METHODS IN MOLECULAR BIOLOGY

Series Editor
John M. Walker
School of Life Sciences
University of Hertfordshire
Hatfield, Hertfordshire, AL10 9AB, UK

For further volumes:
http://www.springer.com/series/7651

Cerebral Angiogenesis

Methods and Protocols

Edited by

Richard Milner

Department of Molecular and Experimental Medicine,
The Scripps Research Institute, La Jolla, CA, USA

Humana Press

Editor
Richard Milner
Department of Molecular and Experimental Medicine
The Scripps Research Institute
La Jolla, CA, USA

ISSN 1064-3745 ISSN 1940-6029 (electronic)
ISBN 978-1-4939-0319-1 ISBN 978-1-4939-0320-7 (eBook)
DOI 10.1007/978-1-4939-0320-7
Springer New York Heidelberg Dordrecht London

Library of Congress Control Number: 2014931262

Printed on acid-free paper

Humana Press is a brand of Springer
Springer is part of Springer Science+Business Media (www.springer.com)

Preface

While blood vessels in the adult central nervous system are extremely stable under normal conditions, it is now accepted that these vessels can be rapidly remodeled in a variety of physiological and pathological situations, including cerebral hypoxia, cerebral ischemia, and tumor development. All of these conditions trigger cerebral angiogenesis, i.e., the sprouting of new vessels. In light of the critical importance of cerebral angiogenesis in promoting physiological adaptation (hypoxia), recovery (cerebral ischemia), or pathogenesis (brain tumors), it becomes a high priority to fully understand the molecular mechanisms underlying this process. In the last 20 years, we have tremendously expanded our working knowledge of how vascular remodeling in the brain occurs and identified many of the key cellular and molecular events underlying this process. From a clinical viewpoint, clear goals are to promote angiogenesis following ischemic stroke or block angiogenic remodeling during brain tumor formation. However, we are still some way off from achieving these noble aims, and it is likely that only after more intensive research and studies, based on the kind of principles and experimental approaches outlined in this volume, will these goals be realistically attainable.

In light of the rapidly expanding field of cerebral angiogenesis, the aim of *Cerebral Angiogenesis: Methods and Protocols* is to provide a selection of the key techniques that are used in characterizing cerebral angiogenesis and to define the cellular and molecular mechanisms underlying this important process. Following the objective of the highly successful *Methods in Molecular Biology* series, the aim of this book is to provide a comprehensive step-by-step guide for the many models and techniques for studying cerebral angiogenesis as well as more specialized approaches for examining endothelial cell behavior, both in vitro and in vivo. As such, *Cerebral Angiogenesis: Methods and Protocols* should help all researchers, both those new to this field as well as those looking to use more specialized and sophisticated techniques, to examine blood vessel growth in the brain. *Cerebral Angiogenesis: Methods and Protocols* covers a wide range of protocols, which provide a useful resource for vascular biologists, cell biologists, molecular biologists, and neuroscientists. This book is divided into six parts. Part I provides an overview of cerebral angiogenesis occurring in different scenarios. Part II describes a variety of different models in which cerebral angiogenesis can be studied as well as includes one chapter describing cerebral arteriogenesis, the formation of new functional arteries. Part III presents a number of methods to characterize and quantify angiogenic events as well as several different approaches to measure changes in cerebral blood flow. Part IV describes different approaches to investigate the role of specific candidate genes in cerebral angiogenesis as well as an important method to determine the contribution of blood-derived hematopoietic stem cells. Part V presents several methods of therapeutically manipulating cerebral angiogenesis by gene delivery. Finally, Part VI outlines some important approaches to examine cerebral angiogenic mechanisms in vitro, including primary culture of brain endothelial cells from different species, assays to examine different aspects of endothelial cell function, and cell culture systems to examine blood vessel formation in vitro. It also includes two chapters that describe in vitro systems of the blood–brain barrier.

I hope that all readers, both first year graduate students and experienced investigators, will find *Cerebral Angiogenesis: Methods and Protocols* to be both timely and useful in their research endeavours. I would like to thank all contributors for their valuable time and effort spent in preparing these excellent chapters and for passing on the critical steps and snippets of experimental knowledge, painstakingly acquired over many years in the laboratory, that can make all of our lives that much easier. My sincere thanks also go to John Walker for giving me the opportunity to put this collection together and for providing editorial assistance during the preparation of this book.

La Jolla, CA, USA *Richard Milner*

Contents

Contributors

YVONNE ALEXANDER • *School of Healthcare Science, Manchester Metropolitan University, Manchester, UK*
LINA BADIMON • *Institut Català de Ciències Cardiovasculars, Hospital de la Santa Creu i Sant Pau, Barcelona, Spain*
ANDREW BAIRD • *Department of Surgery, University of California, San Diego, La Jolla, CA, USA*
MARIBEL BALDELLOU • *Institut Català de Ciències Cardiovasculars, Hospital de la Santa Creu i Sant Pau, Barcelona, Spain*
PETER BAUMGARTEN • *Neurological Institute (Edinger Institute), Frankfurt University Medical School, Frankfurt, Germany; University Cancer Center Frankfurt (UCT), Frankfurt University Medical School, Frankfurt, Germany*
GIRRISO F. BENDERRO • *Department of Physiology and Biophysics, Case Western Reserve University School of Medicine, Cleveland, OH, USA*
KIERSTEN L. BERGGREN • *Department of Psychology, University of Wisconsin-Milwaukee, Milwaukee, WI, USA*
GREGORY BIX • *Sanders-Brown Center on Aging, University of Kentucky, Lexington, KY, USA*
AMIN BOROUJERDI • *Department of Molecular and Experimental Medicine, The Scripps Research Institute, La Jolla, CA, USA*
ANDREW BRESNEN • *Department of Radiology, Research Imaging Institute, University of Texas Health Science Center, San Antonio, TX, USA*
BIRGER BRODIN • *Department of Pharmacy, The Faculty of Health and Medical Sciences, University of Copenhagen, Copenhagen, Denmark*
ALASTAIR M. BUCHAN • *Acute Stroke Programme, Radcliffe Department of Medicine, University of Oxford, Oxford, UK*
MARION S. BUCKWALTER • *Department of Neurology and Neurological Sciences, Stanford University School of Medicine, Stanford, CA, USA; Department of Neurosurgery, Stanford University School of Medicine, Stanford, CA, USA*
IVO R. BUSCHMANN • *Center for Cardiovascular Research (CCR), Universitaetsmedizin Berlin, Berlin, Germany*
MICHAEL CARROLL • *School of Healthcare Science, Manchester Metropolitan University, Manchester, UK*
CLAUDIE CHARLEBOIS • *ADME & Imaging Team, Department of Translational Biosciences, National Research Council of Canada, Ottawa, ON, Canada*
WANQIU CHEN • *Department of Anesthesia and Perioperative Care, Center for Cerebrovascular Research, University of California, San Francisco, San Francisco, CA, USA*
MOHAMMAD CHOWDHURY • *Cardiovascular Research Group, School of Medicine, University of Manchester, Manchester, UK*
CATHRIN J. CZUPALLA • *Institute of Neurology (Edinger Institute), Johann Wolfgang Goethe-University Frankfurt Medical School, Frankfurt, Germany*

OLEKSANDR DASHKIN • *Department of Neurology, University of Miami, Miami, FL, USA*
RICHARD ANTHONY DEFAZIO • *Department of Molecular and Integrative Physiology, University of Michigan, Ann Arbor, MI, USA*
HANS DEGENS • *School of Healthcare Science, Manchester Metropolitan University, Manchester, UK*
KAVI DEVRAJ • *Institute of Neurology (Edinger Institute), Johann Wolfgang Goethe-University Frankfurt Medical School, Frankfurt, Germany*
PAULA DORE-DUFFY • *Division of Neuroimmunology, Department of Neurology, Wayne State University School of Medicine, Detroit, MI, USA*
KRISTIAN P. DOYLE • *Department of Neurology and Neurological Sciences, Stanford University School of Medicine, Stanford, CA, USA; Department of Neurosurgery, Stanford University School of Medicine, Stanford, CA, USA*
ANDRÉ DUELSNER • *Center for Cardiovascular Research (CCR), Universitaetsmedizin Berlin, Berlin, Germany*
TIMOTHY Q. DUONG • *Department of Ophthalmology, Research Imaging Institute, University of Texas Health Science Center, San Antonio, TX, USA; Department of Radiology, Research Imaging Institute, University of Texas Health Science Center, San Antonio, TX, USA; Medicine/Cardiology Division, Department of Physiology, Research Imaging Institute, University of Texas Health Science Center, San Antonio, TX, USA*
BRIAN P. ELICEIRI • *Department of Surgery, University of California, San Diego, La Jolla, CA, USA*
MOISES FREITAS-ANDRADE • *Department of Cellular and Physiological Science, Life Sciences Institute, University of British Columbia, Vancouver, BC, Canada*
NORA GATZKE • *Center for Cardiovascular Research (CCR), Universitaetsmedizin Berlin, Berlin, Germany*
PAULA GRAMMAS • *Department of Neurology, Garrison Institute on Aging, Texas Tech University Health Sciences Center, Lubbock, TX, USA*
DAVID A. GREENBERG • *Buck Institute for Research on Aging, Novato, CA, USA*
SHUZHEN GUO • *Neuroprotection Research Laboratory, Department of Radiology and Neurology, Harvard Medical School, Massachusetts General Hospital, Boston, MA, USA*
GINA HADLEY • *Acute Stroke Programme, Radcliffe Department of Medicine, University of Oxford, Oxford, UK*
PATRICK N. HARTER • *Neurological Institute (Edinger Institute), Frankfurt University Medical School, Frankfurt, Germany; University Cancer Center Frankfurt (UCT), Frankfurt University Medical School, Frankfurt, Germany*
KAZUHIDE HAYAKAWA • *Neuroprotection Research Laboratory, Department of Radiology and Neurology, Harvard Medical School, Massachusetts General Hospital, Boston, MA, USA*
JANA HEDRICH • *Institute of Physiology and Pathophysiology, University Medical Center of the Johannes Gutenberg University Mainz, Mainz, Germany*
HANS C. HELMS • *Department of Pharmacy, The Faculty of Health and Medical Sciences, University of Copenhagen, Copenhagen, Denmark*
ELSPETH HILL • *Department of Vascular Surgery, Central Manchester University Hospitals NHS Foundation Trust, Manchester, UK*
ROBERT M. HOFFMAN • *Department of Surgery, University of California San Diego Medical Center, San Diego, CA, USA*
SHEA HUGHES • *Department of Neurology, University of Miami, Miami, FL, USA*

MASAFUMI IHARA • *Department of Stroke and Cerebrovascular Diseases, National Cerebral and Cardiovascular Center, Osaka, Japan*
XUNMING JI • *Cerebrovascular Research Center, Xuanwu Hospital, Capital Medical University, Beijing, China*
JACOB J.M. KAY • *Department of Psychology, University of Wisconsin-Milwaukee, Milwaukee, WI, USA*
JERZY KRUPINSKI • *Division of Cerebrovascular Diseases, Department of Neurology, Hospital Universitari, Barcelona, Spain*
JOSEPH C. LAMANNA • *Department of Physiology and Biophysics, Case Western Reserve University School of Medicine, Cleveland, OH, USA*
HYE SHIN LEE • *Department of Cancer Biology, University of Texas M.D. Anderson Cancer Center, Houston, TX, USA*
JISOOK LEE • *Department of Surgery, University of California, San Diego, La Jolla, CA, USA*
WENDY LEUNG • *Neuroprotection Research Laboratory, Department of Radiology and Neurology, Harvard Medical School, Massachusetts General Hospital, Boston, MA, USA*
YANING LI • *Neuroscience and Neuroengineering, Research Center, Med-X Research Institute and School of Biomedical Engineering, Shanghai, China*
STEFAN LIEBNER • *Institute of Neurology (Edinger Institute), Johann Wolfgang Goethe-University Frankfurt Medical School, Frankfurt, Germany*
XIAOJIE LIN • *Neuroscience and Neuroengineering, Research Center, Med-X Research Institute and School of Biomedical Engineering, Shanghai, China*
FUDONG LIU • *Department of Neuroscience, University of Connecticut Health Center, Farmington, CT, USA*
YI LIU • *Neuroprotection Research Laboratory, Department of Radiology and Neurology, Harvard Medical School, Massachusetts General Hospital, Boston, MA, USA*
ENG H. LO • *Neuroprotection Research Laboratory, Department of Radiology and Neurology, Harvard Medical School, Massachusetts General Hospital, Boston, MA, USA*
JOSEPHINE LOK • *Department of Pediatrics, Harvard Medical School, Massachusetts General Hospital, Boston, MA, USA*
HEIKO J. LUHMANN • *Institute of Physiology and Pathophysiology, University Medical Center, Johannes Gutenberg University Mainz, Mainz, Germany*
JINHUA LUO • *Department of Neurology, Garrison Institute on Aging, Texas Tech University Health Sciences Center, Lubbock, TX, USA*
MARCIA REGINA MACHEIN • *Department of Neurosurgery, Freiburg University Medical School, Freiburg, Germany*
TAKAKUNI MAKI • *Department of Neurology, Kyoto University Graduate School of Medicine, Kyoto, Japan*
AILEEN MARCELO • *Sanders-Brown Center on Aging, University of Kentucky, Lexington, KY, USA*
JOSEPH MARTINEZ • *Department of Neurology, Garrison Institute on Aging, Texas Tech University Health Sciences Center, Lubbock, TX, USA*
JOSEPH H. MCCARTY • *Department of Cancer Biology, University of Texas M.D. Anderson Cancer Center, Houston, TX, USA*
LOUISE D. MCCULLOUGH • *Department of Neuroscience, University of Connecticut Health Center, Farmington, CT, USA*
GARRY MCDOWELL • *University of Edge Hill, Ormskirk, UK*
PENG MIAO • *Neuroscience and Neuroengineering, Research Center, Med-X Research Institute and School of Biomedical Engineering, Shanghai, China*

RICHARD MILNER • *Department of Molecular and Experimental Medicine, The Scripps Research Institute, La Jolla, CA, USA*
MICHEL MITTELBRONN • *Neurological Institute (Edinger Institute), Frankfurt University Medical School, Frankfurt, Germany; University Cancer Center Frankfurt (UCT), Frankfurt University Medical School, Frankfurt, Germany*
MARIA J. MORENO • *ADME & Imaging Team, Department of Translational Biosciences, National Research Council of Canada, Ottawa, ON, Canada*
ERIC R. MUIR • *Department of Ophthalmology, Research Imaging Institute, University of Texas Health Science Center, San Antonio, TX, USA*
CHRISTOPHER MURGATROYD • *School of Healthcare Science, Manchester Metropolitan University, Manchester, UK*
MICHALIS PAPADAKIS • *Acute Stroke Programme, Radcliffe Department of Medicine, University of Oxford, Oxford, UK*
ANJA BONDKE PERSSON • *Institute of Physiology, Universitaetsmedizin Berlin, Berlin, Germany*
KARL H. PLATE • *Institute of Neurology (Edinger Institute), Frankfurt University Medical School, Frankfurt, Germany*
TAMER RABIE • *Radcliffe Department of Medicine, University of Oxford, Oxford, UK*
NORMA ROVIRA • *Division of Cerebrovascular Diseases, Department of Neurology, Hospital Universitari, Barcelona, Spain*
ALMA SANCHEZ • *Department of Neurology, Garrison Institute on Aging, Texas Tech University Health Sciences Center, Lubbock, TX, USA*
FABIAN T. SCHNEIDER • *Department of Biosciences and Nutrition, Karolinska Institute, Huddinge, Sweden*
FERDINAND SERRACINO-INGLOTT • *Cardiovascular Research Group, School of Medicine, University of Manchester, Manchester, UK*
MARK SLEVIN • *School of Healthcare Science, Manchester Metropolitan University, Manchester, UK*
QIANG SHEN • *Department of Ophthalmology, Research Imaging Institute, University of Texas Health Science Center, San Antonio, TX, USA*
JACQUELINE SLINN • *ADME & Imaging Team, Department of Translational Biosciences, National Research Council of Canada, Ottawa, ON, Canada*
HUA SU • *Department of Anesthesia and Perioperative Care, Center for Cerebrovascular Research, University of California, San Francisco, San Francisco, CA, USA*
XIAOYAN SUN • *Department of Physiology and Biophysics, Case Western Reserve University School of Medicine, Cleveland, OH, USA*
BRAD A. SUTHERLAND • *Radcliffe Department of Medicine, University of Oxford, Oxford, UK*
RODNEY A. SWAIN • *Department of Psychology, University of Wisconsin-Milwaukee, Milwaukee, WI, USA*
AKIHIKO TAGUCHI • *Department of Regenerative Medicine and Research, Institute of Biomedical Research and Innovation, Kobe, Japan*
YAOHUI TANG • *Neuroscience and Neuroengineering, Research Center, Med-X Research Institute and School of Biomedical Engineering, Shanghai, China*
ULRICH TIGGES • *Department of Molecular and Experimental Medicine, The Scripps Research Institute, La Jolla, CA, USA*
YASH VARDHAN TIWARI • *Department of Biomedical Engineering, Research Imaging Institute, University of Texas Health Science Center, San Antonio, TX, USA*

Hidekazu Tomimoto • *Department of Neurology, Mie University Graduate School of Medicine, Mie, Japan*
Debjani Tripathy • *Department of Neurology, Garrison Institute on Aging, Texas Tech University Health Sciences Center, Lubbock, TX, USA*
Constantinos P. Tsipis • *Department of Physiology and Biophysics, Case Western Reserve University School of Medicine, Cleveland, OH, USA*
Yongting Wang • *Neuroscience and Neuroengineering, Research Center, Med-X Research Institute and School of Biomedical Engineering, Shanghai, China*
Kazuo Washida • *Department of Neurology, Kobe University Graduate School of Medicine, Kobe, Japan*
Lora Talley Watts • *Department of Cellular and Structural Biology, Research Imaging Institute, University of Texas Health Science Center, San Antonio, TX, USA; Department of Neurology, Research Imaging Institute, University of Texas Health Science Center, San Antonio, TX, USA*
Jennifer V. Welser-Alves • *Department of Molecular and Experimental Medicine, The Scripps Research Institute, La Jolla, CA, USA*
Robin White • *Institute of Physiology and Pathophysiology, University Medical Center, Johannes Gutenberg University Mainz, Mainz, Germany*
Ian R. Winship • *Centre for Neuroscience, University of Alberta, Edmonton, AB, Canada*
Ina M. Wittko-Schneider • *Institute for Stroke and Dementia Research, Klinikum der Universität München, München, Germany*
Changhong Xing • *Neuroprotection Research Laboratory, Department of Radiology and Neurology, Harvard Medical School, Massachusetts General Hospital, Boston, MA, USA*
Kui Xu • *Department of Physiology and Biophysics, Case Western Reserve University School of Medicine, Cleveland, OH, USA*
Guo-Yuan Yang • *Neuroscience and Neuroengineering, Research Center, Med-X Research Institute and School of Biomedical Engineering, Shanghai, China*
Xiangling Yin • *Department of Neurology, Garrison Institute on Aging, Texas Tech University Health Sciences Center, Lubbock, TX, USA*
William L. Young • *Department of Anesthesia and Perioperative Care, Center for Cerebrovascular Research, University of California, San Francisco, San Francisco, CA, USA*
Christoph M. Zehendner • *Institute of Physiology and Pathophysiology, University Medical Center of the Johannes Gutenberg University Mainz, Mainz, Germany*
Nicole Ziegler • *Institute of Neurology, Edinger-Institute, Johann Wolfgang Goethe-University Frankfurt Medical School, Frankfurt, Germany*

Part I

Overview of Cerebral Angiogenesis

Chapter 1

Cerebral Angiogenesis During Development: Who Is Conducting the Orchestra?

Ina M. Wittko-Schneider, Fabian T. Schneider, and Karl H. Plate

Abstract

Blood vessels provide the brain with the oxygen and the nutrients it requires to develop and function. Endothelial cells (ECs) are the principal cell type forming the vascular system and driving its development and remodeling. All vessels are lined by a single EC layer. Larger blood vessels are additionally enveloped by vascular smooth muscle cells (VSMCs) and pericytes, which increase their stability and regulate their perfusion and form the blood–brain barrier (BBB). The development of the vascular system occurs by two processes: (1) vasculogenesis, the de novo assembly of the first blood vessels, and (2) angiogenesis, the creation of new blood vessels from preexisting ones by sprouting from or by division of the original vessel. The walls of maturing vessels produce a basal lamina and recruit pericytes and vascular smooth muscle cells for structural support. Whereas the process of vasculogenesis seems to be genetically programmed, angiogenesis is induced mainly by hypoxia in development and disease. Both processes and the subsequent vessel maturation are further orchestrated by a complex interplay of inhibiting and stimulating growth factors and their respective receptors, many of which are hypoxia-inducible. This chapter intends to give an overview about the array of factors directing the development and maintenance of the brain vasculature and their interdependent actions.

Key words Brain angiogenesis, Development, Tip cell, Stalk cell, Signaling pathways, Guidance molecules, Blood–brain barrier

1 The Development of the Vascular System

To ensure their proper functionality, all tissues must be continuously provided with adequate amounts of oxygen, nutrients, and signaling molecules and simultaneously need to be cleared of cellular and metabolic waste. As the oxygen diffusion distance in tissues is very limited, multicellular organisms were forced to develop a long-distance transportation structure. Whereas invertebrates developed simpler systems, vertebrates established an extremely complex circuit of blood and lymphatic vessels, namely, the vascular system.

Ina M. Wittko-Schneider and Fabian T. Schneider have contributed equally to this work.

Richard Milner (ed.), *Cerebral Angiogenesis: Methods and Protocols*, Methods in Molecular Biology, vol. 1135,
DOI 10.1007/978-1-4939-0320-7_1, © Springer Science+Business Media New York 2014

Endothelial cells (ECs) are the primary cell type forming the vascular system and driving its development and remodeling. A monolayer of ECs lines all blood vessels and provides a non-coagulant surface for blood. Endothelial tubes are enwrapped by vascular smooth muscle cells (VSMCs) and pericytes which increase their stability and regulate their perfusion and, together with astrocyte end-feet, form the blood–brain barrier (BBB) [1] (*see* Fig. 1).

1.1 Vasculogenesis and Angiogenesis

In vertebrates vascular system development starts shortly after gastrulation and is divided into two distinct processes. The initial process has been termed *vasculogenesis* by Werner Risau and is defined as the de novo assembly of the first blood vessels. Vasculogenesis starts early in life, i.e., embryonic day (E) 6–6.5 in mice and E18 in humans. During gastrulation three germ layers, namely, the ectoderm, mesoderm and endoderm, are formed [2]. Initiated by mesodermal activation signals, the primary vascular plexus emerges by differentiation of splanchnopleuric mesodermal cells into hemangioblasts, which then differentiate further into hematopoietic precursors or endothelial precursor cells, so-called angioblasts [2–4]. Guided by endodermal signals, these angioblasts migrate to sites of vascularization, i.e., firstly the extraembryonic visceral mesoderm of the yolk sac. There the angioblasts proliferate, differentiate into ECs, and subsequently aggregate into blood islands, which then coalesce into tubelike structures to form the primary vascular plexus [5, 6]. In mammals the two dorsal aortas are the first vessels that develop on both sides of the neural tube. These fuse at the midline during later embryonic development [1]. Several blood islands from the vascular plexus migrate to the head region to produce cranial vessels [7].

Later (~E9 in mice) new blood vessels are formed by a process named *angiogenesis*, which is defined as the formation of capillaries from preexisting vessels. Angiogenesis is the process whereby ectodermal–mesodermal organs, e.g., the brain, are vascularized [7, 8]. Angiogenesis can occur either by the proliferation of ECs (sprouting angiogenesis) or by division of the original vessel (intussusception; non-sprouting angiogenesis) [7, 9, 10]. In response to the increasing metabolic need of the growing tissue, the vascular network expands and evolves constantly by remodeling and pruning of additional vessels and matures into the complex, hierarchical organized adult circulatory system. Under physiological conditions sprouting angiogenesis is the major process driving embryonic blood vessel development, particularly during brain development [7, 11, 12].

Angiogenesis is typically initiated in response to angiogenic signals (e.g., growth factors) that are predominantly released under hypoxic conditions. The formation of new vessels starts with the construction of a multicellular EC cord, is followed by lumen creation (tubologenesis), and concludes with the initiation of blood

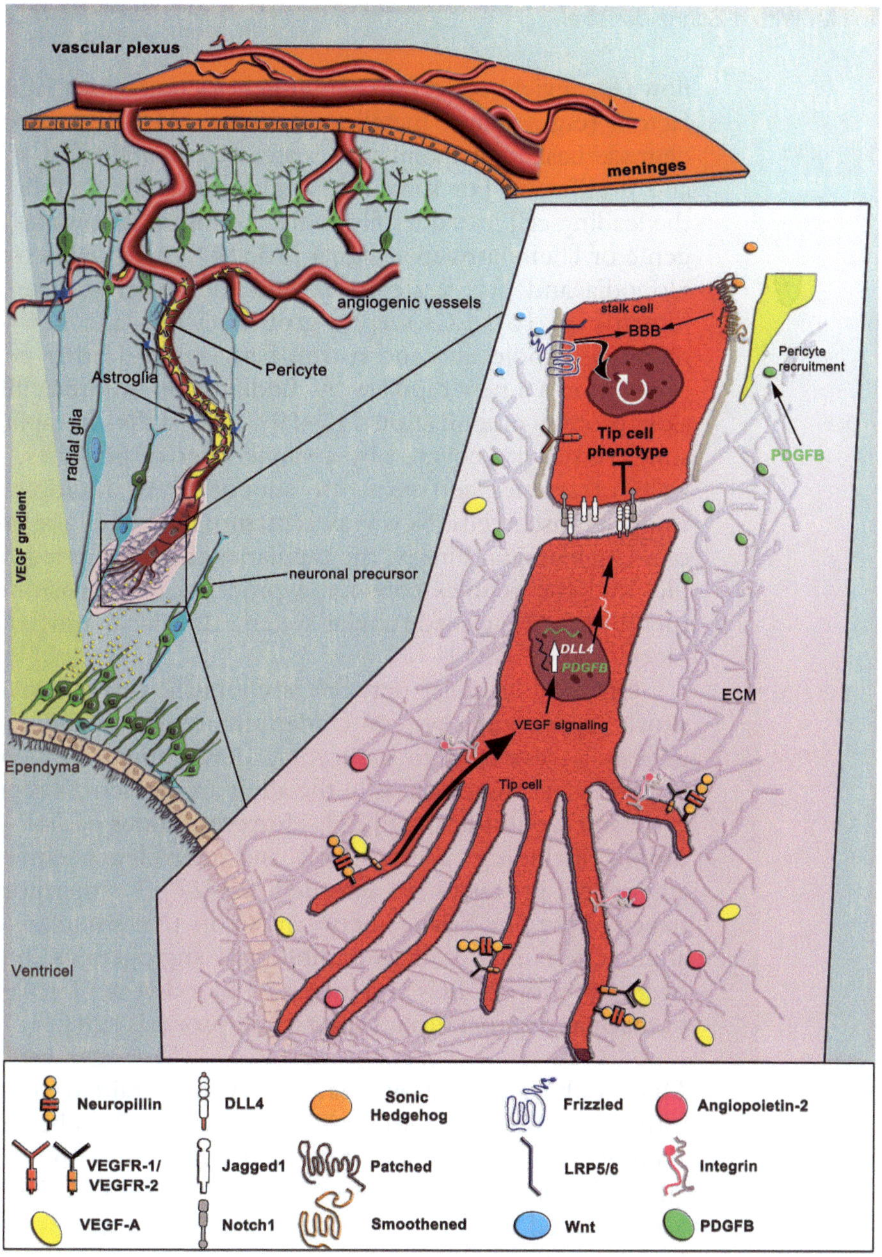

Fig. 1 Scheme of angiogenic outgrowth of vessels from the vascular plexus into the brain and towards the ventricular zone of the developing mammalian brain: during brain development vessels are growing into the parenchyma from the vascular plexus on the meninges following a chemotactic gradient of VEGF. Maturing vessels become covered with pericytes (*yellow* cells). The VEGF gradient is produced mainly by neuronal progenitor cells, which migrate along radial glia cells towards cortical regions where they differentiate into neurons (*dark* and *light green* cells). VEGF binds to VEGFR-1 that is expressed on tip cells (see close-up) and induces the expression of PDGFB (*green*) and DLL4 (*white*). Another important tip cell guidance factor is angiopoietin-2, which binds to integrins on tip cells. The expression of DLL4 in tip cells maintains its phenotype by binding to the Notch1 receptor (*gray*), expressed on stalk cells, which inhibits the tip cell phenotype in the later. While PDGFB is important for the recruitment and guidance of pericytes (*yellow*), which cover the endothelial cells and establish a functional blood–brain barrier. The main signaling pathways, which generate the functional blood–brain barrier, are Wnt (*blue*) and Sonic hedgehog signaling pathways (*orange*). Furthermore, Wnt signaling controls proliferation of endothelial stalk cells (*white curved arrow*)

flow (for review, see ref. 1). Angiogenesis starts with ECs releasing factors (e.g., matrix metalloproteinases (MMPs)), which locally degrade basement membrane structures, enabling ECs to detach and proliferate. For sprouting angiogenesis the so-called tip cell, the leading cell, extrudes filopodia and migrates towards an angiogenic or chemoattractive signal [13, 14]. Stalk cells extend fewer filopodia and proliferate to expand the vessel stalk and form a lumen [15, 16]. ECs secrete growth factors and new basement membrane molecules and recruit pericytes and other perivascular cells to form a new capillary by modification of intercellular adhesion protein organization [17–19]. The walls of capillaries and small vessels are covered by a single layer of pericytes, while the walls of arteries and veins are supported by multiple layers of smooth muscle cells. New vessels mature further by the production of a basal lamina. Finally, the capillaries that originate from arterioles and the venules connect, a process called anastomosis, and thereby create a closed tubular system, thus facilitating continuous blood flow [20, 21].

Once the vascular network is established, it is a stable and a slowly regenerating system. Under physiological conditions angiogenesis is restricted to embryonic development. Exceptions for physiological angiogenesis in the adult are the female reproductive cycle and wound healing [22]. However, under pathological conditions, including cancer, stroke, and neurodegenerative diseases, angiogenic processes are activated [23–27]. Furthermore, disturbance of the fine balance between factors that stimulate or inhibit angiogenesis may lead to pathological angiogenesis, resulting in over-activation of ECs and increased blood vessel formation, e.g., in macular degeneration, tumors, and psoriasis [28].

1.2 Brain Angiogenesis

Many publications show that embryonic brain development is intimately associated with brain angiogenesis. This becomes especially evident as treatment of rat embryos (d14) with thalidomide, an inhibitor of angiogenesis, leads to vascular abnormalities accompanied by severe developmental defects in the cortex and hippocampus [29].

The cerebral microvasculature is structurally and functionally unique compared to blood vessels in other organs. Cerebral capillaries form the BBB which is created by junctions between neighboring ECs and between ECs and VSMCs or astrocytes, thus hindering the uncontrolled entry of molecules from the blood into the brain. Disruption of the BBB is an integral part of the pathogenesis of many CNS pathologies, including ischemic stroke, cerebral edema, and brain tumors (reviewed in ref. 30).

Vascularization of the CNS initially follows an intrinsic program [31] and starts during early embryonic development with the migration of angioblasts from the splanchnopleuric mesoderm into the head region, where they create the perineural vascular plexus

(PNVP). This primitive network covers first the surface of the brain (E9 in mice) and is therefore termed extracerebral vascularization. Later by leptomeningeal vascularization, consisting of EC differentiation, extramural cell recruitment, vessel remodeling, and regression, this PNVP evolves into meningeal vessels [32]. Whereas pial vessels develop without a specific temporal and spatial pattern [7], intracerebral vessels sprout from this extracellular network and grow gradually deeper into the neuroectodermal tissue towards the subventricular zone (SVZ) in characteristic phases. These phases coincide with neuronal development [31, 33, 34] and are driven mostly by EC proliferation [35]. First, a capillary plexus is formed next to the PVNP, leaving the cortical plate avascular. Vascularization of the SVZ takes place (phase I) by new branches growing around the ventricles [35]. These periventricular vessels arise from a basal vessel that is most probably situated in the basal ganglia primordium. This vessel presumably branches out from the pharyngeal arch arteries of the cervical region (reviewed in refs. 31, 34). This ingrowth of vessels into brain tissue is guided by chemoattractants secreted by cells of the SVZ, many of which are also regulators of neurogenesis, e.g., VEGF [36–38] (Fig. 1). Only at later stages of development, a second capillary plexus forms in the hemispheric wall, and interestingly, vascularization of the developing white matter (phase II) and the cortical plate (phase III) occurs in an inside-out fashion. As a consequence depending on the time of their formation, the cortical branches are located at specific cortical depths [32, 35, 39].

The extensive vascular branching is based on vast endothelial proliferation and lasts until early postnatal stages, at which point a complex vascular network has been created [32, 35, 39]. Constant pruning of excess vessel branches and widening and merging of endothelial tubes to build large vessels shape the adult vascular tree [7]. Newly formed vessels mature and create the BBB by secreting extracellular matrix (ECM) proteins that construct a basal lamina; by recruiting perivascular support cells, like pericytes or VSMCs and by establishing endothelial–astrocytic contacts and intercellular tight junctions [40]. The BBB is formed partly in response to neural signals, and its development starts as early as E11–E13 in mice [41].

Similar to the rest of the body, the vertebral CNS vasculature consists of arteries and veins. The aortic arch system gives rise to ventral arteries providing blood flow to the face (external carotid arteries) and the frontal part of the skull base (internal carotid arteries). The internal carotid arteries branch intracranially into the anterior cerebral artery and the middle cerebral artery. The latter divides into anterior, medial, and posterior segments. The dorsal arteries extend into the brain and merge at the midline into one unpaired basilar artery. This supplies the ventral surface of the brain stem. At the level of the diencephalon, the internal carotid artery branches fuse with the basilar artery into the posterior communicating arteries, thereby connecting the internal carotid and basilar systems. The anterior communicating artery forms ventrally through the connection of two small branches from the anterior cerebral arteries. Fused together, these branches form the so-called circle of Willis. This vascular ring seems to guarantee continuous blood supply to the brain in the event of the occlusion of a major artery. Specific to the brain is that arteries and veins do not run in parallel. By 10-week gestation in humans, the initial venous plexuses are modified to form major intracerebral and extracerebral veins and sinuses, independent from the arteries (reviewed in ref. 42).

2 Growth Factors Are Major Regulators of Cerebral Vascular Development: A Special Case for VEGF

Whereas the initial process of vasculogenesis seems to be genetically programmed [7, 43], hypoxia is the major stimulus of angiogenesis in development and disease [26, 44]. Both processes are further regulated by a complex interplay of directly or indirectly hypoxia-inducible inhibitory or stimulatory growth factors and receptors [26, 45] (Table 1). The early events in vessel formation and invasive capillary sprouting are triggered by vascular endothelial growth factors (VEGF) and their receptor pathways [46], which act predominantly on ECs. These pathways are indispensable for angiogenesis throughout life and are the most extensively studied pathways in angiogenesis research. Other important angiogenic regulatory systems include the Notch–Jagged/Delta-like pathways for early angiogenic events [47, 48], while later vascular development and lumen formation is regulated by alternative signaling pathways that include the ephrin–Eph system [49–51], the Wnt pathway [30], EGFL7 [52], and pathways that are involved in axonal guidance, e.g., the semaphorins, netrins, and ROBO/Slits [53, 54]. The angiopoietin–Tie [55] and the platelet-derived growth factor (PDGF/PDGFR) and transforming growth factor beta (TGFβ) families [8, 43] control the late steps of angiogenesis that include vessel maturation and vascular remodeling. Heinke et al. (2012) [56] listed the phenotypes of genetic deletion and overexpression studies of numerous vascular growth factors and receptors. Interestingly, many of these have been shown to act in parallel on both vascular and neural cells, demonstrating the interdependence and functional connection of the two systems using shared regulatory mechanisms [57–60].

The VEGF/VEGF receptor signaling system is absolutely essential for vasculogenesis, angiogenesis, and cerebral vascularization. Therefore, as an example, the molecular determinants of this system will be covered here in some detail. The VEGF growth factor family includes six different secreted homologous

Table 1
Overview of the main signaling pathways in cerebral angiogenesis, their major function in brain angiogenesis, and their specific role in tip and stalk cells

Signaling pathway	Effects on brain angiogenesis	Tip cell function	Stalk cell function
VEGF	Guides vessel sprouts	Chemotactic gradient via VEGFR-2 signaling	Vessel integrity VEGFR-1 signaling
AKT	Downstream signal of VEGFR	Induces DLL4 expression	
Wnt	Major angiogenic molecule, endothelial cell differentiation, triggers VEGF expression		Proliferative signal, BBB formation, and maintenance
Notch	Aortic differentiation, vascular branching	Represses tip cell phenotype in stalk cells via DLL4/ Notch1 signaling	The Notch target Nrarp enhances b-catenin signaling in retina vessels. Proliferative signal
Hedgehog	Capillary/tube formation, migration, triggers Vegf and Ang-2 expression	Shh signaling triggers Ang-2 and VEGF expression in glial cells to	BBB induction
Eph/Ephrin	Repulsion from preexisting neuronal cells, vessel sprouting		
Angiopoietins/ tie receptors	Ang-1/Tie-2 vessel stabilization, antagonistic to Ang-2; Ang-2/Tie-2: vessel destabilization, tip cell formation	Guidance molecule	Ang-1, stabilization; Ang-2, pericyte detaching
EGFL7	Vessel lumen formation and stabilization	Enhances Dll4/ Notch1-mediated stalk cell phenotype	
PDGF	Pericyte coverage of ECs	Express PDGF to create a gradient for pericyte migration	
TGF	Vessel stabilization		Inhibits proliferation
Netrin	Vessel guidance, arterial expression	Expression in arterial tip cells	
Slits	EC migration and chemotaxis, vessel repulsion, proliferation		
Hif	Initiator of angiogenic cytokine transcription for the formation of new vessels		

Abbreviations: *AKT* protein kinase B, *EGFL7* EGF-like domain-containing protein 7, *Hif* hypoxia-inducible factor, *PDGF* platelet-derived growth factor, *TGF* transforming growth factor, *VEGF* vascular endothelial growth factor, *WNT* combination of wingless and integration 1 dimeric glycoproteins: VEGF-A–E and placenta growth factor (PlGF) [61, 62]. While VEGF-A, VEGF-B, VEGF-E, and PlGF act predominantly on blood vessels, VEGF-C and VEGF-D regulate the development and homeostasis of lymphatic vessels [63]. In the last two decades many studies described additional effects on neural cells, but it is largely unclear if these are mediated directly or indirectly through vascular effects [36, 37, 64]. VEGF family growth factors transmit their angiogenic signals mainly via EC membrane-bound tyrosine kinase receptors: VEGFR-1 (Fms-related tyrosine kinase 1; Flt-1) [65, 66] and VEGFR-2 (fetal liver kinase 1; Flk-1) [67]. VEGFR signaling is modified by the co-receptors neuropilin-1(Nrp-1) and neuropilin-2 (Nrp-2) [61, 62]. The great importance of the VEGF family and their receptors for proper vascular development becomes most obvious in genetic loss-of-function studies. The phenotypes of all knockout mice of the VEGF family members and VEGFRs in mice have been nicely reviewed and portrayed by Olsson et al. (2006) [68].

VEGF-A, which has the highest angiogenic activity of the VEGF family, induces and regulates the formation of new blood vessels during development and is a mitogen and survival-promoting factor for ECs and controls EC migration and differentiation during vascular development [7, 38, 69]. The deletion of a single allele of the *vegf-a* gene leads to embryonic lethality on day E9, as a result of major abnormalities of the vascular system both in the whole embryo and the yolk sac [70, 71]. VEGF-A occurs in six isoforms that are generated by alternative splicing and range in length from 121 to 206 amino acids (VEGF121–206) (reviewed in ref. 72). The isoforms differ in their binding affinity to ECM proteins, which results in a VEGF-A gradient with concentrations decreasing with distance from the target region. These gradients of VEGF-A serve as guidance cues for forming vessels [15, 16, 73].

In the developing brain, VEGF-A is secreted by neuroectodermal cells close to the ventricles and by the choroid plexus. The resulting VEGF-A gradient induces the growth of developing vessels towards the ventricles [74, 75] (Fig. 1). Specific deletion of VEGF-A in nestin+ cells (mainly neural progenitor cells) of the SVZ hinders intracerebral vascularization in mice. Additionally, the loss of secreted VEGF-A results in extensive apoptosis of subventricular cells, diminishes overall cerebral growth, and leads to a strong flattening of the skull, a phenotype resembling human microencephaly [38, 76]. Comparable to the brain, in the retina VEGF gradients have been demonstrated to guide blood vessels during angiogenesis [15].

The tip cell of growing vessels highly expresses VEGFR-2 and organizes a filopodia to migrate towards the area of the higher VEGF-A expression, while the stalk cells express only low levels of VEGFR-2 and react to VEGF-A binding by proliferation and consequently by vessel growth [15, 16, 73] (Fig. 1). In the adult, VEGF-A has been shown to play an autocrine role in vascular homeostasis [77].

The expression of VEGF-A protein is regulated by low oxygen concentration, growth factors, and hormones. Hypoxia-inducible factors (HIF1α and HIF2α) bind to the hypoxia response element (HRE) in the 5′ *vegf-a* promoter region, thus provoking *vegf* gene expression [78–80]. Consequently, VEGF-A expression is upregulated following hypoxic conditions, as in developmental tissue growth and also in tumors [81, 82] and after various pathological ischemic insults [83, 84].

VEGF-B displays great homology to VEGF-A and exists in two isoforms: VEGF-B167 and VEGF-B186. VEGF-B promotes proliferation of EC in vitro and in vivo, but contrary to VEGF-A, VEGF-B shows only minor angiogenic activity [85] and its activity is restricted to certain conditions [86, 87]. VEGF-B is not essential for the development of the vascular system as homozygous deletion of the *vegf-b* gene (VEGF-B −/−) in mice leads to a viable and fertile phenotype. These animals show only minor disturbances in cardiac development and function [88, 89]. However, VEGF-B is an important survival factor for blood vessels [90, 91] and plays a decisive role in vascular remodeling and protection after stroke [92, 93]. Unlike VEGF-A, VEGF-B expression is not regulated by hypoxia, owing to the lack of an HRE in the promoter region of the *vegf-b* gene [94]. In a similar manner to VEGF-A, direct actions of VEGF-B on neural cells have also been described [95, 96]. *VEGF-C and VEGF-D* share 30 % homology with VEGF-A and are regulators of lymphatic system development [97–100].

All three VEGFRs are typical receptor tyrosine kinases and are comprised of an extracellular immunoglobulin-like domain, a transmembrane domain, and an intracellular split tyrosine kinase (TK) domain. Ligand binding induces receptor homo- or hetero-dimerization and subsequent kinase activation and autophosphorylation of specific tyrosine residues, which in turn activates multiple downstream signaling pathways [61, 62].

VEGFR-1 exists in two different isoforms: a membrane-bound form and a truncated soluble form (soluble VEGFR-1; sVEGFR-1; sFlt-1), which lacks the intracellular TK domain and is therefore incapable of transducing a VEGF signal. In contrast to VEGFR-2, the expression of VEGFR-1 is induced by hypoxia, similar to that of VEGF-A. Despite its much higher binding affinity to VEGF-A compared to VEGFR-2, VEGFR-1 shows only weak autophosphorylation of its TK domains. This leads to the general assumption that VEGFR-1, especially sVEGFR-1, functions mainly by trapping VEGF-A, thus preventing ligand binding to VEGFR-2 and Nrps and thereby negatively regulating VEGFR-2 signaling. This inhibitory effect of VEGFR-1 was demonstrated by homozygous deletion of *vegfr-1*, which results in early embryonic lethality (E8.5) due to a disorganized vasculature, vast proliferation of ECs, and vascular overgrowth [101, 102]. This effect is subsequent to an increased mesenchymal–hemangioblast transition [102]. Accordingly, the overexpression of sVEGFR-1 diminishes VEGFR-2 signaling. In contrast to the general deletion of *vegfr-1*, the specific deletion of the intracellular domain (Flt-1TK−/−) of the receptor does not obviously affect angiogenesis [103].

VEGFR-2 is indispensable for embryonic blood vessel formation and is regarded as the main receptor mediating the mitogenic, angiogenic, and permeability-increasing signals of VEGF-A in ECs and neural cells. Activation of VEGFR-2 leads to an amplification of EC proliferation, reduction of EC apoptosis, and EC assembly into vascular systems and blood vessel sprouting (reviewed in refs. 37, 104). Homozygous deletion of *vegfr-2* leads to embryonic death between E8.5 and 9.5 owing to an inability to develop blood islands and the incapacity to form a functioning vascular system [6, 105]. The necessity for VEGFR-1 and VEGFR-2 specifically for CNS vascularization has not been studied in detail due to the early embryonic lethality of full homozygous deletion of these genes. Similar to sVEGFR-1, a shortened soluble form of VEGFR-2 (soluble VEGFR-2, sVEGFR-2) exists physiologically. This form functions as a decoy receptor and serves as an endogenous antagonist of VEGF-C, thus negatively regulating lymphangiogenesis [106]. In contrast to VEGFR-1 and VEGF-A, the expression of VEGFR-2 is not controlled by hypoxia but is upregulated secondarily by its ligand VEGF-A [107].

The responses of VEGFRs are modulated by the non-TK transmembrane receptors *Nrp-1 and Nrp-2*. Neuropilins are non-tyrosine kinase receptors that bind to VEGFR-2 thereby reinforcing VEGF-A signaling [108–110]. Nrp-1 further forms complexes with VEGFR-1, reducing the binding of VEGF-A and Nrp-1 to VEGFR-2 and further diminishing VEGF-A signaling [110]. Nrp-2 interacts also with VEGFR-3 and has affinity for the VEGFR-3 ligands VEGF-C and VEGF-D [109]. In Nrp-1 knockout mice, a defective neural patterning and vascular regression lead to embryonic lethality [111, 112]. Correspondingly, overexpression of Nrp-1 induces excess blood vessel formation, provoking cardiovascular and heart malformations, dilated blood vessels, and hemorrhages. Mutations of the Nrp-2 gene result in 40 % embryonic lethality, and the surviving mice are smaller and show a severe impairment of lymphangiogenesis [113]. Simultaneous deletion of both Nrp homologs induces a block of vascular development causing early embryonic death (E8.5) [114]. Nrps are expressed with specificity to arterious–venous vessel identity, which reflects their role in arterious–venous specification [113, 115, 116].

3 Molecular Mechanisms of Brain Angiogenesis

Genetic gain- and loss-of-function studies have identified a number of signaling molecules as important regulators of the development and maintenance of brain vascularization, each regulating different processes and at different developmental stages (Table 1). In early embryonic CNS vascularization, VEGF-A seems to be the major regulator. VEGF-A is expressed in the lateral neural tube when the PNVP is formed, and its inhibition has been shown to block neural tube vascularization in explant cultures and in vivo [117]. Besides VEGF-A, Shh and Ang-1 are expressed by neural progenitors of the neural tube at E8.5–9, when vessel sprouting occurs [118]. The VEGFRs and the Tie receptors are expressed at the same time by invading ECs, and genetic deletion of Tie-2 inhibits vascular sprouts from the PVNP [119].

Similar to the neural tube, VEGF is the main developmental regulator of vascularization of the brain. In the developing brain, VEGF-A is secreted from neuroectodermal cells close to the ventricles and by the choroid plexus. The resulting VEGF-A gradient induces the growth of vessels towards the highest VEGF-A concentration at the ventricles (Fig. 1). The tip cells of growing vessels express high levels of VEGFR-2 and sprout filopodia. Thereby tip cells migrate towards the area of the highest VEGF-A expression [15, 16]. ECs located behind tip cells (i.e., the stalk cells) express low levels of VEGFR-2 and react to VEGF-A with proliferation and subsequent vessel growth [15, 16]. Besides VEGFR-2, Nrp-1 also seems to be crucial to transduce VEGF-mediated effects on ECs as Nrp-1 knockout mice show impaired brain angiogenesis. Angiopoietins seem to have a special role in cerebral development. Ang-1 and Ang-2 and VEGF-A are primarily expressed in astrocytes and Purkinje cells. Their expression is in a temporal and spatial pattern coherent with vessel growth, suggesting a coordinated interplay of VEGF-A and Angs in the regulation of cerebral vascular system development and a cross talk between astrocytes and invading vascular cells [120].

During vessel branching and maturation, the Ang/Tie-2 receptor system plays the most crucial role in the interaction of ECs with VSMCs and pericytes. Whereas Ang-1 appears to be essential for pericyte coverage and vessel stability, Ang-2 together with VEGF might be important for the initiation of tip cells [121]. Importantly, both ligands are direct target genes of Shh signaling, which is expressed by NPCs in lateral somites and the VZ. Tie-2, similar to VEGFR-2, has been detected specifically at the tip of invading capillaries [121]. Besides Tie-2 and VEGF, the Notch ligand DLL4 is crucial for tip cell identity, and its expression is induced by VEGF-A signaling in tip cells [122, 123]. Notch-1 is expressed on stalk cells, and Notch-1 signaling suppresses VEGFR-2 expression in these cells, further accentuating the molecular differences between tip and stalk cells [124]. Nrps which are expressed on the tip cell further reinforce VEGF signaling by increasing the affinity of VEGF-A to VEGFR-2 [108, 112] (Fig. 1). These findings suggest that multiple factors are important for the initiation of vascular sprouting and vessel growth.

The emerging sprouts need to be directed to their target area. Besides tip cell navigation via VEGF gradients and Notch ligands [125], axonal guidance molecules have been found to control EC migration and vascular guidance via repellence from the cells surrounding the vascular bed [126]. Ephrins, Netrins, and Slits and their respective receptors have been shown to play an important role in these processes [126]. However, the detailed interplay of these factors is not yet fully understood. Maturing blood vessels acquire stabilization by the recruitment of intramural cells, the formation of inter-EC tight junctions as well as with astrocytic end feet, and the deposition of ECM proteins. In the brain, pericytes and SMCs appear directly after initial sprouting of ECs, and their recruitment to newly formed vessels is regulated by multiple growth factors, i.e., the angiopoietins, PlGF, TGF-β, and PDGF [127]. VEGF-A and Angs cooperate on vascular remodeling [121]. Ang-1 stabilizes preexisting vessels under physiological conditions [128]. When Ang-2 blocks Ang-1 signaling via Tie-2 activation, perivascular cells detach and vessels destabilize and ECs become more receptive to other angiogenic growth factors and stimuli. Depending on the extent of available VEGF-A, ECs proliferate and vessels grow. In the case of low VEGF-A protein levels, Ang-2/Tie-2 signaling facilitates vessel regression and EC apoptosis [129].

BBB development that begins as early as E11–13 is regulated by factors secreted partially by astrocytes such as bFGF, TGF-α GDNF, Ang-1, Wnt7a/b, and Shh [30]. ECM proteins are secreted by ECs, and ECM-integrin interactions play an important role in promoting and maintaining brain vessel maturation [130]. In the rat, brain angiogenesis peaks early postnatally until P10 and declines thereafter [39]. Simultaneously, VEGF-A and VEGFR-2 are gradually downregulated between P10 and P20, and VEGFR-2 expression is barely detectable at P30 [33]. This goes together with a reduction of canonical Wnt/β-catenin signaling in CNS ECs after birth [30]. By contrast, VEGFR-1 expression stays unchanged after birth [33, 74, 131]. VEGF at very low levels seems to be essential to ensure vessel stability and avert regression [132]. More recently, an autocrine role in vascular homeostasis has been demonstrated in adult mice [77]. As an exception, VEGF expression stays high in epithelial cells and in ECs of the choroid plexus throughout life, where it might be necessary for the maintenance of a fenestrated phenotype [75]. Similar to VEGF, Ang-1 is constitutively expressed in the adult brain and might also be important for preservation of the cerebral vasculature [121, 133].

Like VEGF, Ang-1 seems to be a survival factor for ECs as it inhibits EC apoptosis [134]. Like VEGFRs, Tie-1 and Tie-2 mRNA levels are present at high levels in the vasculature of the developing brain and are downregulated in the adult. However, Tie-2 is expressed at low levels in the adult vasculature [121, 135].

Although the adult vasculature is quiescent, under pathological conditions adult ECs are mitotic and can generate new vessels, when activated by hypoxia or angiogenic stimuli. Whereas VEGF and its receptors have long been thought to be the major angiogenic regulator, it becomes clear that the development of the complex vascular network is controlled by a plethora of interacting signaling molecules and that modifications in this molecular network can lead to dynamic changes in the angiogenic state. On the other hand, further understanding of the fine-tuning of vessel development, repair, and BBB formation and maintenance opens the possibility for targeted therapeutic interventions during cancer and tissue repair.

References

1. Lammert E, Axnick J (2012) Vascular lumen formation. Cold Spring Harb Perspect Med 2:a006619
2. Risau W, Flamme I (1995) Vasculogenesis. Annu Rev Cell Dev Biol 11:73–91
3. Choi K (1998) Hemangioblast development and regulation. Biochem Cell Biol 76: 947–956
4. Eichmann A, Corbel C, Pardanaud L, Breant C, Moyon D, Yuan L (2000) Hemangioblastic precursors in the avian embryo. Curr Top Microbiol Immunol 251:83–90
5. Yancopoulos GD, Davis S, Gale NW, Rudge JS, Wiegand SJ, Holash J (2000) Vascular-specific growth factors and blood vessel formation. Nature 407:242–248
6. Shalaby F, Rossant J, Yamaguchi TP, Gertsenstein M, Wu XF, Breitman ML, Schuh AC (1995) Failure of blood-island formation and vasculogenesis in Flk-1-deficient mice. Nature 376:62–66
7. Risau W (1997) Mechanisms of angiogenesis. Nature 386:671–674
8. Patel-Hett S, D'Amore PA (2011) Signal transduction in vasculogenesis and developmental angiogenesis. Int J Dev Biol 55: 353–363
9. Djonov V, Schmid M, Tschanz SA, Burri PH (2000) Intussusceptive angiogenesis: its role in embryonic vascular network formation. Circ Res 86:286–292
10. Patan S (2000) Vasculogenesis and angiogenesis as mechanisms of vascular network formation, growth and remodeling. J Neurooncol 50:1–15
11. Pardanaud L, Dieterlen-Lievre F (2000) Ontogeny of the endothelial system in the avian model. Adv Exp Med Biol 476:67–78
12. Carmeliet P (2000) Mechanisms of angiogenesis and arteriogenesis. Nat Med 6:389–395
13. Davis GE, Senger DR (2005) Endothelial extracellular matrix: biosynthesis, remodeling, and functions during vascular morphogenesis and neovessel stabilization. Circ Res 97:1093–1107
14. Senger DR, Davis GE (2011) Angiogenesis. Cold Spring Harb Perspect Biol 3:a005090
15. Gerhardt H, Golding M, Fruttiger M, Ruhrberg C, Lundkvist A, Abramsson A, Jeltsch M, Mitchell C, Alitalo K, Shima D, Betsholtz C (2003) VEGF guides angiogenic sprouting utilizing endothelial tip cell filopodia. J Cell Biol 161:1163–1177
16. Ruhrberg C, Gerhardt H, Golding M, Watson R, Ioannidou S, Fujisawa H, Betsholtz C, Shima DT (2002) Spatially restricted patterning cues provided by heparin-binding VEGF-A control blood vessel branching morphogenesis. Genes Dev 16:2684–2698
17. Lindahl P, Johansson BR, Leveen P, Betsholtz C (1997) Pericyte loss and microaneurysm formation in PDGF-B-deficient mice. Science 277:242–245
18. Lindahl P, Hellstrom M, Kalen M, Betsholtz C (1998) Endothelial-perivascular cell signaling in vascular development: lessons from knockout mice. Curr Opin Lipidol 9:407–411
19. Hellstrom M, Kalen M, Lindahl P, Abramsson A, Betsholtz C (1999) Role of PDGF-B and PDGFR-beta in recruitment of

vascular smooth muscle cells and pericytes during embryonic blood vessel formation in the mouse. Development 126:3047–3055
20. Kurz H (2000) Physiology of angiogenesis. J Neurooncol 50:17–35
21. Geudens I, Gerhardt H (2011) Coordinating cell behaviour during blood vessel formation. Development 138:4569–4583
22. Hoeben A, Landuyt B, Highley MS, Wildiers H, Van Oosterom AT, De Bruijn EA (2004) Vascular endothelial growth factor and angiogenesis. Pharmacol Rev 56:549–580
23. Plate KH, Risau W (1995) Angiogenesis in malignant gliomas. Glia 15:339–347
24. Plate KH, Scholz A, Dumont DJ (2012) Tumor angiogenesis and anti-angiogenic therapy in malignant gliomas revisited. Acta Neuropathol 124:763–775
25. Beck H, Plate KH (2009) Angiogenesis after cerebral ischemia. Acta Neuropathol 117: 481–496
26. Acker T, Plate KH (2003) Role of hypoxia in tumor angiogenesis-molecular and cellular angiogenic crosstalk. Cell Tissue Res 314: 145–155
27. Reiss Y, Machein MR, Plate KH (2005) The role of angiopoietins during angiogenesis in gliomas. Brain Pathol 15:311–317
28. Folkman J (2007) Angiogenesis: an organizing principle for drug discovery? Nat Rev Drug Discov 6:273–286
29. Hallene KL, Oby E, Lee BJ, Santaguida S, Bassanini S, Cipolla M, Marchi N, Hossain M, Battaglia G, Janigro D (2006) Prenatal exposure to thalidomide, altered vasculogenesis, and CNS malformations. Neuroscience 142:267–283
30. Liebner S, Plate KH (2010) Differentiation of the brain vasculature: the answer came blowing by the Wnt. J Angiogenes Res 2:1
31. Vasudevan A, Long JE, Crandall JE, Rubenstein JL, Bhide PG (2008) Compartment-specific transcription factors orchestrate angiogenesis gradients in the embryonic brain. Nat Neurosci 11:429–439
32. Bar T (1983) Patterns of vascularization in the developing cerebral cortex. Ciba Found Symp 100:20–36
33. Ogunshola OO, Stewart WB, Mihalcik V, Solli T, Madri JA, Ment LR (2000) Neuronal VEGF expression correlates with angiogenesis in postnatal developing rat brain. Brain Res Dev Brain Res 119:139–153
34. Vasudevan A, Bhide PG (2008) Angiogenesis in the embryonic CNS: a new twist on an old tale. Cell Adh Migr 2:167–169
35. Bar T (1980) The vascular system of the cerebral cortex. Adv Anat Embryol Cell Biol 59:I–VI, 1–62
36. Wittko IM, Schanzer A, Kuzmichev A, Schneider FT, Shibuya M, Raab S, Plate KH (2009) VEGFR-1 regulates adult olfactory bulb neurogenesis and migration of neural progenitors in the rostral migratory stream in vivo. J Neurosci 29:8704–8714
37. Raab S, Plate KH (2007) Different networks, common growth factors: shared growth factors and receptors of the vascular and the nervous system. Acta Neuropathol 113:607–626
38. Raab S, Beck H, Gaumann A, Yuce A, Gerber HP, Plate K, Hammes HP, Ferrara N, Breier G (2004) Impaired brain angiogenesis and neuronal apoptosis induced by conditional homozygous inactivation of vascular endothelial growth factor. Thromb Haemost 91: 595–605
39. Robertson PL, Du Bois M, Bowman PD, Goldstein GW (1985) Angiogenesis in developing rat brain: an in vivo and in vitro study. Brain Res 355:219–223
40. Engelhardt B (2003) Development of the blood–brain barrier. Cell Tissue Res 314: 119–129
41. Bauer HC, Bauer H (2000) Neural induction of the blood–brain barrier: still an enigma. Cell Mol Neurobiol 20:13–28
42. Plate KH (1999) Mechanisms of angiogenesis in the brain. J Neuropathol Exp Neurol 58: 313–320
43. Li H-Y, Zhou X-F (2007) Potential conversion of adult clavicle-derived chondrocytes into neural lineage cellsin vitro. J Cell Physiol 214:630–644
44. Stamataki D, Ulloa F, Tsoni SV, Mynett A, Briscoe J (2005) A gradient of Gli activity mediates graded Sonic Hedgehog signaling in the neural tube. Genes Dev 19:626–641
45. Isohata N, Aoyagi K, Mabuchi T, Daiko H, Fukaya M, Ohta H, Ogawa K, Yoshida T, Sasaki H (2009) Hedgehog and epithelial-mesenchymal transition signaling in normal and malignant epithelial cells of the esophagus. Int J Cancer 125:1212–1221
46. Maury JJP, Choo AB-H, Chan KK-K (2011) Technical advances to genetically engineering human embryonic stem cells. Integr Biol 3: 717–723
47. Krebs LT, Xue Y, Norton CR, Shutter JR, Maguire M, Sundberg JP, Gallahan D, Closson V, Kitajewski J, Callahan R, Smith GH, Stark KL, Gridley T (2000) Notch signaling is essential for vascular morphogenesis in mice. Genes Dev 14:1343–1352
48. Uyttendaele H, Ho J, Rossant J, Kitajewski J (2001) Vascular patterning defects associated with expression of activated Notch4 in embryonic endothelium. Proc Natl Acad Sci U S A 98:5643–5648

49. Cheng N, Brantley DM, Chen J (2002) The ephrins and Eph receptors in angiogenesis. Cytokine Growth Factor Rev 13:75–85
50. Wang HU, Chen ZF, Anderson DJ (1998) Molecular distinction and angiogenic interaction between embryonic arteries and veins revealed by ephrin-B2 and its receptor Eph-B4. Cell 93:741–753
51. Wang Y, Nakayama M, Pitulescu ME, Schmidt TS, Bochenek ML, Sakakibara A, Adams S, Davy A, Deutsch U, Luthi U, Barberis A, Benjamin LE, Makinen T, Nobes CD, Adams RH (2010) Ephrin-B2 controls VEGF-induced angiogenesis and lymphangiogenesis. Nature 465:483–486
52. Nikolic I, Plate KH, Schmidt MH (2010) EGFL7 meets miRNA-126: an angiogenesis alliance. J Angiogenes Res 2:9
53. Klagsbrun M, Eichmann A (2005) A role for axon guidance receptors and ligands in blood vessel development and tumor angiogenesis. Cytokine Growth Factor Rev 16:535–548
54. Kise Y, Morinaka A, Teglund S, Miki H (2009) Sufu recruits GSK3beta for efficient processing of Gli3. Biochem Biophys Res Commun 387:569–574
55. Sellheyer K (2011) Basal cell carcinoma: cell of origin, cancer stem cell hypothesis and stem cell markers. Br J Dermatol 164:696–711
56. Dakubo GD, Mazerolle C, Furimsky M, Yu C, St-Jacques B, McMahon AP, Wallace VA (2008) Indian hedgehog signaling from endothelial cells is required for sclera and retinal pigment epithelium development in the mouse eye. Dev Biol 320:242–255
57. Eberl M, Klingler S, Mangelberger D, Loipetzberger A, Damhofer H, Zoidl K, Schnidar H, Hache H, Bauer H-C, Solca F, Hauser-Kronberger C, Ermilov AN, Verhaegen ME, Bichakjian CK, Dlugosz AA, Nietfeld W, Sibilia M, Lehrach H, Wierling C, Aberger F (2012) Hedgehog-EGFR cooperation response genes determine the oncogenic phenotype of basal cell carcinoma and tumour-initiating pancreatic cancer cells. EMBO Mol Med 4:218–233
58. Puppo F, Thomé V, Lhoumeau A-C, Cibois M, Gangar A, Lembo F, Belotti E, Marchetto S, Lécine P, Prébet T, Sebbagh M, Shin W-S, Lee S-T, Kodjabachian L, Borg J-P (2011) Protein tyrosine kinase 7 has a conserved role in Wnt/β-catenin canonical signalling. EMBO Rep 12:43–49
59. Choy SW, Cheng SH (2012) Hedgehog signaling. Vitam Horm 88:1–23
60. Cheng X, Huber TL, Chen VC, Gadue P, Keller GM (2008) Numb mediates the interaction between Wnt and Notch to modulate primitive erythropoietic specification from the hemangioblast. Development 135:3447–3458
61. Robel S, Berninger B, Götz M (2011) The stem cell potential of glia: lessons from reactive gliosis. Nat Rev Neurosci 12:88–104
62. Paces-Fessy M, Boucher D, Petit E, Paute-Briand S, Blanchet-Tournier M-F (2004) The negative regulator of Gli, Suppressor of fused (Sufu), interacts with SAP18, Galectin3 and other nuclear proteins. Biochem J 378: 353–362
63. Okano H, Imai T, Okabe M (2002) Musashi: a translational regulator of cell fate. J Cell Sci 115:1355–1359
64. Aubry L, Bugi A, Lefort N, Rousseau F, Peschanski M, Perrier AL (2008) Striatal progenitors derived from human ES cells mature into DARPP32 neurons in vitro and in quinolinic acid-lesioned rats. Proc Natl Acad Sci 105:16707–16712
65. (2007) Untitled-14338:1–7
66. Habela CW, Ernest NJ, Swindall AF, Sontheimer H (2008) Chloride accumulation drives volume dynamics underlying cell proliferation and migration. J Neurophysiol 101:750–757
67. Lenth RV (2007) Statistical power calculations. J Anim Sci 85:E24–E29
68. de Bont JM, Packer RJ, Michiels EM, den Boer ML, Pieters R (2008) Biological background of pediatric medulloblastoma and ependymoma: a review from a translational research perspective. Neuro-Oncology 10: 1040–1060
69. Ferrara N (2004) Vascular endothelial growth factor: basic science and clinical progress. Endocr Rev 25:581–611
70. Kalin TV, Ustiyan V, Kalinichenko VV (2011) Multiple faces of FoxM1 transcription factor: lessons from transgenic mouse models. Cell Cycle 10:396–405
71. Fodde R, Brabletz T (2007) Wnt/β-catenin signaling in cancer stemness and malignant behavior. Curr Opin Cell Biol 19:150–158
72. Kittappa R, Chang WW, Awatramani RB, McKay RDG (2007) The foxa2 gene controls the birth and spontaneous degeneration of dopamine neurons in old age. PLoS Biol 5: e325
73. Duan Y, Fan M (2011) Lentivirus-mediated gene silencing of beta-catenin inhibits growth of human tongue cancer cells. J Oral Pathol Med 40:643–650
74. Breier G, Clauss M, Risau W (1995) Coordinate expression of vascular endothelial growth factor receptor-1 (flt-1) and its ligand suggests a paracrine regulation of murine vascular development. Dev Dyn 204:228–239
75. Vaid M, Prasad R, Sun Q, Katiyar SK (2011) Silymarin targets β-catenin signaling in blocking migration/invasion of human melanoma cells. PLoS ONE 6:e23000

76. Shimizu K, Chiba S, Hosoya N, Kumano K, Saito T, Kurokawa M, Kanda Y, Hamada Y, Hirai H (2000) Binding of Delta1, Jagged1, and Jagged2 to Notch2 rapidly induces cleavage, nuclear translocation, and hyperphosphorylation of Notch2. Mol Cell Biol 20: 6913–6922
77. Emami KH, Nguyen C, Ma H, Kim DH, Jeong KW, Eguchi M, Moon RT, Teo J-L, Oh SW, Kim HY, Moon SH, Ha JR, Kahn M (2004) A small molecule inhibitor of beta-catenin/CREB-binding protein transcription [corrected]. Proc Natl Acad Sci U S A 101: 12682–12687
78. Forsythe JA, Jiang BH, Iyer NV, Agani F, Leung SW, Koos RD, Semenza GL (1996) Activation of vascular endothelial growth factor gene transcription by hypoxia-inducible factor 1. Mol Cel Biol 16:4604–4613
79. Yang L, Jackson E, Woerner BM, Perry A, Piwnica-Worms D, Rubin JB (2007) Blocking CXCR4-mediated cyclic AMP suppression inhibits brain tumor growth in vivo. Cancer Res 67:651–658
80. Samuelov L, Sprecher E, Tsuruta D, Bíró T, Kloepper JE, Paus R (2012) P-cadherin regulates human hair growth and cycling via canonical Wnt signaling and transforming growth factor-β2. J Invest Dermatol 132(10): 2332–2341
81. Grigoryan T, Wend P, Klaus A, Birchmeier W (2008) Deciphering the function of canonical Wnt signals in development and disease: conditional loss- and gain-of-function mutations of beta-catenin in mice. Genes Dev 22: 2308–2341
82. Tihan T, Pekmezci M, Karnezis A (2011) Neural stem cells and their role in the pathology and classification of central nervous system tumors. Türk Patoloji Derg 27:1–11
83. Plate KH, Beck H, Danner S, Allegrini PR, Wiessner C (1999) Cell type specific upregulation of vascular endothelial growth factor in an MCA-occlusion model of cerebral infarct. J Neuropathol Exp Neurol 58:654–666
84. Amoh Y, Aki R, Hamada Y, Niiyama S, Eshima K, Kawahara K, Sato Y, Tani Y, Hoffman RM, Katsuoka K (2012) Nestin-positive hair follicle pluripotent stem cells can promote regeneration of impinged peripheral nerve injury. J Dermatol 39:33–38
85. Hockemeyer D, Jaenisch R (2010) Gene targeting in human pluripotent cells. Cold Spring Harb Symp Quant Biol 75:201–209
86. Pannuti A, Foreman K, Rizzo P, Osipo C, Golde T, Osborne B, Miele L (2010) Targeting Notch to target cancer stem cells. Clin Cancer Res 16:3141–3152
87. Kovacs JJ, Whalen EJ, Liu R, Xiao K, Kim J, Chen M, Wang J, Chen W, Lefkowitz RJ (2008) Arrestin-mediated localization of smoothened to the primary cilium. Science 320:1777–1781
88. Mohapel P, Frielingsdorf H, Haggblad J, Zachrisson O, Brundin P (2005) Platelet-derived growth factor (PDGF-BB) and brain-derived neurotrophic factor (BDNF) induce striatal neurogenesis in adult rats with 6-hydroxydopamine lesions. Neuroscience 132:767–776
89. Rahnama F, Shimokawa T, Lauth M, Finta C, Kogerman P, Teglund S, Toftgård R, Zaphiropoulos PG (2006) Inhibition of GLI1 gene activation by Patched1. Biochem J 394: 19–26
90. Zhang F, Tang Z, Hou X, Lennartsson J, Li Y, Koch AW, Scotney P, Lee C, Arjunan P, Dong L, Kumar A, Rissanen TT, Wang B, Nagai N, Fons P, Fariss R, Zhang Y, Wawrousek E, Tansey G, Raber J, Fong G-H, Ding H, Greenberg DA, Becker KG, Herbert J-M, Nash A, Ylä-Herttuala S, Cao Y, Watts RJ, Li X (2009) VEGF-B is dispensable for blood vessel growth but critical for their survival, and VEGF-B targeting inhibits pathological angiogenesis. Proc Natl Acad Sci 106: 6152–6157
91. Li X, Lee C, Tang Z, Zhang F, Arjunan P, Li Y, Hou X, Kumar A, Dong L (2009) VEGF-B: a survival, or an angiogenic factor? Cell Adh Migr 3:322–327
92. Mould AW, Tonks ID, Cahill MM, Pettit AR, Thomas R, Hayward NK, Kay GF (2003) Vegfb gene knockout mice display reduced pathology and synovial angiogenesis in both antigen-induced and collagen-induced models of arthritis. Arthritis Rheum 48:2660–2669
93. Sun Y, Jin K, Childs JT, Xie L, Mao XO, Greenberg DA (2004) Increased severity of cerebral ischemic injury in vascular endothelial growth factor-B-deficient mice. J Cereb Blood Flow Metab 24:1146–1152
94. Enholm B, Paavonen K, Ristimaki A, Kumar V, Gunji Y, Klefstrom J, Kivinen L, Laiho M, Olofsson B, Joukov V, Eriksson U, Alitalo K (1997) Comparison of VEGF, VEGF-B, VEGF-C and Ang-1 mRNA regulation by serum, growth factors, oncoproteins and hypoxia. Oncogene 14:2475–2483
95. Kele J, Andersson ER, Villaescusa JC, Cajanek L, Parish CL, Bonilla S, Toledo EM, Bryja V, Rubin JS, Shimono A, Arenas E (2012) SFRP1 and 2 dose-dependently regulate midbrain dopamine neuron development in vivo and in embryonic stem cells. Stem Cells 30(5):865–875

96. Barakat MT, Scott MP (2009) Tail wags dog: primary cilia and tumorigenesis. Cancer Cell 16:276–277
97. Androutsellis-Theotokis A, Rueger MA, Park DM, Boyd JD, Padmanabhan R, Campanati L, Stewart CV, LeFranc Y, Plenz D, Walbridge S, Lonser RR, McKay RDG (2010) Angiogenic factors stimulate growth of adult neural stem cells. PLoS ONE 5:e9414
98. Athar M, Tang X, Lee JL, Kopelovich L, Kim AL (2006) Hedgehog signalling in skin development and cancer. Exp Dermatol 15: 667–677
99. Guo A, Owens WA, Coady M, Liu D, Jahoda CAB (2012) An in vivo mouse model of human skin replacement for wound healing and cell therapy studies. J Plast Reconstr Aesthet Surg 65:1129–1131
100. Lim S-H, Choi SA, Lee JY, Wang K-C, Phi JH, Lee D-H, Song SH, Song JH, Jin X, Kim H, Lee HJ, Lim I, Kim SU, Kim S-K (2011) Therapeutic targeting of subdural medulloblastomas using human neural stem cells expressing carboxylesterase. Cancer Gene Ther 18:817–824
101. Blanpain C, Lowry WE, Pasolli HA, Fuchs E (2006) Canonical notch signaling functions as a commitment switch in the epidermal lineage. Genes Dev 20:3022–3035
102. Fong GH, Zhang L, Bryce DM, Peng J (1999) Increased hemangioblast commitment, not vascular disorganization, is the primary defect in flt-1 knock-out mice. Development 126:3015–3025
103. Fitch MJ, Campagnolo L, Kuhnert F, Stuhlmann H (2004) Egfl7, a novel epidermal growth factor-domain gene expressed in endothelial cells. Dev Dyn 230:316–324
104. Whitman M (1997) Signal transduction. Feedback from inhibitory SMADs. Nature 389:549–551
105. Chinchilla P, Xiao L, Kazanietz MG, Riobo NA (2010) Hedgehog proteins activate pro-angiogenic responses in endothelial cells through non-canonical signaling pathways. Cell Cycle 9:570–579
106. Klein RS, Rubin JB, Gibson HD, DeHaan EN, Alvarez-Hernandez X, Segal RA, Luster AD (2001) SDF-1 alpha induces chemotaxis and enhances Sonic hedgehog-induced proliferation of cerebellar granule cells. Development 128:1971–1981
107. Kremer C, Breier G, Risau W, Plate KH (1997) Up-regulation of flk-1/vascular endothelial growth factor receptor 2 by its ligand in a cerebral slice culture system. Cancer Res 57:3852–3859
108. Aberger F, Kern D, Greil R, Hartmann TN (2012) Canonical and noncanonical Hedgehog/GLI signaling in hematological malignancies. Vitam Horm 88:25–54
109. Jiang J, Hui CC (2008) Hedgehog signaling in development and cancer. Dev Cell 15: 801–812
110. Rizzo P, Miao H, D'Souza G, Osipo C, Yun J, Zhao H, Mascarenhas J, Wyatt D, Antico G, Hao L, Yao K, Rajan P, Hicks C, Siziopikou K, Selvaggi S, Bashir A, Bhandari D, Marchese A, Lendahl U, Qin JZ, Tonetti DA, Albain K, Nickoloff BJ, Miele L (2008) Cross-talk between notch and the estrogen receptor in breast cancer suggests novel therapeutic approaches. Cancer Res 68:5226–5235
111. Clement V, Sanchez P, de Tribolet N, Radovanovic I, Ruiz i Altaba A (2007) HEDGEHOG-GLI1 signaling regulates human glioma growth, cancer stem cell self-renewal, and tumorigenicity. Curr Biol 17: 165–172
112. Bergstein I (2002) In vivo enhanced expression of patched dampens the sonic hedgehog pathway. Mol Ther 6:258–264
113. Silver DJ, Steindler DA (2009) Common astrocytic programs during brain development, injury and cancer. Trends Neurosci 32: 303–311
114. Xu J, Krebs LT, Gridley T (2010) Generation of mice with a conditional null allele of the Jagged2 gene. Genesis 48:390–393
115. Rodig SJ, Payne EG, Degar BA, Rollins B, Feldman AL, Jaffe ES, Androkites A, Silverman LB, Longtine JA, Kutok JL, Fleming MD, Aster JC (2008) Aggressive Langerhans cell histiocytosis following T-ALL: clonally related neoplasms with persistent expression of constitutively active NOTCH1. Am J Hematol 83:116–121
116. Kuraguchi M, Wang X-P, Bronson RT, Rothenberg R, Ohene-Baah NY, Lund JJ, Kucherlapati M, Maas RL, Kucherlapati R (2006) Adenomatous polyposis coli (APC) is required for normal development of skin and thymus. PLoS Genet 2:e146
117. Pierfelice T, Alberi L, Gaiano N (2011) Notch in the vertebrate nervous system: an old dog with new tricks. Neuron 69: 840–855
118. Pola R, Ling LE, Silver M, Corbley MJ, Kearney M, Blake Pepinsky R, Shapiro R, Taylor FR, Baker DP, Asahara T, Isner JM (2001) The morphogen Sonic hedgehog is an indirect angiogenic agent upregulating two families of angiogenic growth factors. Nat Med 7:706–711

119. Patan S (1998) TIE1 and TIE2 receptor tyrosine kinases inversely regulate embryonic angiogenesis by the mechanism of intussusceptive microvascular growth. Microvasc Res 56:1–21
120. Kelly BD, Hackett SF, Hirota K, Oshima Y, Cai Z, Berg-Dixon S, Rowan A, Yan Z, Campochiaro PA, Semenza GL (2003) Cell type-specific regulation of angiogenic growth factor gene expression and induction of angiogenesis in nonischemic tissue by a constitutively active form of hypoxia-inducible factor 1. Circ Res 93:1074–1081
121. Heiser PW, Lau J, Taketo MM, Herrera PL, Hebrok M (2006) Stabilization of beta-catenin impacts pancreas growth. Development 133:2023–2032
122. Roca C, Adams RH (2007) Regulation of vascular morphogenesis by Notch signaling. Genes Dev 21:2511–2524
123. Hellström M, Phng L-K, Hofmann JJ, Wallgard E, Coultas L, Lindblom P, Alva J, Nilsson A-K, Karlsson L, Gaiano N, Yoon K, Rossant J, Iruela-Arispe ML, Kalén M, Gerhardt H, Betsholtz C (2007) Dll4 signalling through Notch1 regulates formation of tip cells during angiogenesis. Nature 445:776–780
124. Hofmann JJ, Iruela-Arispe ML (2007) Notch signaling in blood vessels: who is talking to whom about what? Circ Res 100:1556–1568
125. Blanco R, Gerhardt H (2012) VEGF and notch in tip and stalk cell selection. Cold Spring Harb Perspect Med
126. Adams RH, Eichmann A (2010) Axon guidance molecules in vascular patterning. Cold Spring Harb Perspect Biol 2:a001875
127. Kurz H, Gartner T, Eggli PS, Christ B (1996) First blood vessels in the avian neural tube are formed by a combination of dorsal angioblast immigration and ventral sprouting of endothelial cells. Dev Biol 173:133–147
128. Armulik A, Abramsson A, Betsholtz C (2005) Endothelial/pericyte interactions. Circ Res 97:512–523
129. Stefater JA, Lewkowich I, Rao S, Mariggi G, Carpenter AC, Burr AR, Fan J, Ajima R, Molkentin JD, Williams BO, Wills-Karp M, Pollard JW, Yamaguchi T, Ferrara N, Gerhardt H, Lang RA (2011) Regulation of angiogenesis by a non-canonical Wnt-Flt1 pathway in myeloid cells. Nature 474:511–515
130. Berger AC, Wang X-Q, Zalatoris A, Cenna J, Watson JC (2004) A murine model of ex vivo angiogenesis using aortic disks grown in fibrin clot☆. Microvasc Res 68:179–187
131. Benchabane H, Xin N, Tian A, Hafler BP, Nguyen K, Ahmed A, Ahmed Y (2011) Jerky/Earthbound facilitates cell-specific Wnt/Wingless signalling by modulating β-catenin-TCF activity. EMBO J 30: 1444–1458
132. Park HJ, Carr JR, Wang Z, Nogueira V, Hay N, Tyner AL, Lau LF, Costa RH, Raychaudhuri P (2009) FoxM1, a critical regulator of oxidative stress during oncogenesis. EMBO J 28:2908–2918
133. Thurston G, Rudge JS, Ioffe E, Zhou H, Ross L, Croll SD, Glazer N, Holash J, McDonald DM, Yancopoulos GD (2000) Angiopoietin-1 protects the adult vasculature against plasma leakage. Nat Med 6:460–463
134. Lim KJ, Bisht S, Bar EE, Maitra A, Eberhart CG (2011) A polymeric nanoparticle formulation of curcumin inhibits growth, clonogenicity and stem-like fraction in malignant brain tumors. Cancer Biol Ther 11: 464–473
135. Lemjabbar-Alaoui H, Dasari V, Sidhu SS, Mengistab A, Finkbeiner W, Gallup M, Basbaum C (2006) Wnt and Hedgehog are critical mediators of cigarette smoke-induced lung cancer. PLoS ONE 1:e93

Chapter 2

Cerebral Angiogenesis: A Realistic Therapy for Ischemic Disease?

David A. Greenberg

Abstract

Angiogenesis, the sprouting of new capillaries from existing blood vessels, accompanies clinical and experimental stroke and is focused particularly in the salvageable ischemic border zone. As this endogenous angiogenic response correlates positively with clinical prognosis, a more complete understanding of the underlying molecular mechanisms and timing of these events may help in the design of novel therapies for vascular regeneration after stroke. In this review we discuss endogenous protective mechanisms, including angiogenesis and vasculogenesis, and underlying molecular mechanisms. We also consider the feasibility of angiogenic therapy for stroke and its optimal timing.

Key words Angiogenesis, Arteriogenesis, Collaterals, Endothelial progenitor cells, Ischemia, Stroke, Thrombolytics, Vasculogenesis, Vascular endothelial growth factor

1 Introduction

Angiogenesis, the sprouting of new capillaries from existing blood vessels, typically in response to hypoxia, is an integral feature of physiological (e.g., development) and pathological (e.g., cancer) processes [1]. Angiogenesis also accompanies clinical [2] and experimental [3] stroke, wherein the expression of angiogenic factors is induced within hours and new capillaries are formed within days of the onset of ischemia [3, 4]. Although angiogenesis in this setting is focused in the salvageable ischemic border zone and correlates with prognosis [5], the precise relationship between angiogenesis and outcome from stroke is uncertain. Unsettled in particular are whether this relationship is causal and, if so, what mechanisms are responsible.

Richard Milner (ed.), *Cerebral Angiogenesis: Methods and Protocols*, Methods in Molecular Biology, vol. 1135, DOI 10.1007/978-1-4939-0320-7_2, © Springer Science+Business Media New York 2014

2 Endogenous Mechanisms of Improving Blood Flow to Ischemic Brain Tissue

Several scenarios have the potential for improving blood flow to ischemic brain tissue. Immediately upon the establishment of a pressure differential between normally perfused and occluded arteries, flow occurs through preexisting collateral channels, such as the leptomeningeal anastomoses connecting the anterior, middle, and posterior cerebral arteries [6]. These collaterals may then enlarge over time to support increased flow through a process termed arteriogenesis which, like the initiation of collateral flow, is pressure—rather than hypoxia—driven [7]. Collateral blood flow to ischemic brain regions is associated with reduced stroke risk [8] and improved outcome [9], and also with a better response to thrombolytic therapy [10], perhaps due to enhanced access of thrombolytics to the target thrombus through collateral channels. Because the onset of collateral flow is virtually instantaneous in ischemia, it is easy to imagine how this process could salvage tissue and prevent or reduce neurological deficits.

Vasculogenesis, defined as the de novo generation of blood vessels from progenitor cells, is usually associated with development, but may also occur postnatally [11]. Endothelial progenitor cells in blood-forming or local tissues appear to be mobilized by ischemia and may contribute, directly or indirectly, to vessel repair and growth following stroke [12]. Some studies suggest that this process might help promote a more favorable clinical outcome [13]. However, vasculogenesis is unlikely to occur quickly enough to prevent tissue damage from acute cerebral ischemia.

The role of spontaneous or therapeutically induced angiogenesis in recovery from stroke is also unclear, partly because of the delay involved in constructing new, functional vessels. Moreover, demonstrating a functional benefit of angiogenesis would require that angiogenesis be selectively ablated or stimulated, and the consequences observed, but such selectivity is difficult to achieve. For example, angiogenic stimuli like hypoxia, growth factors, and cell transplants can trigger additional, potentially protective or restorative processes, such as neurogenesis, gliogenesis, and trophic effects.

3 Potential Mechanisms of Angiogenic Protection Following Ischemic Stroke

At least three potential mechanisms have been invoked as possible bases for the salutary effects of angiogenesis in stroke [14]. One possibility is that angiogenesis protects against the acute effects of cerebral ischemia by enhancing cell survival, but this is difficult to reconcile with the observation that ischemic cell death proceeds much more rapidly than angiogenesis. On the other hand, angiogenic factors induced by brain ischemia have acute cytoprotective effects that are evident well before the appearance of new vessels [15], so in this sense angiogenesis could be viewed as acutely protective. Another way in which angiogenesis might afford protection from ischemia is in the setting of recurrent transient ischemic attacks. Thus, ischemia-induced angiogenesis might contribute to ischemic preconditioning leading to ischemic tolerance, as proposed for exercise-induced tolerance [16].

Another theory, the clean-up hypothesis [17, 18], posits that angiogenesis contributes to recovery from stroke by providing macrophages with access to necrotic brain tissue and thereby facilitating its removal. Supportive evidence includes the transiency of many vessels induced by ischemia and their extension into the ischemic core.

Finally, angiogenesis may be an important factor in the regeneration of brain tissue after stroke. First, any regenerated tissue presumably requires a new vascular supply. Second, angiogenesis is thought to provide a niche for the proliferation of new neurons (neurogenesis) and their migration to sites of ischemic brain injury [19]. Third, as noted in relation to acute neuroprotection, signaling by angiogenic factors may have non-angiogenic effects that promote recovery. For example, vascular endothelial growth factor (VEGF-A) [20] and its principal angiogenesis-related receptor (VEGFR-2) [21] are induced at remote sites of brain plasticity after experimental stroke in primates.

4 Potential Clinical Use of Angiogenic Therapy

Is cerebral angiogenesis a realistic therapy for ischemic disease? Probably not, in the sense that angiogenesis is unlikely to be inducible quickly enough at the onset of ischemia to protect brain cells from acute ischemic death. However, if the definition of angiogenic therapy is extended to include treatment with drugs or cells that are angiogenic but also have other (e.g., neuroprotective or neuroregenerative) effects, then preclinical studies suggest that this approach may be effective [22, 23]. Purely angiogenic treatments could also be effective if administered preemptively, such as for recurrent transient ischemic attacks, where cerebral ischemia can be anticipated for at least several days before its occurrence. This presupposes that the cause of ischemia in a particular situation is amenable to correction by an increase in capillary flow, which may or may not be the case. Selective therapeutic enhancement of angiogenesis might also contribute to brain repair after stroke, if this involves processes, like neurogenesis, that depend on a vascular niche, or if regenerating tissue outgrows the existing blood supply. In each instance, an underlying assumption is that postischemic induction of angiogenesis by endogenous factors is not already optimal.

References

1. Ribatti D, Crivellato E (2012) Sprouting angiogenesis, a reappraisal. Dev Biol 372:157–165
2. Liu HM (1988) Neovasculature and blood–brain barrier in ischemic brain infarct. Acta Neuropathol 75:422–426
3. Chen HH, Chien CH, Liu HM (1994) Correlation between angiogenesis and basic fibroblast growth factor expression in experimental brain infarct. Stroke 25:1651–1657
4. Kovacs Z, Ikezaki K, Samoto K et al (1996) VEGF and flt: expression time kinetics in rat brain infarct. Stroke 27:1865–1873
5. Krupinski J, Kaluza J, Kumar P et al (1994) Role of angiogenesis in patients with cerebral ischemic stroke. Stroke 25:1794–1798
6. Weidner W, Hanafee W, Markham CH (1965) Intracranial collateral circulation via leptomeningeal and rete mirabile anastomoses. Neurology 15:39–48
7. Liebeskind DS (2003) Collateral circulation. Stroke 34:2279–2284
8. Liebeskind DS, Cotsonis GA, Saver JL et al (2011) Collaterals dramatically alter stroke risk in intracranial atherosclerosis. Ann Neurol 69:963–974
9. Bang OY, Saver JL, Buck BH et al (2008) Impact of collateral flow on tissue fate in acute ischaemic stroke. J Neurol Neurosurg Psychiatry 79:625–629
10. Bang OY, Saver JL, Kim SJ et al (2011) Collateral flow predicts response to endovascular therapy for acute ischemic stroke. Stroke 42:693–699
11. Pearson JD (2010) Endothelial progenitor cells: an evolving story. Microvasc Res 79:162–168
12. Tongers J, Roncalli JG, Losordo DW (2010) Role of endothelial progenitor cells during ischemia-induced vasculogenesis and collateral formation. Microvasc Res 79:200–206
13. Sobrino T, Hurtado O, Moro MA et al (2007) The increase of circulating endothelial progenitor cells after acute ischemic stroke is associated with good outcome. Stroke 38:2759–2764
14. Ergul A, Alhusban A, Fagan SC (2012) Angiogenesis: a harmonized target for recovery after stroke. Stroke 43:2270–2274
15. Sun Y, Jin K, Xie L et al (2003) VEGF-induced neuroprotection, neurogenesis, and angiogenesis after focal cerebral ischemia. J Clin Invest 111:1843–1851
16. Zhang F, Wu Y, Jia J (2011) Exercise preconditioning and brain ischemic tolerance. Neuroscience 177:170–176
17. Manoonkitiwongsa PS, Jackson-Friedman C, McMillan PJ et al (2001) Angiogenesis after stroke is correlated with increased numbers of macrophages: the clean-up hypothesis. J Cereb Blood Flow Metab 21:1223–1231
18. Yu SW, Friedman B, Cheng Q et al (2007) Stroke-evoked angiogenesis results in a transient population of microvessels. J Cereb Blood Flow Metab 27:755–763
19. Ohab JJ, Fleming S, Blesch A et al (2006) A neurovascular niche for neurogenesis after stroke. J Neurosci 26:13007–13016
20. Stowe AM, Plautz EJ, Eisner-Janowicz I et al (2007) VEGF protein associates to neurons in remote regions following cortical infarct. J Cereb Blood Flow Metab 27:76–85
21. Stowe AM, Plautz EJ, Nguyen P et al (2008) Neuronal HIF-1α protein and VEGFR-2 immunoreactivity in functionally related motor areas following a focal M1 infarct. J Cereb Blood Flow Metab 28:612–620
22. Zhang ZG, Zhang L, Jiang Q et al (2000) VEGF enhances angiogenesis and promotes blood–brain barrier leakage in the ischemic brain. J Clin Invest 106:829–838
23. Horie N, Pereira MP, Niizuma K et al (2011) Transplanted stem cell-secreted vascular endothelial growth factor effects poststroke recovery, inflammation, and vascular repair. Stem Cells 29:274–285

Chapter 3

Vascular Normalization in Cerebral Angiogenesis: Friend or Foe?

Jisook Lee, Andrew Baird, and Brian P. Eliceiri

Abstract

Current antiangiogenic therapies have led to the observation that such agents can lead to improved tumor vessel structure and function termed "vascular normalization" which reduces tumor burden. However, vessel normalization is a transient process, and patients often develop resistance/poor response to anti-vascular strategies that remains an important clinical challenge. Therefore, increasing effort has been made to better understand the cellular and molecular mechanisms of vascular normalization and its contribution to immunomodulation. Herein, we summarize the recent effort to better understand the cellular and molecular mechanisms of vascular normalization with a focus on preclinical genetic models. These studies remain important directions for a mechanistic understanding of the complexities of the maintenance of BBB integrity and the impact of its breakdown on tumor dissemination and pharmaco-distribution of therapeutics.

Key words Vascular normalization, Angiogenesis, Blood–brain barrier, Glioma, Antiangiogenic therapy, Bevacizumab

1 Introduction

Cerebral vascular normalization refers to the process in which the central nervous system (CNS) microvasculature, when damaged in disease and/or by aging, is no longer compromised by surrounding metabolic, pharmacological, or immunological challenges. Under normal circumstances, an intact blood–brain barrier (BBB) protects the brain by limiting access of blood-borne molecules and circulating immune cells [1, 2]. While angiogenesis and remodeling of the BBB occur in normal development and retain many of the essential cellular and molecular parameters of an adult BBB [3], breakdown of the BBB is generally associated with CNS injury, ischemia, and brain cancers. The tight regulation and organization of brain endothelia, pericytes, astrocytes, and neurons, is termed the neurovascular unit (NVU) and functions within the extracellular matrix (ECM) microenvironment [4]. Because dysregulation of

Richard Milner (ed.), *Cerebral Angiogenesis: Methods and Protocols*, Methods in Molecular Biology, vol. 1135,
DOI 10.1007/978-1-4939-0320-7_3, © Springer Science+Business Media New York 2014

the NVU occurs in response to various CNS injury pathologies particularly in CNS tumors, understanding the mechanism of cerebral angiogenesis and BBB dysfunction is essential for the delivery of therapeutics and restoration of BBB homeostasis. Furthermore, the mobilization of local (i.e., microglia) vs. circulating immune cells (i.e., blood monocytes/macrophages and lymphocytes) to CNS tumors can influence the vascular endothelium and BBB integrity to regulate tumor progression.

2 Antiangiogenesis Therapy and Vascular Normalization in CNS Tumors

While many growth factors and cytokines are known to mediate various responses to CNS injury, vascular endothelial growth factor (VEGF/VPF) is unique among these factors as both a potent growth factor and inducer of vascular permeability (VP) of blood vessels. With the discovery of its central role in vascular biology and the development of VEGF antagonists to starve tumors of their blood supply (i.e., antiangiogenesis), the efficacy of VEGF blockade on reducing angiogenic blood vessel density, VP, and vessel diameter has proven therapeutically valuable. Tumor-associated blood vessels have elevated VP, vessel tortuosity, and decreased patency which lead to a hypoxic tumor microenvironment with high interstitial fluid pressure (IFP) [5], that there is also a disruption of efficient drug delivery and reduced immunosurveillance by circulating leukocytes. Targeting tumor vessels with antiangiogenic therapies (i.e., bevacizumab) has led to the paradoxical observation that such therapies can have the effect of restoring blood vessel integrity (i.e., vascular normalization) [6–9]. Vascular normalization has been identified as a novel mechanism for balancing several molecular and physiological properties of angiogenesis-related vascular remodeling. For example, Goel et al. detail the specific effects of the mouse anti-human VEGF monoclonal antibody (mAb, A.4.6.1), the humanized variant of A.4.6.1 (i.e., bevacizumab), an anti-mouse VEGF mAb, and a VEGF trap (i.e., aflibercept) as well as various anti-VEGFR agents (DC101 and tyrosine kinase inhibitors) [10]. Dosing of these anti-VEGF agents, primarily in tumor-bearing animal models, typically results in a reduction in vessel diameter, tortuosity, and permeability thus underscoring the link between VEGF and the vascular normalization phenotype. While the vascular normalization phenotype is also observed in colorectal, melanoma, breast, and prostate models, glioma-associated tumor vessels have been the focus of the vascular normalization phenotype based in part on the availability of preclinical models with intravital imaging and the number and scope of clinical reports of anti-VEGF therapy in brain tumors. In fact, with the FDA approval of bevacizumab for glioma and the increasing numbers of breast and melanoma cancer patients with brain metastases secondary to managed primary neoplasms, there has been an increased effort to better understand the mechanisms of action of vascular normalization and develop strategies to exploit vascular normalization for reducing tumor burden.

For example, in the brain, preclinical models of orthotopic glioma treated with agents that induce the vascular normalization phenotype are associated with a recruitment of perivascular pericytes, changes in endothelial proteins regulating endothelial stability/permeability (VEGF, bFGF, Ang-1/Ang-2, TSP-1, iNOS, and Src/FAK), and changes in the ECM microenvironment of blood vessels [6]. The effect of antiangiogenic therapies of increasing tumor oxygenation further complicates approaches designed to starve a tumor of a blood supply where there may be only a narrow time frame in which therapy-induced vascular normalization occurs. The transient nature of pharmacologically induced vascular normalization contrasts with genetic models and presents important opportunities for a better understanding of experimental models and a translational path.

Interestingly, metronomic dosing of sustained yet lower doses of cytotoxic chemotherapy agents leads to reduced tumor burden [11, 12]. Such dosing strategies are generally associated with an increase in better tumor vascular perfusion and tumor oxygenation and reductions in IFP. While the mechanisms of action of such approaches are poorly understood, the efficacy of combined antiangiogenic/chemotherapeutic strategies may involve vascular normalization and be even more relevant to emerging drug delivery technologies. For example, even with adequate tumor perfusion, a significant limitation of targeted nanoparticle delivery across the BBB remains due to the complexity of multicellular NVU, ECM, and elevated IFP, all three being barriers that prevent efficient drug delivery.

3 Cell Biology of Vascular Normalization

Tumor angiogenesis is a multicellular process that is coordinated by factors that are produced from tumor and stromal cells that influence the proliferation and maturation of endothelial cells. Accordingly, the vascular normalization process also involves endothelial cells and stromal cells that are responsible for regulating the steps of vascular remodeling.

3.1 Endothelial Cells

Prolonged exposure to angiogenic factors (e.g., VEGF) secreted from tumor and stroma induces structurally and functionally abnormal vessels in the tumor. Tumor vessels are tortuous and have irregular diameter and length and discontinuous lining with sometimes multiple layers of endothelial cells. Functionally, they are poorly perfused and leaky, allowing tumor cells to intravasate and metastasize and also inhibiting trafficking of leukocytes. In contrast, normalized vessels appear to have reduced vessel diameter, less tortuous with regular branching, and the endothelial cell layer is continuous with smooth lining, and there is improved perfusion and oxygenation. Genetic modification of oxygen sensor protein prolyl hydroxylase domain-containing protein 2 (PHD2) [13] or Sox17 transcription factor [14] in endothelial cells is associated with a vascular normalization phenotype, indicating the importance of these particular pathways in endothelial cells. In the brain, endothelial-specific deletion of focal adhesion kinase (FAK) demonstrated vascular normalization in a model of glioma [6].

3.2 Pericytes and Basement Membrane

Maturation of the new vessels involves pericyte recruitment and deposition of extracellular matrix. In normal vessels, pericyte-deficient mice display BBB breakdown and neuronal defects [15, 16]. Angiogenic tumor vessels often lack sufficient coverage of pericytes or the endothelial–pericyte interaction is lost, leaving "empty sleeves"—pericytes and basement membrane that are devoid of endothelial cells. The thickness of the basement membrane is also irregular, inhibiting the appropriate matrix binding of growth factors that are important for vascular maturation. This leads to irregular flow and perfusion, causing uneven distribution of oxygen and nutrients, and hence, hypoxia in the tumor. Normalized tumor vessels demonstrate improved pericyte coverage and perfusion, which can also help with the delivery of chemotherapeutic agents to the tumor core. For example, Hamzah et al. demonstrated that loss of regulator of G-protein signaling 5 (*Rgs5*) protein induces maturation of pericytes that cover endothelial cells and induce vascular normalization [17]. Another study demonstrated that Ras [18] is important for pericyte association and maturation of tumor vessels, and upregulation of Ras signaling induced vascular normalization, indicating pericytes as alternative targets for inducing normalization of tumor vasculature.

3.3 Astrocytes

In the brain, the recruitment of pericytes and astrocytes is an essential process for the formation of tight junctions that are unique to the BBB. The tight junctions play an essential role in controlling the permeability of the BBB. Astrocytes form perivascular astrocytic end feet to maintain the astrocyte–endothelial interaction, whereas in tumors, this contact is lost, which correlates with increased permeability. Restoration of the astrocyte contact is found in a vascular normalization model in glioma as shown by the upregulation of aquaporin-4 surrounding the tumor vasculature [6].

4 Genetic Models of Vascular Normalization

Knockout mouse models have been used to establish the functional relevance of specific molecules in vascular normalization where in some models deletions are cell-type restricted. For example, a role for the Ang/Tie2 pathway in endothelium is supported by knockout animal models and pharmacological approaches showing a vascular normalization phenotype, while Ang/Tie2 also functions in tumor-infiltrating monocytes [19, 20]. Notably, antagonists of placental growth factor (PlGF), angiopoietins, and integrins have also been shown to exhibit a vascular normalization phenotype in various models [21].

Using a glioma model, our lab has shown that a conditional and endothelial-restricted deletion of FAK in the endothelium leads to a vascular normalization phenotype in tumor-associated brain–blood vessels [6]. The conditional nature of this model enables testing of the possible temporal restriction of the vascular normalization phenotype at different developmental stages, in a specific cell type. The fundamental importance of the FAK pathway in regulating endothelial barrier integrity is further demonstrated by knock-in mutation of the kinase domain in response to recombinant VEGF [22]. The importance of tumor hypoxia in the vascular normalization phenotype is demonstrated functionally in knockout mice lacking the Siah family of ubiquitin ligases [23]. In this model, the loss of Siah led to vascular normalization and reduced breast tumor burden, but with tumors having increased perfusion and pericyte coverage of blood vessels. While Siah2 was shown to attenuate the hypoxia-responsive transcription factor HIF1α, consistent with a link between vascular normalization and hypoxia, the Siah2$^{-/-}$ mice also exhibited an increased sensitivity to chemotherapy. Recently, Yang et al. demonstrated that mice lacking the Sox17 transcription factor in endothelial cells have reduced tumor burden associated with a vascular normalization phenotype associated with increased number of infiltrating myeloid cells [14]. Thus, while these genetic examples (summarized in Table 1) define the functional relevance of specific signaling/transcriptional pathways in vascular normalization, these models also demonstrate that the vascular normalization phenotype observed in preclinical and clinical paradigms following antiangiogenic therapy can also be mediated by the inflammatory response.

5 Immunomodulation of Vascular Normalization

Tumor-associated macrophages (TAMs) are an abundant myeloid-derived cell type of the innate immune system that exist in both quiescent and activated states to regulate vascular remodeling and

Table 1
Host-restricted loss-of-function models for vascular normalization

Target gene	Cell type	Model	Reference
Rgs5	General	RIP1-Tag5 pancreatic tumor	Hamzah et al. [17]
Ang2	General	s.c. lung tumor and melanoma	Nasarre et al. [20]
PlGF	General	Skin tumor, liver tumor, RIP1-Tag2 pancreatic tumor, s.c. mammary tumor, fibrosarcoma, Panc02 pancreatic tumor	Van de Veiri et al. [21], Rolny et al. [25]
R-Ras	General	s.c. lung tumor, melanoma, prostate tumor	Sawada et al. [18]
Siah2	General	PyMT mammary tumor	Wong et el. [23]
PHD2	Endothelial	Pancreatic tumor, s.c. lung tumor, and melanoma	Mazzone et al. [13]
FAK	Endothelial	Glioma	Lee et al. [6]
Sox17	Endothelial	s.c. lung tumor and melanoma	Yang et al. [14]
VEGF-A	Myeloid cell	MMTV-PyVT, s.c. lung tumor	Stockmann et al. [24]

tumor progression. Extensive infiltration of myeloid-derived cells (i.e., $CD45^+$, $CD11b^+$) contributes to a significant percentage of the overall tumor burden (i.e., 10–20 %) of tumor progression. Cytokines secreted by infiltrating myeloid-derived cells (i.e., VEGF, PDGF, PlGF) influence endothelial cell proliferation and angiogenesis. For example, deletion of myeloid-cell-derived VEGF-A induced vascular normalization in lung tumors [24]. Therefore, regulation of TAM activation to affect tumor vessel physiology is emerging as an important intersection of tumor immunology and vascular biology. Strategies to switch these tumor macrophages into becoming tumor-inhibiting phenotypes (i.e., M1) with host-produced factors such as histidine-rich glycoprotein [25] establish experimental paradigms for regulating tumor vessel permeability, patency, and therefore perfusion of chemotherapeutics. Establishing a link between switching TAM phenotypes and vascular patency for the purpose of enhancing drug delivery is an important translational research area in tumor biology. For example, in a screen of antifibrotic drugs, losartan was identified as a potent enhancer of nanoparticle penetration into solid tumors of the breast, pancreas, and skin [26]. Regulating the balance of growth factors and cytokines holds promise in obtaining a better understanding of the mobilization/switching of macrophages and other myeloid and stem cell-like cell populations to the tumor microenvironment. Across various tumor models, there is a strong correlation of TAMs with angiogenesis and tumor progression, and brain tumors, with gliomas in particular, are no exception [27]. In the CNS, microglia, the resident macrophage of the brain, provide a dynamic and abundant innate immune response to tumors, infection, and trauma; however, the recruitment of inflammatory myeloid-derived monocytes augments the initial immune response. With time, this local and mobilized immune response converts to an overall immunosuppressive tumor microenvironment. The immunosuppressive, proangiogenic state of myeloid cells in tumors can be altered in response to the secreted factors in the microenvironment. Therefore, a better understanding of the mechanisms that regulate the switching to the immunosuppressive phenotype and identification of therapeutic strategies to reverse the phenotype will provide novel therapeutic approaches to treat CNS cancers.

6 Drug Delivery to Normalized Tumor Vasculature

Drug delivery to the brain remains to be a challenge as the BBB prevents endocytosis or transport of various molecules to the brain parenchyma. Recent progress has been made on strategies to overcome this barrier by utilizing vehicle-mediated transport of drugs, for example, nanoparticles [28–30]. In brain pathologies, much of the current CNS drug delivery dogma is based either on their hydrophobicity or the capacity of larger molecules (i.e., antibodies) to gain access to CNS injury sites through overtly compromised BBB. As such, it has been well established that IFP and tissue fibrosis impede drug delivery to brain tumors that are considered to have a "leaky" barrier. Furthermore, as oppose to the dysregulated tumor vessels in the hypoxic, necrotic tumor core, endothelial cells at the tumor–stroma interface display characteristics similar to those of normal BBB. This type of heterogeneity of tumor vessels causes the delivery of chemotherapeutic agents to be uneven throughout the tumor. Yet vascular normalization has been associated with a decrease in vascular tortuosity, increased patency, and increased perfusion and drug delivery. Increasing evidence in preclinical models of vascular normalization demonstrates that these changes in the tumor vasculature accompany improvement in the delivery of chemotherapeutic drugs [23, 31, 32].

7 Resistance and Escape from Antiangiogenesis Therapy

Patients subjected to antiangiogenic therapy have shown vascular normalization to be only a transient process, therefore only providing a short window of time when combined therapy gives the best outcome [33]. Another serious problem in the clinical management of gliomas is the diffuse and infiltrative nature of the CNS tumor burden that is generally refractory to complete surgical resection. Moreover, glioma patients develop resistance/poor response to anti-vascular strategies, which is accompanied with enhanced invasion of tumor cells [34]. Therefore, there is a critical need to determine optimal dosing of vascular normalization agents to promote more sustained/prolonged vessel normalization. Developing alternative vessel normalization targets and approaches can also overcome resistance to current antiangiogenesis therapies to control invasiveness of glioma cells. For example, targeting of the PI3K and mTOR pathways with the inhibitor NVP-BEZ235 [35] or EGFR inhibitors [36] to induce prolonged vascular normalization could be used as an alternative to VEGF targeting.

8 Conclusions

Although antiangiogenic strategies reduce tumor burdens by starving tumors of their blood supply in preclinical models, clinical studies indicate that several compensatory mechanisms exist in humans that generally result in only a modest improvement in disease-free survival [37, 38]. Therefore, a better mechanistic understanding of the "adaptive" vascular mechanisms of antiangiogenesis therapies is essential. The concept of enhancing perfusion of tumors to yield "normalized" vessels is paradoxical in the sense that one would assume that a "better" vascular supply would enhance, rather than diminish, tumor progression. So while an enhanced vascular supply improves access to nutrients and more efficient removal of toxins, efflux pumps may be more efficient in the removal of chemotherapeutic agents. Thus, mechanistic studies of the cellular and molecular basis of neovascularization and regulation of VP remain important directions for further research to understand the complexities of BBB integrity and their role on tumor dissemination and pharmaco-distribution of therapeutics.

References

1. Jain RK, di Tomaso E, Duda DG et al (2007) Angiogenesis in brain tumours. Nat Rev Neurosci 8:610–622
2. Liebner S, Czupalla CJ, Wolburg H (2011) Current concepts of blood–brain barrier development. Int J Dev Biol 55:467–476
3. Ek CJ, Dziegielewska KM, Stolp H et al (2006) Functional effectiveness of the blood–brain barrier to small water-soluble molecules in developing and adult opossum (Monodelphis domestica). J Comp Neurol 496:13–26
4. Abbott NJ, Ronnback L, Hansson E (2006) Astrocyte-endothelial interactions at the blood–brain barrier. Nat Rev Neurosci 7:41–53
5. Thompson EM, Frenkel EP, Neuwelt EA (2011) The paradoxical effect of bevacizumab in the therapy of malignant gliomas. Neurology 76:87–93
6. Lee J, Borboa AK, Chun HB et al (2010) Conditional deletion of the focal adhesion kinase FAK alters remodeling of the blood–brain barrier in glioma. Cancer Res 70:10131–10140
7. de Groot JF, Fuller G, Kumar AJ et al (2010) Tumor invasion after treatment of glioblastoma with bevacizumab: radiographic and pathologic correlation in humans and mice. Neuro Oncol 12:233–242

8. McGee MC, Hamner JB, Williams RF et al (2010) Improved intratumoral oxygenation through vascular normalization increases glioma sensitivity to ionizing radiation. Int J Radiat Oncol Biol Phys 76:1537–1545
9. von Baumgarten L, Brucker D, Tirniceru A et al (2011) Bevacizumab has differential and dose-dependent effects on glioma blood vessels and tumor cells. Clin Cancer Res 17:6192–6205
10. Goel S, Duda DG, Xu L et al (2011) Normalization of the vasculature for treatment of cancer and other diseases. Physiol Rev 91:1071–1121
11. Folkins C, Man S, Xu P et al (2007) Anticancer therapies combining antiangiogenic and tumor cell cytotoxic effects reduce the tumor stem-like cell fraction in glioma xenograft tumors. Cancer Res 67:3560–3564
12. Kesari S, Schiff D, Doherty L et al (2007) Phase II study of metronomic chemotherapy for recurrent malignant gliomas in adults. Neuro Oncol 9:354–363
13. Mazzone M, Dettori D, Leite de Oliveira R et al (2009) Heterozygous deficiency of PHD2 restores tumor oxygenation and inhibits metastasis via endothelial normalization. Cell 136:839–851
14. Yang H, Lee S, Kim K et al (2013) Sox17 promotes tumor angiogenesis and destabilizes tumor vessels in mice. J Clin Invest 123:418–431
15. Armulik A, Genove G, Mae M et al (2010) Pericytes regulate the blood–brain barrier. Nature 468:557–561
16. Bell RD, Winkler EA, Sagare AP et al (2010) Pericytes control key neurovascular functions and neuronal phenotype in the adult brain and during brain aging. Neuron 68:409–427
17. Hamzah J, Jugold M, Kiessling F et al (2008) Vascular normalization in Rgs5-deficient tumours promotes immune destruction. Nature 453:410–414
18. Sawada J, Urakami T, Li F et al (2012) Small GTPase R-Ras regulates integrity and functionality of tumor blood vessels. Cancer Cell 22:235–249
19. Falcon BL, Hashizume H, Koumoutsakos P et al (2009) Contrasting actions of selective inhibitors of angiopoietin-1 and angiopoietin-2 on the normalization of tumor blood vessels. Am J Pathol 175:2159–2170
20. Nasarre P, Thomas M, Kruse K et al (2009) Host-derived angiopoietin-2 affects early stages of tumor development and vessel maturation but is dispensable for later stages of tumor growth. Cancer Res 69:1324–1333
21. Van de Veire S, Stalmans I, Heindryckx F et al (2010) Further pharmacological and genetic evidence for the efficacy of PlGF inhibition in cancer and eye disease. Cell 141:178–190
22. Chen XL, Nam JO, Jean C et al (2012) VEGF-induced vascular permeability is mediated by FAK. Dev Cell 22:146–157
23. Wong CS, Sceneay J, House CM et al (2012) Vascular normalization by loss of Siah2 results in increased chemotherapeutic efficacy. Cancer Res 72:1694–1704
24. Stockmann C, Doedens A, Weidemann A et al (2008) Deletion of vascular endothelial growth factor in myeloid cells accelerates tumorigenesis. Nature 456:814–818
25. Rolny C, Mazzone M, Tugues S et al (2011) HRG inhibits tumor growth and metastasis by inducing macrophage polarization and vessel normalization through downregulation of PlGF. Cancer Cell 19:31–44
26. Diop-Frimpong B, Chauhan VP, Krane S et al (2011) Losartan inhibits collagen I synthesis and improves the distribution and efficacy of nanotherapeutics in tumors. Proc Natl Acad Sci U S A 108:2909–2914
27. Chen P, Bonaldo P (2013) Role of macrophage polarization in tumor angiogenesis and vessel normalization: implications for new anticancer therapies. Int Rev Cell Mol Biol 301:1–35
28. Dilnawaz F, Singh A, Mewar S et al (2012) The transport of non-surfactant based paclitaxel loaded magnetic nanoparticles across the blood brain barrier in a rat model. Biomaterials 33:2936–2951
29. Kong SD, Lee J, Ramachandran S et al (2012) Magnetic targeting of nanoparticles across the intact blood–brain barrier. J Control Release 164:49–57
30. Wohlfart S, Gelperina S, Kreuter J (2011) Transport of drugs across the blood–brain barrier by nanoparticles. J Control Release 161:264–273
31. Liu J, Liao S, Huang Y et al (2011) PDGF-D improves drug delivery and efficacy via vascular normalization, but promotes lymphatic metastasis by activating CXCR4 in breast cancer. Clin Cancer Res 17:3638–3648
32. Leite de Oliveira R, Deschoemaeker S, Henze AT et al (2012) Gene-targeting of Phd2 improves tumor response to chemotherapy and prevents side-toxicity. Cancer Cell 22:263–277
33. Winkler F, Kozin SV, Tong RT et al (2004) Kinetics of vascular normalization by VEGFR2 blockade governs brain tumor response to radiation: role of oxygenation, angiopoietin-1, and matrix metalloproteinases. Cancer Cell 6:553–563
34. Sennino B, McDonald DM (2012) Controlling escape from angiogenesis inhibitors. Nat Rev Cancer 12:699–709
35. Fokas E, Im JH, Hill S et al (2012) Dual inhibition of the PI3K/mTOR pathway increases

tumor radiosensitivity by normalizing tumor vasculature. Cancer Res 72:239–248

36. Cerniglia GJ, Pore N, Tsai JH et al (2009) Epidermal growth factor receptor inhibition modulates the microenvironment by vascular normalization to improve chemotherapy and radiotherapy efficacy. PLoS One 4:e6539

37. Jain RK, Duda DG, Clark JW et al (2006) Lessons from phase III clinical trials on anti-VEGF therapy for cancer. Nat Clin Pract Oncol 3:24–40

38. Ellis LM, Hicklin DJ (2008) VEGF-targeted therapy: mechanisms of anti-tumour activity. Nat Rev Cancer 8:579–591

Chapter 4

Pericytes and Adaptive Angioplasticity: The Role of Tumor Necrosis Factor-Like Weak Inducer of Apoptosis (TWEAK)

Paula Dore-Duffy

Abstract

The TNF superfamily member TWEAK has emerged as a pleiotropic cytokine that regulates many cellular functions that include immune/inflammatory activity, angiogenesis, cell proliferation, and fate. TWEAK through its inducible receptor, FGF-inducible molecule 14 (Fn14), can induce both beneficial and deleterious activity that has a profound effect on cell survival. Thus it is highly likely that TWEAK and Fn14 expressed by cells of the neurovascular unit help regulate and maintain vascular and tissue homeostasis. In this chapter we discuss the expression of TWEAK and Fn14 signaling in the cerebral microvascular pericyte. Pericytes are a highly enigmatic population of microvascular cells that are important in regulatory pathways that modulate physiological angiogenesis in response to chronic mild hypoxic stress. A brief introduction will identify the microvascular pericyte. A more detailed discussion of pericyte TWEAK signaling during adaptive angioplasticity will follow.

Key words Pericytes, Hypoxia, Physiological angiogenesis, Adaptation, Migration, Tumor necrosis family, TWEAK, Fn14

1 Introduction

The French scientist Jean Marie Benjamin Rouget discovered the microvascular pericyte in the late nineteenth century [1]. It was many years before his observations were confirmed and the first review was written [2]. It was over 100 years before scientists began to unravel the complex web surrounding the role of this enigmatic cell in tissue physiology. However, to this day navigating the vast increase in publications dealing with the pericyte still remains a confusing undertaking. Following brief introductory comments on pericytes in general, we will focus our discussion on the role played by the pericyte in vascular adaptation to chronic mild hypoxia. In particular, emphasis will be given to the possible role played by members of the tumor necrosis factor superfamily of signaling molecules and their receptors. Emphasis will be given to angioplasticity in the brain.

Richard Milner (ed.), *Cerebral Angiogenesis: Methods and Protocols*, Methods in Molecular Biology, vol. 1135,
DOI 10.1007/978-1-4939-0320-7_4, © Springer Science+Business Media New York 2014

2 What Is a Pericyte?

Pericytes have been called contractile motile cells of the vasculature that surround vessels in a tunic-like fashion [1], smooth muscle cells of the capillaries [3], the mural cells of blood microvessels, adventitial cells [4, 5], regulatory cells of the microvasculature [5, 6], and a source of adult stem cells in most tissues [7]. Pericytes are found abluminal to endothelial cells (EC) in precapillary arterioles, capillaries, postcapillary venules, and collecting venules [2–6]. However, pericytes have also been identified in association with arteries as well as microvessels [8]. They are totally surrounded by the basal lamina including even the smallest of cellular projections. Only migrating pericytes have minimal contact with the basal lamina [9, 10]. In its vascular location, the basal lamina may regulate/restrict local signaling pathways [11, 12].

By virtue of their ability to express a number of muscle-specific proteins, pericytes were originally called vascular smooth muscle cells [13–15]. Alpha smooth muscle-specific actin (αSMA) was considered a pericyte marker [14]. However, with continued investigation, it became clear that the majority of pericytes do not express αSMA (Fig. 1a). Induction of smooth muscle actin in vivo is related to a functional change associated with focal regulation of vascular diameter and capillary blood flow [16–19]. For example, under conditions of stress, a subset of vascular pericytes are induced to express αSMA at injury sites (Fig. 1b). The expression of alpha

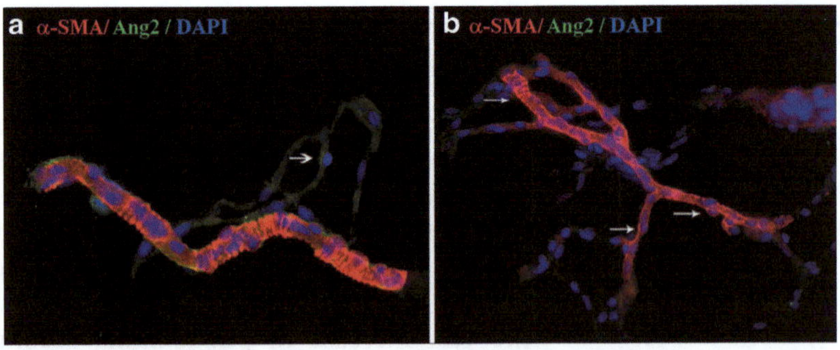

Fig. 1 Pericyte expression of alpha smooth muscle actin in CNS capillaries. Capillaries were isolated from normal Sprague Dawley rat brain and rats exposed to traumatic brain injury using the weight drop method of Marmarou [132]. Thin sections were fixed with paraformaldehyde and stained for the expression of angiopoietin 2 (endothelial cell marker) and alpha smooth muscle actin (αSMA). Nuclei were stained with DAPI. In (**a**) small round nuclei in microvessels adjacent to a smooth muscle-positive precapillary arteriole were found to be αSMA−. Following injury, there was a significant increase in the number of αSMA+ cells associated with round DAPI+ nuclei (**b**). This is consistent with an upregulation of αSMA in microvascular pericytes following TBI [20]

muscle actin is variable in normal vascular beds. Transition pericytes located in precapillary arterioles are in contact with smooth muscle cells and are routinely muscle actin positive. Following traumatic brain injury (TBI), αSMA upregulation is mediated by endothelin-1 and associated with a decrease in capillary diameter [20]. Increased alpha actin may be responsible for transient hypoperfusion seen following TBI. In tissue culture in the presence of serum, pericytes differentiate along the mesenchymal lineage to mesenchymal stem cell (MSC)-like cells [18, 19]. During this process cells are induced to express αSMA.

A number of other markers have proved useful in pericyte identification in defined systems. Pericytes are proteoglycan NG2 positive [21]. NG2 chondroitin sulfate proteoglycan (high-molecular-weight melanoma-associated antigen) [21–24] is expressed at high levels during development and in tumor pericytes [21]. In adult pericytes, NG2 is heterogeneously expressed at lower levels. Pericytes located in postcapillary venules are NG2 negative [24]. Brain pericytes express intercellular adhesion molecule-1 (ICAM-1) [25] and express vimentin but not desmin [23, 26]. One of the most useful markers is the protein platelet-derived growth factor β receptor (PDGFβR). In our hands antibodies directed toward PDGFβR label the majority of non-differentiated pericytes in all tissues tested thus far. Other markers expressed include the 140 kDa aminopeptidase N [27], nestin [28], and a 3G5-defined ganglioside [29]. Using expression analysis, additional developmental markers such as the RGS-5 protein have been identified in knockout mice [30]. Not all markers are expressed in adult pericytes and in CNS pericytes. Pericytes have also been reported to express CD146 [19, 31, 32], annexin 5A [33], and alkaline phosphatase [34].

Not all reported markers have been validated in both pure pericyte cultures or in tissue sections. For example, expression of endoglin (CD105) in combination with negative expression of CD31 has been used to identify pericytes versus endothelial cells in some studies [35]. The vast majority of studies have reported pericytes as CD31 negative. However, an antibody clone that recognizes PECAM is available commercially with a claim that it labels both endothelial cells and pericytes. Thus care must be taken not to draw conclusions from data dependent on the use of a single marker. We have repeatedly recommended that in tissue section dual staining techniques be used. In section pericytes should be identified as peri-endothelial by labeling the basal lamina. For example, PDGFβR coupled with an endothelial marker such as factor VIII or the binding of the Griffonia symplicifolia agglutinin (GSA) lectin [36] is quite useful. Confirmation with a second pericyte marker is advisable. The lack of a definitive pan-marker for pericytes may be due to the fact that these cells are multipotent self-renewing cells [7, 9]. As pericytes begin to differentiate immediately upon isolation, they have functional and phenotypic heterogeneity depending on their tissue location; it may be impossible to definitively identify and validate a unique marker.

3 Pericyte Function

Many functions have been attributed to pericytes both within the microvessels and in the perivascular space following their migration from a microvascular location. From a simple structural standpoint, pericytes provide a protective vascular support structure that maintains vascular integrity [37–40]. Pericyte projections wrapping around the abluminal surface of the vessels provide stability and resistance to injury. Loss of pericyte coverage due to injury or migration of the pericyte results in increased susceptibility to mechanical injury [41, 42] and vascular instability. Genetically manipulated knockout animals that have decreased pericyte coverage show enhanced susceptibility to trauma with evidence of micro-hemorrhage [43, 44]. Astrocytes aid the pericyte in maintenance of vascular integrity. Glial fibrillary acidic protein (GFAP) knockout mice that lack astrocyte-vascular contacts also exhibit increased susceptibility to trauma [45]. However, they have nearly ten times the number of pericytes as wild-type mice in what appears to be an attempt to compensate for the lack of astrocytes [46].

Within their microvascular location, pericytes function as regulatory cells. In response to changes in the microenvironment, they make fine-tuned adjustments and adaptations that promote vascular and tissue homeostasis and thus tissue survival [40]. At the blood–brain barrier (BBB), pericytes help maintain barrier function by regulating, in part, the movement of nutrients and hormones across the endothelial barrier [21, 44, 47, 48]. Pericytes express P-glycoprotein [49]. They have been shown to modulate multidrug resistance protein-6 (MRP6) messenger RNA (mRNA) and protein in coculture [50, 51]. MRP6 is the only transporter whose endothelial mRNA expression is influenced and possibly dependent on the pericyte [50]. MRP6 deficiency leads to the connective tissue disorder pseudoxanthoma elasticum (PXE) [52] associated with increased vascular permeability and by inference may involve a pericyte abnormality.

Pericytes help regulate barrier function by the synthesis and release of bioactive substances that maintain junctional proteins [46, 48, 53]. Pericyte induction of junctional proteins is thought to underlie its role in differentiation of the blood–brain barrier [44, 46, 48]. Loss of pericyte function or coverage during development or due to injury or migration results in increased permeability, vascular regression, and loss of metabolic homeostasis [54]. Pericyte regulation of vascular and metabolic homeostasis underscores its role in new vessel formation (angiogenesis) [54].

4 Pericytes Regulate Angiogenesis

The role of the pericyte in new vessel formation is continually being redefined. New vessel formation (vasculogenesis during development and angiogenesis in the adult) or new vessel formation under pathological conditions (pathological angiogenesis), while sharing some parameters, may be fundamentally different. Gene expression profiles have been shown to be dissimilar [55]. In general however our understanding of the role of pericytes in angiogenesis is a compendium of data derived from both in vivo and in vitro studies. Pericytes are involved in three phases of vascular remodeling: (1) initiation of angiogenesis, (2) sprout formation and migration, and (3) maturation or termination. Following exposure to proangiogenic stimuli, there is a significant reduction in the area of contact between pericytes and the basal lamina on the capillary abluminal surface and withdrawal of pericyte processes. This coincides with initiation of sprout formation and pericyte migration [46, 56–58]. A variety of proteinases are involved in pericyte direction of sprout formation [46, 56, 58–62]. During angiogenesis endothelial cells proliferate and new vessels migrate into the tissue. Pericytes proliferate adjacent to sites of new vessel formation. This is particularly apparent in tissue culture angiogenesis systems. New vessels ultimately undergo canalization and secrete basement membrane material. It is unclear whether lumen formation occurs before or is associated with onset of pericyte coverage.

Pericytes migrate during the early phases of angiogenesis [62]. Pericyte migration was initially described as "dying back" [63]. Following exposure to a proangiogenic signal or environmental induction, pericytes mount a HIF-1 response within hours. In physiological angiogenesis, we find that pericytes are the first cell to produce HIF-1-dependent proteins such as vascular endothelial growth factor (VEGF). They then migrate by 2 days [54]. Migration is related to their specialized role during adaptive angiogenesis [54]. Cat pericytes were shown to be the first cell responding to more acute levels of hypoxia and were found to begin migration by 2 hours [64], confirming our studies in the traumatic brain injury (TBI) model [65]. Following TBI, pericyte migration is associated with changes in the basal lamina. The migratory phenotype was associated with the expression of cell surface proteases at the leading tip of the migrating cell. Of interest is that stress-induced migration in the TBI model results in survival of the migrating pericyte [65]. Pericytes remaining in the vessel became apoptotic. Similar degenerative changes in pericytes were reported in human brain edema [62], but in this model, there was no evidence of migration. Changes to the basal lamina leading to migration have also been observed in a model of focal brain compression and ischemia [66]. Four days after compression injury, the basement membrane was blurred and thickened, and pericytes were found to migrate through the basal lamina. Migration was interpreted to be important to microvessel repair and plasticity. Pericyte migration in association with altered basal lamina has also been observed following middle cerebral artery occlusion (MCAO) [67].

Pericytes also migrate during developmental angiogenesis [62] in response to PDGFβ and is thought to be the signal that recruits pericytes into newly formed vessels. Two modes of recruitment have been suggested: (1) de novo differentiation of vascular smooth muscle cells (vSMC) and pericytes by induction of undifferentiated perivascular MSC and (2) co-migration of vSMC and pericytes from a preexisting pool of differentiated cells. Data by Hellström and colleagues [68] support both models. They also proposed that PDGFβR-positive pericyte progenitors initially form around certain vessels in a PDGFβ-independent induction. Subsequent angiogenic sprouting and vessel enlargement involves PDGFβ-dependent progenitor co-migration and proliferation, and/or PDGFβ-independent new induction of vSMC or pericytes, depending on the specific tissue. Pericyte migration in wild-type mice requires heparin sulfate (HS) [69]. From studies of liver hypoxia, we know that engagement of PDGFβR results in activation of Ras/Erk, which differentially regulates proliferation and migration of liver pericytes [70, 71]. Quantitative morphological analysis showed that PDGFβ KO mice have normal microvessel density, length, and branch points. However, absence of pericytes correlates with endothelial hyperplasia [68], increased capillary diameter, abnormal EC shape and ultrastructure, changes in cellular localization of junctional proteins, and increased permeability [62].

In developmental as well as adult new vessel formation, termination of angiogenesis is associated with renewed pericyte coverage. This is seen during physiological angiogenesis but is absent in some forms of pathological angiogenesis (Dore-Duffy unpublished observations). In adults, PDGF stimulates migration of smooth muscle cells but not pericytes, whereas purified platelet factor IV (PF4) stimulated migration of pericytes but not smooth muscle cells nor endothelial cells [72]. In our hands and in the hands of other investigators, CXCR3-B and/or pericyte-synthesized heparin sulfate proteoglycans are important in pericyte recoverage stabilizing the newly formed vessels and terminating angiogenesis [71]. That angiogenesis is inhibited may be due to pericyte synthesis of other angiostatic substances in addition to the heparin sulfate proteoglycans. With the presence of pericytes in new vessels, the basal lamina appears on the inner luminal surface of the endothelial layer. Pericytes secrete components of the extracellular matrix (ECM) [73]. The basal lamina is deposited on the parenchymal side of the pericyte and the endothelial cell until surrounded by the basement membrane. It is unclear whether a similar sequence occurs in all organs. Further, while the majority of data suggests that capillaries form locally, it cannot be ruled out that they may also differentiate from wandering mesenchymal cells, as first suggested by Clark and Clark [74]. It is also unclear whether the mechanisms governing vasculogenesis/angiogenesis are similar to those defined in adult adaptive physiological angiogenesis or pathological angiogenesis. For further information on this subject, please see the following reviews [59, 62, 75–77].

5 The Role of the Pericyte in Adaptive Angioplasticity

Rodents exposed to stress stimuli such as moderate low oxygen (simulated high-altitude environment) are induced to mount both a systemic- and tissue-specific adaptation that promotes tissue survival [54, 78–81]. Nowhere is this regulatory system more important than in the adult brain where bioenergetic balance is integral to neuronal survival and tissue repair. Endogenous adaptive vascular remodeling or "angioplasticity" is a natural healing process.

Pericytes have an important role in adaptive vascular remodeling. Loss of pericyte coverage or pericyte dysfunction inhibits endogenous adaptation [9, 46, 54, 82, 83]. The ability to respond to environmental changes underscores the role of the pericyte as a regulatory cell. The pericyte response to low oxygen mediates induction of angioplasticity. During the initiation of angiogenesis in response to chronic mild hypoxia, pericytes very rapidly become HIF-1 positive and synthesize VEGF [54, 81]. Pericytes are the first cell to mount a HIF-1 response and release HIF-1-dependent gene products such as VEGF within 4 h exposure to 10 % oxygen. With continued exposure to the hypoxic stimulus, astrocytes, oligodendrocyte progenitor cells (OPC), and oligodendrocytes become HIF-1 positive (Fig. 2). Exactly how pericytes detect changes in the microenvironment is still unknown and under investigation. In response to these changes, the pericyte may functionally and phenotypically differentiate, migrate, and/or proliferate. During adaptive physiological angiogenesis, increased permeability is characteristic of newly formed vessels and thought to be the result of pericyte migration. Vascular instability is, however, only transient as it is terminated with renewed pericyte coverage [54].

Angioplasticity in response to chronic exposure to mild hypoxia involves the coordinated synthesis and release of a number of hypoxia-/stress-driven signaling molecules that are both HIF-1 dependent and independent [54, 80, 81]. These extracellular and intracellular signaling molecules tightly regulate homeostasis and are potential mediators or targets for therapeutic intervention. However, the design of targeted therapeutic strategies requires understanding of the molecular regulation of a particular gene at the cellular level. In this regard, the expression of these molecules may vary in a species, as well as cell-type- and stimulus-specific

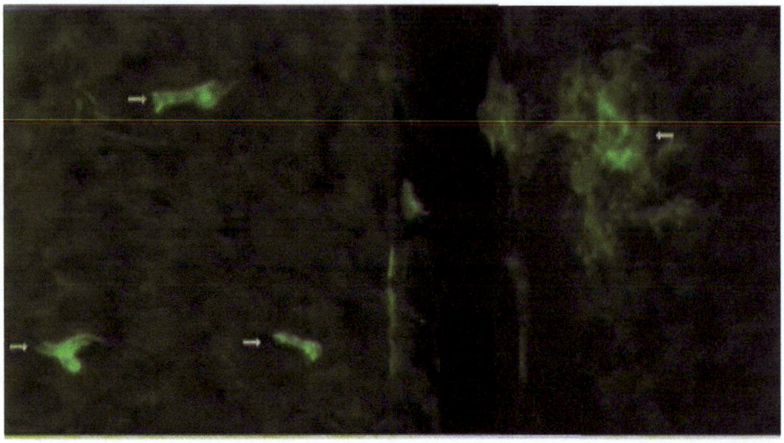

Fig. 2 Cellular HIF-1 response to mild hypoxia. Transgenic animals genetically engineered to fate map HIF-1 responding cells were used to investigate the cellular response to moderate hypoxia in vivo (Gow and Dore-Duffy unpublished observations). Animals were exposed to 10 % oxygen for 24 h. Tissue samples were taken from the cerebellum and whole mounts examined for the presence of reporter gene expression. In these animals, HIF-1+ cells were *green*. Results show that the majority of fluorescent cells were found in microvessels in what appeared to be a vascular pericyte location. HIF-1+ astrocytes and/or oligodendrocyte progenitor cells were also observed

manner. In our studies, we have investigated signaling molecules released by the pericyte that are involved in regulation of adaptive angioplasticity. In this chapter we discuss novel members of the TNF superfamily (TNFSF): TWEAK and APRIL.

6 The Tumor Necrosis Factor (TNF) Ligand Superfamily: TNF-Like Weak Inducer of Apoptosis (TWEAK)

The tumor necrosis factor (TNF) ligand superfamily and their receptors are a cytokine-receptor subgroup that has attracted considerable interest as potential therapeutic targets for the management of human diseases [84] and may prove useful targets to regulate vascular remodeling. Members of the TNF superfamily regulate several cell responses, including proliferation, differentiation, and apoptotic cell death [85]. The TNF family members TNFα and TNFβ have been extensively studied in CNS injury models and are known to be involved in the pathophysiology of neurodegenerative disease [86, 87]. TNF-like weak inducer of apoptosis (TWEAK) (TNFSF12) is a relatively new member of the TNF superfamily [88–90]. TWEAK was previously dubbed APO3L and CD255 [91, 92]. TWEAK functions both as a type II transmembrane protein and as a soluble secreted molecule that retains biological function [91, 92]. Full-length TWEAK is proteolytically processed in the C-terminal region by the serine protease furin. TWEAK amino acid residues 90–93 were identified by the Winkles laboratory [93] as the predominant furin recognition site. In addition TWEAK may undergo intergenic splicing to produce hybrid molecules with other TNF family members. Under these conditions TWEAK provides the anchor for other predominantly secreted TNF family members [94]. Both the anchored and soluble TWEAK molecules contain a C-terminal TNF homology domain that mediates trimerization and TWEAK receptor binding. Cells may co-express full-length plasma membrane-anchored TWEAK, soluble TWEAK, and theoretically intergenic splice variants TWE-PRIL [88–90], but the mechanism governing the relative production of these forms within a given cell is not entirely understood.

TNF superfamily members bind to one or more members of the TNF receptor (TNFR) superfamily. TNFR are type I or type III transmembrane proteins [95]. Fibroblast growth factor-inducible 14 (Fn14) is the primary TWEAK receptor [88, 95, 96] and is also known as TNFRSF12A, TWEAKR, or CD266. Fn14 is a 102-amino acid type I transmembrane protein described as a serum- and growth factor-inducible gene product in fibroblasts and smooth muscle cells [97, 98]. TWEAK signaling is through engagement of Fn14. Soluble TWEAK and full-length, membrane-anchored TWEAK both bind the Fn14 receptor [93]. Cell-associated TWEAK can act on neighboring cells and activate the NF-κB signaling pathway. Therefore, TWEAK can act in a juxtacrine manner [93]. In addition to autocrine/paracrine and juxtacrine signaling, TWEAK may also signal within the cell of origin. Nuclear localization sequences have been described suggesting that endogenously synthesized TWEAK may enter the cell nucleus [91, 99]. However, more studies are necessary to determine whether TWEAK nuclear localization occurs in all cell types.

TWEAK was initially described as a pro-apoptotic factor for tumor cell lines [100], but we now know that TWEAK can stimulate cell proliferation, survival, and differentiation [101, 102]. The TWEAK–Fn14 signaling axis has been implicated in wound repair and in the pathogenesis of ischemic stroke, chronic inflammatory diseases [103–106], and inflammatory bowel disease [107]. Circulating TWEAK levels are significantly increased in other human autoimmune diseases such as multiple sclerosis (MS) and systemic lupus erythematosus (SLE) [92, 107–109], and TWEAK-neutralizing antibody ameliorates collagen-induced arthritis in mice [110, 111]. Overexpression of TWEAK leads to increased CCL2 and pro-inflammatory cytokines including TNFα, IL-1β, and IL-6 [92].

Like many other TNFSF members, TWEAK–Fn14 signaling mediates pleiotropic effects. TWEAK signaling is summarized in Fig. 3. TWEAK attenuates the transition from innate to adaptive

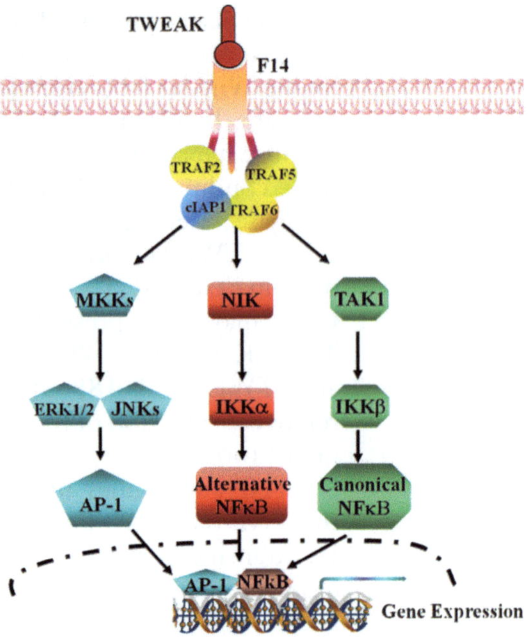

Fig. 3 TWEAK signaling cascade: the TWEAK–Fn14 signaling cascade. Binding of TWEAK to Fn14 leads to the recruitment of cIAP1, TRAF2, TRAF5, and/or TRAF6 proteins leading to downstream activation of transforming growth factor β-activated kinase 1 (TAK1), NF-κB-inducing kinase (NIK), and various mitogen-activated protein kinase kinases (MKK). Activation of TAK1 stimulates IB kinase (IKKβ) leading to the activation of canonical NF-κB. NIK phosphorylates and activates IKKβ leading to downstream activation of NF-κB through alternative pathways. Various MKKs activate c-Jun N-terminal kinase 1 (JNK1) and p38 MAPK, which in turn activates transcription factors such as transcription factor activator protein-1 (AP-1). Increased activation of NF-κB and AP-1 leads to the expression of specific genes involved in TWEAK-mediated responses

immunity by suppressing the production of interferon-γ and IL-12 [112]. TWEAK has been shown to activate p44/p42 MAPK, c-Jun N-terminal kinase (JNK), transcription factor activator protein-1 (AP-1), and NF-κB signaling pathway in various cell types including skeletal muscle [113–116]. Binding of TWEAK to the extracellular domain of Fn14 leads to receptor trimerization, association of TRAF2/cIAP1 (cellular inhibitor of apoptosis protein 1) complex to cytoplasmic domain, and subsequent activation of various signaling proteins including TRAF6, transforming growth factor-β-activated kinase1 (TAK1), I kappa B kinase (IKK), and MAPKs leading to altered expression of genes involved in various cellular responses [91, 92, 114–118].

TWEAK and Fn14 appear to have a minimal role in embryonic development or postnatal growth because mice null for TWEAK or Fn14 are viable and show no major lethal abnormalities [119, 120]. However, TWEAK signaling through the Fn14 does have an effect on cell differentiation [121] and by inference on developmental differentiation. In vitro, TWEAK inhibits the differentiation of myoblasts to myotubes [122]. TWEAK knockout mice exhibit augmented muscle tissue regeneration while overexpression inhibits myofiber regeneration. For an excellent review on this subject, please see Bhatnager and colleagues [122]. In tissue culture, TWEAK has been found to stimulate proliferation and expansion of progenitor cells and bone marrow-derived MSC [99, 120, 122–124], but this may ultimately inhibit terminal differentiation [111, 120]. Thus dysregulation or overproduction of TWEAK may lead to inhibition of adult stem cell activity. This concept has been put forward to explain mechanisms that lead to skeletal muscle wasting [122].

7 TWEAK and Angiogenesis

Soluble TWEAK was shown in 1999 to induce proliferation in cultured human endothelial cells and in aortic smooth muscle cells [96, 125, 126]. Blocking antibodies to VEGF did not inhibit TWEAK-induced proliferation, indicating that TWEAK does not function through upregulation of VEGF. However, TWEAK treatment activated several endothelial cell intracellular signaling pathways and potentiated FGF-2- and VEGF-A-stimulation of proliferation [127]. We have observed similar results in pericytes (Dore-Duffy, unpublished observations). TWEAK induction of endothelial proliferation can be inhibited by antibodies that block TWEAK binding to Fn14 and by the use of an Fn14 decoy receptor [127]. These results suggest that TWEAK may affect angiogenesis alone or in combination with other proangiogenic factors [127]. Studies have shown that an angiogenic factor, *VG5Q*, co-immunoprecipitates with TWEAK and is expressed in microvessels early in after induction of angiogenesis. Two genetic defects of *VG5Q* are found in patients with the vascular disease Klippel–Trenaunay syndrome (KTS) [128]. VG5Q protein acts as a potent angiogenic factor in promoting angiogenesis, and suppression of VG5Q expression inhibits vessel formation. This data further suggest that proangiogenic factors stimulate angiogenesis by increasing TWEAK-dependent Fn14 signaling [91, 119]. Numerous other studies have confirmed the importance of TWEAK-mediated signaling in tube formation in culture models [129, 130].

TWEAK also regulates vascular remodeling in vivo. Pellets containing TWEAK implanted in rat cornea were found to induce a strong angiogenic response [125]. Jakubowski and colleagues reported similar results [119]. They further substantiated a dual role of TWEAK in vascular remodeling, by showing that TWEAK is involved in mitogenic activity but not cell survival. Silencing TWEAK suppressed invasion and new vessel formation in tumors in vivo.

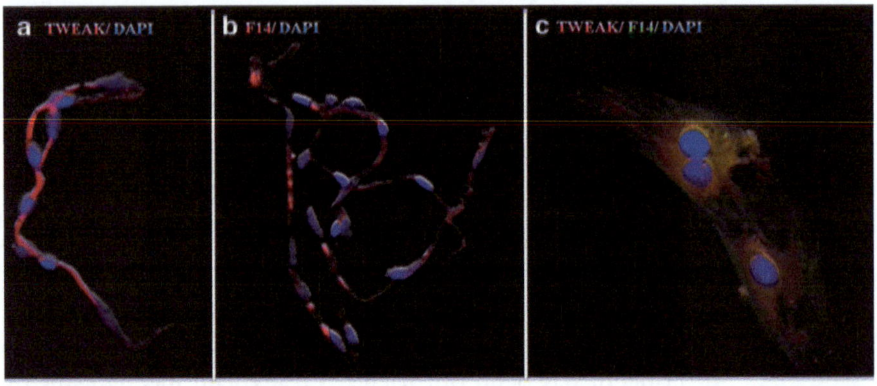

Fig. 4 Microvessel and pericyte expression of TWEAK and Fn14 in brain. Microvessels were isolated from 6- to 8-week-old rats. Pericytes were subcultured after isolation with enzymatic disruption and sequential sieving [9]. Pericytes were cultured for 3–4 days before staining. Pericytes and microvessels were fixed with paraformaldehyde and permeabilized with Triton X-100 before staining. Antibodies directed against TWEAK (Santa Cruz [FL-249 Sc-5558] and R+D [AF1090]) and Fn14 [ITEM-4, Sc-56250] were used. Dual staining was performed sequentially

Tumor cells and vascular endothelial cells within cancer foci expressed Fn14. Fn14 expression is reduced in TWEAK knockdowns. Taken together, this data suggests that expression of TWEAK stabilizes Fn14 and that coordinated expression of these molecules is important for vascular remodeling.

Using immunohistochemical techniques on purified brain microvessel fragments, we have confirmed significant expression of both TWEAK and Fn14. Positive staining was observed in both endothelial cells and pericytes (Fig. 4) as well as vascular smooth muscle cells seen in precapillary arterioles. Primary pericytes in culture also express both TWEAK and Fn14. The ratio of TWEAK and Fn14 in pericytes changes following exposure to pro-angiogenic stimuli in vivo as well as in vitro (data not shown). Following exposure to in vitro hypoxia, TWEAK and Fn14 protein was increased (Fig. 4). The increase in Fn14 is greater than that of TWEAK. Western analysis of TWEAK protein is difficult without some effort to localize to specific cells. Whole brain homogenates are not successful. We have had better results with protein isolates from capillary preparations or pericyte cultures. Increased TWEAK correlated with the HIF-1 response in pericytes and to VEGF protein release. It is unclear at present whether the hypoxia-driven pericyte HIF-1 response and release of mitogens are dependent on pericyte Fn14. However, a blocking antibody did alter angioplasticity in vivo (Dore-Duffy and LaManna unpublished data). Care must be given to antibody selection as most anti-TWEAK molecules are directed against specific TWEAK recombinant peptides. Not all are available for purchase and many have debatable efficacy. The antibody we have used was directed toward the intracellular anchoring portion of the

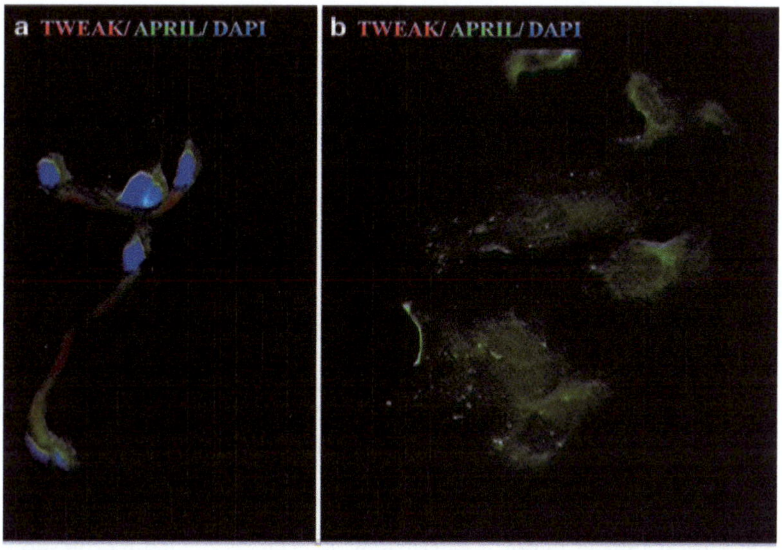

Fig. 5 Microvessel and pericyte expression of TWEAK and APRIL. Microvessels were isolated from 6- to 8-week-old rats and pericytes isolated following enzymatic disruption [9]. Pericytes were cultured for 4 days before staining. Pericytes and microvessels were fixed with paraformaldehyde and permeabilized with Triton X-100. Antibodies directed against TWEAK and APRIL (Santa Cruz [R-15, Sc-5739]) were used at appropriate dilutions. Dual staining was performed sequentially

TWEAK molecule. As such it does not distinguish between TWEAK expressed as a whole molecule and as an anchor for secreted members of the TNFSF such as APRIL (TNFSF13a). We therefore examined pericytes for immunocytochemical detection of APRIL. Results in Fig. 5 show that primary normoxic pericytes have detectable cell-associated APRIL. Under normal conditions, APRIL is synthesized and rapidly released [131]. Thus the appearance of detectable APRIL by immunofluorescence and FACs analysis (data not shown) suggests that TWEAK may not be expressed as a full-size molecule in normoxic pericytes but rather is present as the anchor for APRIL. Additional studies are currently underway to determine whether APRIL is associated with specific subsets of pericytes and whether there is a role for TWE-PRIL as well as TWEAK in pericyte regulation of angiogenesis in both culture models and in acclimatization to low oxygen in vivo.

References

1. Rouget C (1874) Note sur le developpement de la tunique contractile des vaisseaux. Compt Rend Acad Sci 59:559–562
2. Dorè SE (1923) On the contractility and nervous supply of the capillaries. Brit J Dermatol 35:398–404
3. Fabry Z, Fitzsimmons KM, Herlein JA et al (1993) Production of the cytokines interleukin 1 and 6 by murine brain microvessel endothelium and smooth muscle pericytes. J Neuroimmunol 47:23–34
4. Spraycar M (ed) (1995) Stedman's medical dictionary, 6th edn. Williams and Wilkins, Baltimore, MD
5. Ding R, Darland DC, Parmacek MS et al (2004) Endothelial-mesenchymal interactions

in vitro reveal molecular mechanisms of smooth muscle/pericyte differentiation. Stem Cells Dev 13:509–520
6. Hautekeete ML, Geerts A (1997) The hepatic stellate (Ito) cell: its role in human liver disease. Virchows Arch 430:195–207
7. Dore-Duffy P, Katychev A, Wang X et al (2006) CNS microvascular pericytes exhibit multipotential stem cell activity. J Cereb Blood Flow Metab 26:613–624
8. Andreeva ER, Pugach IM, Gordon D et al (1998) Continuous subendothelial network formed by pericyte-like cells in human vascular bed. Tissue Cell 30:127–135
9. Dore-Duffy P (2008) Pericytes: pluripotent cells of the blood brain barrier. Curr Pharm Des 14:1581–1593
10. Sims DE (1991) Recent advances in pericyte biology–implications for health and disease. Can J Cardiol 7:431–443
11. Dore-Duffy P, Owen C, Balabanov R et al (2000) Pericyte migration from the vascular wall in response to traumatic brain injury. Microvasc Res 60:55–69
12. Pfister F, Feng Y, vom Hagen F (2008) Pericyte migration: a novel mechanism of pericyte loss in experimental diabetic retinopathy. Diabetes 57:2495–2502
13. Joyce NC, Haire MF, Palade GE (1985) Contractile proteins in pericytes. II. Immunocytochemical evidence for the presence of two isomyosins in graded concentrations. J Cell Biol 100:1387–1395
14. Herman IM, D'Amore PA (1985) Microvascular pericytes contain muscle and nonmuscle actins. J Cell Biol 101:43–52
15. Skalli O, Pelte MF, Peclet MC et al (1989) Alpha-smooth muscle actin, a differentiation marker of smooth muscle cells, is present in microfilamentous bundles of pericytes. J Histochem Cytochem 37:315–321
16. Hamilton NB, Attwell D, Hall CN (2010) Pericyte-mediated regulation of capillary diameter: a component of neurovascular coupling in health and disease. Front Neuroenergetics. doi:10.3389/fnene.2010.00005
17. Kennedy-Lydon TM, Crawford C, Wildman SS et al (2013) Renal pericytes: regulators of medullary blood flow. Acta Physiol (Oxf) 207:212–225
18. Kinner B, Zaleskas JM, Spector M (2002) Regulation of smooth muscle actin expression and contraction in adult human mesenchymal stem cells. Exp Cell Res 278:72–83
19. Dore-Duffy P, Mehedi A, Wang X et al (2011) Immortalized CNS pericytes are quiescent smooth muscle actin-negative and pluripotent. Microvasc Res 82:18–27
20. Dore-Duffy P, Wang S, Mehedi A et al (2011) Pericyte-mediated vasoconstriction underlies TBI-induced hypoperfusion. Neurol Res 33:176–186
21. Balabanov R, Dore-Duffy P (1998) Role of the CNS microvascular pericyte in the blood–brain barrier. J Neurosci Res 53:637–644
22. Schlingemann RO, Rietveld FJ, de Waal RM et al (1990) Expression of the high molecular weight melanoma-associated antigen by pericytes during angiogenesis in tumors and in healing wounds. Am J Pathol 136:1393–1405
23. Murfee WL, Skalak TC, Peirce SM (2005) Differential arterial/venous expression of NG2 proteoglycan in perivascular cells along microvessels: identifying a venule-specific phenotype. Microcirculation 12:151–160
24. Stallcup WB (2002) The NG2 proteoglycan: past insights and future prospects. J Neurocytol 31:423–435
25. Verbeek MM, Otte-Holler I, Wesseling P et al (1994) Induction of alpha-smooth muscle actin expression in cultured human brain pericytes by transforming growth factor-beta 1. Am J Pathol 144:372–382
26. Bandopadhyay R, Orte C, Lawrenson JG et al (2001) Contractile proteins in pericytes at the blood–brain and blood–retinal barriers. J Neurocytol 30:35–44
27. Kunz J, Krause D, Kremer M et al (1994) The 140-kDa protein of blood–brain barrier-associated pericytes is identical to aminopeptidase N. J Neurochem 62:2375–2386
28. Alliot F, Rutin J, Leenen PJ et al (1999) Pericytes and periendothelial cells of brain parenchyma vessels co-express aminopeptidase N, aminopeptidase A, and nestin. J Neurosci Res 58:367–378
29. Nayak RC, Berman AB, George KL et al (1988) A monoclonal antibody (3G5)-defined ganglioside antigen is expressed on the cell surface of microvascular pericytes. J Exp Med 167:1003–1015
30. Cho H, Kozasa T, Bondjers C et al (2003) Pericyte-specific expression of Rgs5: implications for PDGF and EDG receptor signaling during vascular maturation. FASEB J 17: 440–442
31. Crisan M, Yap S, Casteilla L et al (2008) A perivascular origin for mesenchymal stem cells in multiple human organs. Cell Stem Cell 3:301–313
32. Paquet-Fifield S, Schluter H, Li A et al (2009) A role for pericytes as microenvironmental regulators of human skin tissue regeneration. J Clin Invest 119:2795–2806
33. Brachvogel B, Moch H, Pausch F et al (2005) Perivascular cells expressing annexin A5 define a novel mesenchymal stem cell-like population with the capacity to differentiate into multiple mesenchymal lineages. Development 132: 2657–2668

34. Schor AM, Allen TD, Canfield AE et al (1990) Pericytes derived from the retinal microvasculature undergo calcification in vitro. J Cell Sci 97(Pt 3):449–461
35. Dar A, Domev H, Ben-Yosef O et al (2012) Multipotent vasculogenic pericytes from human pluripotent stem cells promote recovery of murine ischemic limb. Circulation 125:87–99
36. Dore-Duffy P (2003) Isolation and characterization of cerebral microvascular pericytes. Methods Mol Med 89:375–382
37. Rucker HK, Wynder HJ, Thomas WE (2000) Cellular mechanisms of CNS pericytes. Brain Res Bull 51:363–369
38. Tilton RG (1991) Capillary pericytes: perspectives and future trends. J Electron Microsc Tech 19:327–344
39. Shepro D, Morel NM (1993) Pericyte physiology. FASEB J 7:1031–1038
40. Fisher M (2009) Pericyte signaling in the neurovascular unit. Stroke 40:S13–S15
41. Bergers G, Song S (2005) The role of pericytes in blood-vessel formation and maintenance. Neuro Oncol 7(4):452–464
42. Siao CJ, Lorentz CU, Kermani P et al (2012) ProNGF, a cytokine induced after myocardial infarction in humans, targets pericytes to promote microvascular damage and activation. J Exp Med 209:2291–2305
43. Armulik A, Abramsson A, Betsholtz C (2005) Endothelial/pericyte interactions. Circ Res 97:512–523
44. Armulik A, Genove G, Mae M et al (2010) Pericytes regulate the blood–brain barrier. Nature 468:557–561
45. Nawashiro H, Messing A, Azzam N et al (1998) Mice lacking GFAP are hypersensitive to traumatic cerebrospinal injury. Neuroreport 9:1691–1696
46. Bonkowski D, Katyshev V, Balabanov RD et al (2011) The CNS microvascular pericyte: pericyte-astrocyte crosstalk in the regulation of tissue survival. Fluids Barriers CNS 8(1):8
47. Daneman R, Zhou L, Kebede AA et al (2010) Pericytes are required for blood–brain barrier integrity during embryogenesis. Nature 468:562–566
48. Al Ahmad A, Taboada CB, Gassmann M et al (2011) Astrocytes and pericytes differentially modulate blood–brain barrier characteristics during development and hypoxic insult. J Cereb Blood Flow Metab 31:693–705
49. Berezowski V, Landry C, Dehouck MP et al (2004) Contribution of glial cells and pericytes to the mRNA profiles of P-glycoprotein and multidrug resistance-associated proteins in an in vitro model of the blood–brain barrier. Brain Res 1018:1–9
50. Dohgu S, Takata F, Yamauchi A et al (2005) Brain pericytes contribute to the induction and up-regulation of blood–brain barrier functions through transforming growth factor-beta production. Brain Res 1038:208–215
51. Bendayan R, Ronaldson PT, Gingras D et al (2006) In situ localization of P-glycoprotein (ABCB1) in human and rat brain. J Histochem Cytochem 54:1159–1167
52. Quaglino D, Boraldi F, Annovi G et al (2011) The multifaceted complexity of genetic diseases: a lesson from pseudoxanthoma elasticum. In: Kenji I (ed) Advances in the study of genetic disorders. Intech, Croatia. doi:10.5772/22161. ISBN 978-953-307-305-7
53. Lonigro AJ, McMurdo L, Stephenson AH et al (1996) Hypotheses regarding the role of pericytes in regulating movement of fluid, nutrients, and hormones across the microcirculatory endothelial barrier. Diabetes 45(Suppl 1):S38–S43
54. Dore-Duffy P, LaManna JC (2007) Physiologic angiodynamics in the brain. Antioxid Redox Signal 9(9):1363–1371
55. Seaman S, Stevens J, Yang MY et al (2007) Genes that distinguish physiological and pathological angiogenesis. Cancer Cell 11:539–554
56. Diaz-Flores L, Gutierrez R, Valladares F et al (1994) Intense vascular sprouting from rat femoral vein induced by prostaglandins E1 and E2. Anat Rec 238:68–76
57. Lai CH, Kuo KH (2005) The critical component to establish in vitro BBB model: pericyte. Brain Res Brain Res Rev 50:258–265
58. Virgintino D, Girolamo F, Errede M et al (2007) An intimate interplay between precocious, migrating pericytes and endothelial cells governs human fetal brain angiogenesis. Angiogenesis 10:35–45
59. Egginton S, Zhou AL, Brown MD et al (2000) The role of pericytes in controlling angiogenesis in vivo. Adv Exp Med Biol 476:81–99
60. Reynolds LP, Grazul-Bilska AT, Redmer DA (2000) Angiogenesis in the corpus luteum. Endocrine 12:1–9
61. Ozerdem U, Stallcup WB (2003) Early contribution of pericytes to angiogenic sprouting and tube formation. Angiogenesis 6:241–249
62. Gerhardt H, Betsholtz C (2003) Endothelial-pericyte interactions in angiogenesis. Cell Tissue Res 314:15–23
63. Diaz-Flores L, Gutierrez R, Gonzalez P et al (1991) Inducible perivascular cells contribute to the neochondrogenesis in grafted perichondrium. Anat Rec 229:1–8

64. Gonul E, Duz B, Kahraman S (2002) Early pericyte response to brain hypoxia in cats: an ultrastructural study. Microvasc Res 64(1):116–119
65. Dore-Duffy P, Wang X, Mehedi A et al (2007) Differential expression of capillary VEGF isoforms following traumatic brain injury. Neurol Res 29:395–403
66. Melgar MA, Rafols J, Gloss D et al (2005) Postischemic reperfusion: ultrastructural blood–brain barrier and hemodynamic correlative changes in an awake model of transient forebrain ischemia. Neurosurgery 56:571–581
67. Duz B, Oztas E, Erginay T et al (2007) The effect of moderate hypothermia in acute ischemic stroke on pericyte migration: an ultrastructural study. Cryobiology 55: 279–284
68. Hellstrom M, Kalen M, Lindahl P et al (1999) Role of PDGF-B and PDGFR-beta in recruitment of vascular smooth muscle cells and pericytes during embryonic blood vessel formation in the mouse. Development 126:3047–3055
69. Jin K, Mao XO, Sun Y et al (2002) Heparin-binding epidermal growth factor-like growth factor: hypoxia-inducible expression in vitro and stimulation of neurogenesis in vitro and in vivo. J Neurosci 22:5365–5373
70. Rovida E, Navari N, Caligiuri A et al (2008) ERK5 differentially regulates PDGF-induced proliferation and migration of hepatic stellate cells. J Hepatol 48:107–115
71. Bonacchi A, Romagnani P, Romanelli RG et al (2001) Signal transduction by the chemokine receptor CXCR3: activation of Ras/ERK, Src, and phosphatidylinositol 3-kinase/Akt controls cell migration and proliferation in human vascular pericytes. J Biol Chem 276:9945–9954
72. Bernstein LR, Antoniades H, Zetter BR (1982) Migration of cultured vascular cells in response to plasma and platelet-derived factors. J Cell Sci 56:71–82
73. Tigges U, Boroujerdi A, Welser-Alves JV et al (2013) TNF-alpha promotes cerebral pericyte remodeling in vitro, via a switch from alpha1 to alpha2 integrins. J Neuroinflammation 10:33
74. Clark ER, Clark EL (2005) The development of adventitial (Rouget) cells on the blood capillaries of amphibian larvae. Am J Anat 35:239–264. doi:10.1002/aja.1000350205
75. Carmeliet P (2004) Manipulating angiogenesis in medicine. J Intern Med 255(5):538–561
76. Distler O, Neidhart M, Gay RE et al (2002) The molecular control of angiogenesis. Int Rev Immunol 21:33–49
77. Hall AP (2006) Review of the pericyte during angiogenesis and its role in cancer and diabetic retinopathy. Toxicol Pathol 34: 763–775
78. Harik SI, Hritz MA, LaManna JC (1995) Hypoxia-induced brain angiogenesis in the adult rat. J Physiol 485(Pt 2):525–530
79. Chavez JC, Agani F, Pichiule P et al (2000) Expression of hypoxia-inducible factor-1alpha in the brain of rats during chronic hypoxia. J Appl Physiol 89:1937–1942
80. LaManna JC, Vendel LM, Farrell RM (1992) Brain adaptation to chronic hypobaric hypoxia in rats. J Appl Physiol 72:2238–2243
81. Pichiule P, LaManna JC (2002) Angiopoietin-2 and rat brain capillary remodeling during adaptation and deadaptation to prolonged mild hypoxia. J Appl Physiol 93:1131–1139
82. Vandenhaute E, Culot M, Gosselet F et al (2012) Brain pericytes from stress-susceptible pigs increase blood–brain barrier permeability in vitro. Fluids Barriers CNS 9:11
83. Manea A, Constantinescu E, Popov D et al (2004) Changes in oxidative balance in rat pericytes exposed to diabetic conditions. J Cell Mol Med 8:117–126
84. Paepe BD, Creus KK, Bleecker JLD (2012) The tumor necrosis factor superfamily of cytokines in the inflammatory myopathies: potential targets for therapy. Clin Dev Immunol 2012:369–432
85. Ando T, Ichikawa J, Wako M et al (2006) TWEAK/Fn14 interaction regulates RANTES production, BMP-2-induced differentiation, and RANKL expression in mouse osteoblastic MC3T3-E1 cells. Arthritis Res Ther 8:R146
86. Tweedie D, Sambamurti K, Greig NH (2007) TNF-alpha inhibition as a treatment strategy for neurodegenerative disorders: new drug candidates and targets. Curr Alzheimer Res 4:378–385
87. Frankola KA, Greig NH, Luo W et al (2011) Targeting TNF-α to elucidate and ameliorate neuroinflammation in neurodegenerative diseases. CNS Neurol Disord Drug Targets 10:391–403
88. Locksley RM, Killeen N, Lenardo MJ (2001) The TNF and TNF receptor superfamilies: integrating mammalian biology. Cell 104:487–501
89. Chicheportiche Y, Bourdon PR, Xu H et al (1997) TWEAK, a new secreted ligand in the tumor necrosis factor family that weakly induces apoptosis. J Biol Chem 272:32401–32410
90. Schneider P, Schwenzer R, Haas E et al (1999) TWEAK can induce cell death via endogenous TNF and TNF receptor 1. Eur J Immunol 29:1785–1792
91. Winkles JA (2008) The TWEAK-Fn14 cytokine-receptor axis: discovery, biology and

therapeutic targeting. Nat Rev Drug Discov 7:411–425
92. Burkly LC, Michaelson JS, Hahm K et al (2007) TWEAKing tissue remodeling by a multifunctional cytokine: role of TWEAK/ Fn14 pathway in health and disease. Cytokine 40:1–16
93. Brown SA, Ghosh A, Winkles JA (2010) Full-length, membrane-anchored TWEAK can function as a juxtacrine signaling molecule and activate the NF-kappaB pathway. J Biol Chem 285:17432–17441
94. Pradet-Balade B, Medema JP, Lopez-Fraga M et al (2002) An endogenous hybrid mRNA encodes TWE-PRIL, a functional cell surface TWEAK-APRIL fusion protein. EMBO J 21:5711–5720
95. Wiley SR, Winkles JA (2003) TWEAK, a member of the TNF superfamily, is a multi-functional cytokine that binds the TweakR/ Fn14 receptor. Cytokine Growth Factor Rev 14:241–249
96. Wiley SR, Cassiano L, Lofton T et al (2001) A novel TNF receptor family member binds TWEAK and is implicated in angiogenesis. Immunity 15:837–846
97. Bossen C, Ingold K, Tardivel A et al (2006) Interactions of tumor necrosis factor (TNF) and TNF receptor family members in the mouse and human. J Biol Chem 281:13964–13971
98. Nakayama M, Harada N, Okumura K et al (2003) Characterization of murine TWEAK and its receptor (Fn14) by monoclonal anti-bodies. Biochem Biophys Res Commun 306:819–825
99. Dogra C, Hall SL, Wedhas N et al (2007) Fibroblast growth factor inducible 14 (Fn14) is required for the expression of myogenic regulatory factors and differentiation of myo-blasts into myotubes. Evidence for TWEAK-independent functions of Fn14 during myogenesis. J Biol Chem 282:15000–15010
100. Tran NL, McDonough WS, Savitch BA et al (2005) The tumor necrosis factor-like weak inducer of apoptosis (TWEAK)-fibroblast growth factor-inducible 14 (Fn14) signaling system regulates glioma cell survival via NFkappaB pathway activation and BCL-XL/ BCL-W expression. J Biol Chem 280:3483–3492
101. Tran NL, McDonough WS, Donohue PJ et al (2003) The human Fn14 receptor gene is up-regulated in migrating glioma cells in vitro and overexpressed in advanced glial tumors. Am J Pathol 162:1313–1321
102. Tran NL, McDonough WS, Savitch BA et al (2006) Increased fibroblast growth factor-inducible 14 expression levels promote gli-oma cell invasion via Rac1 and nuclear factor-kappaB and correlate with poor patient outcome. Cancer Res 66:9535–9542
103. Nakayama M, Ishidoh K, Kayagaki N et al (2002) Multiple pathways of TWEAK-induced cell death. J Immunol 168:734–743
104. Chang HY, Sneddon JB, Alizadeh AA et al (2004) Gene expression signature of fibro-blast serum response predicts human cancer progression: similarities between tumors and wounds. PLoS Biol 2:E7
105. Chang HY, Nuyten DS, Sneddon JB et al (2005) Robustness, scalability, and integra-tion of a wound-response gene expression sig-nature in predicting breast cancer survival. Proc Natl Acad Sci U S A 102:3738–3743
106. Yepes M, Brown SA, Moore EG et al (2005) A soluble Fn14-Fc decoy receptor reduces infarct volume in a murine model of cerebral ischemia. Am J Pathol 166:511–520
107. Lembo AJ, Neri B, Tolley J et al (2009) Use of serum biomarkers in a diagnostic test for irritable bowel syndrome. Aliment Pharmacol Ther 29:834–842
108. Serafini B, Magliozzi R, Rosicarelli B et al (2008) Expression of TWEAK and its recep-tor Fn14 in the multiple sclerosis brain: impli-cations for inflammatory tissue injury. J Neuropathol Exp Neurol 67:1137–1148
109. Michaelson JS, Wisniacki N, Burkly LC et al (2012) Role of TWEAK in lupus nephritis: a bench-to-bedside review. J Autoimmun 39:130–142
110. Kamata K, Kamijo S, Nakajima A et al (2006) Involvement of TNF-like weak inducer of apoptosis in the pathogenesis of collagen-induced arthritis. J Immunol 177:6433–6439
111. Perper SJ, Browning B, Burkly LC et al (2006) TWEAK is a novel arthritogenic mediator. J Immunol 177:2610–2620
112. Maecker H, Varfolomeev E, Kischkel F et al (2005) TWEAK attenuates the transition from innate to adaptive immunity. Cell 123:931–944
113. Han S, Yoon K, Lee K et al (2003) TNF-related weak inducer of apoptosis receptor, a TNF receptor superfamily member, activates NF-kappa B through TNF receptor-associated factors. Biochem Biophys Res Commun 305:789–796
114. Saitoh T, Nakayama M, Nakano H et al (2003) TWEAK induces NF-kappaB2 p100 processing and long lasting NF-kappaB acti-vation. J Biol Chem 278:36005–36012
115. Perkins ND (2007) Integrating cell-signalling pathways with NF-kappaB and IKK function. Nat Rev Mol Cell Biol 8:49–62
116. Hayden MS, West AP, Ghosh S (2006) NF-kappaB and the immune response. Oncogene 25:6758–6780

117. Fotin-Mleczek M, Henkler F, Hausser A et al (2004) Tumor necrosis factor receptor-associated factor (TRAF) 1 regulates CD40-induced TRAF2-mediated NF-kappaB activation. J Biol Chem 279:677–685
118. Hauer J, Puschner S, Ramakrishnan P et al (2005) TNF receptor (TNFR)-associated factor (TRAF) 3 serves as an inhibitor of TRAF2/5-mediated activation of the noncanonical NF-kappaB pathway by TRAF-binding TNFRs. Proc Natl Acad Sci U S A 102:2874–2879
119. Jakubowski A, Browning B, Lukashev M et al (2002) Dual role for TWEAK in angiogenic regulation. J Cell Sci 115:267–274
120. Girgenrath M, Weng S, Kostek CA et al (2006) TWEAK, via its receptor Fn14, is a novel regulator of mesenchymal progenitor cells and skeletal muscle regeneration. EMBO J 25:5826–5839
121. Polek TC, Talpaz M, Darnay BG et al (2003) TWEAK mediates signal transduction and differentiation of RAW264.7 cells in the absence of Fn14/TweakR. Evidence for a second TWEAK receptor. J Biol Chem 278:32317–32323
122. Bhatnagar S, Kumar A (2012) The TWEAK-Fn14 system: breaking the silence of cytokine-induced skeletal muscle wasting. Curr Mol Med 12:3–13
123. Bird T, Knight B, Boulter L et al (2010) P55 Syngenic bone marrow transfer stimulates hepatic progenitor cell expansion via TWEAK/Fn14 Signalling: implications for human autologous cell therapy. Gut 59:A33
124. Hamill CA, Michaelson JS, Hahm K et al (2007) Age-dependent effects of TWEAK/Fn14 receptor activation on neural progenitor cells. J Neurosci Res 85:3535–3544
125. Lynch CN, Wang YC, Lund JK et al (1999) TWEAK induces angiogenesis and proliferation of endothelial cells. J Biol Chem 274:8455–8459
126. Biancone L, Martino AD, Orlandi V et al (1997) Development of inflammatory angiogenesis by local stimulation of Fas in vivo. J Exp Med 186:147–152
127. Donohue PJ, Richards CM, Brown SA et al (2003) TWEAK is an endothelial cell growth and chemotactic factor that also potentiates FGF-2 and VEGF-A mitogenic activity. Arterioscler Thromb Vasc Biol 23:594–600
128. Tian XL, Kadaba R, You SA et al (2004) Identification of an angiogenic factor that when mutated causes susceptibility to Klippel-Trenaunay syndrome. Nature 427:640–645
129. Kawakita T, Shiraki K, Yamanaka Y et al (2005) Functional expression of TWEAK in human colonic adenocarcinoma cells. Int J Oncol 26:87–93
130. Shimada K, Fujii T, Tsujikawa K et al (2012) ALKBH3 contributes to survival and angiogenesis of human urothelial carcinoma cells through NADPH oxidase and tweak/Fn14/VEGF signals. Clin Cancer Res 18:5247–5255
131. Hahne M, Kataoka T, Schroter M et al (1998) APRIL, a new ligand of the tumor necrosis factor family, stimulates tumor cell growth. J Exp Med 188:1185–1190
132. Marmarou A, Foda MA, van den Campbell BW et al (1994) A new model of diffuse brain injury in rats. Part I: pathophysiology and biomechanics. J Neurosurg 80:291–300

Part II

Animal Models of Cerebral Angiogenesis

Chapter 5

Analysis of Angiogenesis in the Developing Mouse Central Nervous System

Nicole Ziegler, Karl H. Plate, and Stefan Liebner

Abstract

In order to study basic mechanisms of sprouting angiogenesis, researchers worldwide rely on the use of model tissues such as rodent retina, which becomes vascularized postnatally, to study the growth of blood vessels. By definition, models have to be simple, recapitulating angiogenic processes in a stereotyped and relatively easy accessible manner, allowing the application of standardized analyses. These criteria also apply in an ideal manner to the embryonic mouse hindbrain, which becomes vascularized by sprouting angiogenesis from a preformed perineural vascular plexus, leading to the stereotypical formation of a capillary subventricular plexus. Similar to the retina model, between embryonic days 10.5 and 13.5, the hindbrain can be flat-mounted in an "open-book" preparation, allowing the analysis of the vascular bed in two-dimensional extension, of parameters like vessel density, morphology, and remodeling including branching and sprouting. In addition to sprouting angiogenesis, the hindbrain is a suitable model for investigating inductive mechanisms towards the blood–brain barrier phenotype of microvessels in the central nervous system. In this chapter, we describe how to fix, dissect, stain, and analyze the developing hindbrain vasculature.

Key words Angiogenesis, Mouse hindbrain, High-resolution imaging, Transgenic mice

1 Introduction

Because of the direct or indirect involvement of blood or lymphatic vessels in disease initiation and/or progression [1], angiogenesis has become one of the major topics in biomedical research throughout the last few decades. To understand the basic principles of vessel growth and differentiation, transgenic rodent model systems have become a valuable tool to investigate vasculo- and angiogenesis.

Here we describe the mouse embryonic hindbrain (HB) as a model organ to elucidate the quality and properties of the developing vasculature, providing an experimental protocol for imaging and quantifying different parameters including vessel morphology, mural cell coverage, as well as branching and sprouting angiogenesis in this organ.

Richard Milner (ed.), *Cerebral Angiogenesis: Methods and Protocols*, Methods in Molecular Biology, vol. 1135,
DOI 10.1007/978-1-4939-0320-7_5, © Springer Science+Business Media New York 2014

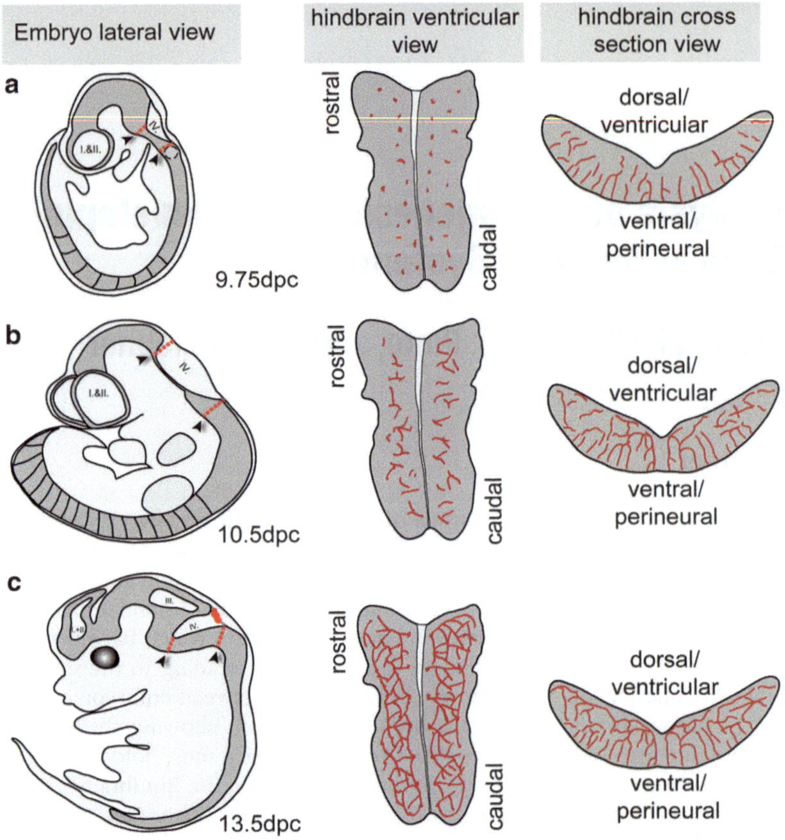

Fig. 1 Schematics of a complete embryo as well as embryo hindbrains in the ventricular and cross-sectional view for different developmental stages. In the *left panel*, the localization of the hindbrain for each developmental stage is indicated by *red dashed lines*. The *middle* and *left panels* illustrate the time course of hindbrain vascularization. (**a**) After the establishment of the perineural vascular plexus (PVP) at embryonic day 9.5 (9.5dpc), first vessel sprouting into the embryonic hindbrain occurs around 9.75dpc. The cross-sectional view thereby illustrates that the invading vessels lineally grow towards the ventricular zone. (**b**) The onset of radial vessel outgrowth, meaning that the vessel sprouts start growing laterally (around 10.0dpc), is followed by their anastomosis. This process initiates 10.5dpc and subsequently leads to the establishment of the subventricular vascular plexus (SVP). (**c**) These primary angiogenic courses are followed by vessel remodeling and maturation starting around 12.5dpc, including the recruitment of perivascular cells, leading to a functional vascular network. Arteriovenous differentiation in the hindbrain has been suggested to take place from E14.5 onwards [6]. *Arrowheads* point to the borders of the hindbrain

1.1 Vascularization of the Mouse Embryonic Hindbrain

In mammals, the brain becomes exclusively vascularized by angiogenic sprouting from a preformed perineural vascular plexus (PVP), which in the mouse is established by embryonic day 9.5 (9.5dpc, Fig. 1). Vascularization of the avascular neuroectoderm starts from caudal to rostral, and therefore the hindbrain as a phylogenetically ancient part of the central nervous system (CNS), located adjacent to the spinal cord, is vascularized early in development. Vessels first sprout into the mouse hindbrain around 9.75dpc and subsequently grow perpendicular to the PVP towards the ventricular zone (Fig. 1a) [2, 3]. Starting from 10.0dpc, radial vessels form sprouts parallel to the ventricular hindbrain surface, which anastomose, thereby generating a subventricular vascular plexus (SVP) (Fig. 1b). This ventral-to-dorsal scheme of HB angiogenesis is highly reproducible and evolutionarily conserved, starting from the basilar artery as the major supplying vessel in all vertebrates [3].

Following the initial angiogenic processes, starting from 12.5dpc vessels remodel which includes the recruitment of perivascular cells. Pericytes (PCs) enter the neuroectoderm together with the first vessels and contribute to vessel maturation [4]. In particular, most vessels in the CNS acquire specific barrier properties, known as the blood–brain barrier (BBB), via the cellular interaction in the neurovascular unit (NVU), comprising endothelial cells, pericytes, astrocytes, some interneurons, and perivascular microglia [5].

Arteriovenous differentiation in the hindbrain has been suggested to take place from E14.5 onwards [6].

Owing to its unique architecture and vascularization early in development, the mouse embryo hindbrain has become a key model system for studying sprouting angiogenesis in vivo.

1.2 The Mouse Embryo Hindbrain as a Model for Angiogenesis and Vessel Differentiation

The mouse, but also the zebrafish hindbrain, has extensively been used to study angiogenic processes [3, 6]. As most of the studies utilize HB angiogenesis as a representative model system for sprouting angiogenesis, we have focused on vessel differentiation towards the BBB phenotype in the HB model [7]. In particular, we evaluated activation of the Wnt/β-catenin pathway in ECs entering the HB, employing the BAT-gal reporter mouse for canonical Wnt signaling (Fig. 2 [8]). This pathway has been shown by us, and subsequently and independently by two other groups, to control brain angiogenesis on the one hand and BBB differentiation on the other [9]. Accordingly, Tie-2-Cre:βCat$^{fl/del\text{-}mediated}$ endothelial deletion of β-catenin is embryonically lethal around 11.5dpc and leads to impaired vascularization of the neuroectoderm. Specifically, vessel morphology and caliber of the PVP was altered, suggesting a vascular remodeling defect upon β-catenin deletion in ECs [10]. Regarding brain vascularization, endothelial deletion of β-catenin was demonstrated to reduce vessel penetration from the PVP into the avascular neuroectoderm [11, 12], which can be visualized at the HB (Fig. 2).

1.3 Pros and Cons of the Hindbrain Model

Pros: *First*, the mouse HB is suited for the analysis of early developmental stages starting from 10.5dpc, whereas other model organs, such as the rodent retina, are limited to postnatal stages.

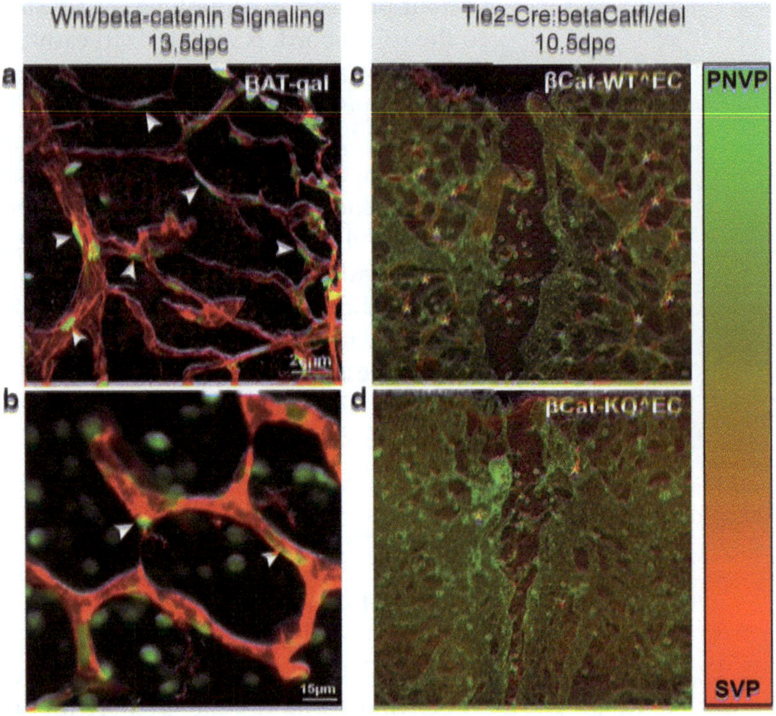

Fig. 2 (**a**, **b**) Whole-mount HB staining for nuclear β-gal (*green*) and IB4 (*red*) of BAT-gal embryos (13.5dpc) analyzed by confocal microscopy. After image or Z-stack acquisition, files are processed using appropriate software (in our case: Nikon NISElements AR, Nikon, Japan; alternatively freeware such as ImageJ can be used). Quantitative analysis of positive nuclei revealed a high incidence of Wnt pathway activation during embryonic HB angiogenesis [7, 11, 12]. *Arrowheads* point to β-gal-positive nuclei in the developing hindbrain endothelium. (**c**, **d**) Staining of a mouse HB from wild type (**c**) or endothelial-specific β-catenin knockout mice at 10.5dpc. The HB specimens were stained for IB4 and after confocal Z-stack acquisition, vessels of the PVP (*green*) and inside the brain parenchyma and the SVP (*red*, *asterisks*) were color coded in ImageJ to visualize differential progression of HB angiogenesis. Note also the altered organization and morphology of the PVP in the KO HB. See online version of the article for complete color information

This is particularly important for transgenic mice in which an early lethal phenotype is observed or is expected (e.g., *see* refs. 7, 10). *Second*, preparation of the HB is relatively easy and can be applied to entirely fixed and stored embryos, opening the possibility to carry out staining and subsequent analysis at the same time on a statistically relevant number of embryos. *Third*, the mouse HB is well suited to quantify angiogenic sprouting and vascular complexity in ventro-dorsal growth direction from the PVP (1D) and in mediolateral direction in the SVP (2D). *Fourth*, arteriovenous specialization starts relatively late during HB vascularization, leaving a large window of opportunity to study vessel density and morphology in capillary-like structures (*see* also ref. 6).

Cons: First, lethality of transgenic mouse lines occurring before 10.5dpc prevents the use of the HB model to investigate vascular defects. Instead, entire embryos or organs such as the yolk sac have to be considered for analysis. However, conditional and inducible transgenic techniques, using the Cre/lox system, can help circumvent lethality before 10.5dpc. *Second,* the embryonic mouse HB is not as suitable for manipulation as the postnatal mouse retina, which is accessible to intravitreal injections of drugs and viruses carrying engineered genes [13]. Nevertheless, in principle it is possible to apply intrauterine manipulation to embryos and to the HB in particular, such as transfection/transduction by electroporation [14]. Although this technique is quite challenging and laborious, it might open the possibility to tackle the HB vasculature by pharmacologic compounds. The limitation of intrauterine manipulation is most likely the stage of the embryos that can be treated by this technique, which is described as 15.5dpc. *Third,* with increasing age of the embryos (>14.5dpc), the HB gets thicker, preventing flat-mounting of the sample. Furthermore, whole-mount staining procedures become increasingly difficult due to limited penetration of the antibodies. However, we have successfully stained and mounted HBs up to stage 17.5dpc. For larger samples, we also describe an alternative method that specifically accounts for the size of older staged HBs, employing the light sheet or ultramicroscopy. *Last,* as discussed in Fantin et al., ex vivo applications such as organotypic cultures of the HB are not possible [6].

2 Materials

2.1 Sample Collection and Hindbrain Preparation

2.1.1 Equipment

1. Scissors for dissection of pregnant mouse female.
2. Forceps.
3. Petri dishes (10 cm).
4. Stereomicroscope with bright light illumination.
5. Orbital/plate shaker (preferentially in a cold room).
6. Round-bottom safe-lock tubes (2 ml).
7. Scalpel.

2.1.2 Reagents/Buffer

1. Phosphate buffered saline (PBS) 1×.
2. 4 % paraformaldehyde (PFA) in PBS, pH 7.4.
3. Blocking buffer (1 % BSA) 0.3 % Triton X-100, in 1× PBS.
4. Permeabilization/block buffer (1 % BSA), 0.5 % Triton X-100, PBS, pH 7.4–7.6.
5. 100 % methanol.
6. Isoflurane (1-chloro-2, 2,2-trifluoroethyl difluoromethyl ether) 100 % (V/V).

Table 1
Examples of antibodies used for HB whole-mount staining

			Dilution	Company	#	
Vessels	IB4		1:50	Sigma	L 2140	
	CD31 (PECAM-1)		1:100	BD Pharmingen	553370	
	Podocalyxin		1:200	R&D systems	AF1556	
	Endomucin			eBioscience	14-5851	
Vessel-associated cell types	SMA	Smooth muscle cells	1:200	Sigma	C6198	
	NG2 PDGFRβ Desmin	Pericytes	1:200 1:100	Millipore Dako Cytomation	AB5320 M 0760	
	Collagen IV	Basement membrane	1:200	AbD Serotec	2150-1470	
	Aquaporin 4 GFAP	Astrocytes	1:200	Millipore Dako Cytomation	AB2218	
Junctions	ZO-1	Tight junctions	1:200	Zymed	40-2300	[a]
	VE-Cadherin	Adherens junctions	1:100	Santa Cruz	sc-6458	
	Claudin3	Tight junctions	1:200	Zymed	34-1700	[a]
	Claudin5	Tight junctions	1:200	Invitrogen	35-2500	[a]
	Occludin	Tight junctions		Santa Cruz	sc-8144	

[a]Do not use on PFA-fixed tissue. Dissect hindbrain first and fix afterwards with methanol (*see* Subheading 3)

2.2 Isolectin and Antibody Incubations and Mounting

2.2.1 Equipment

1. Orbital/plate shaker (at room temperature/ 4 °C).
2. Glass slides and coverslips.
3. Parafilm.

2.2.2 Reagents/Buffer

1. Phosphate buffered saline (PBS).
2. Pblec buffer: 1 mM $CaCl_2$, 1 mM $MgCl_2$, 0.1 mM $MnCl_2$, 1 % Triton X-100 in PBS.
3. Biotinylated isolectin B4.
4. Primary (*see* Table 1) and secondary antibodies (Molecular Probes/Invitrogen, dilution 1:200).
5. 4′,6-Diamidino-2-phenylindole (DAPI).
6. Mounting medium.
7. Blocking buffer: 1 % BSA, 0.3 %Triton X-100 in PBS.

8. Antibody incubation buffer: 0.5 % BSA, 0.25 % Triton X-100 in PBS, pH 7.2.
9. 4 % PFA in PBS.
10. Serum appropriate for the dilution of secondary antibodies (e.g., goat serum).

2.3 Clearing

2.3.1 Equipment

1. Plate shaker (horizontal).
2. Glass cuvette.
3. Parafilm.

2.3.2 Reagents/Buffer

1. 100 % methanol.
2. Benzyl alcohol.
3. Benzyl benzoate.
4. Entellan (Merck, #107961).
5. Nail polish.

3 Methods

3.1 Day 1: Sample Collection

General note: Mouse embryos from 10.5 to 15.5 are suitable for isolation of HBs.

1. Sacrifice a pregnant female by brief isoflurane inhalation and subsequent cervical dislocation, dissect both cones of the uterus containing the embryos, and transfer them into a 10 cm Petri dish with ice-cold PBS (keep on ice).
2. Cut off single embryos and dissect one after the other under a stereomicroscope.
3. Remove the uterus and carefully peel off the decidua with two fine forceps; open the yolk sac to expose the embryo, cut the umbilical cord, transfer the embryo in a 2 ml round-bottom safe-lock tube with PBS, and keep on ice until all the embryos are prepared (*see* **Note 1**).
4. Remove the PBS and fix the embryos in 4 %PFA (1.5–2 ml per 2 ml tube) overnight in an orbital or on a plate shaker at 4 °C (*see* **Note 2**). *Important:* It has to be taken into consideration that some staining procedures have to be performed with alternatively fixed tissue due to the properties of the particular primary antibodies (*see* Table 1). In that case, directly proceed with the hindbrain preparation.

3.2 Day 2: Hindbrain Preparation (see Fig. 3)

1. To remove PFA, wash the embryos twice with cold 1× PBS for 5 min at room temperature (RT).
2. Sever the head with forceps and expose the brain of the embryo.
3. Carefully separate the brain and remove the skin, the primordial skull bones, and the meninges from the brain (*see* **Note 3**).

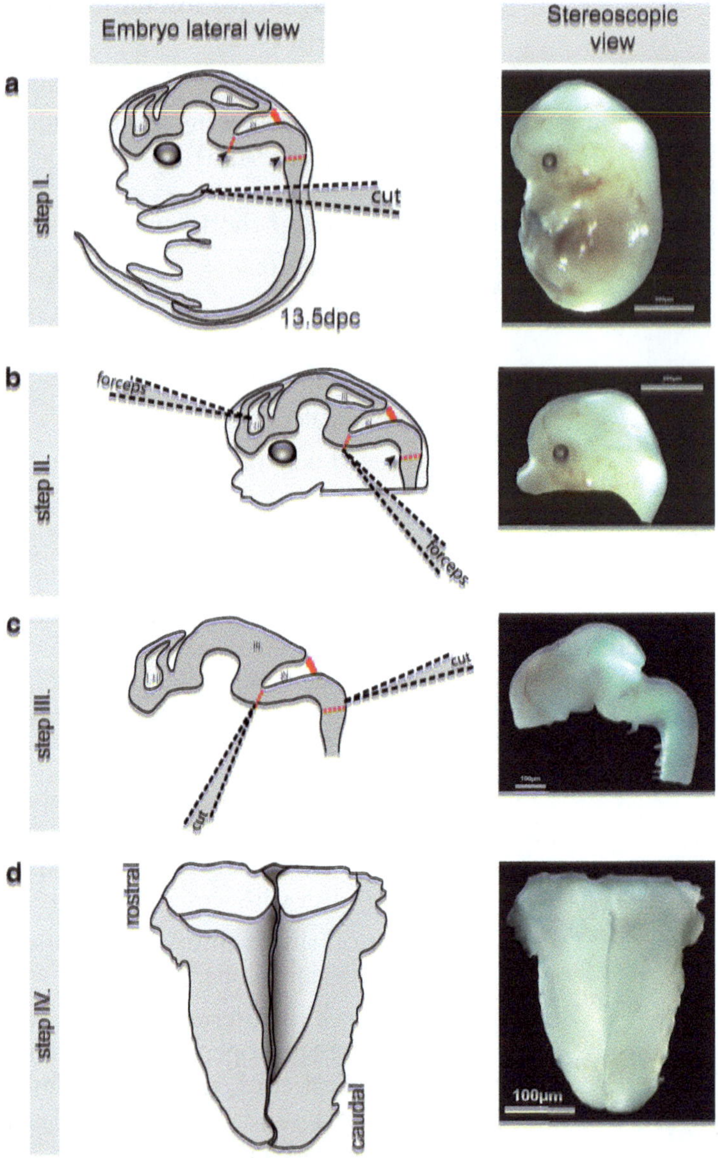

Fig. 3 "Open-book" preparation of an E13.5 mouse embryo hindbrain. Schemes of the preparation procedure including some accessory information are depicted in the *left panel*, pictures on the *right* show the corresponding stereoscopic view of the tissue. Remove the uterus and expose the embryos from the yolk sac. Remove the decidua and yolk sac. (**a**) Separate the head from the body by severing with forceps as indicated in step I (bar, 250 µm). (**b**) In step II, carefully dissect the brain of the embryo by removing the skin, the skull anlagen as well as the meninges (bar, 250 µm). (**c**) To dissever the hindbrain, tweeze with forceps or carefully cut with a scalpel at the indicated sites to get rid of the fore- and midbrain as well as the spinal cord tissue. Clean the hindbrain from remaining meninges and nerve fibers (bar, 100 µm). (**d**). Use fine forceps to open the hindbrain longitudinally so that it can completely unfurl (bar, 100 µm). At that point, the preparation is ready for fixation in methanol (*see* Subheading 3.2) further staining procedures (*see* Subheading 3). *Arrowheads* point to the borders of the hindbrain

4. Detach the hindbrain with a scalpel and remove remaining meninges and nerve fibers.
5. For embryonic stages 13.5–15.5, open the hindbrain longitudinally with fine forceps.

   Important: For junctional proteins, fix the native tissue (*see* Table 1 for appropriate antibodies) with ice-cold 100 % methanol for 5 min on ice and wash once with PBS for 5 min; this is to be done on day 1.
6. Transfer the hindbrains into 2 ml round-bottom tubes and incubate in blocking buffer (1.5–2 ml per 2 ml tube) overnight in an orbital or on a plate shaker at 4 °C (*see* **Note 4**).
7. For isolectin B4 staining, use blocking buffer containing 0.3 % Triton X-100. For all other primary antibodies, blocking can be performed in "permeabilization/block buffer" containing 0.5 % Triton X-100 (*see* Subheading 2).

3.3 Day 3: Isolectin B4 and Primary Antibody Incubation (See Note 5)

1. For isolectin B4 staining, wash the hindbrains three times for 20 min in Pblec buffer at RT [13].
2. Incubate with primary antibodies overnight at 4 °C on a plate shaker, and use 400 μl of antibody solution per hindbrain (*see* **Note 6**).
3. Dilute isolectin B4, and if required also primary antibodies, in Pblec buffer. For primary antibodies without isolectin B4, incubation can be performed in "antibody incubation buffer" (*see* Subheading 2).

3.4 Day 4: Secondary Antibody Incubation and Mounting/Tissue Clearing

1. Wash the hindbrains five times, 20 min at RT. For isolectin B4 staining in blocking buffer 1:1 with PBS. For all other primary antibodies with PBS.
2. Incubate with appropriate secondary antibodies (e.g., streptavidin-AlexaFluor 488 for visualizing isolectin B4 (1:200) and goat α-rabbit 568 (1:200) for visualizing β-galactosidase) for 2 h at RT on a plate shaker (*see* **Notes 7** and **8**). Dilution buffers differ depending on the antibody used. For isolectin B4 staining, use blocking buffer 1:1 with PBS, and for all other antibodies use antibody incubation buffer.
3. For isolectin B4 staining, wash 3 times 20 min in blocking buffer 1:1 with PBS at RT.
4. Wash twice for 5 min in PBS at RT.
5. For nuclear counterstaining, incubate with DAPI (1:1,000) in PBS for 10 min at RT.
6. Fix in 4 % PFA for 5 min on ice.
7. Wash 5 min in PBS at RT.

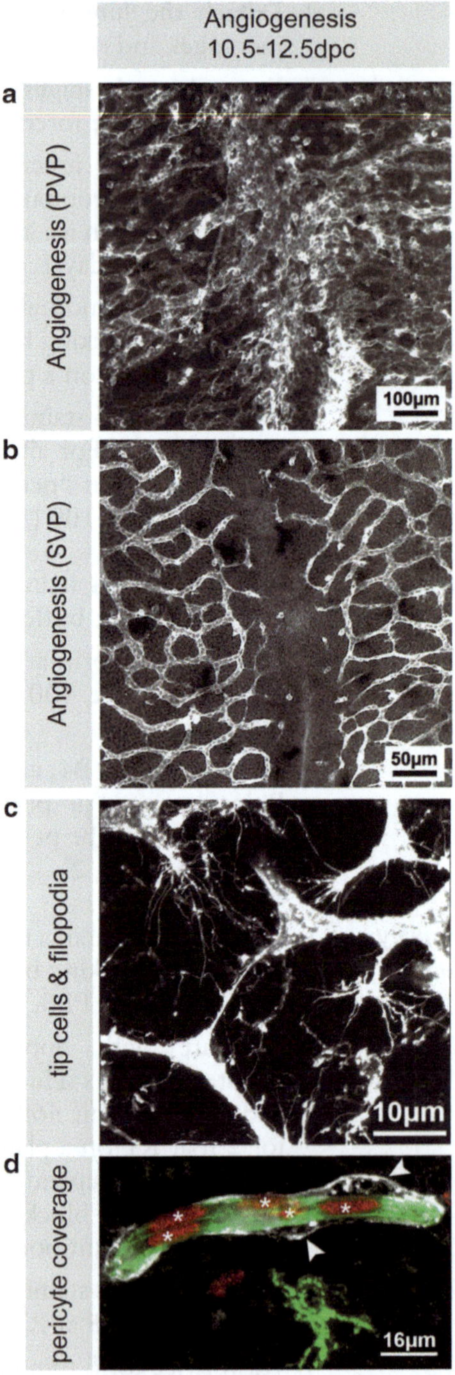

Fig. 4 Representative fluorescent staining for visualization of different parameters that can be analyzed using the mouse embryo hindbrain as a model system for angiogenesis and vessel differentiation. (**a**, **b**) The vascular network of the perineural vascular plexus (PVP) as well as the subventricular vascular plexus can easily be visualized by any vessel marker, for example, IB4 (*see* also Table 1). Pictures with low and medium magnification are particularly suitable for the analysis of different angiogenic parameters like vessel density as well as the

8. For direct mounting *without* tissue clearing, mount the hindbrains in an appropriate mounting medium (e.g., Aqua-Poly/Mount) and let mounting medium cure for several hours before analyzing. If tissue clearing is intended, omit **step 8** and proceed directly to **step 9** (below).
9. As dehydration of the tissue leads to its hardening, we recommend arranging the hindbrain tissue on the slide before starting the procedure. Arrange the hindbrain on a slide in the appropriate orientation and cover with a cover slip.
10. Important: To avoid squeezing the tissue, use Parafilm as a spacer between the slide and the coverslip.
11. Fix the coverslip at its anterior and posterior edge by wrapping Parafilm around the slide and the coverslip.
12. Put the slide upright into a glass cuvette filled with 100 % methanol and place on a horizontal plate shaker at room temperature overnight (*see* **Note 9**) (for review, *see* refs. 15, 16).
13. Substitute the methanol with benzyl alcohol and benzyl benzoate solution (mixing ratio 1:2 benzyl alcohol/ benzyl benzoate) and incubate for at least 6 h on a horizontal plate shaker at RT (*see* **Note 10**).
14. Seal the spaces between the slide and the coverslip by coating the edges with nail polish so that the tissue remains in the benzyl alcohol and benzyl benzoate solution. Alternatively, replace this solution by Entellan.

3.5 Analysis of Vascular Parameters (See Fig. 4)

1. Analysis of whole-mount stained, uncleared or cleared, and mounted HB specimens is carried out using a conventional confocal microscope. Alternatively, HBs can also be stained, embedded in low melting agarose, and subsequently cleared as described above. This preparation method, however, can only be analyzed by light sheet/ultramicroscopy (for further information, *see* ref. 17).
2. For quantifying angiogenic parameters in HB preparations, it is advisable to take pictures from randomly chosen fields

Fig. 4 (continued) number of branch points or intersections. Furthermore, this staining can give some indication of vascular abnormalities or defects in transgenic animals. (**c**) Higher magnification images of the same vascular staining are applicable for investigating the cellular mechanisms of angiogenesis, including the number of endothelial tip and stalk cells as well as filopodia protrusions. (**d**) Double or triple stainings are ideally suitable to illustrate the attachment of perivascular cells as a measure of vessel maturation. As indicated in our example, vessel staining (IB4, *green*) was combined with a pericyte marker (NG2, *white, arrowheads*) and a nuclear Wnt reporter (β-gal, *red, asterisks*) to evaluate cell type-specific activation of the Wnt/β-catenin pathway. Other mural cell markers can be combined with the vascular staining according to requirements. For further information concerning antibodies, *see* Table 1 (number of pictures, $n^p = 3–5$) from each side of the hindbrain and average the numeric values. Depending on the strength of the phenotype, the range for the number of individual HB specimens (n^s) varies between 3 and 15.

3. Counting of angiogenic parameters such as vascular branch points, intersections, endothelial tip cells, or pericyte coverage can be carried out either manually or semiautomated, depending on the quality of staining and the parameter to be determined. For example, vascular density can be evaluated by binarizing the IB4 image and by automatically measuring the stained area. A prerequisite for this form of analysis is staining with a high signal-to-noise ratio, allowing for a relatively painless adjustment of threshold levels. These kinds of measurements can be carried out in open-source software such as ImageJ (http://rsb.info.nih.gov/ij/) that provides a large number of potentially suitable plug-ins. But also many commercially available software packages bundled with your microscope, or stand-alone software such as Openlab and Volocity (PerkinElmer, USA) or Imaris (Bitplane, CH), offer appropriate analysis tools.

   Similarly, the latter commercially available software allow for the 3D reconstruction of vessels, given that the observed phenotype benefits from a 3D presentation for didactic and explanation reasons. Again, open-source software as ImageJ on which ImageSurfer (University of North Carolina, USA) is also based provide extensive possibilities which are beyond the scope of this article.

4 Notes

1. A small piece of each yolk sac can be kept to determine the genotype of the embryos if transgenic mice are investigated.
2. If necessary, embryos can be stored in PBS at 4 °C after fixation.
3. Removal of the meninges is facilitated if the brain is still intact.
4. In our hands, the staining works better with freshly prepared blocking buffer.
5. All washing and antibody incubation steps are performed on a plate shaker.
6. Double and triple stainings of vessels and other cell types, basement membrane, or junction proteins can be achieved by utilizing primary antibodies of different species. Primary and secondary antibody incubation can be performed simultaneously in the same buffer – Pblec and blocking buffer, respectively, for stainings where IB4 is included, and antibody incubation buffer for stainings with alternative vascular markers (*see* Table 1).

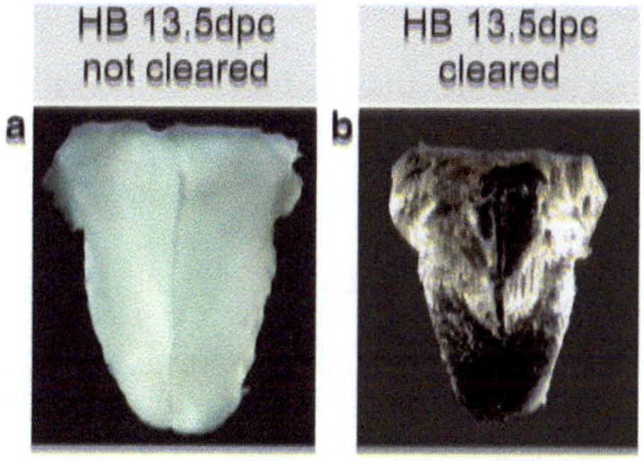

Fig. 5 Stereoscopic view of an uncleared or cleared hindbrain preparation of an E13.5 mouse embryo. Chemical clearing requires tissue dehydration, implemented by incubation with 100 % methanol overnight at room temperature. One important aspect which has to be considered in this regard is that dehydration leads to hardening of the tissue, which could impede proper orientation of the hindbrain for mounting and further microscopic analysis. We provide a strategy to circumvent these difficulties by primarily arranging the specimen on the slide and the subsequent dehydration (*see* Subheading 3). The chemical clearing is performed by incubation with benzyl alcohol/benzyl benzoate (1:2 ratio) for at least 6 h at room temperature. This results in the complete transparency of the tissue, which especially in lower magnification microscopy clearly reduces background staining. Depending on the accessible equipment for specific analyses by high-resolution imaging, the dehydrated tissue can alternatively be embedded in agarose and analyzed by light sheet/ultramicroscopy [17]

7. Keep samples protected from light from this step onwards.
8. Secondary antibody incubation can be performed simultaneously in the same buffer by combining appropriate antibodies labeled with different fluorescent dyes, but cross-reactivity should be considered.
9. To ensure complete soak of the tissue, the orientation of the slide in the cuvette should be parallel to the movement of the shaker.
10. Tissue will become completely translucent (*see* Fig. 5).

References

1. Risau W (1997) Mechanisms of angiogenesis. Nature 386:671–674
2. Lee HS, Han J, Bai H-J et al (2009) Brain angiogenesis in developmental and pathological processes: regulation, molecular and cellular communication at the neurovascular interface. FEBS J 276:4622–4635
3. Fujita M, Cha YR, Pham VN et al (2011) Assembly and patterning of the vascular network of the vertebrate hindbrain. Development (Cambridge, England) 138:1705–1715
4. Quaegebeur A, Segura I, Carmeliet P (2010) Pericytes: blood–brain barrier safeguards against neurodegeneration? Neuron 68:321–323
5. Abbott NJ, Patabendige AAK, Dolman DEM et al (2010) Structure and function of the blood–brain barrier. Neurobiol Dis 37:13–25

6. Fantin A, Vieira JM, Plein A et al (2013) The embryonic mouse hindbrain as a qualitative and quantitative model for studying the molecular and cellular mechanisms of angiogenesis. Nat Protoc 8:418–429
7. Liebner S, Corada M, Bangsow T et al (2008) Wnt/beta-catenin signaling controls development of the blood–brain barrier. J Cell Biol 183:409–417
8. Maretto S, Cordenonsi M, Dupont S et al (2003) Mapping Wnt/beta-catenin signaling during mouse development and in colorectal tumors. Proc Natl Acad Sci U S A 100: 3299–3304
9. Liebner S, Plate KH (2010) Differentiation of the brain vasculature: the answer came blowing by the Wnt. J Angiogenes Res 2:1
10. Cattelino A, Liebner S, Gallini R et al (2003) The conditional inactivation of the beta-catenin gene in endothelial cells causes a defective vascular pattern and increased vascular fragility. J Cell Biol 162:1111–1122
11. Stenman JM, Rajagopal J, Carroll TJ et al (2008) Canonical Wnt signaling regulates organ-specific assembly and differentiation of CNS vasculature. Science (New York, NY) 322:1247–1250
12. Daneman R, Agalliu D, Zhou L et al (2009) Wnt/beta-catenin signaling is required for CNS, but not non-CNS, angiogenesis. Proc Natl Acad Sci 106:641–646
13. Pitulescu ME, Schmidt I, Benedito R et al (2010) Inducible gene targeting in the neonatal vasculature and analysis of retinal angiogenesis in mice. Nat Protoc 5:1518–1534
14. Walantus W, Castaneda D, Elias L et al (2007) In Utero Intraventricular Injection and Electroporation of E15 Mouse Embryos. J Vis Exp 6:e239
15. Becker K, Jährling N, Saghafi S et al (2012) Chemical clearing and dehydration of GFP expressing mouse brains. PLoS ONE 7:e33916
16. Hama H, Kurokawa H, Kawano H et al (2011) Scale: a chemical approach for fluorescence imaging and reconstruction of transparent mouse brain. Nat Neurosci 14:1481–1488
17. Hägerling R, Pollmann C, Andreas M et al (2013) A novel multistep mechanism for initial lymphangiogenesis in mouse embryos based on ultramicroscopy. EMBO J 32:629–644

Chapter 6

Hypoxia-Induced Angiogenesis and Capillary Density Determination

Constantinos P. Tsipis, Xiaoyan Sun, Kui Xu, and Joseph C. LaManna

Abstract

Chronic exposure to moderate hypoxia elicits structural and functional changes in the microvascular network of the mammalian CNS. Hypoxia-induced angiogenesis can be elicited and studied by a relatively simple experimental method. Rats or mice can be exposed to mild hypoxia in a hypobaric chamber, or alternatively in a normobaric hypoxia chamber. After 3 weeks, the animals are perfused and fixed, the brain removed, and paraffin embedded and sectioned at 5 μm. The sections are stained for the blood-brain barrier glucose transporter (GLUT-1) by immunohistochemistry, capillary profiles identified and counted as a measure of angiogenesis.

Key words Hypoxia, Hypoxic adaptation, Hypobaric hypoxia, Brain, Angiogenesis, Immunohistochemistry, GLUT-1, Capillary density

1 Introduction

Chronic exposure to moderate hypoxia elicits structural and functional changes in the microvascular network of the mammalian CNS [1]. The extent of mammalian brain microvascular plasticity is becoming better appreciated. The significance of "angioplasticity" [2] to the pathophysiology of cerebrovascular and neurodegenerative diseases stimulates the need for methods that allow the study of the mechanisms that regulate this dynamic homeostatic property.

Mild hypobaric hypoxia in rats results in nearly doubling of the cerebral cortical capillary density with a time course of about 3 weeks [3–5]. Both hypo- and normobaric approaches are practical methods for exposing small animals to chronic moderate hypoxia. Computer-controlled normobaric chambers are particularly adept in performing intermittent hypoxia protocols where precise temporal control of oxygen is required. Normobaric chambers are also appropriate for situations where multiple gas mixtures are used, such as CO_2 for studies involving hypercapnia. Hypobaric chambers

Richard Milner (ed.), *Cerebral Angiogenesis: Methods and Protocols*, Methods in Molecular Biology, vol. 1135,
DOI 10.1007/978-1-4939-0320-7_6, © Springer Science+Business Media New York 2014

are not well suited for intermittent studies, but are more practical for experiments requiring a longer exposure period. Although there has been extensive discussion on the potential differences between decreasing oxygen partial pressure by decreasing atmospheric pressure (P_BO_2) and by decreasing the fraction inspired oxygen (FiO_2) [6], for the most part, both methods yield indistinguishable results with respect to brain angiogenesis.

GLUT-1 immunohistochemistry provides a direct and quantitative method for estimating the dimensions of the brain capillary network because it is present at all blood-brain barrier sites at 20 times the density found in astrocytes, thus eliminating the arterioles where the endothelial cells do not have the transporter. Other sites of high GLUT-1 concentration (e.g., tanicytes, glia limitans, choroid plexus, vaso vasorum) are easily distinguished from capillaries in the stained sections. The GLUT-1 transporter serves as a reliable marker for functioning, post-arteriolar microvessels [7].

2 Materials

All solutions must be prepared using ultrapure water (purified to 18 MΩ cm sensitivity at 25 °C) and reagent grade chemicals at room temperature. 1× PBS and 4 % PFA should be stored at 4 °C after preparation until use.

2.1 Hypoxia Chamber

1. Modified Wright chamber for hypobaric hypoxic exposure (*see* **Note 1**).
2. Oxycycler™ model A44C (BioSpherix, Ltd., Lacona, NY, USA) normobaric chamber connected to a computer-controlled gas mixing apparatus (*see* **Note 2**).

2.2 Perfusion and Fixation Solutions

1. 1× Phosphate buffered saline (PBS) with heparin: 0.01 M PBS, pH 7.4. Add 1 mL of heparin sodium to 1.0 L of 0.01 M PBS. Store at 4 °C.
2. 4 % paraformaldehyde in PBS: Heat 100 mL of 0.01 M PBS, pH 7.4 in the fume hood to 50 °C. Weigh 4.0 g of paraformaldehyde, and transfer to PBS (*see* **Note 3**). Allow the solution to completely dissolve and pH to 7.4 if necessary. Bring to room temperature and filter the solution through a 0.45 μm Corning filter. Store at 4 °C.

2.3 Immunohistochemistry Solutions

Immunohistochemistry solutions are to be prepared at the time of staining procedure and cannot be made and stored ahead of time.

1. Antigen retrieval buffer: 0.01 M sodium citrate, pH 6.0. Weigh 2.941 g of sodium citrate and transfer to 1 L of ultrapure water. Allow the solution to completely dissolve and titrate with 12 N HCl to pH 6.0.

2. 1 % H_2O_2 in 0.01 M PBS: Add 8 mL of 30 % hydrogen peroxide to 242 mL of 0.01 M PBS.
3. Blocking solution: 0.01 M PBS, pH 7.4, 10 % normal horse serum, 0.3 % Triton X-100 (*see* **Note 4**). Store on ice for duration of preparation.
4. ABC solution: 0.01 M PBS, pH 7.4, Vectastain ABC kit (Vector Laboratories, Burlingame, CA, USA). Add two drops of stock solutions A and B to 5.0 mL of 0.01 M PBS solution. Allow the solution to sit at room temperature for 30 min before use.
5. DAB solution: 0.01 M PBS, pH 7.4, 3,3′-diaminobenzidine tetrahydrochloride. Add 1 tablet of DAB (10 mg) to 30 mL of 1× PBS. Allow the tablet to completely dissolve before filtering through Whatman 41 ashless filter paper. Add 30 μL of 30 % H_2O_2 right before use (*see* **Note 5**). This procedure must be performed under a chemical hood.
6. EUKITT mounting medium (Calibrated Instruments, Inc., Hawthorne, NY, USA).

2.4 Antibodies and Conjugates

1. The Vectastain ABC kit can be purchased from Vector Laboratories, Burlingame, CA, USA.
2. Glucose transporter 1 (GLUT-1) goat polyclonal primary antibody (C-20, SC-1605) can be purchased from Santa Cruz Biotechnology, Dallas, TX, USA.
3. Biotinylated anti-goat made in horse secondary antibody can be purchased from Vector Laboratories.

3 Methods

3.1 Exposure to Hypoxia

1. Rats and mice, at appropriate cage density, are placed in either chamber (Fig. 1) (*see* **Note 6**).
2. Individual animals are weighed and maintained for food and water on a regular basis, and clean bedding is provided every

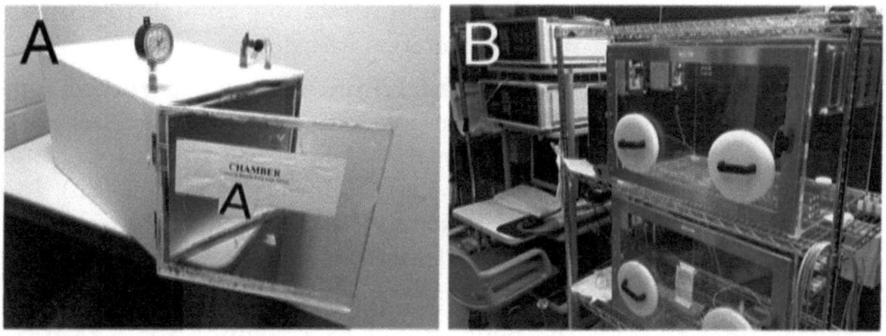

Fig. 1 (**a**) Modified Wright hypobaric hypoxia chamber connected to a facility vacuum line. (**b**) Biospherix OxyCycler™ normobaric hypoxia chamber (gas cylinders not shown)

day or two, necessitating return to normobaric conditions for up to 1 h (*see* Note 7).

3.2 Hypobaric Hypoxia

1. Wistar rats are exposed to 0.5 ATM (380 mmHg) for the designated period of time, usually 3 weeks, by adjusting the bleeder valve accordingly (*see* **Note 8**).
2. C57BL/6 mice are exposed to 0.4 ATM (300 mmHg) for the designated period of time, usually 3 weeks, by adjusting the bleeder valve (*see* **Note 8**).

3.3 Normobaric Hypoxia

1. Computer: Software provided by BioSpherix™ is designed to receive input from sensors and appropriately control the valve apparatus (OxyCycler™) mixing the gasses. For rats use 10 % O_2, for mice 8 % O_2.
2. Controller: Sets the overall flow rate for the system and is connected to N_2, O_2, and CO_2 tanks that are connected directly to the chamber.
3. Chamber: Sensors require calibration and open holes are necessary to facilitate airflow and excessive CO_2 accumulation during exposure (*see* **Note 9**).

3.4 Perfusion, Fixation

1. Body weight: Remove animals from chamber and record body weight in g.
2. Deeply anesthetize the animals with isofluorane and create a midsagittal thoracic incision exposing the heart and carefully clip the right atrium. Collect blood with a capillary tube for hematocrit (HCT) determination (*see* **Note 10**).
3. Insert a 21 gauge butterfly needle into the apex of the heart and perfuse transcardially (1 mL/g body weight) with ~30 mL for mouse and ~200 mL for rat with ice-cold 1× PBS with heparin placed at about 100 cm above heart level, and allow for perfusion to continue for at least 10 min, or until the liver is completely devoid of blood (*see* **Note 11**).
4. Close the PBS line and start 4 % PFA flow immediately through the butterfly needle and allow fixation to occur for 5 min, again, keeping the flow rate low enough to ensure structural integrity of the cerebral vasculature.
5. Decapitate at the level of the shoulders and carefully remove the skin and underlying cranium with sharp scissors by making an incision down the midline and carefully peeling back the cranium using forceps for mouse (*see* **Note 12**). In the rat, 2 bilateral incisions are made and skull is subsequently removed exposing the underlying tissue. Remove the brain by carefully lifting it from the base of the skull.
6. Place the brain in 4 % PFA solution for a 24-h postfixation period at 4 °C. Brains can be stored for a long-term period at this step.

3.5 Tissue Processing

1. Place the brain in a labeled tissue-processing cassette and wash under cold running tap water for 1 h (*see* **Note 13**).
2. Remove cassettes from running water and place in 80 % ethanol overnight (~14 h) at 4 °C.
3. Transfer the cassettes to 95 % ethanol for 90 min at room temperature and repeat this step twice with fresh 95 % ethanol each time for a total of 3 h.
4. Transfer the cassettes to 100 % ethanol for 90 min at room temperature and repeat this step twice with fresh 100 % ethanol each time for a total of 3 h.
5. Transfer the cassettes to xylene for 4 h at room temperature (*see* **Note 14**).
6. Remove the cassettes and transfer to molten paraffin overnight (60 °C).
7. Transfer the cassettes to a vacuum oven and apply pressure for 2 h at 80 °C (*see* **Note 15**).
8. Remove the brains from the tissue-processing cassette and embed them in paraffin using an embedding mold (*see* **Note 16**). Store at 4 °C overnight to allow paraffin to harden. Tissue blocks can be stored at room temperature forever.

3.6 Sectioning

1. Cut 5 μm coronal sections of cerebral cortex between Bregma 1.2 mm and 0.7 mm (*see* **Note 17**).
2. Using a hot water bath set to a temperature of 45 °C, mount each section onto Fisherbrand Superfrost Plus™ slides (Fisher Scientific) serially (*see* **Note 18**).
3. Allow slides to dry on slide rack overnight. Slides can be stored in a box for extended periods of time.

3.7 GLUT-1 Immunohistochemistry

All slides should be dried on a slide warmer overnight at 40–45 °C prior to starting staining preparation. All steps are to be performed at room temperature unless otherwise noted.

1. Deparaffinize: Place the slides in xylene for 5 min. Repeat this step twice for a total of 15 min in three separate xylene baths.
2. Rehydrate with graded alcohols: Transfer the slides to 100 % ethanol for 3 min. Repeat this step for a total of 6 min in two separate 100 % ethanol baths. Transfer the slides to 95 % ethanol for 3 min. Repeat this step for a total of 6 min in two separate 95 % ethanol baths. Transfer the slides to 70 % ethanol for 3 min, followed by 50 % ethanol for 3 min.
3. Transfer the slides to ultrapure deionized water for 3 min. Repeat this step twice for a total of 9 min in three separate washes (*see* **Note 19**).

4. Antigen retrieval: Transfer the slides to 0.01 M Antigen retrieval buffer, pH 6.0 in a 1.5 L beaker and heat to 100 °C. Once the target temperature is reached, allow the slides to remain in 100 °C for 15 min. Remove the beaker from heat and allow the slides to reach room temperature before continuing to the next step (*see* **Note 20**).
5. Wash: Transfer the slides to 1× PBS, pH 7.4 for 3 min. Repeat this step twice for a total of 9 min in three separate washes (*see* **Note 21**).
6. Incubate the slides in a 1 % H_2O_2 in PBS solution for 10 min to quench endogenous peroxidase activity (*see* **Note 22**). Prepare the 10 % blocking buffer during this step.
7. Wash: Transfer the slides to 1× PBS, pH 7.4 for 3 min. Repeat this step twice for a total of 9 min in three separate washes.
8. Incubate the slides in the 10 % blocking buffer for 1 h at room temperature inside of a humid chamber (*see* **Note 23**).
9. Remove excess blocking buffer from each slide carefully using a Kimwipe™, confirming that the tissue section has not become dry. Carefully pipet 250 μL of diluted primary antibody solution onto each slide and store in a humid chamber overnight at 4 °C (*see* **Note 24**). Store the remaining blocking buffer at 4 °C.
10. Wash: Transfer the slides to 1× PBS, pH 7.4 for 3 min. Repeat this step twice for a total of 9 min in three separate washes. Prepare the secondary antibody solution by adding an aliquot of biotinylated anti-goat IgG antibody at a dilution of 1:250 to 10 % blocking buffer, and immediately store on ice.
11. Carefully apply 250 μL of the diluted secondary antibody solution to each slide and incubate for 1 h in humid chamber. Prepare the ABC solution 45 min into the hour of incubation.
12. Wash: Transfer the slides to 1× PBS, pH 7.4 for 3 min. Repeat this step twice for a total of 9 min in three separate washes.
13. Apply 250 μL of ABC solution to each slide and incubate for 30 min at room temperature. Prepare the DAB solution in the chemical fume hood, but do not add H_2O_2 until right before use.
14. Wash: Transfer the slides to 1× PBS, pH 7.4 for 3 min. Repeat this step twice for a total of 9 min in three separate washes.
15. Add 30 μL of 30 % H_2O_2 to the filtered DAB solution. Develop by applying 250 μL of DAB solution to each slide. Color change will occur within seconds to a minute. Once sections have started to darken, gently tap off the DAB solution onto a Kimwipe™ and transfer the slides to ultrapure water to stop the reaction.
16. Dehydrate with graded alcohols: Transfer the slides to 50 % ethanol for 1 min, then transfer to 70 % ethanol for 1 min.

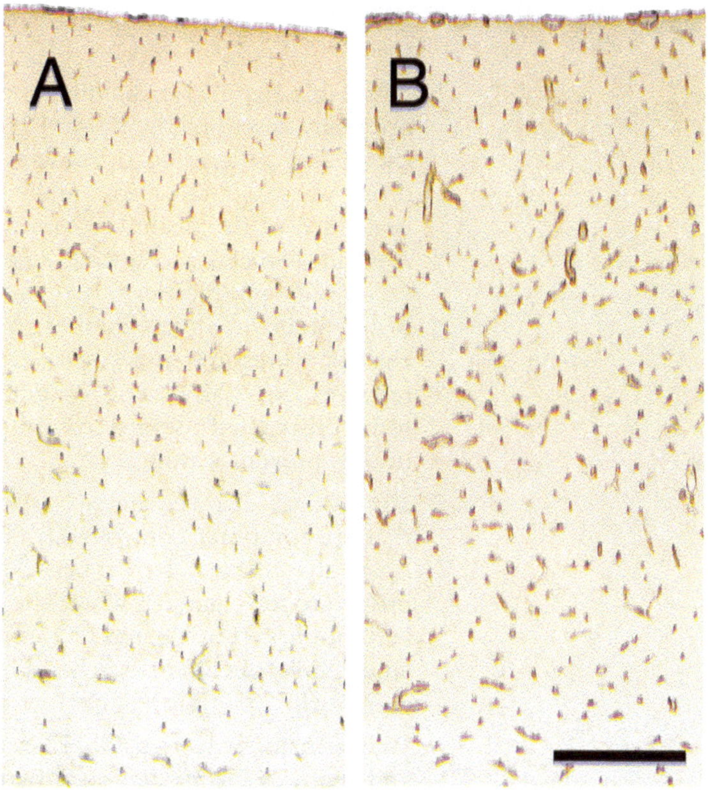

Fig. 2 GLUT-1-positive capillaries illustrating hypoxia-induced angiogenesis in the cerebral cortex. (**a**) Normoxia. (**b**) 21 days hypoxia. Scale bar: 125 μm

Transfer the slides to 95 % alcohol for 1 min. Repeat this step for a total of 2 min in two separate 95 % ethanol baths. Transfer the slides to 100 % ethanol for 1 min. Repeat this step for a total of 2 min in two separate 100 % ethanol baths.

17. Transfer the slides to xylene for 5 min. Repeat this step twice for a total of 15 min in three separate xylene baths.
18. Coverslip the slides with EUKITT media and dry overnight. This step is to be performed under the fume hood.

3.8 Capillary Density Determination

1. Capture digital photographs at 200× magnification spanning the entire depth of the cerebral cortex (Fig. 2) (*see* **Note 25**).
2. Mark each cerebral capillary per section using image-processing software of choice (*see* **Note 26**).
3. Measure the entire area of interest (mm^2) and count the total number of dots (N) per photographed section (*see* **Note 27**).
4. Report density value as the number of capillaries per unit area (N/mm^2) for each individual section. Sections can then be averaged for an aggregate value of each specific slide (*see* **Note 28**).

4 Notes

1. A suitable hypobaric chamber with a fairly simple design was originally described by Wright [8] and was constructed making use of a 55 gallon steel drum with a thick plexiglass cover. The drum was mounted horizontally with a grating placed about one third of the distance from the bottom to act as a floor. A small-diameter access hole was made for a vacuum line and vacuum gauge/bleeder valve. The bleeder valve allows the leak of room air into the chamber to provide a control to adjust chamber pressure and also provides the air turnover that removes CO_2. Modified "Wright chambers" can also be custom constructed using steel and plexiglass of sufficient thickness to prevent warping.
2. Normobaric hypoxia units exist as large plexiglass chambers that are computer-controlled by an apparatus containing multiple valves (Oxycycler™). Gasses are appropriately mixed and pushed to the chamber, where the gas concentration is detected and monitored by O_2 and CO_2 sensors. Air turnover volume is adjusted by opening small holes in the side of the chamber to allow airflow out of the chamber.
3. Paraformaldehyde is a OSHA-regulated hazardous chemical and requires specific approval by the institutional safety office for laboratory use. Follow appropriate safety guidelines and always handle paraformaldehyde under the fume hood.
4. Each slide requires 0.75 mL of blocking solution throughout the entire staining preparation. Modify the final volume based on the number of slides stained per protocol. 10 mL of blocking buffer is sufficient to stain 12 slides.
5. The developing step of this protocol (with slight modifications) can be adapted for fluorescence staining as well, which allows for co-localization studies. For the purpose of capillary density determination, the DAB method is preferred due to the long-lasting stability of the signal. Fluorescent stains require dark, refrigerated storage and are subject to rapid signal degradation. DAB is also a hazardous chemical and it requires proper handling and safety clearance.
6. There are significant differences between strains of mice and rats in their response to hypoxia, both physiologically and structurally [9].
7. The normal body weight for 3-month-old Wistar rats is between 180 and 220 g. Rats lose 10 % of their body weight initially during the first week of exposure as part of the normal physiological adaptation to hypoxia [10, 11]. A normal body weight range for a 3-month-old C57BL/6 mouse is between 23 and 30 g. Mice can lose even more, as much as 20 % of

initial body weight [12]. Loss of body weight is part of the normal physiological acclimatization response to hypoxia [5].

8. 0.5 ATM is 380 mmHg at sea level. If the altitude of the laboratory is significantly higher, the barometric pressure must be known and the appropriate gauge value calculated to give the residual 380 mmHg. For example, if barometric pressure is 680 mmHg, then the gauge should read 300 mmHg to arrive at 0.5 ATM. It is useful to report the hypoxic exposure to include the partial pressure for oxygen. Vacuum chambers are subject to fluctuations in facility vacuum pressure so the gauge needs to be checked often. These chambers can be costly to produce, but are cost-effective in the long run and generally stable over extended periods of time with relatively low maintenance.
9. Long-term management of the system during prolonged exposures requires frequent gas replenishment and is therefore costly. For example, a 1 m chamber will use1 size 200 tank of USP O_2 (Airgas) and 3–4 tanks of N_2 per week of 10 % O_2 exposure. The latter might be supplied by liquid N2 tanks to avoid having to change the tank as often and at inconvenient time of day. Sensors also need to be replaced every few months. The effect of altitude also applies to the "normobaric" chambers that are at ambient pressures and the partial pressure of oxygen should be calculated by multiplying the ambient pressure times the percent gas.
10. Hematocrit (packed red cell volume percent in whole blood) values under normal conditions are about 45–50 % in the control mouse or rat and rise to around 75–80 % by the end of 3 weeks as part of the normal physiological response to prolonged hypoxia.
11. Cerebral capillaries are delicate and can be severely damaged by high flow rate during this step. Perfusion should take no less than 10 min. Complete absence of blood from the liver is a good indicator of a complete perfusion.
12. Take specific care to make midline incisions (mouse) and bilateral (rat) as superficial as possible as to avoid damaging the underlying brain tissue with scissors or forceps.
13. Washing the brains removes residual paraformaldehyde. One hour is sufficient for samples that have not been in storage for longer than 2 days postfixation. A longer washing period should be used for samples that have been in storage for longer, but not exceeding 2 h of total wash time.
14. Xylene hardens tissue rapidly, which can make the brain brittle and difficult to section without excessive artifact during preparation. Some procedures allow for this to occur overnight, but it is suggested to not allow this step to exceed 5 h. Xylene is a hazardous chemical and requires proper handling under a fume hood and clearance from institutional safety offices.

15. Applying vacuum pressure allows for paraffin to better infiltrate the tissue, which is a critical step for producing quality sections with less histological artifact.
16. Depending on desired area of interest, embedding the brains anteriorly with the prefrontal cortex situated downwards towards the bottom of the mold is usually the most effective way to orient the tissue. Reverse orientation in the mold (caudal end pointed downwards) may be beneficial when sectioning posterior structures of the brain such as the brainstem, cerebellum, and hippocampus.
17. This area can be most easily visualized at the level where each lateral ventricle elongates and the corpus callosum becomes continuous and connects both hemispheres. Bregma values corresponding to Figs. 22–28 in *the Mouse Brain in Stereotaxic Coordinates* [13]. Five microns sections are required to see clear cross sections of each capillary. Sections of greater thickness include capillary depth and can confound capillary determination. The equivalent sections from the rat brain can be found in [14].
18. Floating sections flatten and begin to expand rapidly during this step. Special care is required to prevent sections from becoming wrinkled or discontinuous when mounted onto slides. Superfrost Plus™ slides are recommended because they inhibit dissociation of tissue from the slide during the staining procedure. Mounting sections serially achieves a minimum 150 μm distance between capillary beds on the first and last section of each slide as to ensure variability of capillaries counted.
19. It is critical to keep slides from drying out after rehydration or the current preparation must be discarded and restarted with new tissue sections.
20. A water bath or a microwave can be used during this step for bringing the solution to temperature. Both methods are equally effective, but the microwave allows for reaching the target temperature faster, which is more time efficient. After the slides have incubated in retrieval buffer, it is suggested to allow them to return within 10 °C of room temperature before continuing to the next step. This is to ensure that artifacts are not imposed on the tissue from drastic temperature change between buffers.
21. Although not a strict requirement, using fresh 1× PBS for each specific wash appears to decrease excessive background staining and yield cleaner results.
22. Methanol can be substituted for PBS during this step.
23. Some applications of this step include using a hydrophobic "PAP pen" to localize staining reagents on specific sections,

but this is unnecessary. The humid chamber prevents slides from drying out and can be simply constructed using damp paper towels at the bottom of a chamber with a sealable lid.

24. Primary antibody incubation can also be performed at room temperature for 1 h. Although this is more time efficient, a higher incidence of nonspecific background staining is observed if this method is used, which ultimately leads to less contrast between microvessels and background. High visual contrast is crucial for capillary detection.

25. Depending on imaging hardware and optical field, stitching together multiple photographs may be necessary in creating a montage that captures the entire depth of the cerebral cortex from the glia limitans to the corpus callosum. Imaging with a 20× objective on an Eclipse E600 microscope (Nikon Instruments Inc., Melville, NY, USA) and an Aquos EXi camera (QImaging, Surrey, BC, CA) requires 2–3 photos to capture the entire cortical depth. Processing software such as Photoshop (Adobe, Mountain View, CA, USA) can be useful for automating construction of image montages.

26. Cerebral capillaries are defined as GLUT-1-positive microvascular profiles that do not exceed a luminal diameter of 20 μm and do not possess an oval profile where the major axis is twice the size of the minor one in 5 μm sections. Image-processing software of choice can be used for scoring individual images; again Photoshop (Adobe) is preferred.

27. Determining area and counting can be performed using ImageJ software (NIH, Bethesda, MD, USA). This software is easy to use, programmable, and free of charge.

28. For cerebral cortex, one full field from the right and left parietal cortex on each of three sections separated by 150 μm are counted, and the six data samples averaged to generate each data point. The average cortical column is about 300 μm across so this separation includes more than one column. There are differences in capillary densities across the cortical layers, so either the layers have to be independently studied as individual regions of interest or the entire depth of the cortex must be included to minimize sampling error.

References

1. LaManna JC et al (2004) Structural and functional adaptation to hypoxia in the rat brain. J Exp Biol 207(Pt 18):3163–3169
2. Dore-Duffy P, LaManna JC (2007) Physiologic angiodynamics in the brain. Antioxid Redox Signal 9(9):1363–1371
3. LaManna JC et al (1992) Brain adaptation to chronic hypobaric hypoxia in rats. J Appl Physiol 72(6):2238–2243
4. Harik N et al (1996) Time-course and reversibility of the hypoxia-induced alterations in cerebral vascularity and cerebral capillary glucose

transporter density. Brain Res 737(1–2): 335–338
5. LaManna JC et al (2004) Structural and functional adaptation to hypoxia in the rat brain. J Exp Biol 207(Pt 18):3163–3169
6. Millet GP et al (2012) Last word on point: counterpoint: hypobaric hypoxia induces different responses from normobaric hypoxia. J Appl Physiol 112(10):1795
7. Harik SI et al (1990) Immunocytochemical localization of the erythroid glucose transporter: abundance in tissues with barrier functions. J Neurosci 10:3862–3872
8. Wright BM (1964) Apparatus for exposing animals to reduced atmospheric pressure for long periods. Br J Haematol 10:75–77
9. Ward NL et al (2007) Cerebral angiogenic factors, angiogenesis, and physiological response to chronic hypoxia differ among four commonly used mouse strains. J Appl Physiol 102(5): 1927–1935
10. Puchowicz MA et al (2005) Adaptation to chronic hypoxia during diet-induced ketosis. Adv Exp Med Biol 566:51–57
11. Puchowicz MA et al (2007) Diet-induced ketosis increases capillary density without altered blood flow in rat brain. Am J Physiol Endocrinol Metab 292(6):E1607–E1615
12. Benderro GF, LaManna JC (2011) Hypoxia-induced angiogenesis is delayed in aging mouse brain. Brain Res 1389:50–60
13. Paxinos G, Franklin KBJ (2001) The mouse brain in stereotaxic coordinates. Academic, San Diego
14. Paxinos G, Watson C (1986) The rat brain in stereotaxic coordinates. Academic, San Diego

Chapter 7

The Middle Cerebral Artery Occlusion Model of Transient Focal Cerebral Ischemia

Fudong Liu and Louise D. McCullough

Abstract

Transient middle cerebral artery occlusion (tMCAO) in rodents is one of the most widely utilized models in experimental stroke studies on focal cerebral ischemia. tMCAO can be modeled in different ways, all aimed at mimicking the clinical scenario of early reperfusion after an ischemic infarct. Some models utilize mechanical occlusion to transiently occlude blood flow with an intraluminal suture, others use "humanized" clot with adjunctive thrombolytic use. This chapter will focus on these two models; the intraluminal suture and thromboembolic MCAO, as they are widely used in stroke research. In addition, several methods of cerebral blood flow (CBF) monitoring during a tMCAO procedure including laser Doppler flowmetry (LDF), laser speckle flowmetry (LSF), and carbon-14 Iodoantipyrine Autoradiography (^{14}C-IAP) will be described.

Key words C-14 IAP autoradiography, Cerebral blood flow, Ischemic stroke, Laser Doppler flowmetry, Laser speckle flowmetry, Middle cerebral artery occlusion

1 Introduction

Various ischemic stroke models have been developed and can be summarized into four major categories [1]: (1) models not requiring craniectomy (thromboembolic model, intraluminal suture MCAO model, photothrombosis model, endothelin-1-induced stroke model), (2) models requiring craniectomy (Tamura's model), (3) posterior cerebral circulation stroke models, and (4) cerebral venous thrombosis models. Among these, thromboembolic and intraluminal suture MCAO models enable transient occlusion for reperfusion and mimic recombinant tissue plasminogen activator (rtPA) therapy that induces recanalization of occluded intracranial arteries. Therefore, they are more clinically relevant compared to other models and are frequently utilized in stroke research.

The intraluminal suture MCAO model was first described by Koizumi et al. [2] in rats and has been modified for different strains

Richard Milner (ed.), *Cerebral Angiogenesis: Methods and Protocols*, Methods in Molecular Biology, vol. 1135,
DOI 10.1007/978-1-4939-0320-7_7, © Springer Science+Business Media New York 2014

of rodents, and even for other species such as rabbits. Many specific factors in this model, such as suture diameter, coating of the suture, insertion length, etc., can affect lesion reproducibility and size. Currently the technique using silicone-coated sutures has led to larger and more consistent infarcts with good reliability even among investigators of varying experience [3]. The model does not require craniotomy, produces focal occlusion of a large cerebral artery as seen in human stroke, and can be done in a high-throughput manner [4]. Disadvantages of the suture occlusion method include (1) an approximately 12 % rate of subarachnoid hemorrhage, (2) transection of the external carotid artery (ECA) leading to paralysis of muscles of mastication and swallowing producing feeding difficulty and weight loss, (3) spontaneous hyperthermia associated with hypothalamic injury, and (4) tracheal edema [3–6].

Hill et al. [7] were the first group to use a thromboembolic model in animals by injecting homologous blood clots into the carotid artery to induce cerebral infarction. The emboli in this model can either be spontaneous clots induced from blood of a donor animals of the same strain [8, 9], or thrombin-induced clots from autologous blood [10, 11]. This model closely mimics large vessel human stroke which is mostly caused by thromboembolism. The model has the advantage that thrombolytic agents can be tested in conjunction with putative neuroprotective agents. However, the infarction produced by this model is smaller and more variable than the intraluminal suture-induced MCAO due to the continuous process of endogenous thrombolysis that is initiated soon after the clots are injected into the carotid artery. To reduce the chance of spontaneous recanalization, the use of autolysis-resistant fibrin-rich emboli, or so-called humanized clots [12], has been increasingly utilized by mixing arterial blood with human fibrinogen.

In this chapter, we focus on these two most widely used tMCAO models in rodents and give step-by-step operational procedures. In addition, several CBF monitoring methods frequently used to confirm the effect of tMCAO will also be outlined.

2 Materials

2.1 Intraluminal Suture MCAO

1. Stereo microscope.
2. Microscope light source.
3. Isofluorane vaporizer.
4. Plexiglas chamber.
5. Heating pad.
6. Temperature controller.

7. Silicon rubber-coated monofilaments.
8. 6-0 silk suture.
9. Microsurgery tools.
10. Variable temperature cautery (Beaver-Visitec, Waltham, MA, USA).

2.2 Thromboembolic MCAO

In addition to the materials and instruments listed under intraluminal suture MCAO, you will need:

1. Petri dishes.
2. 1 ml and 3 ml syringes.
3. Gauge #21, #27, and #30 needles.
4. PE08, PE10, and PE50 polyethylene tubing.
5. Recombinant tissue plasminogen activator (rtPA).
6. Syringe infusion pump.

2.3 Monitoring of Regional CBF Peri-MCAO

2.3.1 Laser Doppler Flowmetry (LDF)

1. Laser Doppler flowmetry (Moor Instruments Ltd, England).
2. Microsurgical tools.

2.3.2 Laser Speckle Flowmetry (LSF)

1. MoorFLPI full-field laser perfusion imager (Moor Instruments, Devon, UK).
2. Computer installed with image processing software (MoorFLPI Measurement Software Version 2.0; Moor Instruments).
3. Microsurgery tools.

2.3.3 Carbon-14 Iodoantipyrine Autoradiography (^{14}C-IAP)

1. ^{14}C-IAP solution (PerkinElmer, Waltham, MA), 4 μCi in 81 μl isotonic saline (1 mCi/10 ml stock diluted in equal volume of saline to obtain 5 μCi/100 μl).
2. Liquid scintillation spectroscopy (Beckman, model 3801, CA, USA).
3. An infusion pump (Harvard Apparatus, Mode 11, south Natick, MA, USA).
4. Cryostat for brain sectioning.
5. Tissue homogenizing solution (Soluene-350, Packard Instrument Co., Downers Grove, IL, USA).
6. Liquid scintillation cocktail (Packard, Formula 989).
7. Carbon-14 standard microscale (Amersham, San Diego, CA, USA).
8. X-ray film (Kodak, Rochester, NY, USA).

9. Image analysis software (Inquiry, Loats, Westminster, MD, USA).
10. 2-Methylbutane (2-MB).
11. A wide-mouthed scintillation vial for storage of frozen brain.
12. Paraffin wells.

3 Methods

3.1 Intraluminal Suture MCAO (Fig. 1)

3.1.1 Pre-MCAO Preparations

1. Sterilize all surgical tools by autoclaving (minimum 121 °C, 15 PSI, for 15 min). Sanitize the surgery area and associated equipment using 70 % ethanol.
2. Before anesthesia, examine the animals for physical abnormalities (e.g., checking for tumors, fight wounds, etc.) and any obvious behavior deficits, and then weigh the animals (*see* **Note 1**).
3. Anesthetize the animals with isoflurane using a mixture of gas and oxygen (30 % O_2/70 % N_2O) in a small Plexiglas chamber at a flow rate of 5 L/min (induction of anesthesia).
4. After induction, reduce the flow rate to 1 L/min of anesthetic mixture for the remainder of the surgical procedure

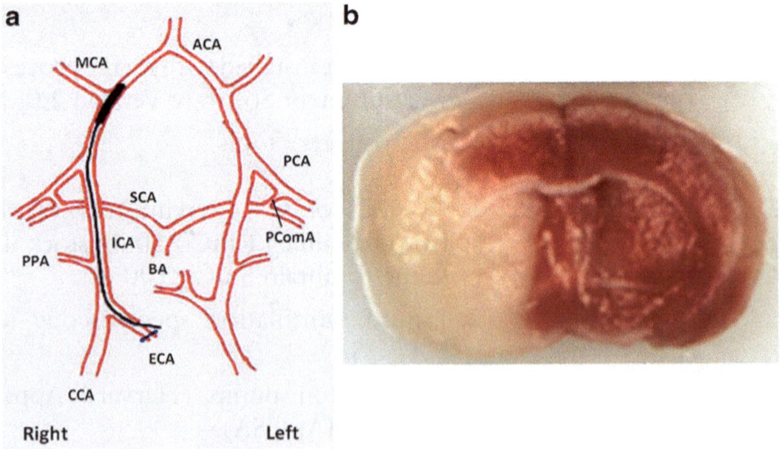

Fig. 1 Suture MCAO model. (**a**). Simplified diagram of the cranial circulatory system of the mouse is shown with a silicon-coated intraluminal suture occluding the origin of the MCA. The right ECA has been ligated and cauterized to a stump, where the suture goes through into the ICA. ACA, anterior cerebral artery; PCA, posterior cerebral artery; PComA, posterior communicating artery; SCA, superior cerebellar artery; BA, basilar artery; PPA, pterygopalatine artery. (**b**). Coronal section of the mouse brain after suture MCAO modeling (90 min MCAO followed by 24 h survival). TTC staining; white area indicates the infarct and red area is the normal tissue

(maintenance of anesthesia). Then remove the animal from the chamber and place in the left lateral position with a nosecone for continued anesthesia and for placement of the laser Doppler flowmetry (LDF) probe (for procedures *see* Subheading 3.3.1).

5. After placement of the LDF probe, place the animal in the supine position. Insert the rectal probe of the temperature controller and aim to maintain the body temperature of the animal at 36.5 °C ± 0.5 °C during surgery with a heating pad connected to the feedback temperature controller. Additional probes can be placed in the masseter muscle.

3.1.2 MCAO Procedure

1. Disinfect the fur and skin of the throat and right neck region with an appropriate agent (e.g., 70 % ethanol or Betadine). In rats, it is necessary to shave off fur.
2. Make a 1 cm long midline incision on the neck and gently separate the soft tissues under the stereo microscope.
3. Dissect the right common carotid artery (CCA) free from the surrounding nerves and place a tight slipknot on the CCA using 6-0 silk suture to temporarily reduce the arterial pressure in the operative field (*see* **Note 2**).
4. Dissect the external carotid artery (ECA) distally and apply a dead knot with silk suture; cauterize the ECA and its superior thyroid artery (STA) branch using a temperature cautery. Cut the ECA and STA at the cauterized segment leaving a residual ECA stump (*see* **Note 3**).
5. Isolate the internal carotid artery (ICA); make a slipknot at the ICA to block the blood flow.
6. Place a suture loop at the ECA stump near the bifurcation with the ICA; cut a small hole in the vessel wall of the ECA stump.
7. Introduce a silicon rubber-coated monofilament (*see* **Note 4**) into the ECA lumen down toward the bifurcation and then gently pull back the ECA stump to allow the filament to slide into the ICA. Once the tip of the filament meets the slipknot at the ICA, partially tighten the suture loop at the ECA stump to secure the filament (*see* **Note 5**).
8. Undo the slipknot at the ICA and advance the filament inside the ICA lumen until mild resistance is felt (*see* **Note 6**). The regional CBF (rCBF) monitored by LDF should drop below 80 % of the baseline at this time (*see* **Note 7**).
9. Redo the slipknot at the ICA to secure the filament.
10. Close the incision (dermis, panniculus carnosus, and subcutaneous tissue layers) with silk suture; apply 0.25 % Bupivacaine and topical antibiotic cream to the incision.
11. Place the animal in a recovery cage; make sure that the cage floor around the nose and mouth are free of bedding material.

The animal should recover from anesthesia in 5–10 min. The animal with a successful MCAO should exhibit neurological deficits during the ischemic period which are scored (NDS) as follows [13]: (0) no deficit, (1) forelimb weakness and torso turning to the ipsilateral side when held by tail, (2) circling to affected side, (3) unable to bear weight on affected side, and (4) no spontaneous locomotor activity or barrel rolling.

3.1.3 Reperfusion

1. Re-anesthetize the animal and disinfect the incision.
2. Reopen the incision; undo the slipknot at the ICA and loosen the suture loop at the ECA stump.
3. Withdraw the monofilament and make a dead knot below the hole at the ECA stump.
4. Undo the slipknot at the CCA. Confirm adequate hemostasis.
5. Close the incision and apply 0.25 % Bupivacaine and topical antibiotic cream; return the animal to the recovery cage (*see* **Note 8**).

3.2 Thromboembolic MCAO (Figs. 2 and 3)

3.2.1 Preparation of Emboli

1. Draw arterial blood from a donor mouse (*see* **Note 9**) and supplement this with human fibrinogen (2 mg/ml).
2. Transfer the fibrinogen-supplemented blood immediately into PE50 polyethylene tubing for 2 h at room temperature (RT) followed by storage at 4 °C overnight (less than 24 h).
3. Before use, transfer the clot (~5 cm) into a modified PE10 tube filled with sterile PBS and retract.

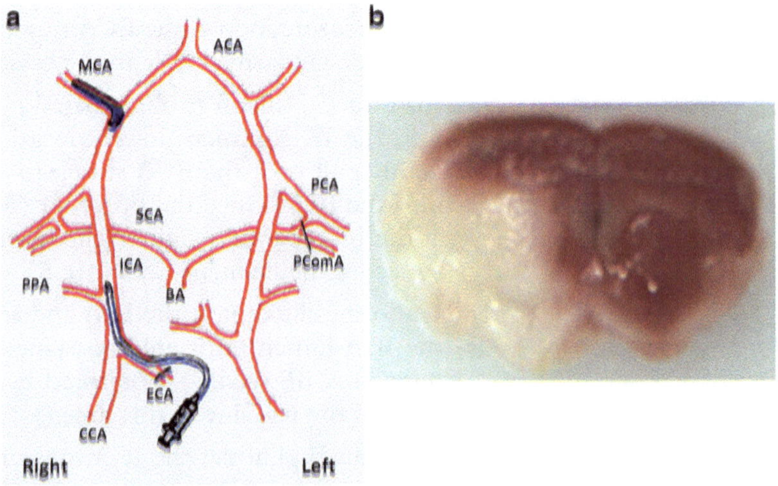

Fig. 2 Thrombotic MCAO model. (**a**). Simplified diagram of the cranial circulatory system of the mouse is shown with a human fibrinogen-treated blood clot inside the MCA. The right ECA has been ligated and cauterized to a stump, where the catheter goes through into the ICA for the clot delivery. (**b**). Coronal section (TTC staining) of the mouse brain 24 h after thrombotic MCAO modeling

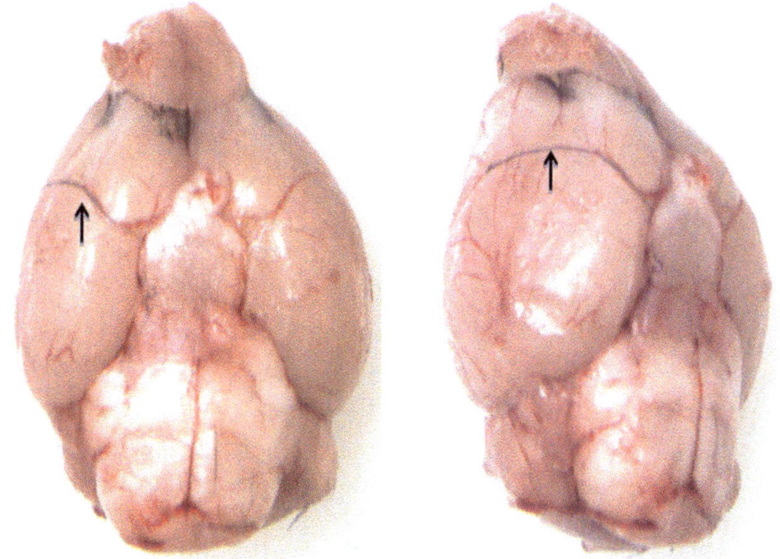

Fig. 3 A human fibrinogen-treated blood clot stained by Indian ink is shown in the MCA (*black line* in front of the *arrow*) after the clot delivery

4. Transfer the clot to a Petri dish containing 1× PBS and leave for further retraction at RT for 4 h.
5. Transfer a single 9 ± 0.5 mm long clot to a modified PE10 or PE08 catheter for embolization.

3.2.2 Pre-MCAO Preparations

Same procedures as in the intraluminal suture MCAO model.

3.2.3 MCAO Procedure

1. Disinfect the fur and skin of the throat and right neck region with an appropriate agent (e.g., 70 % ethanol or Betadine).
2. Make a 1 cm long midline incision on the neck and gently separate the soft tissues under the stereo microscope.
3. Carefully dissect the right common carotid artery (CCA) free from the surrounding nerves and place a tight slipknot on the CCA using 6-0 silk suture to temporarily reduce the arterial pressure in the operative field.
4. Dissect the external carotid artery (ECA) further distally and apply a dead knot with silk suture; cauterize the ECA and its superior thyroid artery (STA) branch using cautery. Cut the ECA and STA at the cauterized segment to make an ECA stump.
5. Isolate the internal carotid artery (ICA); make a slipknot at the ICA to occlude blood flow.
6. Place a suture loop at the residual ECA stump near the bifurcation with the ICA; cut a small hole in the vessel wall of the ECA stump.

7. Introduce a modified PE10 or PE08 catheter containing a clot into the ECA lumen and advance into the ICA [12]. Once the tip of the catheter meets the slipknot at the ICA, partially tighten the suture loop at the ECA stump to secure the catheter.
8. Undo the slipknot at the ICA and advance the catheter inside the ICA lumen until the resistance is felt (*see* **Note 10**).
9. The clot is gently injected with 100 μl of the sterile saline. The rCBF monitored by LDF should drop below 80 % at this time (*see* **Note 11**).
10. After thromboembolization remove the catheter and make a dead knot below the hole at the ECA stump.
11. Undo the slipknot at the CCA.
12. Close the incision and apply 0.25 % Bupivacaine and topical antibiotic cream; return the animal to the recovery cage.

3.2.4 Thrombolytic Procedures

1. Re-anesthetize the animal and disinfect the incision.
2. Reopen the incision and expose the right jugular vein.
3. Place a catheter (PE-10) in the jugular vein and connect to a syringe infusion pump.
4. Administer a total dose of 10 mg/kg rtPA (diluted in saline at 0.1 mg/100 μl): 10 % bolus at the beginning followed by a continuous infusion over a 30 min interval using a syringe infusion pump for the reminder [14]. Inject control mice with the same volume of saline.
5. After administration of the drug, ligate the right jugular vein. Close the incision and apply 0.25 % Bupivacaine and topical antibiotic cream; return the animal to the recovery cage.

3.3 Monitoring of Regional CBF Peri-MCAO

3.3.1 Laser Doppler Flowmetry (LDF)

1. Calibrate the LDF probe using PFS flux standard (Moor Instruments).
2. After induction of anesthesia (*see* Subheading 3.2.4), remove the animal from the chamber and place in the left lateral position.
3. Make a small skin incision (2 mm) in the midpoint of the right eyeball and external auditory canal.
4. Apply blunt dissection to the connective tissue and muscle until the coarse surface of the skull can be felt with forceps; attach the LDF probe to the skull perpendicularly (*see* **Note 12**).
5. Place the animal in the supine position; fix the LDF probe with tape. During the surgery, the probe should stay undisturbed and stable.
6. LDF gives the CBF value (arbitrary unit) pre-, intra-, and post-ischemias during MCAO procedure. Measure the rCBF before and after MCAO while keeping the animal anesthetized.
7. After LDF monitoring, close the incision by a very small amount of cyanoacrylate.

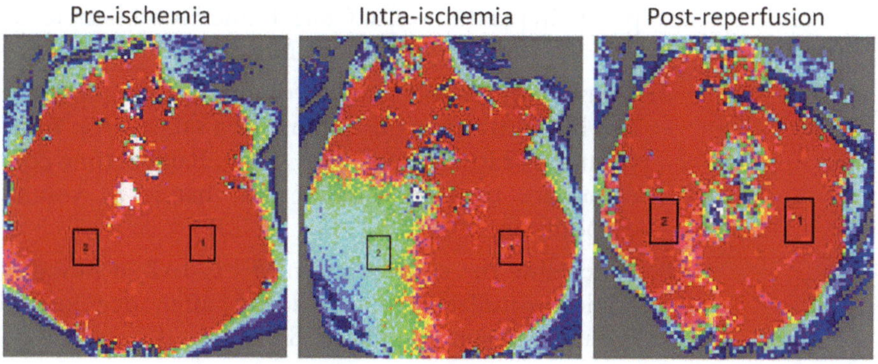

Fig. 4 LSF imaging of a mouse brain subjected to suture MCAO (90 min MCAO followed by 30 min reperfusion). *Orange* and *blue*, low CBF area; *red*, high CBF area

3.3.2 Laser Speckle Flowmetry (LSF) (Fig. 4)

1. Fix the arms of the LSF probe very firmly to a solid and unmovable object; connect the LSF to the computer.
2. Calibrate the LSF with the provided calibrating box.
3. After induction of anesthesia (*see* Subheading 3.2.4), remove the animal from the chamber and place in the prone position.
4. Perform a midline incision and expose the skull. Prevent the skull from drying with a thin layer of mineral oil.
5. Put the probe approximately 25 cm right above the skull; zoom in/out and focus the probe to achieve the optimal picture (*see* **Note 13**).
6. Speckle imaging and CBF flux data can be obtained before and during MCAO and after reperfusion by the computer system.

3.3.3 Carbon-14 Iodoantipyrine Autoradiography (^{14}C-IAP)

1. Subject the animal to MCAO.
2. Set the rate of infusion pump at 6.48 ml/h; attach a 30-gauge adapter to the 2 ml glass syringe with 8.92 mm diameter and draw up 81 μl of 5 μCi/100 μl ^{14}C-IAP.
3. Mount the syringe on the pump and check readiness of infusion.
4. Instrument the animals with femoral vascular catheters (both the femoral vein and artery). Close the artery line temporarily with a hemostat on the catheter; connect the venous line to the infusion pump.
5. 5–10 min before the end of ischemic period, take an arterial blood sample for determination of PO_2, pH, and PCO_2.
6. At the end of ischemia (*see* **Note 14**), start the infusion pump and intravenously infuse the ^{14}C-IAP solution over 45 s. At the same time, open the artery line and allow the blood to spontaneously flow into heparinized saline drops of known volume

placed in paraffin wells. Collect nine blood samples at 5 s intervals and promptly cover (*see* **Note 15**).

7. Measure the blood sample volume using a pipette and calculate by subtracting the volume of a saline drop from the total volume of blood sample plus saline. Parallel time control saline drops are used to account for change in volume due to evaporation.
8. Determine the concentration of ^{14}C-IAP in blood samples by liquid scintillation spectroscopy after decolorization with 0.2 ml of tissue solubilizer (Soluene).
9. Decapitate the animal immediately after the 45 s period of blood collection. Remove the brain and transfer to pre-chilled 2-MB on dry ice for 10–15 min. Then place the brain in a 20 ml cryovial and store in a –80 °C freezer to be sectioned (*see* **Note 16**).
10. Section each brain into 20 μm-thick coronal sections at –20 °C and thaw-mount onto cover glasses using a standard cryostat (*see* **Note 17**). Sections are apposed for 1 week to X-ray film with a carbon-14 standard microscale.
11. Autoradiographic images representing seven coronal levels (+4, +3, +2, +1, 0, –1, and –2 mm from bregma, three images each) are digitalized and rCBF determined by image analysis software.
12. Two methods of analysis are usually used. First, the CBF is sampled at discrete 0.08 mm^2 regions within the cortical and striatal regions vulnerable to MCAO (the parietal and frontal cerebral cortices and lateral and medial striata), averaged over three to seven consecutive coronal slices [15]. Second, the images of each coronal slice are scanned and pixels stratified according to corresponding blood flow rates. Pixels with flow rates falling within a range of blood flow are summed and converted to volume units [16].

4 Notes

1. This is especially important when using aged animals because they are more likely to develop tumors.
2. When separating carotid arteries (CCA, ECA, ICA), be cautious that they are dissected free from the surrounding nerves (without harming the vagal nerve).
3. To better facilitate insertion of the monofilament into the ICA, the connective tissue at the bifurcation of the ECA and ICA should be cleared so that the stump of the ECA can be pulled back to form an obtuse angle with the ICA (>135°).

4. A 6-0 nylon suture monofilament is usually used for mice (20 g–30 g) with Φ0.21 mm silicon rubber-coated tip; for rats (250 g–350 g), 4-0 nylon suture should be chosen and the size of the silicon-coated tip is 0.35–0.37 mm.
5. During suture MCAO modeling, there are two occasions when inadvertent bleeding is most likely: (1). immediately after insertion of the filament into the ICA (strategy: make sure the slipknot at the stump is relatively tight; the filament should barely fit through this loose knot) and (2). Immediately after withdrawal of the filament (strategy: tighten the slipknot at the ICA immediately after the filament passes through).
6. For mice, when the resistance is felt after inserting the filament, the length of the filament from the bifurcation of the ICA and ECA to the tip should be 9 ± 0.5 mm; for rats, the length should be 19 ± 0.5 mm.
7. For rats, LDF drops below 60 % of the baseline when the filament is inserted to the MCA.
8. For chronic survival cohort, fresh and wet food mash is usually served everyday in the cage after MCAO until the animal gains weight. Saline (100 μl/10 g, s.c.) can also be considered to administer depending on the experiment design.
9. The femoral artery in the mouse is not a good source for drawing blood to make clots due to the artery's small size. Drawing blood from the carotid artery guarantees enough blood for an entire cohort of MCAO mice. The blood must be drawn from a donor mouse of the same strain (homologous blood).
10. In order to make sure that the clot will be injected into the MCA, the tip of the catheter should at least pass the opening of pterygopalatine artery (a large branch of the ICA).
11. After injection, the pressure on the syringe should be kept for 1–2 min to induce a firm attachment of clot with the interior wall of the MCA.
12. Since the skull of rats is thicker than that of mice, a Dremel tip is usually applied to thin the skull to a level just above the dura for accurate CBF reading by the LDF probe.
13. The laser beam from the probe reflects on the surface of the skull and produces a patch of white color on the speckle image. To avoid this, the position of the mouse head can be adjusted to remove the white patch to a less important area (e.g., cerebellum).
14. The ^{14}C-IAP experiment could also be done after reperfusion if desired.
15. Since timing is crucial for this step, two people should be available with one in charge of the artery line and the other taking care of the vein line.

16. Injection of ^{14}C-IAP makes the animal tissue and diapers radioactive. After removal of the brain, discard the remaining part of the body and contaminated materials as appropriate for a radioactive waste. Not that the cryostat that is used to section the brain samples must also be considered contaminated.
17. To achieve a high quality of autoradiography, all the brains should be sectioned within 30 days of ^{14}C-IAP injection.

Acknowledgement

This work was supported by the NIH/NINDS (grants NS050505 and NS055215 to LDM), and by AHA (grant 12SDG9030000 to FL).

References

1. Liu F, McCullough LD (2011) Middle cerebral artery occlusion model in rodents: methods and potential pitfalls. J Biomed Biotechnol 2011:464701
2. Koizumi J, Yoshida Y, Nakazawa T, Ooneda G (1986) Experimental studies of ischemic brain edema: 1. A new experimental model of cerebral embolism in rats in which recirculation can be introduced in the ischemic area. Jpn Stroke J 8:1–8
3. Durukan A, Tatlisumak T (2007) Acute ischemic stroke: overview of major experimental rodent models, pathophysiology, and therapy of focal cerebral ischemia. Pharmacol Biochem Behav 87(1):179–197
4. Carmichael ST (2005) Rodent models of focal stroke: size, mechanism, and purpose. NeuroRx 2(3):396–409
5. Li F, Omae T, Fisher M (1999) Spontaneous hyperthermia and its mechanism in the intraluminal suture middle cerebral artery occlusion model of rats. Stroke 30(11):2464–2470, discussion 2470-2461
6. Schmid-Elsaesser R, Zausinger S, Hungerhuber E, Baethmann A, Reulen HJ (1998) A critical reevaluation of the intraluminal thread model of focal cerebral ischemia: evidence of inadvertent premature reperfusion and subarachnoid hemorrhage in rats by laser-Doppler flowmetry. Stroke 29(10):2162–2170
7. Hill NC, Millikan CH, Wakim KG, Sayre GP (1955) Studies in cerebrovascular disease. VII. Experimental production of cerebral infarction by intracarotid injection of homologous blood clot; preliminary report. Proc Staff Meet Mayo Clin 30(26):625–633
8. Jiang Q, Zhang RL, Zhang ZG, Ewing JR, Divine GW, Chopp M (1998) Diffusion-, T2-, and perfusion-weighted nuclear magnetic resonance imaging of middle cerebral artery embolic stroke and recombinant tissue plasminogen activator intervention in the rat. J Cereb Blood Flow Metab 18(7):758–767
9. Zhang Z, Chopp M, Zhang RL, Goussev A (1997) A mouse model of embolic focal cerebral ischemia. J Cereb Blood Flow Metab 17(10):1081–1088
10. Zhang Z, Zhang RL, Jiang Q, Raman SB, Cantwell L, Chopp M (1997) A new rat model of thrombotic focal cerebral ischemia. J Cereb Blood Flow Metab 17(2):123–135
11. Takano K, Carano RA, Tatlisumak T, Meiler M, Sotak CH, Kleinert HD, Fisher M (1998) Efficacy of intra-arterial and intravenous prourokinase in an embolic stroke model evaluated by diffusion-perfusion magnetic resonance imaging. Neurology 50(4):870–875
12. Hoda MN, Li W, Ahmad A, Ogbi S, Zemskova MA, Johnson MH, Ergul A, Hill WD, Hess DC, Sazonova IY (2011) Sex-independent neuroprotection with minocycline after experimental thromboembolic stroke. Exp Transl Stroke Med 3(1):16
13. Liu F, Schafer DP, McCullough LD (2009) TTC, fluoro-Jade B and NeuN staining confirm evolving phases of infarction induced by middle cerebral artery occlusion. J Neurosci Methods 179(1):1–8

14. Jia L, Chopp M, Zhang L, Lu M, Zhang Z (2010) Erythropoietin in combination of tissue plasminogen activator exacerbates brain hemorrhage when treatment is initiated 6 hours after stroke. Stroke 41(9):2071–2076
15. Alkayed NJ, Harukuni I, Kimes AS, London ED, Traystman RJ, Hurn PD (1998) Gender-linked brain injury in experimental stroke. Stroke 29(1):159–165, discussion 166
16. Rusa R, Alkayed NJ, Crain BJ, Traystman RJ, Kimes AS, London ED, Klaus JA, Hurn PD (1999) 17beta-estradiol reduces stroke injury in estrogen-deficient female animals. Stroke 30(8):1665–1670

Chapter 8

A Mouse Model of Chronic Cerebral Hypoperfusion Characterizing Features of Vascular Cognitive Impairment

Masafumi Ihara, Akihiko Taguchi, Takakuni Maki, Kazuo Washida, and Hidekazu Tomimoto

Abstract

Vascular dementia or vascular cognitive impairment occurs as a result of persistently compromised blood flow to the brain and represents the second most common type of dementia after Alzheimer's disease. In order to investigate its underlying mechanisms, a mouse model of chronic cerebral hypoperfusion has been developed, which involves the narrowing of the bilateral common carotid arteries with newly designed microcoils. This mouse model provides a unique platform to investigate the mechanisms of angiogenesis following chronic cerebral hypoperfusion and to explore potential drugs or cell therapies designed to enhance angiogenesis as a preclinical step toward developing novel treatments for dementia of vascular origin.

Key words Chronic cerebral hypoperfusion, Vascular cognitive impairment, Dementia, Bilateral common carotid artery occlusion, Microcoil, Mouse, Cerebral blood flow, Laser speckle flowmetry

1 Introduction

Cerebral blood flow (CBF) is decreased diffusely, or at least focally, in elderly patients with vascular cognitive impairment [1, 2]. To mimic such persistent cerebral ischemia in humans, a chronic cerebral hypoperfusion model has been established in rats, gerbils, and mice [3–5]. The model can be generated by bilateral common carotid artery (CCA) occlusion in rats [3, 6], bilateral CCA stenosis in mice (BCAS) [4, 7–9] or in gerbils [5], and unilateral CCA occlusion in mice [10]. Although nonhuman primates appear to represent the best model for the study of vascular cognitive impairment, due to their similarities in cerebral vascular architectures with humans [11], most experiments studying chronic cerebral hypoperfusion have been performed in rodents because of the ease of handling and greater ethical acceptability.

Among rodent models of chronic cerebral hypoperfusion, the rat model is most widely used [12, 13], resulting in cognitive impairment and cholinergic deficits in the animal [3, 6, 14].

Richard Milner (ed.), *Cerebral Angiogenesis: Methods and Protocols*, Methods in Molecular Biology, vol. 1135,
DOI 10.1007/978-1-4939-0320-7_8, © Springer Science+Business Media New York 2014

The animals also develop white matter rarefaction [3, 9, 15], which appears very similar to that found in human cerebrovascular white matter lesions. However, the rat model does possess inherent drawbacks. For example, the visual pathway is invariably damaged by the occlusion of the ophthalmic arteries, thus potentially compromising behavioral assessment. Furthermore, genetic studies may be hampered because of limited accessibility to molecular technologies in the rat.

To circumvent such limitations, we have established a mouse model of chronic cerebral hypoperfusion, which is subjected to various degrees of CBF reduction by the narrowing of the bilateral CCAs with newly designed microcoils [4, 7–9]. The severity of ischemia/hypoperfusion can be easily controlled by internal diameter regulation of the microcoils [4]. The model demonstrates good reproducibility in terms of glial activation, blood–brain barrier disruption, white matter lesion appearance, and vascular cognitive impairment, which appear within a month after the surgery. In the longer-term model, significant hippocampal changes (atrophy and cell death) are documented 8 months after surgery, providing evidence linking chronic cerebral hypoperfusion with neurodegeneration [16]. The apparent advantage of using this surgical technique is that it provides a unique platform to investigate the mechanisms of angiogenesis and to explore potential drugs [17, 18] or cell therapies [19] designed to enhance angiogenesis as a preclinical step toward developing novel treatments for dementia of vascular origin. Another important advantage is its easy application to genetically engineered mice; for instance, β amyloid-overexpressing mice have been subjected to BCAS to clarify the wider question of whether and how chronic cerebral hypoperfusion has detrimental effects on Alzheimer's disease [20–22].

The aims of the current chapter are therefore to provide technical details that have accumulated since establishment of this mouse BCAS model, focusing particularly on a method of CBF monitoring using laser speckle flowmetry in order to consider the particular strengths and pitfalls of the method.

2 Materials

2.1 Animals and Surgical Appliances

1. Male C57BL/6J mice (weight, 24–29 g) (*see* **Note 1**).
2. Animal restraining device.
3. Heating pad.
4. Rectal thermometer.
5. Chemical depilatory.
6. Surgical scrub solutions: povidone-iodine scrub (Betadine scrub) or chlorhexidine scrub.

7. Anesthetics: halothane or isoflurane (*see* **Note 2**).
8. Surgical knife.
9. Silk suture.
10. Forceps: fine-tip and blunt-tip forceps.
11. Gauze.
12. Stainless steel wound clips.
13. Microcoils: manufactured in the Sawane Spring Co., Ltd. (Hamamatsu, Japan) (*see* **Note 3**).
14. Operating microscope.

2.2 CBF Monitoring Device

1. Laser speckle blood flow imager (Omegawave, Inc., Tokyo, Japan).
2. A calibration reference device (Calibrator S/N 080715-5, Omegawave, Inc.)
3. Gel (Aquasonic, Parker Laboratories, Inc., Fairfield, NJ).

3 Methods

Before performing surgery, one should be aware that there are legal and ethical requirements regarding the use of animals in research. The animal must be maintained in a surgical plane of anesthesia, and its vital signs monitored and regulated throughout the procedure.

3.1 Surgical Procedures

1. Anesthetize the mouse with gas anesthetics (*see* **Note 4**).
2. Place the anesthetized mouse in a dorsal recumbent position on the operating board with the tail toward the surgeon.
3. Shave the ventral neck area with a chemical depilatory and swab with surgical scrub.
4. Make a 1.0–1.5 cm midline skin incision from the base of the neck to the point below the lower jaw.
5. Remove the underlying fat and move the salivary glands laterally or upwards using forceps to maximize the operating field.
6. Expose and free both common carotid arteries (CCAs) from their sheaths under an operating microscope (*see* **Note 5**) (Fig. 1).
7. Place two 4–0 silk sutures around the distal and proximal parts of the CCA.
8. Gently lift the artery by the sutures and place between the loops of the microcoil just proximal to the carotid bifurcation (Fig. 1) (*see* **Note 6**).
9. Twine the microcoil by rotating it around the CCA (*see* **Note 7**).
10. 30 min later, twine another microcoil of the same size around the other CCA (Fig. 1) (*see* **Note 8**).

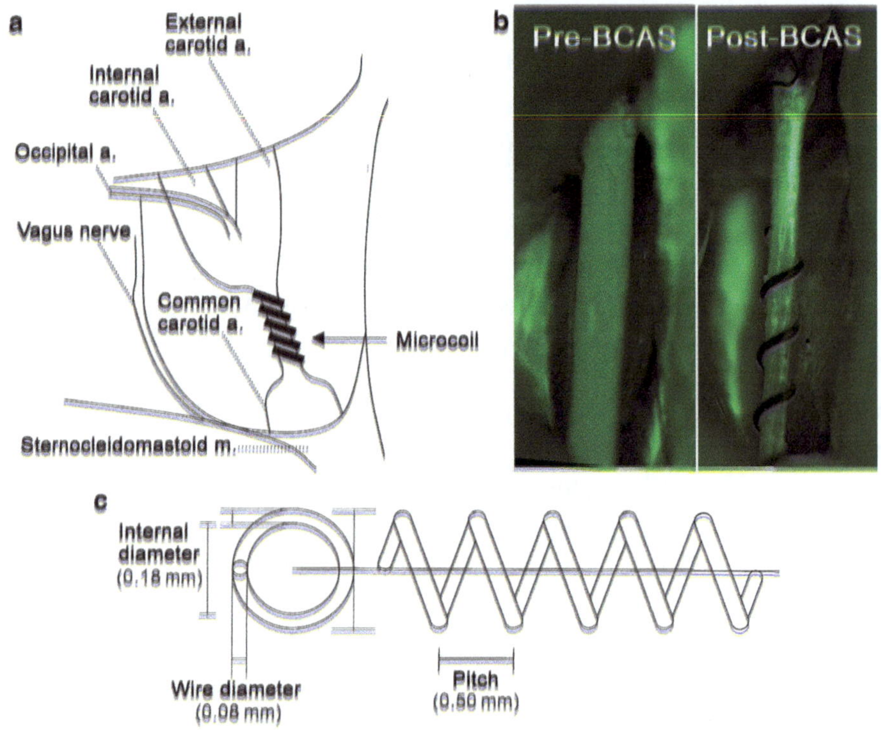

Fig. 1 The procedure for BCAS and microcoil placement. The microcoil is twined by rotating it around the CCA just proximal to the carotid bifurcation of a C57BL/6J mouse (**a**). Representative photographs of a FITC-perfused common carotid artery before (*left*) and after (*right*) placement of a microcoil (**b**). The most frequently used microcoil is made from piano wire (wire diameter of 0.08 mm) with an inner diameter of 0.18 mm, pitch 0.50 mm, and total length 2.5 mm (**c**)

11. Close the incision with stainless steel wound clips or fine sutures.
12. Return the mouse to the animal holding area after they appear normal (*see* **Note 9**).

3.2 Cerebral Blood Flow (CBF) Monitoring

1. Place the anesthetized mouse in a ventral recumbent position with the tail toward the surgeon (*see* **Note 4**).
2. Surgically remove the scalp to expose the skull.
3. Using fine-tip forceps, remove the periosteum, which adheres to the skull (*see* **Note 10**).
4. Wipe the skull surface with saline-soaked gauze and then cover with a thin layer of gel to prevent drying (*see* **Note 11**).
5. Perform calibration with a calibration reference device (*see* **Note 12**).
6. Perform CBF recordings through the skull using the laser speckle blood flow imager, (*see* **Note 13**) (Fig. 2). Define circular regions of interest on the image for quantitative measurement (*see* **Note 14**).

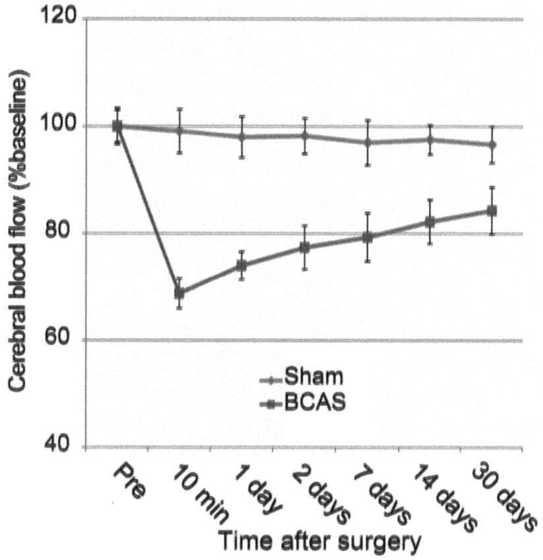

Fig. 2 Cerebral blood flow after BCAS. This figure shows temporal profiles of cerebral blood flow evaluated with laser speckle imager in mice at 2.5 months of age after the surgery using microcoils with diameter of 0.18 mm (*square*) and in sham-operated mice (*diamond*). The data represent mean values ± standard deviation expressed as a percentage of the preoperative value. The CBF values decreased significantly from the preoperative baseline after the surgery with the 0.18 mm diameter microcoils. Immediately after surgery (10 min), there was a significant reduction in CBF values to 68.8 ± 2.8 %, with gradual recovery to 84.3 ± 4.3 % at 1 month ($n=7$), probably due to compensatory mechanisms involving angiogenesis and arteriogenesis [4, 17]. Note that the CBF values, as a percentage of the preoperative value in the sham-operated mice ($n=7$), tended to decrease, although not significantly, probably as a result of minimal fibrous scar tissue build up and bone opacification regardless of appropriate treatment (*see* **Note 10**)

4 Notes

1. This model should be applied exclusively to the C57BL/6J strain, as CBF in other strains may have a greater variability after BCAS [4, 8, 23].
2. Although anesthetics are known to provide varying degrees of neuroprotection against ischemic injury, the selection of anesthesia did not appreciably affect the mortality rates, temporal profile of CBF, and ischemic white matter changes after BCAS.
3. Four types of microcoils are made from piano wire with varying inner diameters from 0.16 mm to 0.22 mm. Researchers in Japan may obtain the microcoils directly from the manufacturer (Sawane Spring Co., Ltd, Hamamatsu, Japan), but those outside Japan may purchase the microcoils from Invitrotech Co., Ltd. (Kyoto, Japan). The microcoils should be thoroughly disinfected with alcohol and air dried just before use.

4. Anesthesia is usually induced with 4 % isoflurane and maintained with 1.5 % isoflurane, via a face mask. Alternatively, halothane may be used with 4 % for induction and 1.5 % for maintenance.
5. Separate the carotid from the vagus nerve, which is a white, string-like object directly lateral to the carotid artery. Particular care should be taken to avoid damage to the vagus nerve.
6. Cessation of CBF for >1 min should be avoided.
7. For this manipulative procedure, one should avoid piercing the artery with either end of the microcoil. Instead of rotating the microcoil around the CCA, one may alternatively wind the artery along the groove of the coil.
8. In our original report, we used 30 min intervals between manipulations on the left and right CCAs to avoid early mortality [4]. However, no intervals may be required to generate this model. During the surgery, rectal temperature should be maintained between 36.5 °C and 37.5 °C.
9. The mortality rates range from 3 % to 5 % in mice with microcoils of 0.18 mm in diameter (unpublished data), although earlier studies suggested higher mortality rates, 13 % in mice with microcoils of 0.22 mm in diameter, 17 % in those of 0.20 mm, and 15–19 % in those of 0.18 mm [4, 24]. In contrast, 75 % (15/20) of mice with microcoils of 0.16 mm diameter administered died within 14 days after the surgery, most of whom were found to have cerebral infarctions [4]. In another study of a modified model with a 0.16 mm microcoil placed on the left CCA and the 0.18 mm microcoil on the right CCA, the mortality rate is reported to be 18.8 % [25].
10. It is important to remove the periosteum, which adheres tightly to the skull, with fine-tip forceps to minimize fibrous scar tissue buildup without significant changes in flow signals in sham-operated mice [19]. Swabbing the skull surface should be done gently. Rough or inappropriate treatment leads to scar tissue build up along the cranial sutures. It is difficult to reverse fibrous scar tissue buildup and bone opacification. If they do occur, another animal should be used for the experiment. Although most bleeding stops spontaneously, the wound should be checked postoperatively for blood stains on the skull, which, if present, must be removed gently with saline-soaked gauze.
11. For each recording, the skull surface should be wiped with saline-soaked gauze, covered with a thin layer of gel (Aquasonic, Parker Laboratories, Inc.), and immersed for 5–10 min. Care should be taken to ensure the surface is fully wet as indicated by a semitransparent appearance. A dry surface, as indicated by a white skull, or a partially wet surface results in a reduction in flow signal intensity.

12. Laser speckle imaging requires a baseline to anchor for repetitive measurement or comparison between different subjects. We use a calibration reference device (Calibrator S/N 080715-5, Omegawave, Inc.) and assign a value to this reference material (arbitrarily assigned value, 25.0). The value is attributed to the Brownian motion of red-colored particles (0.35 μm, 24 °C). Calibration with this device before each test provides standardized values for comparison.
13. CBF should be measured after CBF is stabilized after anesthesia. During the CBF recordings, the rectal temperature should be maintained between 36.5 °C and 37.5 °C. One can successfully image through the skull repeatedly up to 30 days after the operation [17, 19, 26]. Success is mostly dependent on the degree of removal of the periosteum.
14. It is ideal to measure the mean CBF in identically sized regions of interest (900 pixels) located 1 mm posterior and 2 mm lateral from the bregma.

Acknowledgement

We would like to express our gratitude to Dr. Ahmad Khundakar for insightful comments and editing. We would also like to thank Dr. Shibata, Dr. Nakaji, Dr. Fujita, Dr. Nishio, and Dr. Yamada (former members of Kyoto University) who have been involved in the establishment, characterization, and optimization of this animal model. This work was partially supported by Banyu Life Science Foundation International and was also supported by the Grant-in-Aid for Scientific Research (B) (M.I. No. 23390233) and by the Grant-in-Aid from the Ministry of Health, Labour and Welfare of Japan (M.I. No. 2450101-001).

References

1. Ihara M, Tomimoto H, Ishizu K et al (2004) Decrease in cortical benzodiazepine receptors in symptomatic patients with leukoaraiosis: a positron emission tomography study. Stroke 35:942–947
2. Yao H, Sadoshima S, Kuwabara Y et al (1990) Cerebral blood flow and oxygen metabolism in patients with vascular dementia of the Binswanger type. Stroke 21:1694–1699
3. Wakita H, Tomimoto H, Akiguchi I et al (1994) Glial activation and white matter changes in the rat brain induced by chronic cerebral hypoperfusion: an immunohistochemical study. Acta Neuropathol 87:484–492
4. Shibata M, Ohtani R, Ihara M et al (2004) White matter lesions and glial activation in a novel mouse model of chronic cerebral hypoperfusion. Stroke 35:2598–2603
5. Kudo T, Tada K, Takeda M et al (1990) Learning impairment and microtubule-associated protein 2 decrease in gerbils under chronic cerebral hypoperfusion. Stroke 21: 1205–1209
6. Sarti C, Pantoni L, Bartolini L et al (2002) Persistent impairment of gait performances and working memory after bilateral common carotid artery occlusion in the adult Wistar rat. Behav Brain Res 136:13–20
7. Ihara M, Tomimoto H (2011) Lessons from a mouse model characterizing features of vascular cognitive impairment with white matter changes. J Aging Res 2011:978761

8. Shibata M, Yamasaki N, Miyakawa T et al (2007) Selective impairment of working memory in a mouse model of chronic cerebral hypoperfusion. Stroke 38:2826–2832
9. Coltman R, Spain A, Tsenkina Y et al (2011) Selective white matter pathology induces a specific impairment in spatial working memory. Neurobiol Aging 32:2324.e7–12
10. Yoshizaki K, Adachi K, Kataoka S et al (2008) Chronic cerebral hypoperfusion induced by right unilateral common carotid artery occlusion causes delayed white matter lesions and cognitive impairment in adult mice. Exp Neurol 210:585–591
11. Stroke Therapy Academic Industry Roundtable (STAIR) (1999) Recommendations for standards regarding preclinical neuroprotective and restorative drug development. Stroke 30: 2752–2758
12. Farkas E, Luiten PG, Bari F (2007) Permanent, bilateral common carotid artery occlusion in the rat: a model for chronic cerebral hypoperfusion-related neurodegenerative diseases. Brain Res Rev 54:162–180
13. Jiwa NS, Garrard P, Hainsworth AH (2010) Experimental models of vascular dementia and vascular cognitive impairment: a systematic review. J Neurochem 115:814–828
14. Ni JW, Matsumoto K, Li HB et al (1995) Neuronal damage and decrease of central acetylcholine level following permanent occlusion of bilateral common carotid arteries in rat. Brain Res 673:290–296
15. Holland PR, Bastin ME, Jansen MA et al (2011) MRI is a sensitive marker of subtle white matter pathology in hypoperfused mice. Neurobiol Aging 32:2325.e1–6
16. Nishio K, Ihara M, Yamasaki N et al (2010) A mouse model characterizing features of vascular dementia with hippocampal atrophy. Stroke 41:1278–1284
17. Maki T, Ihara M, Fujita Y et al (2011) Angiogenic and vasoprotective effects of adrenomedullin on prevention of cognitive decline after chronic cerebral hypoperfusion in mice. Stroke 42:1122–1128
18. Maki T, Ihara M, Fujita Y et al (2011) Angiogenic roles of adrenomedullin through vascular endothelial growth factor induction. Neuroreport 22:442–447
19. Fujita Y, Ihara M, Ushiki T et al (2010) Early protective effect of bone marrow mononuclear cells against ischemic white matter damage through augmentation of cerebral blood flow. Stroke 41:2938–2943
20. Kalaria RN, Akinyemi R, Ihara M (2012) Does vascular pathology contribute to Alzheimer changes? J Neurol Sci 322(1–2):141–147
21. Okamoto Y, Yamamoto T, Kalaria RN et al (2012) Cerebral hypoperfusion accelerates cerebral amyloid angiopathy and promotes cortical microinfarcts. Acta Neuropathol 123:381–394
22. Yamada M, Ihara M, Okamoto Y et al (2011) The influence of chronic cerebral hypoperfusion on cognitive function and amyloid β metabolism in APP overexpressing mice. PloS One 6:e16567
23. Nakaji K, Ihara M, Takahashi C et al (2006) Matrix metalloproteinase-2 plays a critical role in the pathogenesis of white matter lesions after chronic cerebral hypoperfusion in rodents. Stroke 37:2816–2823
24. Duan W, Gui L, Zhou Z et al (2009) Adenosine A2A receptor deficiency exacerbates white matter lesions and cognitive deficits induced by chronic cerebral hypoperfusion in mice. J Neurol Sci 285:39–45
25. Miki K, Ishibashi S, Sun L et al (2009) Intensity of chronic cerebral hypoperfusion determines white/gray matter injury and cognitive/motor dysfunction in mice. J Neurosci Res 87:1270–1281
26. Washida K, Ihara M, Nishio K et al (2010) Nonhypotensive dose of telmisartan attenuates cognitive impairment partially due to peroxisome proliferator-activated receptor-gamma activation in mice with chronic cerebral hypoperfusion. Stroke 41:1798–1806

Chapter 9

A Mouse Model of Permanent Focal Ischemia: Distal Middle Cerebral Artery Occlusion

Kristian P. Doyle and Marion S. Buckwalter

Abstract

Here we provide a standardized protocol for performing distal middle cerebral artery occlusion (DMCAO) in mice. DMCAO is a method of inducing permanent focal ischemia that is commonly used as a rodent stroke model. To perform DMCAO a temporal craniotomy is performed, and the middle cerebral artery (MCA) is permanently ligated at a point downstream of the lenticulostriate branches. The size of the lesion produced by this surgery is strain dependent. In C57BL/6J mice, DMCAO produces an infarct predominantly restricted to the barrel region of the somatosensory cortex, but in BALB/cJ mice, DMCAO generates a much larger lesion that incorporates more of the somatosensory cortex and part of the M1 region of the motor cortex. The larger lesion produced by DMCAO in BALB/cJ mice produces a clearer sensorimotor deficit, which is useful for investigating recovery from stroke. We also describe how to modify DMCAO in C57BL/6J mice with the application of hypoxia to generate a lesion and sensorimotor deficit that are similar in size to those produced by DMCAO alone in BALB/cJ mice. This is extremely useful for stroke experiments that require a robust sensorimotor deficit in transgenic mice created on a C57BL/6J background.

Key words Distal middle cerebral artery occlusion, Permanent focal ischemia, Mouse stroke model, DH stroke

1 Introduction

DMCAO is a model of permanent focal ischemia that is commonly used in the mouse. It requires a craniotomy to reach the distal portion of the middle cerebral artery (MCA), which is defined as the portion downstream of the lenticulostriate branches. It was first developed in 1981 by Tamura et al as a model of focal ischemia in the rat [1] and has since been adapted for mice [2, 3]. Over the years, the original technique has changed slightly. Initially, the point of ligation on the MCA was more proximal and required that a section of the zygomatic arch be removed from the skull. It is more common now to ligate the MCA at a more distal location, leaving the zygomatic arch intact. This makes the surgery less intrusive and speeds up recovery time, although lesion size is smaller [3, 4].

Richard Milner (ed.), *Cerebral Angiogenesis: Methods and Protocols*, Methods in Molecular Biology, vol. 1135, DOI 10.1007/978-1-4939-0320-7_9, © Springer Science+Business Media New York 2014

Having to perform a craniotomy is the main disadvantage of DMCAO. Studies with sham mice show that performing a craniotomy and opening the dura initiate transcriptional changes and a stress response in the underlying brain tissue [5]. Therefore, it is vital to have sham controls when using this model. Performing a craniotomy can also introduce pathogens into the brain and cause immune responses that are not typical of stroke. The use of a prophylactic dose of antibiotic and aseptic surgical technique mitigates this risk.

Despite these disadvantages, performing a craniotomy also confers an advantage to DMCAO. In mice, the distal portion of the MCA is clearly visible through the temporal bone of the skull. This simplifies selecting the site for the craniotomy and makes successful ligation of the MCA easy to ascertain. This is a significant advantage over alternative stroke models that rely on an intraluminal filament to occlude the MCA. In these stroke models occlusion is not directly visible and as such is harder to verify [3, 6].

Another advantage of DMCAO is that within each mouse strain, it creates highly reproducible cortical lesions, and gives rise to easily definable behavioral deficits [4, 7, 8]. Alternative stroke models frequently generate less predictable lesions [3], which means that behavioral deficits following stroke are more variable from mouse to mouse. The higher fidelity of the lesion produced by DMCAO makes it easier to track recovery while simultaneously investigating mechanisms of repair, such as angiogenesis, brain rewiring, and the resolution of inflammation.

Although lesion size after DMCAO is consistent within a mouse strain, it varies significantly across mouse strains (Fig. 1a, b). For instance, in C57BL/6J and 129/SvJ mice, lesion size is approximately 8 % of the ipsilateral hemisphere; however, in BALB/cJ mice typical lesion volume is three times larger at 25 % of the ipsilateral hemisphere [4, 7]. Based on our experience, BALB/cJ mice have hemiparesis of the contralateral forelimb, and a deficit in body proprioception and vibrissae sensitivity following DMCAO. The smaller lesion created by DMCAO in C57BL/6J mice translates to a smaller behavioral deficit, and mice only appear to have a reduction in vibrissae sensitivity, although altered gait dynamics have been reported [9].

It is possible to increase lesion size and consequently behavioral deficits following DMCAO in C57BL/6J mice. This can be done by transient or permanent ligation of one or both carotid arteries during DMCAO, or by the application of hypoxia after DMCAO [3, 4, 10, 11]. The application of hypoxia is our preferred method of increasing lesion size, because based on our experience it does so with greater consistency than carotid artery ligation, the outcome of which depends on the patency of the Circle of Willis, which is notoriously variable in C57BL/6J mice [12–14]. Therefore, below, in addition to describing a protocol for

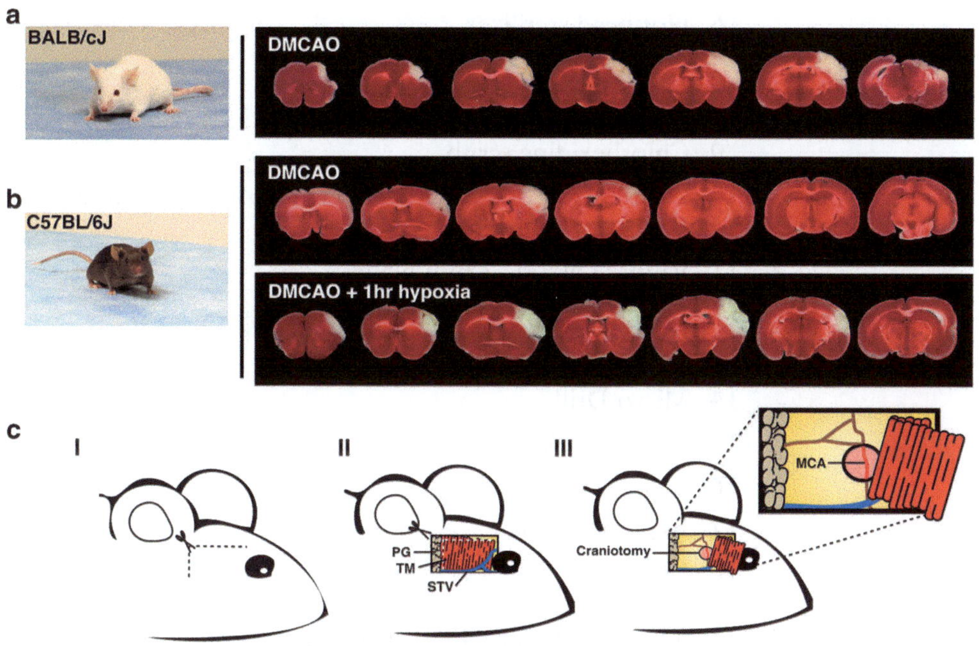

Fig. 1 (**a**) Image of the lesion generated by DMCAO in BALB/cJ mice (Jackson Laboratory, stock number 000651). The lesion incorporates areas of the frontal, parietal, and occipital cortices. (**b**) Image of the lesion generated by DMCAO and DMCAO + 1 h hypoxia in C57BL/6J mice (Jackson Laboratory, stock number 000664). DMCAO produces a small lesion restricted to the frontal cortex and the anterior part of the parietal cortex. The addition of 1 h hypoxia extends the lesion to the occipital cortex. Brains were removed 24 h after stroke and were stained with 1.5 % TTC to visualize the region of infarction. (**c**) Diagram of the DMCAO procedure. *Panel I* depicts the location for the skin incision (dashed line). *Panel II* shows the location for the incision in the temporalis muscle (*dashed line*). PG = parotid gland, TM = temporalis muscle, STV = superficial temporal vein.) *Panel III* shows the location for the craniotomy after the temporalis muscle has been retracted. Caveat: The blood vessels visible through the temporal bone of the skull vary from mouse to mouse and so the diagram only provides a rough guide. To identify the MCA, locate the blood vessel that runs almost vertically from the base of the brain. MCA = middle cerebral artery

performing DMCAO, we also describe our technique for using hypoxia to increase lesion size after DMCAO in C57BL/6J mice (Fig. 1b).

2 Materials

2.1 Materials for DMCAO

1. Air tank for anesthesia.
2. Electric shaver.
3. Binocular surgical microscope.
4. Tabletop anesthesia system.
5. Homeothermic blanket system (Stoelting, Wood Dale, IL, USA).

6. Hot bead sterilizer.
7. Isoflurane.
8. Ophthalmic eye ointment.
9. Chlorhexidine scrub.
10. O'Brien scissors, 3.75 in/angled (Roboz, Gaithersburg, MA, USA).
11. Vannas scissors, 7 mm blades.
12. 2× forceps, 0.05 × 0.01 mm.
13. 1× forceps 4 inch/serrated, 0.5 mm.
14. Micro Drill.
15. Cauterizer kit.
16. Tissue adhesive.
17. Buprenorphine.
18. Cefazolin.

2.2 Additional Materials for DMCAO + Hypoxia (DH Stroke)

1. 8 % oxygen tank (92 % nitrogen).
2. Hypoxia chamber (Coy Lab Products, Grass Lake, Michigan, USA).

2.3 Materials for Infarct Assessment

1. Mouse brain matrix (Ted Pella, Redding, CA, USA).
2. 2,3,5-Triphenyltetrazolium chloride (TTC).
3. Phosphate buffered saline (PBS).
4. 10 % formalin.

3 Methods

Perform all surgery under sterile conditions wearing a lab coat, gloves, and mask. Use sterile instruments and an aseptic tips only technique. Only the sterile working ends of the surgical instruments should come into contact with the animal. If the tips of the surgical instruments touch any surface other than the animal, use a hot bead sterilizer to make them aseptic again. Monitor the respiration and responsiveness of the animal to a toe pinch throughout anesthesia. Reduce the concentration of isoflurane if respiration becomes slow and erratic, or increase concentration if there is a response to toe pinch. Sham-operated mice should be treated in the same way as DMCAO mice, with the single exception that the MCA is not ligated.

3.1 DMCAO

1. Weigh the mouse (*see* **Note 1**) and anesthetize in 2–3 % isoflurane in a warm (37 °C) anesthesia induction box. Verify anesthesia with a toe pinch and then move the mouse to a shaving

station set away from the area where surgery will be performed. This prevents hair from contaminating the surgical area. Shave the side of the mouse's head between the ear and the eye, on the side you wish to occlude. The shaving station and the surgical station should have separate nose cones to deliver 2–3 % isoflurane.

2. Move the mouse to the area designated for surgery and place it on its side on top of a feedback-controlled heating blanket to maintain mouse temperature at 37 °C. Insert a rectal thermometer to control the temperature of the heating blanket (*see* **Note 2**). Inject subcutaneously with a prophylactic antibiotic (e.g., 25 mg/kg cefazolin), and place a drop of ophthalmic eye ointment over each eye to prevent dehydration during anesthesia.
3. Gently stretch the ear towards the tail and tape it in place to keep it out of the way during surgery. Prep the skin for surgery by swabbing with a solution of chlorhexidine, and then rinse the area with sterile saline.
4. Use angled O'Brien scissors to make a 4 mm horizontal incision in the skin between the orbit and the auditory canal. At the auditory canal side of the incision, make a 4–5 mm vertical incision to create a flap of skin that can be retracted. This exposes the temporalis muscle and part of the parotid gland (Fig. 1c, Panel I).
5. Use Vannas scissors to repeat the horizontal and vertical incisions in the temporalis muscle, and then retract the muscle to expose the skull (Fig. 1c, Panel II). Clean residual soft tissue from the surface of the skull using sterile gauze and locate the MCA.
6. Use a micro drill to create a 2 mm diameter hole directly over the MCA (Fig. 1c, Panel III) (*see* **Note 3**). Use forceps (0.05×0.01 mm diameter) to remove the meninges and then cauterize the MCA with a small vessel cauterizer (*see* **Note 4**). If bleeding occurs hold a small piece of sterile gauze in forceps and apply gentle pressure until bleeding stops.
7. Once the MCA has been cauterized, rinse the brain surface with saline and fold the temporalis muscle and skin back into place. Use surgical glue to seal the skin and inject the mouse with buprenorphine (0.1 mg/kg, subcutaneously) to provide analgesia.
8. Place the mouse in a warm cage until recovered from anesthesia. When the mouse is awake and ambulatory, return it to its home cage with soft food.

3.2 DMCAO with Hypoxia

Adding a period of 1 h hypoxia immediately after DMCAO will triple the lesion size and increase the sensorimotor deficit in

C57BL/6J mice. Adding a period of 1 h of hypoxia immediately after DMCAO does not affect lesion size in BALB/cJ mice [4]:

1. Perform DMCAO as above.
2. Immediately after injecting buprenorphine, place the mouse in a hypoxia chamber containing 8 % oxygen for 1 h. Although mice are not anesthetized during the period of hypoxia, respiration should still be regularly checked throughout. Maintain the chamber temperature at 37 °C to prevent hypothermia.
3. After 1 h, place the mouse in a warm cage until it is ambulatory, then return it to its home cage with soft food.

3.3 Follow-Up Care

Analgesia

Administer buprenorphine, 0.1 mg/kg, subcutaneous (SQ), prior to waking from surgery and once more 24 h later.

Antibiotic Prophylaxis

Administer cefazolin, 25 mg/kg, subcutaneous (SQ), once immediately before surgery.

Post-Procedure Monitoring

Mice should be monitored daily for 7 days after surgery, and any mice that appear debilitated longer than 24 h after surgery should be given pain medication in consultation with veterinary staff or immediately euthanized. Early euthanasia criteria should be significant bleeding during the procedure, loss of more than 10 % of body weight, loss of mobility, frequent seizures, poor grooming, abnormal behavior, heightened anxiety, lethargy, extensive abdominal extension, ruffled coat, and respiratory distress. Mice typically tolerate these procedures well and appear healthy by the next day.

3.4 Infarct Assessment

A fast and reliable method to visualize the region of infarction in the first 96 h after DMCAO is to use the metabolic dye 2,3,5-triphenyltetrazolium chloride (TTC) [15]. TTC is a water-soluble salt that is reduced by dehydrogenases within actively respiring mitochondria, changing it to a bright red color:

1. Remove the brain and slice into 1 mm thick sections. Mice do not need to be perfused.
2. Immerse the sections in a 1.5 % solution of TTC in PBS for 15 min.
3. Replace the TTC solution with 10 % formalin.
4. Lay the sections onto a slide and scan using a flatbed scanner with a black background.

4 Notes

1. In choosing animals be aware that age and gender can affect stroke size, as well as other experimental results. We recommend doing pilot studies on your selected age and gender and then utilizing power calculations to determine appropriate group sizes for your experimental question. For most studies with adult mice, groups of 8–10 will be appropriate.
2. During surgery and recovery, maintain the animal temperature at 37 °C to prevent hypothermia, which is neuroprotective and will affect lesion size.
3. To perform the craniotomy, drill over the MCA. Once the skull has been breached, use the tip of the drill bit on the edges of the burr hole to expand its size to a diameter of ~2 mm.
4. The hardest part of DMCAO surgery is to successfully ligate the MCA without causing bleeding. When a fully charged wireless cauterizer is used, the temperature of the wire filament can cause the MCA to burst before the cauterizer makes contact. A simple way to prevent this from occurring is to depress the button on the cauterizer for a minute or so to drain the batteries. This reduces the temperature of the wire filament and prevents the MCA from rupturing before the filament is applied.

Acknowledgements

This work was supported by NIH grants R01NS067132, R21NS078571, and K99NR013593.

References

1. Tamura A, Graham DI, McCulloch J, Teasdale GM (1981) Focal cerebral ischaemia in the rat: 1. Description of technique and early neuropathological consequences following middle cerebral artery occlusion. J Cereb Blood Flow Metab 1:53–60
2. Nawashiro H, Tasaki K, Ruetzler CA, Hallenbeck JM (1997) TNF-alpha pretreatment induces protective effects against focal cerebral ischemia in mice. J Cereb Blood Flow Metab 17:483–490
3. Carmichael ST (2005) Rodent models of focal stroke: size, mechanism, and purpose. NeuroRx 2:396–409
4. Doyle KP, Fathali N, Siddiqui MR, Buckwalter MS (2012) Distal hypoxic stroke: a new mouse model of stroke with high throughput, low variability and a quantifiable functional deficit. J Neurosci Methods 207(1):31–40
5. Glazier SS, O'Rourke DM, Graham DI, Welsh FA (1994) Induction of ischemic tolerance following brief focal ischemia in rat brain. J Cereb Blood Flow Metab 14:545–553
6. Schmid-Elsaesser R, Zausinger S, Hungerhuber E, Baethmann A, Reulen HJ (1998) A critical reevaluation of the intraluminal thread model of focal cerebral ischemia: evidence of inadvertent premature reperfusion and subarachnoid hemorrhage in rats by laser-Doppler flowmetry. Stroke 29:2162–2170
7. Majid A, He YY, Gidday JM, Kaplan SS, Gonzales ER, Park TS, Fenstermacher JD, Wei L, Choi DW, Hsu CY (2000) Differences in vulnerability to permanent focal cerebral ischemia among 3 common mouse strains. Stroke 31:2707–2714
8. Freret T, Bouet V, Leconte C, Roussel S, Chazalviel L, Divoux D, Schumann-Bard P,

Boulouard M (2009) Behavioral deficits after distal focal cerebral ischemia in mice: usefulness of adhesive removal test. Behav Neurosci 123:224–230

9. Lubjuhn J, Gastens A, von Wilpert G, Bargiotas P, Herrmann O, Murikinati S, Rabie T, Marti HH, Amende I, Hampton TG, Schwaninger M (2009) Functional testing in a mouse stroke model induced by occlusion of the distal middle cerebral artery. J Neurosci Methods 184:95–103
10. Sugimori H, Yao H, Ooboshi H, Ibayashi S, Iida M (2004) Krypton laser-induced photothrombotic distal middle cerebral artery occlusion without craniectomy in mice. Brain Res Brain Res Protoc 13:189–196
11. Buchan AM, Xue D, Slivka A (1992) A new model of temporary focal neocortical ischemia in the rat. Stroke 23:273–279
12. Barone FC, Knudsen DJ, Nelson AH, Feuerstein GZ, Willette RN (1993) Mouse strain differences in susceptibility to cerebral ischemia are related to cerebral vascular anatomy. J Cereb Blood Flow Metab 13:683–692
13. Kitagawa K, Matsumoto M, Yang G, Mabuchi T, Yagita Y, Hori M, Yanagihara T (1998) Cerebral ischemia after bilateral carotid artery occlusion and intraluminal suture occlusion in mice: evaluation of the patency of the posterior communicating artery. J Cereb Blood Flow Metab 18:570–579
14. McColl BW, Carswell HV, McCulloch J, Horsburgh K (2004) Extension of cerebral hypoperfusion and ischaemic pathology beyond MCA territory after intraluminal filament occlusion in C57Bl/6J mice. Brain Res 997:15–23
15. Bederson JB, Pitts LH, Germano SM, Nishimura MC, Davis RL, Bartkowski HM (1986) Evaluation of 2,3,5-triphenyltetrazolium chloride as a stain for detection and quantification of experimental cerebral infarction in rats. Stroke 17:1304–1308

Chapter 10

A Method of Inducing Global Cerebral Ischemia

Gina Hadley, Michalis Papadakis, and Alastair M. Buchan

Abstract

The four-vessel occlusion (4-VO) method of global forebrain cerebral ischemia mimics the human clinical condition of cardiac arrest. It results in selective neuronal damage and is a useful experimental system to dissect underlying mechanisms behind ischemic phenomena such as the differential susceptibility of CA1 compared to the CA3 region of the hippocampus. It also provides a "proof-of-principle" system for testing out potential agents for neuroprotection.

Key words Rat, Global, Ischemia, Telemetry, Temperature

1 Introduction

The model of transient global forebrain ischemia is highly successful in rodents [1]. It is a reversible model of forebrain ischemia. In the rat it is relatively easy to control physiological variables. Considering the principles of reduction, replacement, and refinement [2] when conducting animal experiments, the minimum number of animals required should be used to achieve statistical power. Although the focal model of cerebral ischemia more closely approximates to the human clinical condition of stroke [3] and the global model mimics cardiac arrest, the global model can be invaluable in gaining mechanistic insight into potential neuroprotective phenomena [4, 5].

The four-vessel occlusion (4-VO) model was first described in 1979 [6]. Prior to this, solely occluding the carotid arteries did not produce a reliable level of ischemic damage as the rat has an efficient collateral circulation [7]. In addition to transient occlusion of the carotid arteries, the 4-VO model involves cauterization of the vertebral arteries and occlusion of the collateral vessels in the neck muscles. The method was modified in 1983 to allow for the tightness of the suture occluding the collaterals in the neck muscles to be responsive to the condition of the rat [1]. The degree of ischemia in the rat was assessed by measuring the righting response, reaction to painful stimuli, the corneal reflex, and the running

Richard Milner (ed.), *Cerebral Angiogenesis: Methods and Protocols*, Methods in Molecular Biology, vol. 1135,
DOI 10.1007/978-1-4939-0320-7_10, © Springer Science+Business Media New York 2014

response [1]. The success of the model is therefore a balance between achieving the desired complete ischemia versus the adverse outcome of respiratory arrest [8]. The global model is advantageous as the neuronal damage produced is predictable [9] and the occurrence of seizures is low [4]. The incidence of post-ischemic seizures appears to be related to increasing duration of ischemia with 8 % of rats experiencing seizures after 20 min of 4-VO and 40 % following 30 min of 4-VO [10]. In mice, the 4-VO model of ischemia is not practically possible. In these animals, the best model of forebrain ischemia uses two-vessel (carotid artery) occlusion and hypotension, where hypotension controls for the collateral circulation [11].

Core temperature telemetry probe implantation and regulation is vital in these experiments as hypothermia has been shown to be neuroprotective [12, 13]. If temperature is not controlled, it can be a confounding factor, so that 7 days before global ischemia surgery, sterilized core telemetry probes are implanted into the peritoneal cavity under anesthesia. The global model of cerebral ischemia is carried out in three stages completed over an 8-day total experimental period:

1. Implantation of telemetry probes (7 days prior to Pre 4-VO)
2. Pre 4-VO, in which the vertebral arteries are occluded
3. 4-VO (24 h post Pre 4-VO)

2 Materials

2.1 Animals, Anesthesia, and Preparation for Surgery

1. Male Wistar rats 150–200 g from your local laboratory animal supplier. They are allowed free access to chow and tap water. The animal house should be approximately 24 °C with day–night regulation.
2. Weighing scales.
3. Anesthesia induction chamber.
4. Isoflurane delivery unit with flowmeters that can deliver isoflurane carried in other gases.
5. Isoflurane.
6. 100 % oxygen medical grade.
7. 100 % nitrous oxide medical grade (BOC).
8. Buprenorphine: 0.5 ml of 0.3 mg/ml diluted 1 part in 9 parts normal saline.
9. Lidocaine 5 % M/M ointment.
10. Lacri-lube.
11. Shaver.

12. 70 % ethanol.
13. Betadine.
14. Marker pen.

2.2 Surgical Tools

1. 2-0 silk suture.
2. 3-0 silk suture.
3. 2-0 nonabsorbable Mersilk braided suture.
4. 5-0 Monocryl absorbable sutures.
5. Surgical scalpel (sterilized).
6. Surgical clamps (sterilized).
7. Retractor (sterilized).
8. Pasteur pipette.
9. Beaker of water (to clean suction apparatus).
10. Cotton buds.
11. 70 % ethanol.
12. Vaseline or equivalent lubricant.
13. Normal saline 0.9 % w/v injection.
14. Spinal needle (18 GA 3.50 in. 1.2×90 mm).
15. Electrocautery equipment (<1 mm diameter, platinum electrode loop angled tip electrocautery needle connected to the Geiger Thermal Cautery Unit on setting 9).
16. Vascular clips (jaw dimensions 10×2.15 mm, length 22 mm, FST-00325).
17. Rectal thermometer and heat pad system (Harvard Apparatus Homeothermic Blanket Control).
18. Telemetry probe (TA-F40, Data Sciences International).
19. Sterile drapes.
20. Autoclave tape.
21. Sterile gloves.
22. Rig for global ischemia surgery, including nose cone (Fig. 1).
23. Surgical scissors (sterilized).
24. Curved forceps (sterilized) $\times 2$.
25. Stainless steel Stoelting EZ Clip Applier.
26. 9 mm EZ Clips.
27. Wound clip remover.
28. Skin tweezers (sterilized).
29. Telemetry receivers (RLA-1020, Data Sciences International) connected to infrared lamps. Temperature sampled in freely moving rats every 30 s and regulated (37.0 ± 0.5 °C) using the computerized temperature control system ART-2.2.

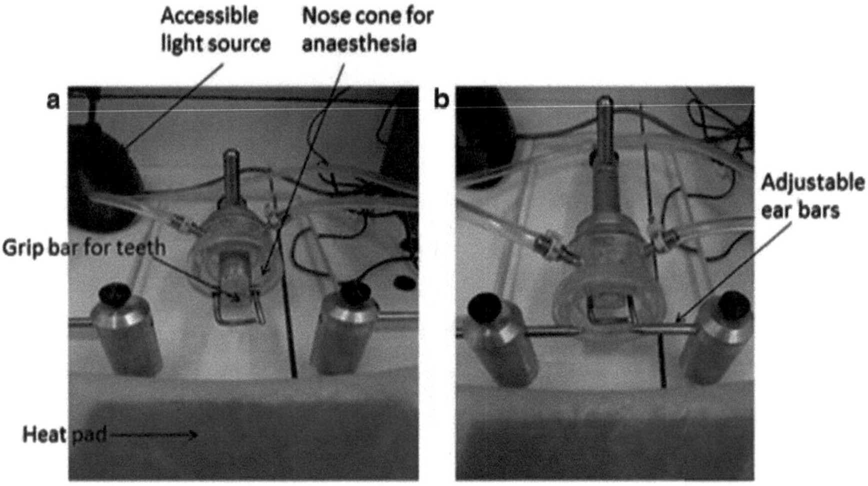

Fig. 1 Apparatus for Pre 4-VO and 4-VO surgery. Panel (**a**) is for when the rat is supine. Panel (**b**) is for when the rat is prone (adjustable ear bars to hold in place)

3 Methods

3.1 Anesthesia

1. Induce anesthesia using isoflurane (3 % v/v) in the induction chamber.
2. Remove the rat from the isoflurane chamber.
3. Shave the fur from the proposed surgical area. For probe implantation, this is the ventral surface of the abdomen. For Pre 4-VO, this area includes the neck (front and back) and the top of the head and at the sides as far as the ears (*see* **Note 1**).
4. Move the rat to the apparatus in Fig. 1 and maintain anesthesia on a draped, sterile heating mat (controlling the rat's body temperature at 37 °C) via a nose cone (isoflurane (3 % v/v), mixed with 30 % (v/v) O_2 and 70 % (v/v) N_2O) connected to a scavenging system.
5. Apply eye ointment (Lacri-lube®) to prevent the eyes from drying out.
6. Insert the rectal temperature probe.
7. Inject buprenorphine 0.5 ml of 0.3 mg/ml diluted in normal saline (1:9) subcutaneously into the scruff of the neck.
8. Rotate the apparatus so that the rat is prone in order to apply the rectal temperature probe with the aid of Vaseline to lubricate.
9. When the rat is stable, reduce the isoflurane from the induction (3 % v/v) to the maintenance level (1–1.5 % v/v), mixed with 30 % (v/v) O_2 and 70 % (v/v) N_2O. *See* Fig. 1 for apparatus.
10. Keep the rat under constant surveillance in order to check that the anesthesia is sufficient, and monitor the breathing rate.

3.2 Telemetry Probe Insertion

1. Soak the telemetry probes in disinfectant overnight prior to implantation. The procedure must be sterile to avoid infection.
2. Anesthetize and prepare the rat, *see* Subheading 3.1.
3. Swab the lower abdomen on either side of the midline using both ends of three cotton wool buds first with ethanol and then with Betadine.
4. Make a midline incision through the skin and peritoneum taking care not to cut any bowel (*see* **Note 2**) and insert the probe small-end first.
5. Close the abdominal wall with 5-0 absorbable suture, taking care to cut the remaining ends short to avoid them being exposed and accessible for the rats to chew.
6. Close the skin with wound staples, ensuring no suture is protruding.
7. Swab the lower abdomen either side of the midline using both ends of three cotton wool buds first with Betadine and then with ethanol.
8. Rub lidocaine into the skin surrounding the stapled wound.
9. Weigh the rat and inspect the wound at least daily.

3.3 Pre 4-VO

1. Anesthetize and prepare the rat, *see* Subheading 3.1.
2. Turn the rat back to a supine position and drape to expose the proposed sterile field.
3. Conduct surgery on a draped heat mat.
4. Swab the neck with Vaseline in order to move excess fur away from the sterile field and then with 70 % ethanol to clean the area.
5. Make an incision using surgical scissors on the ventral surface of the midline of the neck.
6. Use curved forceps to gently dissect out the common carotid arteries. For the left common carotid, carry out blunt dissection along the longitudinal left aspect of the central and adjacent muscular tissues (these are sternocleidomastoid, omohyoid, thyrohyoid, and sternohyoid). It is possible to separate the central muscle from parallel neck muscles and the diagonal thin muscular band, the omohyoid (*see* **Note 3**). The omohyoid lies directly over the carotid vasculature (*see* **Note 4**).
7. Gently separate the vagus nerve away from the common carotids, avoiding overstimulation.
8. Pass approximately 20 cm of 3-0 braided silk suture underneath each common carotid artery when separated from surrounding tissues and nerves (Fig. 2a).
9. Place a knot in this, leaving approximately 5 cm in a loop either side (Fig. 2b).
10. Repeat the procedure for the other common carotid artery.

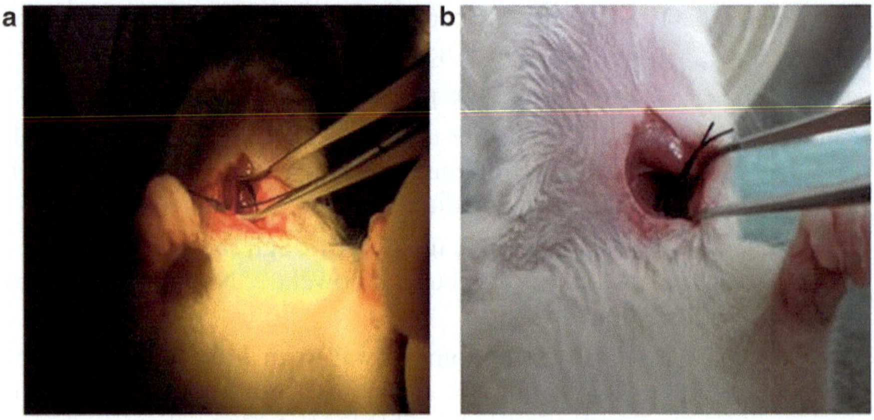

Fig. 2 Dissection of the common carotid artery. (**a**) The common carotid artery is separated from the surrounding tissues and nerves, using gentle dissection with forceps. (**b**) Approximately 20 cm of 3-0 braided silk suture can be passed underneath the artery, and a knot is placed in this leaving approximately 5 cm in a loop on either side

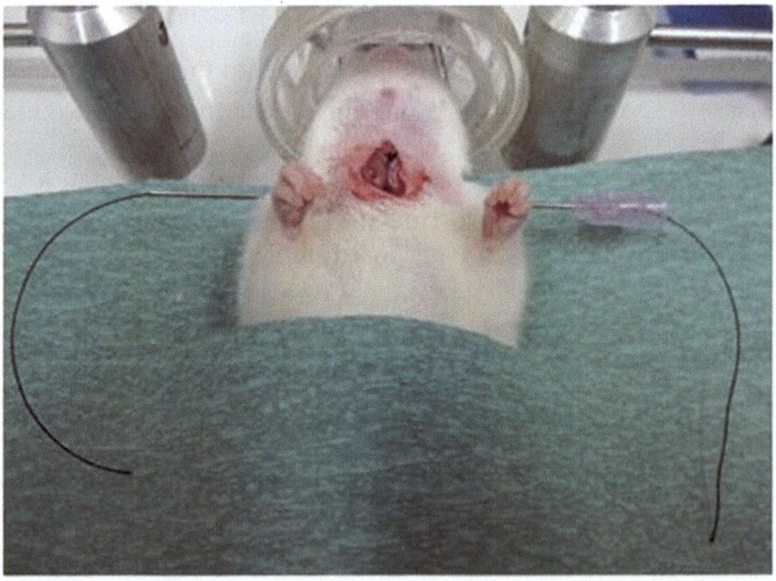

Fig. 3 Preparation for removal of the effect of collateral circulation. A spinal needle (18 GA 3.50 in. 1.2 × 90 mm) is used to pass a 2-0 silk suture through the neck, passing behind the trachea, esophagus, external jugular veins, and common carotid arteries, but in front of the cervical and paravertebral muscles

11. Locate the jugular vein and surrounding tissue and further dissect to expose an area below on both sides.
12. Pass the spinal needle behind the trachea, esophagus, external jugular veins, and common carotid arteries, but in front of the cervical and paravertebral muscles (Fig. 3).
13. Retract the inner part of the needle and pass a length of 2-0 silk suture through it.

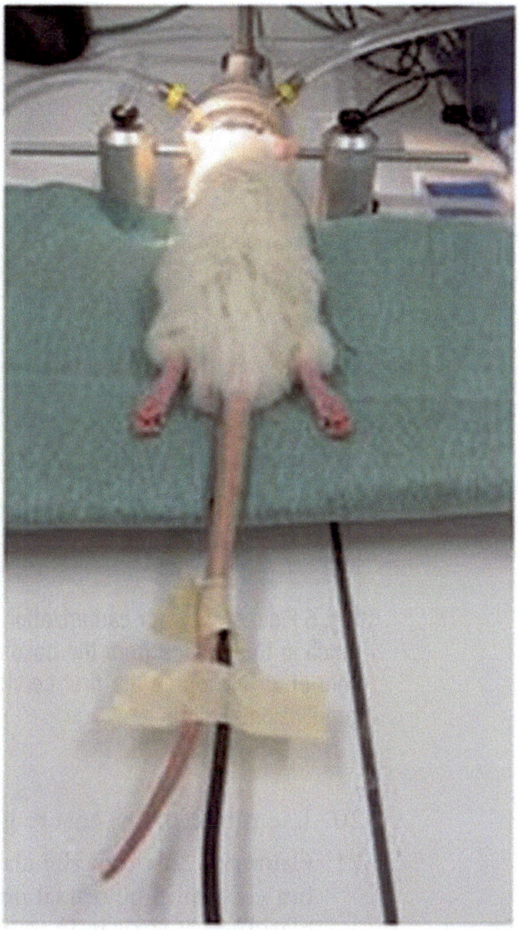

Fig. 4 Positioning for cauterization of the vertebral artery supply. The rat's head is placed in stereotaxic ear bars with the head tilted at approximately 30° to the horizontal. The cervical spine is gently stretched by placing tension on the rat's tail via tape which anchors it to the table

14. Remove the outer part of the needle, taking care to ensure that the suture remains in place at both sides of the neck.
15. Place the two sutures around the common carotids inside the neck.
16. Close the neck wound using wound clips.
17. Place the rat's head in stereotactic ear bars with the head tilted at approximately 30° to the horizontal (Figs. 1b and 4).
18. Gently stretch the cervical spine by putting tension on the rat's tail via autoclave tape which anchors it to the table (Fig. 4).
19. Make an incision in the midline from the base of the skull large enough to provide a good field of vision to see the first cervical vertebra (Fig. 5).

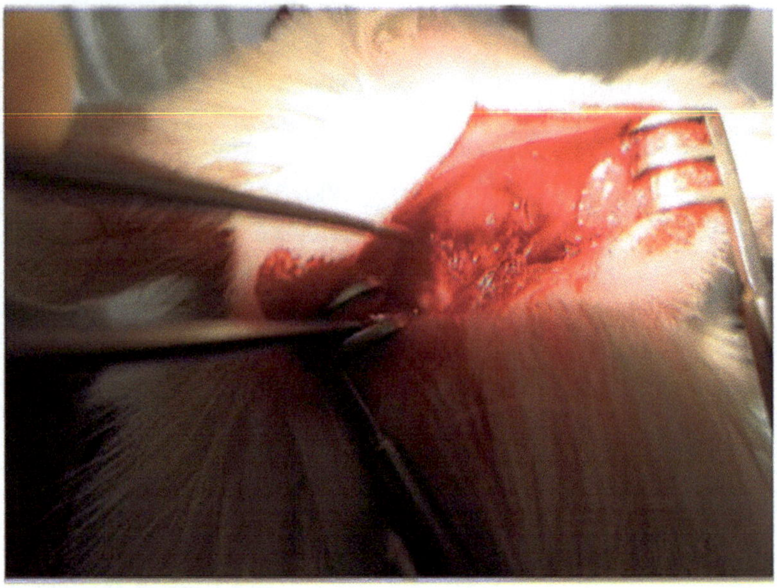

Fig. 5 Field of view for cauterization of the vertebral artery supply. An incision is made in the midline from the base of the skull large enough to provide a good field of vision to see the first cervical vertebra. Retractors are used to ensure good vision at all times

20. Use retractors to ensure good vision at all times (Fig. 5).
21. Cauterize through the alar foramina of the first cervical vertebra via a midline dorsal neck incision using a <1 mm diameter, platinum electrode loop angled tip electrocautery needle with the Geiger Thermal Cautery Unit on setting 9.
22. Keep a vacuum pump with rubber tubing and a Pasteur pipette attached close at hand in case of bleeding (*see* **Note 5**).
23. Use the inner part of the spinal needle to probe the vertebral foramina to check for any remaining collaterals that require further cauterization.
24. Use the Pasteur pipette connected to the vacuum pump to remove any blood to improve the field of view.
25. Close the muscle layer using a 2-0 silk suture, followed by a 2-0 braided suture for the skin.
26. Tape the two lengths of 2-0 silk suture exiting both sides of the neck that were placed for occlusion of the collateral circulation onto the rat's back (*see* **Note 6**).
27. Allow a 24 h interval between the preparatory surgery (Pre 4-VO) and the induction of ischemia (4-VO) to allow the animals time to recover and to permit blood glucose levels to drop. Animals have free access to water.

3.4 Global Ischemia: 4-VO

1. Anesthetize and prepare the rat, *see* Subheading 3.1.
2. This is a two-person procedure. The assistant holds the rat supine, with its teeth attached to the bar of the nose cone apparatus (Fig. 1a) and paws restrained.
3. The surgeon removes the clips from the neck wound and uses forceps to lift the loosely placed sutures to expose the common carotid arteries.
4. Once exposed, place the aneurysm clips to completely occlude blood flow in the carotids (*see* **Note 1**).
5. Tighten the suture thread exiting both sides of the neck to stop the collateral circulation and hold in place with a surgical clamp.
6. Monitor the rat. If global ischemia is successful (complete), the rat should be unconscious. It should display a "running response." Pupils should be dilated and there should be no corneal reflex, response to pain, or righting response (*see* **Note** 7).
7. Carry out the procedure on a draped heat pad and insert a rectal temperature probe.
8. Maintain the ischemia for 5–30 min, depending on the desired protocol, after which the ligature is released and the clips are removed from the carotids.
9. Cut the sutures around the carotids and remove.
10. Staple the neck wound and apply lidocaine cream.
11. Cut the surgical suture used to occlude the collaterals on both sides at the point at which it exits the skin.
12. Observe the rat until it fully recovers post 4-VO in a hot box at 30 °C.
13. Move the rat to telemetry and regulate for the time required for the specific experiment.
14. Observe the rat for the first hour of this period of telemetry to check that it is stable.

4 Notes

1. The rat may need to be returned to the anesthesia box and the procedure completed over a few attempts as the anesthesia may wear off.
2. Use skin tweezers to pinch the skin and hold well away from the abdominal cavity in order to avoid cutting into bowel.
3. Take care not to exert pressure when dissecting out the common carotids as this can impede the rat's ability to breathe.

4. Take care not to overstimulate the vagus nerve as this can have an adverse effect on the heart. The thin white sheath next to the pulsatile carotid artery is the vagus nerve [14].
5. Suction is needed to ensure a good view at all times. Brief intermittent application of electrical current avoids major hemorrhage. Take care not to over-cauterize. Settings that produce minimal local muscle contraction should be used, e.g., level 9.
6. Taping the sutures to the rats back before completing the cauterization step can impede view, so it is best carried out just before recovery from anesthesia. Autoclave tape is used and threads must be secured firmly to the fur to ensure the rat is unable to chew them.
7. The suture should be tightened according to the responsiveness of the rat. If it does not display features of ischemia, it should be tightened; if it stops breathing, it should be loosened immediately.

References

1. Pulsinelli WA, Duffy TE (1983) Regional energy balance in rat brain after transient forebrain ischemia. J Neurochem 40(5):1500–1503
2. The National Centre for the Replacement, Refinement and reduction of animals in research (NC3Rs) http://www.nc3rs.org.uk/landing.asp?id=2
3. Mhairi Macrae I (1992) New models of focal cerebral ischaemia. Br J Clin Pharmacol 34(4): 302–308
4. McBean DE, Kelly PA (1998) Rodent models of global cerebral ischemia: a comparison of two-vessel occlusion and four-vessel occlusion. Gen Pharmacol 30(4):431–434
5. Papadakis M, Hadley G, Xilouri M, Hoyte LC, Nagel S, McMenamin MM, Tsaknakis G, Watt SM, Drakesmith CW, Chen R, Wood MJ, Zhao Z, Kessler B, Vekrellis K, Buchan AM (2013) Tsc1 (Hamartin) confers neuroprotection against ischemia by inducing autophagy. Nat Med 19(3):351–357
6. Pulsinelli WA, Brierley JB (1979) A new model of bilateral hemispheric ischemia in the unanesthetized rat. Stroke 10(3):267–272
7. Eklöf B, Siesjö BK (1972) The effect of bilateral carotid artery ligation upon the blood flow and the energy state of the rat brain. Acta Physiol Scand 86(2):155–165
8. Pulsinelli WA, Buchan AM (1988) The four-vessel occlusion rat model: method for complete occlusion of vertebral arteries and control of collateral circulation. Stroke 19(7):913–914
9. Kirino T (1982) Delayed neuronal death in the gerbil hippocampus following ischemia. Brain Res 239(1):57–69
10. Pulsinelli WA, Brierley JB, Plum F (1982) Temporal profile of neuronal damage in a model of transient forebrain ischemia. Ann Neurol 11:491–498
11. Murakami K, Kondo T, Kawase M, Chan PH (1998) The development of a new mouse model of global ischemia: focus on the relationships between ischemia duration, anesthesia, cerebral vasculature, and neuronal injury following global ischemia in mice. Brain Res 780(2):304–310
12. Morikawa E, Ginsberg MD, Dietrich WD, Duncan RC, Kraydieh S, Globus MY, Busto R (1992) The significance of brain temperature in focal cerebral ischemia: histopathological consequences of middle cerebral artery occlusion in the rat. J Cereb Blood Flow Metab 12(3):380–389
13. Colbourne F, Li H, Buchan AM (1999) Indefatigable CA1 sector neuroprotection with mild hypothermia induced 6 hours after severe forebrain ischemia in rats. J Cereb Blood Flow Metab 19(7):742–749
14. Tulis DA (2007) Rat carotid artery balloon injury model. Methods Mol Med 139:1–30

Chapter 11

Induction of Cerebral Arteriogenesis in Mice

André Duelsner, Nora Gatzke, Anja Bondke Persson, and Ivo R. Buschmann

Abstract

Unilateral common carotid artery occlusion (CCAO) is a standardized method to initiate collateral artery growth (arteriogenesis) in mouse brain. After CCAO is induced, blood circulation in the circle of Willis is changed and increases shear stress, which triggers increased arterial diameter and improvements in cerebrovascular reserve capacity. Functional improvement can be quantified after experimentally induced stroke by external middle cerebral artery occlusion (MCAO). Stroke volume is evaluated by standard tetrazolium chloride (TTC) staining. Here, we describe in vivo methods of CCAO and MCAO in detail and also the evaluation of stroke volume by TTC staining.

Key words Arteriogenesis, CCAO, Blood flow, Stroke volume, MCAO, TTC

1 Introduction

Collateral artery growth (arteriogenesis) is defined as the outgrowth of preexistent collateral arterioles in their diameter to compensate for disturbed blood flow after stenosis or occlusion of a major artery [1, 2]. The altered blood flow through collaterals increases shear stress, the initiator of arteriogenesis. In contrast to angiogenesis, hypoxia and changes in vessel density do not play any role in arteriogenesis. In the case of stenosis, arteriogenesis occurs in arterioles that form preexistent bypasses to the stenotic artery, thus improving blood supply to the distal (hypoxic) area. Angiogenesis occurs in the distal area and improves blood distribution via increased capillary density [3].

Cerebral arteriogenesis can be evaluated in rodents by three vessel occlusions (3-VO) in rats [4, 5] and unilateral common carotid artery occlusion (CCAO) in mice [6]. The proof of concept is performed: (a) morphologically by postmortem angiography, e.g., with colorized latex milk, after maximal vasodilatation, and (b) functionally by measurement of cerebrovascular reserve capacity [4, 5]. Another functional verification is the combination

Richard Milner (ed.), *Cerebral Angiogenesis: Methods and Protocols*, Methods in Molecular Biology, vol. 1135,
DOI 10.1007/978-1-4939-0320-7_11, © Springer Science+Business Media New York 2014

of CCAO with experimentally induced stroke, because the improved reserve capacity as a result of collateral growth, determines stroke volume [6].

Here, we present the necessary methodological details, which were described in short previously: common carotid artery occlusion [6], external permanent middle cerebral artery occlusion (MCAO) [6, 7], and evaluation of stroke volume by tetrazolium chloride (TTC) staining.

2 Materials

1. Ketamine (100 mg/ml injectable solution).
2. Xylazine (2 % injectable solution).
3. Isoflurane.
4. Syringes and cannula.
5. Analgesia: buprenorphine.
6. Hair removal cream: conventional cream.
7. Hot plate.
8. Surgical instruments: scalpel, scissors, forceps, needle holder.
9. Cauterizer.
10. Drill.
11. Isotonic saline (0.9 % solution).
12. Alcohol.
13. Suture material: ETHIBOND EXCEL, Polyester, nonabsorbable, braided, 5/0 USP, 1 metric, 75 cm.
14. Blood flow measurements: Medtronic GmbH, Düsseldorf (Germany).
15. Paraformaldehyde (PFA): 4 % formalin solution.
16. Triphenyl tetrazolium chloride (TTC).
17. Blood flow measurements: Periflux system, Perimed AG, Järfälla, Sweden.

3 Methods

3.1 Common Carotid Artery Occlusion (CCAO)

1. Anesthetize the mice by intraperitoneal injection of ketamine and xylazine (100/10 mg/kg). If the mouse is sleeping, place a mask over the nose, and narcosis is maintained by inhalation of isoflurane in 30 % oxygen/70 % nitrous oxide. Position the mouse on its back on a warming plate. Protect the eyes with eye cream against dehydration (*see* **Note 1**).

2. Apply hair removal cream paramedially to the left neck area and the left temporal area to remove hair. Clear the skin and disinfect the area.
3. In preparation for blood flow measurements, minimally incise the skin at the left temporal area. The skull is prepared at this position.
4. To prepare for the CCA, the skin is incised minimally paramedially at the left side. Expose the left common carotid artery, and prepare and separate from the nerve at the ligation site (*see* **Note 2**).
5. To prepare for CCA occlusion, place a suture filament (USP 5/0) around the CCA and prepare a knot.
6. To make measurements of blood flow, carefully place the laser Doppler flow electrode directly at the left skull. Start to record ipsilateral blood flow at least 1 min before ligation to establish a baseline of blood flow (*see* **Note 3**).
7. One minute after starting the blood flow recording, ligate the suture filament at the CCA. Continue recording for another 10 min.
8. Close the subcutaneous and cutaneous wound sites separately with suture USP 5/0.
9. As part of postsurgical care, apply lidocaine analgesic cream to the wounds.

3.2 Middle Cerebral Artery Occlusion (MCAO)

Unilateral CCAO leads to verifiable collateral growth. To evaluate the reduction in stroke volume resulting from collateral growth, MCAO will be performed 7–10 days after CCAO:

1. As skull drilling is painful, the mouse is premedicated with analgesics by subcutaneous injection with 0.1 mg/kg buprenorphine 20 min before surgery.
2. General anesthesia is induced with 100 mg/kg ketamine/10 mg/kg xylazine by intraperitoneal injection. As MCAO is a very short method, wherefore a short narcosis by injection is optimal. Apply eye cream.
3. Remove the skin at the right temporal area and remove hair using depilation cream and then disinfect.
4. Minimally incise the skin and separate the muscles. Disinfect the skull at this position with alcohol and observe the MCA branches in a Y shape. At this position carefully drill a hole through the skull (*see* **Note 4**).
5. Within the hole carefully incise the pia mater.
6. At the point just proximal to the vessel branching, occlude the MCA by electrocoagulation. If the vessel is occluded, the upper part will turn white (*see* **Note 5**).

7. Close the muscle and cutaneous tissue separately with suture USP 5/0.
8. Following surgery, give buprenorphine (0.1 mg/kg) subcutaneously.

3.3 Quantification of the Ischemic Stroke Volume

Estimation of stroke volume is optimal at 24 h after MCAO. TTC staining is based on a redox reaction, in which tetrazolium chloride in its reduced condition forms the red dye formazan. Viable tissue is stained red, and necrotic tissue remains pale by comparison:

1. To harvest the brain, the mouse is anesthetized with 100 mg/kg ketamine/10 mg/kg xylazine and decapitated and the brain is harvested carefully and placed in cold saline.
2. A 2 % TTC solution is prepared by diluting tetrazolium powder in a phosphate buffer (pH 7.4). The solution should be kept at 37 °C and protected from light (*see* **Note 6**).
3. To prepare the brain sections for staining, the brain is put in an acrylic mouse brain slicer matrix set to 2.0 mm coronal slice intervals and is sectioned into 2 mm slices with a blade (*see* **Note** 7).
4. Incubate the brain slices in TTC solution at 37 °C for 20 min in a water bath. After 20 min aspirate the TTC solution with a pipette and fix the brains in 4 % paraformaldehyde (*see* **Note 8**).
5. To quantify the stroke infarct volume, transfer the brain slices into a culture plate filled with PBS. Under a microscope collect pictures of the slices at the same magnification. Calculate the stroke volume versus total brain volume for each slice and then calculate for the whole brain with conventional photo editing software.

4 Notes

1. Mouse feet should be secured by medical strips.
2. After incision carefully displace the tissue and glands sideways. Separate the musculus sternocephalicus (pars mastoidea et sternohyoideus). Caution: Do not cut the nerve. It should be separated by minimal irritation.
3. Do not move the workstation at this time, otherwise the electrode might move. Alternatively, the electrode can be secured with standardized medical glue.
4. The skull becomes warm by drilling. Therefore, saline at room temperature should be dropped onto the drilling position.
5. Caution: If the vessel is bleeding, press it carefully with a cotton stick for at least 10 s. If not only the vessel is electro coagulated, but rather brain tissue, this leads to a false-positive evaluation of stroke volumes.

6. The solution bottle should be coated with aluminum foil and kept in a water bath at 37 °C.
7. If the brain is warm, it cannot be cut into standard slices. Therefore, brains should be frozen for some minutes and then cut immediately.
8. For light protection again the bottle should be coated with aluminum foil.

Acknowledgements

We thank Prof. James Faber and his colleagues (The University of North Carolina at Chapel Hill School of Medicine) for professionally teaching us the methodological details of the external MCAO procedure.

References

1. Persson AB, Persson PB (2012) Let 'em grow: the Yin and Yang of vessel growth. Acta Physiol (Oxf) 204:466–468
2. Schaper W, Buschmann I (1999) Arteriogenesis, the good and bad of it. Cardiovasc Res 43: 835–837
3. Schirmer SH, van Nooijen FC, Piek JJ, van Royen N (2009) Stimulation of collateral artery growth: travelling further down the road to clinical application. Heart 95:191–197
4. Busch HJ, Buschmann IR, Mies G, Bode C, Hossmann KA (2003) Arteriogenesis in hypoperfused rat brain. J Cereb Blood Flow Metab 23:621–628
5. Duelsner A, Gatzke N, Glaser J, Hillmeister P, Li M, Lee EJ, Lehmann K, Urban D, Meyborg H, Stawowy P, Busjahn A, Nagorka S, Persson AB, Buschmann IR (2012) Acetylsalicylic acid, but not clopidogrel, inhibits therapeutically induced cerebral arteriogenesis in the hypoperfused rat brain. J Cereb Blood Flow Metab 32:105–114
6. Duelsner A, Gatzke N, Glaser J, Hillmeister P, Li M, Lee EJ, Lehmann K, Urban D, Meyborg H, Stawowy P, Busjahn A, Nagorka S, Bondke Persson A, Buschmann IR (2012) Granulocyte colony-stimulating factor improves cerebrovascular reserve capacity by enhancing collateral growth in the circle of willis. Cerebrovasc Dis 33:419–429
7. Clayton JA, Chalothorn D, Faber JE (2008) Vascular endothelial growth factor—a specifies formation of native collaterals and regulates collateral growth in ischemia. Circ Res 103: 1027–1036

Chapter 12

Vessel Painting Technique for Visualizing the Cerebral Vascular Architecture of the Mouse

Shea Hughes, Oleksandr Dashkin, and Richard Anthony DeFazio

Abstract

Vessel painting is a simple, cost-effective way to visualize the vascular architecture of the mouse brain and other organs. DiI is a lipophilic carbocyanine dye that binds to lipid membranes and is commonly used for tract tracing in the brain. After perfusion with PBS to remove the blood, perfusion with a special DiI solution allows direct staining of the vasculature. This step is followed by perfusion fixation and removal of the brain from the skull. Pial vessels can be directly imaged using a standard fluorescent microscope. To acquire images of the whole brain, a montage of images at different focal planes is assembled. Basic cerebral vascular anatomy is reviewed in the context of vessel painting, and examples are presented showing enhanced collateralization in a mouse model of metabolic syndrome. Vessel painting offers a cost-effective and efficient alternative to more complex approaches such as corrosion casting.

Key words Vessel painting, DiI, Cerebral arteries, Fluorescent microscopy, Anastomosis, Collateral, Middle cerebral artery

1 Introduction

The Vessel Painting technique was developed as a method to better visualize the 3D nature of blood vessel networks within the retina [1, 2]. Recent modifications to the approach enabled reliable staining of the cerebral vasculature [3, 4]. The technique depends on the lipophilic properties of carbocyanine dyes such as 1,1′-dioctadecyl-3,3,3′,3′-tetramethylindocarbocyanine perchlorate (DiI) to label cell membranes. DiI is a very bright, relatively photostable fluorescent molecule that permits robust imaging sessions for documentation of vessel structure using either standard fluorescence microscopy or multiphoton laser scanning microscopy. Vessel painting is achieved by intravascular cardiac perfusion with a solution of DiI, bringing the aqueous DiI solution into direct contact with the endothelial cells. The basic procedure is simple: Sacrifice the animal using an institutionally approved method (e.g., CO_2 asphyxiation or isoflurane overdose), cardiac perfuse

Richard Milner (ed.), *Cerebral Angiogenesis: Methods and Protocols*, Methods in Molecular Biology, vol. 1135,
DOI 10.1007/978-1-4939-0320-7_12, © Springer Science+Business Media New York 2014

with phosphate-buffered saline (PBS) to remove blood cells, perfuse with DiI solution to label the vessels, then perfusion fix the animal with 4 % formaldehyde. Visualization of the blood vessel network is achieved by using standard fluorescent microscopy and a 5× objective. Thus, the vessel painting technique can be an affordable and reliable approach to visualize the vascular architecture of the brain in animal models of diseases such as stroke and diabetes.

2 Materials

2.1 Reagents

1. DiI (1,1′-dioctadecyl-3,3,3′,3′-tetramethylindocarbocyanine perchlorate).
2. Sodium nitroprusside.
3. Heparin, sodium injection, 1,000 USP units/mL.
4. 10× phosphate-buffered saline (10× PBS).
5. 4 % paraformaldehyde.

2.2 Preparation of Stock Solutions

1. Prepare the DiI stock solution: We have used the original preparation as described by Li et al. [2]. Add 16.7 mL of 100 % ethanol to 100 mg of DiI. Allow the solution to dissolve slowly on a rocker; overnight is sufficient. If protected from light, this stock is stable for at least 1 year at room temperature.
2. Prepare the SNP stock solution: Add 0.75 mg/mL SNP to sterile-filtered PBS. Protect from light and refrigerate at 3–4 °F. Make this solution on a weekly basis for maximum consistency.
3. Diluent solution: For 50 mL of the diluent, dissolve 2 g of glucose in half amount of distilled water, add 1 mL of PBS 10×, and adjust volume to 50 mL. Filter through a 0.22 m bottle-top filter. Larger quantities can be made in advance. This solution is stable at 3–4 °F for several months.

2.3 Apparatus

1. Two 50 mL conical tubes.
2. PE-50 tubing.
3. Tygon lab tubing, 1/32″ ID × 3/32″ OD.
4. Three 3-way stopcocks (cat# WU-30600-02, Cole-Parmer).
5. 22 gauge Luer stub adapters (to connect tubing to stopcocks; cat# 724439, Harvard Apparatus, Holliston, MA).
6. Male-to-male Luer lock connectors (cat# 12090, Qosina, Edgewood, NY).
7. Pressure gauge and bulb from blood pressure cuff.
8. Pressurized air tank (a 2 L empty plastic bottle with airtight cap will suffice).

9. 27 gauge Luer stub (beveled at 45° for cardiac insertion).
10. 30 gauge needle.
11. 1 mL syringes.
12. 10 mL syringes.
13. Syringe pump (Harvard Pump II Plus single syringe; Harvard Apparatus, Holliston, MA).
14. Basic forceps.
15. Iridectomy scissors (to cut the skull during brain removal).
16. Small hemostat.
17. Blunt dissection scissors.
18. Clamp applying forceps. Alternative tools for clamping the cardiac perfusion needle can be utilized.
19. Standard fluorescence microscope with 4–5× objective compatible with fluorescence, a tetramethylrhodamine excitation/emission filter set, a digital camera for imaging, and image acquisition and manipulation software (e.g., ImageJ or Adobe Photoshop).

3 Methods

3.1 Assembly of Equipment

Figure 1 shows the arrangement of each of the components:

1. Drill one hole in the cap of the air pressure tank (2 L bottle) to accept the tubing from the blood pressure bulb and the pressure gauge. The fittings must be airtight.
2. It may be necessary to seal the tubing connection to the cap using a small amount of aquarium sealant.
3. Drill a 1/16″ hole in the cap and insert a short length of Tygon tubing. This should be a very secure fit, both between the cap and Tygon tubing, as well as between the Tygon tubing and PE50 tubing.
4. Insert a length of PE50 tubing to connect the air pressure tank to the first 50 mL conical tube.
5. Put a stopcock using the Luer stub adapters and male-to-male connectors in this line to seal off the pressure tank from the rest of the apparatus during loading the solutions.
6. Drill three 1/16″ holes in the cap of the first conical tube and insert short lengths of Tygon tubing into each hole. Label this conical tube “PBS.”
7. Insert the connection from the air pressure tank. Insert a longer PE50 length, reaching to the bottom of the conical tube and extending to the stopcock assembly (approximately 25 cm, but this depends on how the apparatus is arranged). The third connection goes to the next conical tube.

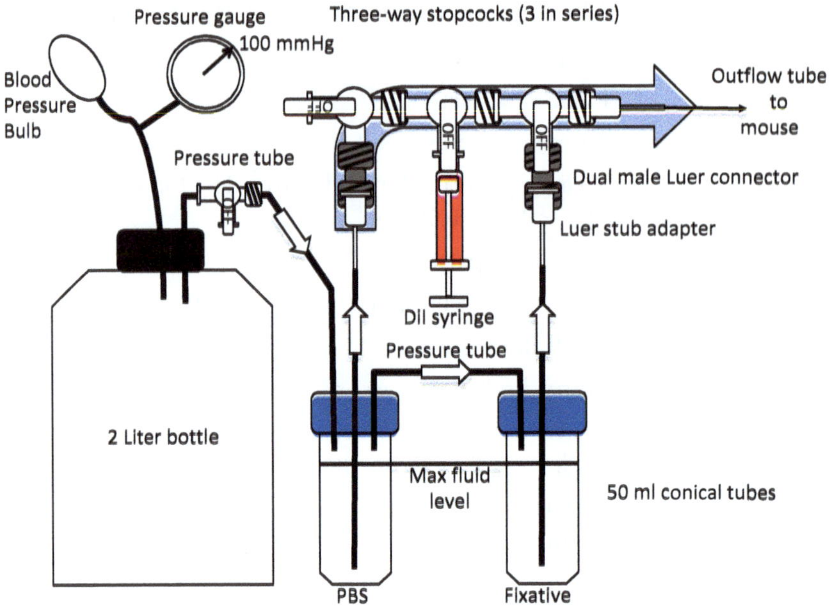

Fig. 1 The experimental apparatus consists of an air tank with pressure gauge, tubes connecting the perfusate to the stopcock assembly, and an outflow port. The blood pressure bulb and gauge are from an inexpensive aneroid blood pressure kit. Pressure in the tank is alternately delivered to the PBS to flush blood from the mouse. DiI is perfused via a syringe pump. Finally, pressure is used to deliver fixative. The route of flow is determined by the stopcock positions (illustrated in Fig. 2)

8. Drill two 1/16″ holes, insert Tygon tubing, and connect the pressure tube from the PBS conical tube.
9. Insert a longer PE50 length, reaching to the bottom of the tube and extending to the stopcock assembly (again about 25 cm). Label the second conical tube “FIX.”
10. Assemble the three stopcocks as shown in Fig. 1. Using the dual male Luer connectors, connect 22 gauge Luer stub adapters to the first and third ports of the stopcock assembly (as shown in Fig. 1). The two PE50 tubes from the conical tubes should fit snugly over the Luer stubs.
11. To assemble the perfusion tube and needle assembly, attach the final Luer stub to the outflow port as indicated in the figure and connect a length of PE50 tubing. This final PE50 tube will carry the experimental reagents to the mouse’s heart during the cardiac perfusion. The PE50 tube should be a bit longer (30–40 cm) to permit some flexibility in arranging the apparatus to ensure reliable perfusion and comfort of the experimentalist.
12. Attach the beveled 27 gauge Luer stub (for cardiac puncture and perfusion) to the PE50 tube using a 22 gauge Luer stub at the end of the PE50 tubing and a dual male-to-male Luer lock connector.

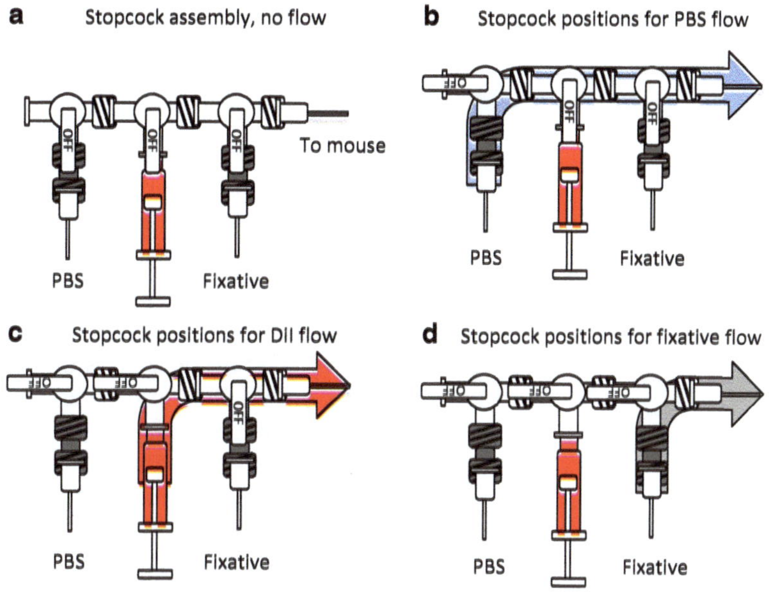

Fig. 2 Stopcock assembly positions for different stages of the experiments. (**a**) The "no flow" position setting is used during setup of the perfusion solutions and pressurization of the system. It is also useful for testing the system for leaks. (**b**) "PBS flow" position is the first setting and permits PBS to flow to the animal under constant pressure. (**c**) "DiI flow" position transfers flow control to the syringe pump to deliver DiI. (**d**) "Fix flow" position switches back to pressurized flow for the final perfusion step

13. Verify that the system is airtight by sealing all stopcocks in the assembly to the "no flow" position (Fig. 2a), and with the PBS and FIX 50 mL conical tubes attached to the caps, first, with the blood pressure bulb and pressure gauge, pressurize the air tank assembly with the pressure tube stopcock in the off position (pointing toward the tank). Confirm that the tank holds 100 mmHg pressure.
14. With the stopcock assembly in the "no flow" position, open the pressure tube stopcock as shown in Fig. 1 to allow pressure to the stopcock assembly. Confirm that the entire assembly is airtight and holds pressure at 100 mmHg.

3.2 Immediately Prior to Sacrifice

1. Prepare a single bolus injection consisting of SNP (final dose, 0.75 mg/kg) and heparin (final dose, 2,000 U/kg). Load into a 1 mL syringe with a 30 gauge needle for cardiac injection.
2. Prepare the DiI working solution. Immediately before use, add 20 μL of the DiI stock solution to 180 μL of 100 % ethyl alcohol in a 15 mL conical tube. Add 10 mL of diluent solution and mix by vigorous shaking. Load the DiI working solution into a 10 mL syringe. A small bowl made from a 50 mL conical tube cut in half can facilitate loading the dye into the 10 mL syringe.

3.3 Procedure

1. Prepare 15 mL conical tubes to store the mouse brains after vessel painting and dissection. If many mice are to be sacrificed, it is advantageous to label the tubes in advance. Add 10 mL of 4 % fixative to each tube and seal the cap.
2. With the stopcock from the pressure tank blocking the flow to the stopcock assembly, pressurize the tank to 100 mmHg using the blood pressure bulb and the pressure gauge (*see* **Note 1**).
3. Load the PBS and FIX 50 mL conical tubes (at least 30 mL each tube for 1–2 mice) and attach the caps with the perfusion tubes as shown in Fig. 1. Set all the stopcocks in the stopcock assembly to the "no flow" position indicated in Fig. 2a. Attach the perfusion needle and tube assembly to the stopcock assembly (use the Luer stub adapter marked "outflow tube to mouse" in Fig. 1). Open the pressure tube stopcock (as in Fig. 1) with the stopcock assembly in the "no flow" position (*please revisit* **Note 1**!).
4. *Flush the lines*: Fill the DiI syringe with the DiI working solution (~10 mL). Flush a small amount of DiI working solution from the syringe to remove any air in the tip of the syringe. Confirm the pump is set for 1 mL/min flow. Mount on the syringe pump and attach the stopcock assembly to the DiI syringe. Open the pressure tube stopcock (Fig. 1). Cycle through the stopcock assembly positions (Fig. 2b–d, in order) to flush a small amount of PBS, DiI working solution (activate the syringe pump briefly to allow flow of the dye into the assembly), and Fix through the lines to remove all air bubbles. Repeat the PBS flow position (Fig. 2b) to flush the remainder of DiI and Fix from the lines (*see* **Note 2**).
5. Sacrifice the mouse using methods approved by the local and national animal care and use authorities. This is usually CO_2 asphyxiation or isoflurane overdose. Similar results can be obtained with an overdose of injectable anesthetic.
6. When the animal has stopped breathing, open the abdomen below the diaphragm. Carefully cut the diaphragm to reveal the chest cavity. Open the chest laterally on each side, being very careful not to cut the lungs or the heart. Clamp the chest at the sternum with a small hemostat and fold toward the head to expose the heart (*see* **Note 3**).
7. Inject the heparin-SNP bolus into the left ventricle using a 30 gauge needle and a 1 mL syringe.
8. Open the pressure tube stopcock and set the stopcock assembly for PBS flow (Fig. 2b). PBS should flow from the outflow tube and there should be no air bubbles. Insert the cardiac perfusion needle into the left atrium via the left ventricle, and clamp with the clip tool (*see* **Note 4**).

9. Carefully cut the right ventricle to allow outflow of the perfusate (*see* **Note 5**).
10. Perfuse with 5 mL of PBS (Fig. 2b, PBS flow position).
11. Activate the syringe pump and quickly switch the stopcock assembly to the DiI flow position (Fig. 2c; *see* **Note 6**).
12. Perfuse with 10 mL of fixative by switching to the Fix flow position (Fig. 2d). Clear fixative should flow from the right ventricle and the mouse's tail and limbs should become stiff (*see* **Note 7**).

3.4 Removing the Brain

After fixation, cut the head off with scissors, remove the skin from the skull, and break the skull between the eyes. Using the iridectomy scissors, cut the skull in the cerebellum area, then from rostral to caudal on both sides at a distance of about 3–5 mm, then from rostral to caudal along the medial line. Use fine forceps to open the skull to the sides. Scoop out the brain using a spatula or similar device and place in the appropriately labeled 15 mL tube containing 4 % fixative (*see* **Notes 8** and **9**).

3.5 Cleanup

Flush the perfusion lines with deionized water followed by 75 % ethanol, then pressurized air to remove remaining liquid. Thoroughly rinse all conical tubes that you plan to reuse. Keep everything clean, dry, and free of dust.

4 Notes

1. Verify that the pressure is stable and that there are no leaks. If the pressure is not stable, check that the connections and fittings are tight and that the cap on the pressure tank is tight. It may be necessary to seal the fittings with aquarium sealant (WARNING: This can take several hours to cure, so test for leaks in advance). Do not proceed if the pressure is not stable. Locate and repair all leaks before proceeding.
2. Confirm that there are no air bubbles visible in the flow lines prior to sacrifice! Repeat the flushing process if bubbles are detected. If this step is troublesome, this can be practiced in advance using water in the PBS and Fix 50 mL tubes.
3. If the lungs are cut/bleeding, it can be very difficult to obtain adequate perfusion of the brain. Continuing after this mistake is unlikely to produce good vessel painting.
4. Although the purpose of this tool is very different, we found the vessel clip application tool is a good device to keep a seal between the heart and the perfusion needle during pressurized perfusion. The handle of the tool can be placed on a spare cap to adjust the height and angle of the perfusion to optimize

perfusion. Alternative approaches to stabilizing the perfusion needle may be equally effective.

5. Blood should flow rapidly from the right ventricle. After a minute or two, blood should run clear to PBS. If PBS is not observed flowing from the right ventricle, double check the position of the perfusion needle, especially for leakage at the site of cardiac puncture. It is possible to reposition the perfusion needle after perfusion has begun, but this must be done quickly to avoid complications and incomplete vessel painting. During PBS perfusion, the nose, ears, and paws should become pale as blood is flushed from the animal.
6. The red dye should be visible in the perfusion line and ultimately outflow from the right ventricle. The nose, ears, and paws should now become pink due to perfusion of the dye. Perfuse at least 9 mL of the DiI solution, but avoid introducing bubbles into the line if there is an air gap in the DiI syringe. Stop the syringe pump before it reaches its maximum position.
7. A major delay between the sacrifice of the animal and the perfusion of the dye will result in poor labeling. After fixation, time can be taken to carefully remove the brain from the skull.
8. It is very easy to damage the pial vessels during removal of the brain. It is advisable to be very patient during this process and take due care to minimize any contact with the brain. Polishing the edges and points of the forceps and spatula can help avoid damage to the surface vessels. After removal of the brain, vessel painting can be rapidly confirmed by visualizing the brain through the 15 mL conical tube at low magnification on the fluorescence microscope.
9. *Troubleshooting incomplete staining and other complications*: Poor dissection technique while removing the brain damages the pial arteries. This will appear as loss of fluorescence with very sharp demarcation where surgical tools have removed the surface vessels. It is crucial to remove the brain with extreme care if the pial arteries are of interest. Once the brain is fixed, the fluorescent labeling is effectively stable in room light for several hours. This allows very careful dissection with no time constraints. If the dorsal surface of the brain is the target, more rapid dissection can proceed from the ventral aspect of the skull. However, it should be noted that with due care, the brain can be removed with minimal damage to the surface arteries. This allows complete tracing of the middle cerebral artery and its branches from the Circle of Willis to its termination in the anterior and posterior cerebral artery territories. Particulates in the perfusate will block perfusion of the microvasculature. Careful attention to preparation of the working solution is absolutely necessary to ensure solubilization of the dye.

Initial attempts using this technique using the original approach of syringes to manually deliver the perfusate resulted in occasional ruptures in the cerebral vasculature. Our belief that this was due to too much pressure led us to use the pressurize air tank to deliver a constant, repeatable perfusion.

10. *Imaging and documentation*: Use of a widefield fluorescence microscope is highly recommended, if such a device is available. This technology can permit visualization of large fluorescent specimens with a single image. However, a standard fluorescence microscope using a 4–5× fluorescence objective can be used to generate composite images of the surfaces of the brain (as shown in Figs. 3 and 4) as well as brain sections [4]. For the dorsal surface, this requires taking 3–4 images at different focal planes for each hemisphere of the brain. The resulting 6–8 images are cropped to remove the out-of-focus regions, then "stitched" together in an image manipulation software, e.g., Adobe Photoshop. If Photoshop or similar software is unavailable, the free NIH ImageJ software package [5] can be used to crop images, which in turn can be assembled for figures in Microsoft PowerPoint. ImageJ also contains many quantitative tools that can permit measurement of the physical characteristics of the vasculature (e.g., see the excellent work of J.E. Faber's group [6]).
11. *Applications of vessel painting*: We have developed vessel painting to quantify the number of collaterals in mouse strains commonly used in experimental models of stroke [3]. An example is shown in Fig. 3. Figure 3a illustrates a cartoon of the Circle of Willis in the C57BL/6J mouse traced from a vessel-painted brain. Figure 3b shows the ventral aspect of a vessel-painted C57BL/6J mouse brain. Figure 3c displays the dorsal aspect, with labels indicating the anterior cerebral artery (ACA) branches, middle cerebral artery (MCA), and posterior cerebral artery (PCA). In Fig. 4, we asked if metabolic syndrome has an effect on the cerebral vasculature. We chose the TALLYHO/JngJ polygenic mouse model [7] to test this hypothesis. In Fig. 4a, we show the vessel-painted brain of an SWR/J mouse (identified as "control" for the TALLYHO model by Jackson Laboratories). The mouse strain has a moderate number of collaterals, compared to BALB/C mice (which have only 5–6 between the dorsal pial arteries) and C57BL/6J (which have >20 between the dorsal pial arteries). An adult TALLYHO mouse shows tremendous collateralization (Fig. 4b). Future studies are required to determine the developmental time course of collaterals relative to the onset of symptoms of metabolic syndrome (hyperinsulinemia, hyperlipidemia, and hypertension). Similar findings were obtained using alternative techniques in diabetic rats [8]. Interestingly,

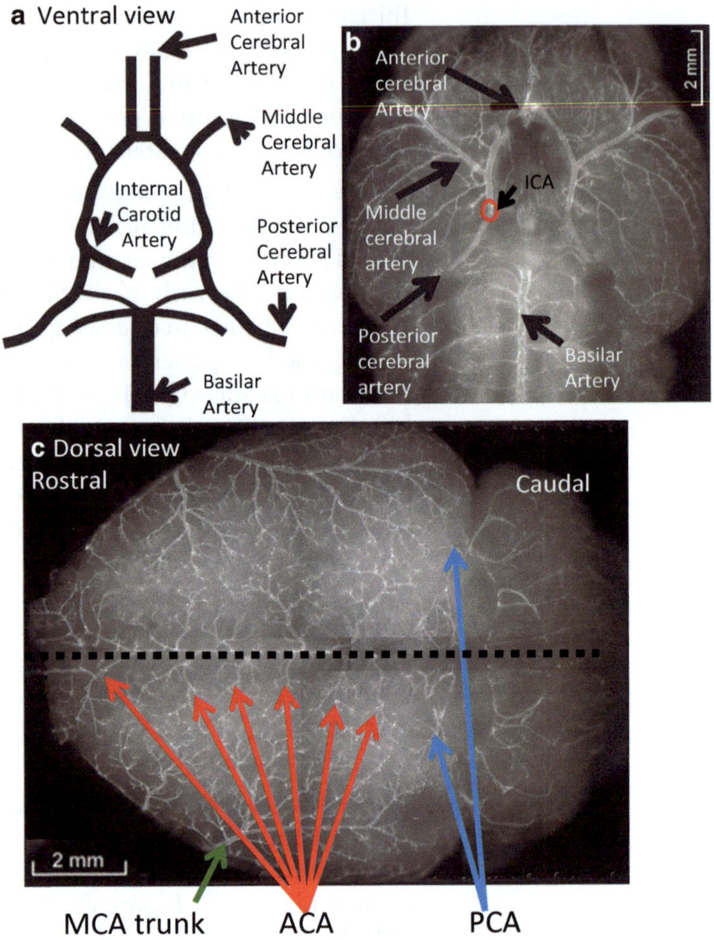

Fig. 3 Basic cerebral vascular anatomy revealed with vessel painting. (**a**) A cartoon of the cerebral vascular architecture viewed from the ventral aspect. These lines are traced from actual vessel painting specimens. The relative diameters and positions are matched to the diameter of the vessels in a single C57BL6/J specimen (shown in **b**). (**c**) The dorsal aspect of the mouse cerebral vascular system. The anterior cerebral artery (ACA), middle cerebral artery (MCA), and posterior cerebral artery (PCA) are identified with *arrows*

in our small study of six mice, not all TALLYHO exhibited hyperglycemia ($n=4$ of 6), yet all showed development of collaterals.

12. *Weaknesses*: DiI preferentially labels arteries, leaving veins only weakly stained. This can be advantageous as it simplifies identification of major arteries and arterioles; however, as a vascular tool it fails to reveal the "other half" of the cerebral vascular system, i.e., the veins and venules. Additional weaknesses arise when attempting to combine this technique with other methods. For example, DiI dissolves in alcohol—a major

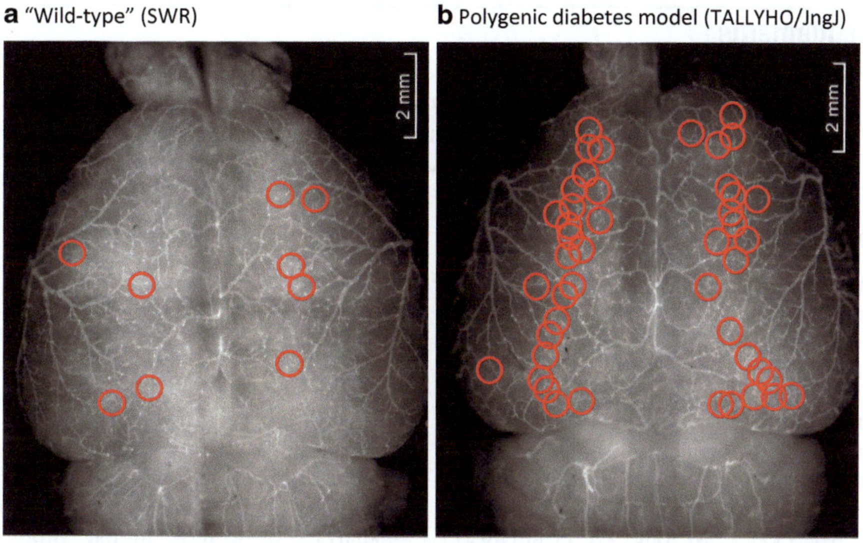

Fig. 4 Comparison of collaterals between ACA, MCA, and PCA territories in a control mouse strain (SWR/J) and a polygenic model of metabolic syndrome (TALLYHO/JngJ). *Red circles* indicate collaterals identified in the two mouse strains

disadvantage if trying to combine vessel painting with immunohistochemical analysis as many standard protocols rely on alcohol dehydration. It may be possible to modify existing immunohistochemistry protocols to maintain the DiI staining of the vasculature.

Additionally, we attempted to combine vessel painting with 2,3,5-triphenyltetrazolium chloride (TTC). TTC is a redox indicator technique that reveals infarct volumes after ischemic stroke. Living cells convert TTC to a red stain, while infarcted tissues lacking cellular respiration fail to convert the stain. The infarct appears as a pale region within the dark red staining of the healthy tissue. TTC can be delivered via the vasculature to stain the brain in an intact animal. When TTC was included in the initial PBS perfusion, vessel painting failed due to severe vasoconstriction. We also tried TTC staining *after* vessel painting by skipping the perfusion of fixative with inconsistent results. This was likely due to the ischemic death of many neurons during the extended time it takes to carefully remove the brain from the skull without damaging the pial vessels.

13. *Application to other organs*: In our hands, cardiac perfusion is the most efficient method to deliver DiI to the brain for vessel painting. However, if other organs are the experimental target, the damage to the chest and heart during cardiac perfusion may limit the applicability of the approach. Alternative approaches to perfusion fixation, e.g., via the femoral vein or direct ex vivo perfusion of individual organs, could be optimized for the purpose of vessel painting the heart and other organs.

Acknowledgements

Supported by the University of Miami Scientific Awards Committee pilot grant mechanism (RAD). We thank Professor Rong Wen at the University of Miami for his assistance in establishing the technique for use in the brain.

References

1. Ravnic DJ, Jiang X, Wolloscheck T et al (2005) Vessel painting of the microcirculation using fluorescent lipophilic tracers. Microvasc Res 70:90–96
2. Li Y, Song Y, Zhao L et al (2008) Direct labeling and visualization of blood vessels with lipophilic carbocyanine dye DiI. Nat Protoc 3:1703–1708
3. Defazio RA, Levy S, Morales CL et al (2011) A protocol for characterizing the impact of collateral flow after distal middle cerebral artery occlusion. Transl Stroke Res 2:112–127
4. Moy AJ, Wiersma MP, Choi B (2013) Optical histology: a method to visualize microvasculature in thick tissue sections of mouse brain. PLoS One 8:e53753
5. Schneider CA, Rasband WS, Eliceiri KW (2012) NIH Image to ImageJ: 25 years of image analysis. Nat Methods 9:671–675
6. Chalothorn D, Faber JE (2010) Formation and maturation of the native cerebral collateral circulation. J Mol Cell Cardiol 49:251–259
7. Kim JH, Sen S, Avery CS et al (2001) Genetic analysis of a new mouse model for non-insulin-dependent diabetes. Genomics 74: 273–286
8. Elgebaly MM, Prakash R, Li W et al (2010) Vascular protection in diabetic stroke: role of matrix metalloprotease-dependent vascular remodeling. J Cereb Blood Flow Metab 30: 1928–1938

Chapter 13

Examining Cerebral Angiogenesis in Response to Physical Exercise

Kiersten L. Berggren, Jacob J.M. Kay, and Rodney A. Swain

Abstract

Capillary growth and expansion (angiogenesis) is a prerequisite for many forms of neural and behavioral plasticity. It is commonly observed in both brain and muscle of aerobically exercising animals. As such, several histological methods have been used to quantify capillary density, including perfusion with India ink, various Nissl stains, and immunohistochemistry. In this chapter, we will describe these histological procedures and describe the stereological analysis used to quantify vessel growth in response to aerobic exercise.

Key words Exercise, Angiogenesis, Immunohistochemistry, Stereology, Nissl staining, India ink

1 Introduction

In recent years, research regarding the effects of exercise on the brain has become increasingly popular. This interest in exercise research is driven partly by findings that exercise results in improved performance in various cognitive tasks in both humans [1–3] and nonhuman animals [4–6]. Furthermore, exercise has been shown to attenuate cognitive decline in older populations [7, 8] and is neuroprotective following brain injury [9–11] and stroke [12, 13].

The mechanisms by which exercise exerts these effects have also been a focus of a large body of research. Several studies have shown that exercise causes significant angiogenesis in the brains of nonhuman animals [14–20]. Findings from our laboratory show that exercise increases blood vessel density in both the cerebellum [18] and hippocampus [21] and that the high-affinity VEGF (vascular endothelial growth factor) receptors are upregulated differentially in response to exercise in the cerebellum and motor cortex [22, 23]. Sikorski et al. [18] found that feeding animals a diet containing conjugated linoleic acid (CLA), rather than standard rat chow, inhibited exercise-induced angiogenesis in the cerebellum. This is relevant because research has shown that increases in

Richard Milner (ed.), *Cerebral Angiogenesis: Methods and Protocols*, Methods in Molecular Biology, vol. 1135,
DOI 10.1007/978-1-4939-0320-7_13, © Springer Science+Business Media New York 2014

capillary density are involved in improved cognitive performance following exercise. Specifically, another recent study from our laboratory demonstrated that blocking angiogenesis resulted in a more severe learning deficit in the Morris water maze (MWM) compared to inhibition of neurogenesis [21].

Here, we outline several methods by which exercise-induced angiogenesis can be quantified in the rat brain. Simple procedures such as the incorporation of India ink during perfusion can be used to visualize capillaries. Furthermore, Nissl stains such as cresyl violet, safranin O, and toluidine blue can be used to accentuate the contrast between brain tissue and vasculature. Additionally, immunohistochemical stains with antibodies against vascular proteins, such as the $\alpha_v\beta_3$ integrin (CD61) and the rat endothelial cell antigen 1 (RECA-1), can be applied to tissue to identify regions of the brain where angiogenesis is occurring. Finally, quantification of angiogenesis can be achieved by way of stereological analysis. It is important to note that most of the visualization techniques mentioned above stain all vasculature generally, so angiogenesis is operationally defined by apparent increases in capillary density or diameter relative to those levels expressed in non-exercising control animals.

2 Materials

All solutions needed for successful transcardial perfusion, Nissl staining, and immunohistochemistry are listed below. Unless otherwise specified, all solutions should be prepared with deionized water (dH_2O) and should be stored as directed. Working solutions should be used for all procedures.

2.1 Solutions for Perfusion

Several different combinations of solutions can be used to perfuse animals. Here we will limit our description to the most commonly used method in our laboratory—working phosphate buffer (PB) followed by 4 % paraformaldehyde. We will also describe perfusion with India ink.

2.1.1 Stock and Working PB and Paraformaldehyde

1. Stock PB: For 4,000 mL, dissolve 428.8 g sodium phosphate dibasic heptahydrate ($Na_2HPO_4 \cdot 7H_2O$) in 3,200 mL dH_2O. After full incorporation, bring volume to 4,000 mL. Dissolve 55.2 g sodium phosphate monobasic monohydrate ($NaH_2PO_4 \cdot H_2O$) in 800 mL dH_2O. After full incorporation bring volume up to 1,000 mL. Mix 3,240 mL of dibasic solution with 760 mL of monobasic solution, add 0.4 g sodium azide (NaN_3; antibiological agent), and adjust the solution's pH to 7.3–7.4. Store at room temperature.
2. Working PB: Dilute one part of stock PB in three parts dH_2O. Store at room temperature.

3. Stock 4 % paraformaldehyde: For 1,000 mL, heat 800 mL dH_2O to 60 °C, and add and dissolve 80 g 95 % paraformaldehyde powder—[$OH(CH_2O)_nH$]. Add several drops of 10 N sodium hydroxide (NaOH) until the solution becomes clear (*see* **Note 1**). Cool the solution in an ice bath until it reaches room temperature. Filter the solution with number 4 qualitative filter paper, and add dH_2O to bring the final volume to 1,000 mL. Adjust the pH to 7.4–7.6. Store the solution at 4 °C.
4. Working 4 % paraformaldehyde: Dilute one part stock paraformaldehyde in one part stock PB solution. Store the solution at 4 °C.

2.1.2 India Ink Incorporation

1. Stock phosphate buffered saline (PBS) with azide: For 1,000 mL, dissolve 41.30 g sodium phosphate dibasic heptahydrate ($Na_2HPO_4{\cdot}7H_2O$) and 5.57 g sodium phosphate monobasic monohydrate ($NaH_2PO_4{\cdot}H_2O$) in 800 mL dH_2O. After complete incorporation bring volume up to 1,000 mL with dH_2O. Add 1 g of sodium azide per 1,000 mL stock solution.
2. Working PBS with azide: Combine 50 mL PBS stock solution and 800 mL dH_2O and stir. Add 9 g sodium chloride (NaCl) and dissolve completely. Add dH_2O to bring the volume to 1,000 mL and adjust the pH to 7.3–7.4.
3. 10 % formalin: Dilute formaldehyde solution to a 10 % concentration.
4. India ink.

2.2 Solutions for Histological Stains

Histological tissue staining requires several alcohol solutions of different concentrations and the actual solutions for the stains themselves. See below for recipes for stains including cresyl violet, toluidine blue, and safranin O.

1. Alcohol solutions: Dilute 100 % ethyl alcohol to various concentrations (50, 70, and 95 %).
2. Cresyl violet: Mix 1 g high purity cresyl violet acetate ($C_{18}H_{15}N_3O_3$), 10 mL glacial acetic acid, and 990 mL dH_2O. Stir and heat on a low setting for several hours. After the solution has cooled to room temperature, filter the portion needed with number 4 qualitative filter paper. Store the remaining/unfiltered stain for later use.
3. Safranin O: For the stock solution, dissolve 2.5 g safranin O ($C_{20}H_{19}ClN_4$) in 100 mL 95 % ethyl alcohol. To prepare the working solution, mix 10 mL stock solution with 90 mL dH_2O.
4. Toluidine blue: Dissolve toluidine blue ($C_{15}H_{16}ClN_3S{\cdot}0.5ZnCl_2$) in dH_2O to a final concentration of 1 %.
5. Xylene or xylene substitute.

6. Glacial acetic acid.
7. Multi-slide carriages.

2.3 Immunohistochemistry

2.3.1 Products and Materials

1. Anti-CD61 ($\alpha_v\beta_3$ integrin) antibody (AbD Serotec, Raleigh, NC).
2. Anti-RECA1 antibody (AbD Serotec, Raleigh, NC).
3. Secondary antibody: Select according to the host species of the primary antibody.
4. ABC Kit: Use a peroxidase-based ABC Kit (e.g., Vectastain ABC Kit—Standard, Vector Laboratories, Burlingame, CA). Prepare according to manufacturer guidelines.
5. 3-3′-diaminobenzidine (DAB, $C_{12}H_{14}N_4$).
6. Perforated and non-perforated multi-well plates.
7. Vessels for washes (e.g., pie plates, disposable trays, etc.)

2.3.2 Solutions

1. Stock PBS: *See* Subheading 2.1.2 above. Two stock PBS solutions should be prepared, one with sodium azide (as described above) and one without sodium azide (prepare the same way; just skip the azide addition step).
2. Working PBS: *See* Subheading 2.1.2 above.
3. 0.3 % hydrogen peroxide (H_2O_2) solution: For 200 mL, mix 2 mL 30 % H_2O_2 in 198 mL working PBS with azide.
4. 2 % serum solutions: For 200 mL, mix 4 mL serum (host will depend on primary and/or secondary antibody hosts; *see* **Note 2**) with 196 mL working PBS with or without azide (depending on the point in the immunohistochemistry protocol).
5. Triton X: Dilute Triton X ($C_2H_4O]_nC_{14}H_{22}O$) to a concentration of 10 % in dH_2O.
6. Blocking solution: 10 % serum, 0.5 % Triton X, and working PBS with azide.
7. Primary antibody solution: 1 % serum, 0.5 % Triton X, and primary antibody (dilution will vary depending on antibody concentration, manufacturer suggestions, and experimental experience; *see* **Note 3**). Add working PBS with azide to volume.
8. Secondary antibody solution: 10 % serum and 0.5–1 % secondary antibody (*see* **Note 2**), bring to volume with PBS with azide.
9. Tris buffer: For 1,000 mL, dilute 6.06 g Trizma HCl and 1.39 g Trizma base in 800 mL dH_2O. Once dissolved, bring volume up to 1,000 mL with dH_2O and adjust the pH to 7.3–7.4.
10. DAB: For 200 mL, dissolve 100 mg DAB (a colorimetric stain) in 150 mL Tris buffer. Dissolve 1.39 g nickel ammonium sulfate hexahydrate ($H_8N_2NiO_8S{\cdot}6H_2O$) in 50 mL Tris buffer.

After both mixtures have gone into solution, add the nickel ammonium sulfate solution to the DAB solution (*see* **Note 4**). Before using the mixture, it must be filtered with number 4 qualitative filter paper. (DAB is a carcinogen and extreme care should be taken to properly dispose of leftover solutions; *see* **Note 5**.)

11. Ethanol solutions in dH_2O: 70, 95, and 100 %.
12. Xylene or xylene substitute.

3 Methods

3.1 Perfusion

3.1.1 Normal Perfusion

1. Euthanize the animal according to standard laboratory protocol (e.g., submersion in a chamber filled with CO_2 or isoflurane vapor).
2. Transcardially perfuse the animal with 200 mL working PB to drain blood, followed by 400 mL 4 % paraformaldehyde (*see* **Note 6**).
3. Carefully extract the brain and postfix overnight in working paraformaldehyde. If obtaining frozen sections, cryoprotect the tissue in a solution of 30 % sucrose in working phosphate buffer.

3.1.2 Perfusion with India Ink (Adapted for Rats from Yu et al. [24]; See Fig. 1)

1. Euthanize the animal according to standard laboratory protocol (e.g., submersion in a chamber filled with CO_2 or isoflurane vapor).
2. Transcardially perfuse the animal with 200 mL working PB to drain blood, followed by 400 mL 10 % formalin. Follow the formalin with 120 mL India ink.
3. Carefully extract the brain and postfix overnight in working paraformaldehyde. If obtaining frozen sections, cryoprotect the tissue in a solution of 30 % sucrose in working phosphate buffer.

3.2 Staining

Prior to staining, tissue should be mounted on gelatin-coated slides and allowed to dry overnight. The procedures described here can be completed using multi-slide carriages. Please *see* Figs. 1 and 2, for examples, of toluidine blue and cresyl violet tissue staining.

3.2.1 Cresyl Violet/ Safranin O

1. Defat and rehydrate the tissue with the following rinses: 100 ethanol (3–5 min), Histoclear/xylene (5 min), 100 % ethanol (3–5 min), 70 % ethanol (3–5 min), 50 % ethanol (3–5 min), and dH_2O (3–5 min).
2. Place the tissue in cresyl violet or safranin O solution for 30 s–5 min (*see* **Note 7**). Dip into two different dH_2O washes for just a few seconds each.

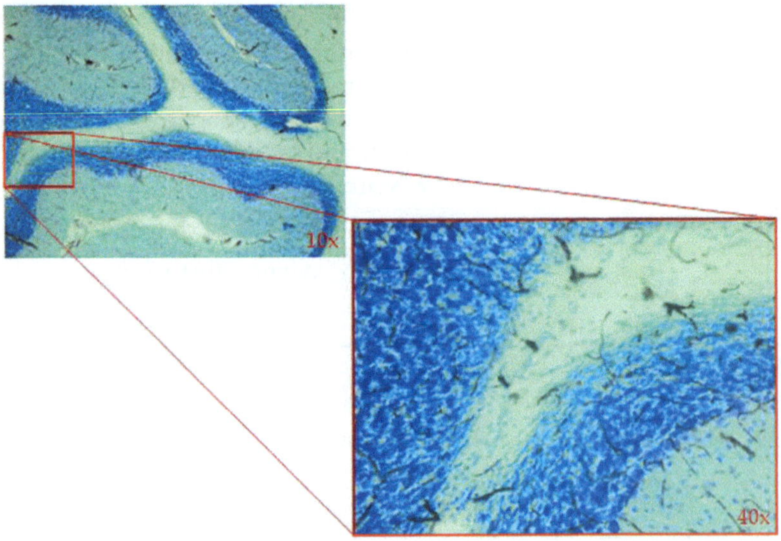

Fig. 1 A cerebellar slice with India ink and a toluidine blue counterstain

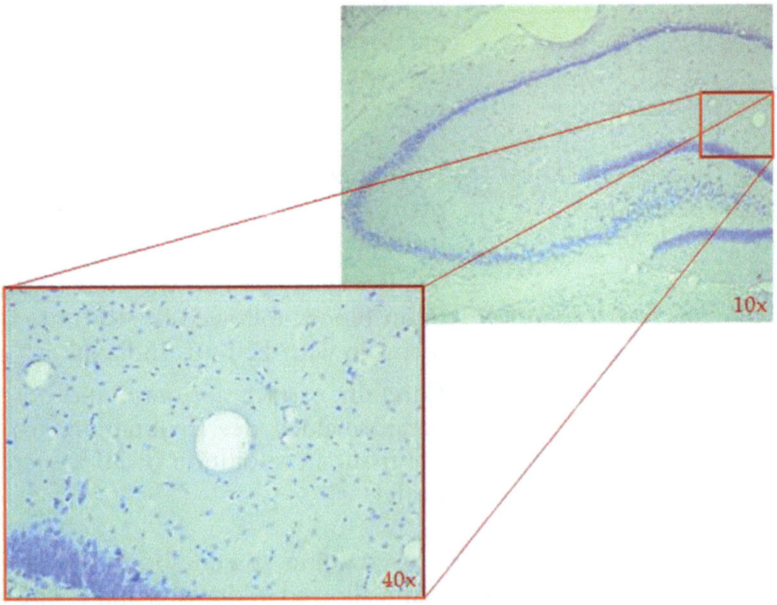

Fig. 2 A hippocampal slice stained with cresyl violet

3. Rehydrate and differentiate the tissue with the following rinses: 50 % ethanol (1–3 min), 70 % ethanol (1–3 min), 95 % ethanol (1–3 min), 95 % ethanol with 2–5 drops of glacial acetic acid (1–3 min), 100 % ethanol (1–3 min), and xylene/xylene substitute (>5 min).
4. After at least 5 min in xylene/xylene substitute, remove slides and coverslip with Permount.

3.2.2 Toluidine Blue

1. Rinse the tissue sections in 75 % ethanol for 30 s, followed by 30 s in a dH_2O rinse.
2. Place the tissue in toluidine blue solution for 30–60 s (*see* **Note 7**).
3. Dehydrate the tissue in a series of ethanol solutions (70, 95, 100 %) for 30 s each. Rinse in xylene for 5 min and again in a fresh container of xylene for another 5 min (this second xylene rinse is optional). Coverslip slides with Permount.
4. Allow slides to dry for at least 20 min before further handling.

3.3 Immunohistochemistry

Use fresh solutions for all rinses, and wash dishware after each rinse. It is advisable to have several different vessels (pie plates, non-perforated wells, etc.) to use throughout the process. This ensures that the timing of each rinse is the same and that there is no delay between rinses (i.e., for washing supplies). All rinses and incubations should be done at room temperature and on a plate shaker unless otherwise specified (*see* **Note 8**). Please *see* Figs. 3 and 4 for illustrations of CD61 and RECA-1 labeling.

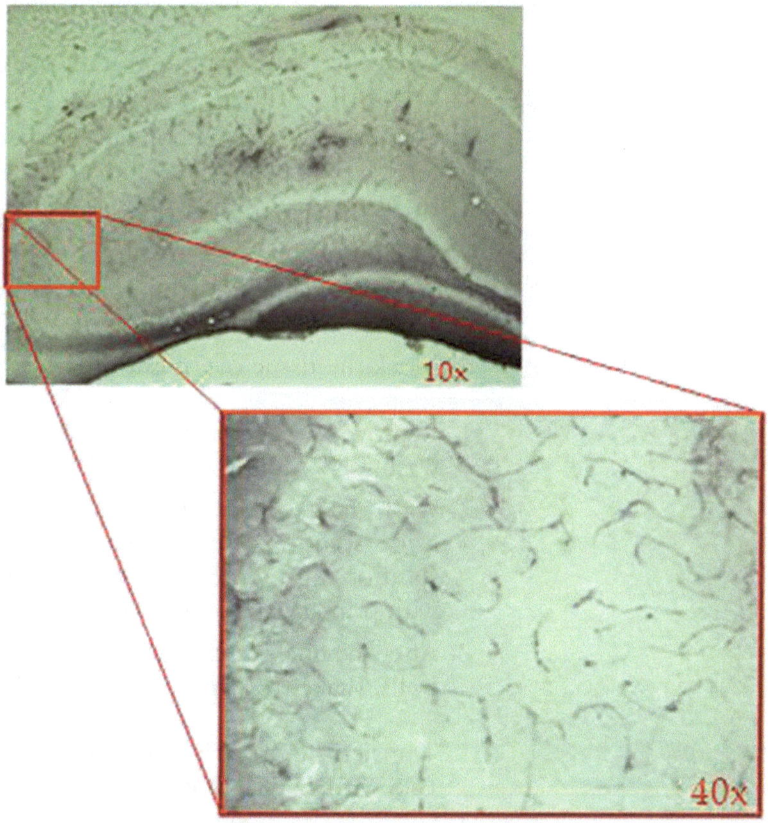

Fig. 3 CD61 immunohistochemistry in the hippocampus

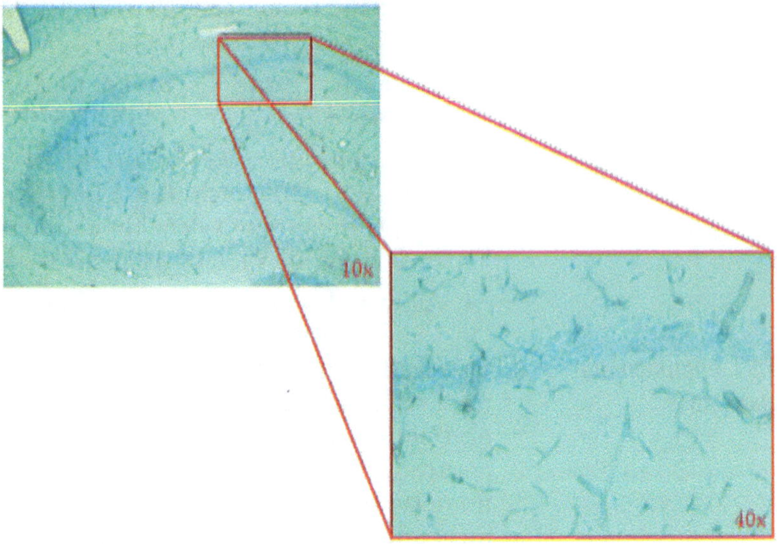

Fig. 4 RECA-1 immunohistochemistry in the hippocampus

3.3.1 Day 1

1. Transfer the brain tissue into perforated wells, making sure that all tissue sections are flat.
2. Wash the tissue for 5 min in working PBS with azide. Repeat once.
3. Catalyze the endogenous peroxidase in 0.3 % H_2O_2 solution for 1 h on a plate shaker.
4. Wash the tissue for 5 min in working PBS with azide. Repeat once.
5. Wash the tissue for 10 min in 2 % serum in working PBS with azide solution. Repeat once.
6. Block the tissue overnight (at 4 °C) in blocking solution.

3.3.2 Day 2

1. Remove the perforated wells from the blocking solution and wash for 5 min in working PBS with azide. Repeat once.
2. Wash the tissue for 10 min in 2 % serum in working PBS with azide solution. Repeat once. Prepare the primary antibody solution during these two washes. Fill the non-perforated wells with primary antibody solution.
3. After the final serum wash, transfer the tissue to non-perforated wells, shake (*see* **Note 9**), and incubate overnight at 4 °C.

3.3.3 Day 3

1. Transfer the tissue into perforated wells and wash for 5 min in PBS with azide. Repeat once.
2. Wash the tissue for 10 min in 2 % serum in working PBS with azide solution. Repeat once. Prepare the secondary antibody

solution during these two washes. Fill the non-perforated wells with the secondary antibody solution. Incubate the tissue in secondary antibody solution at room temperature (on a shaker plate) for 1.5 h.

3. Near the end of the 1.5 h incubation, prepare the ABC according to manufacturer guidelines. Prepare the DAB and nickel ammonium sulfate solutions. Once incorporated, pour the nickel ammonium sulfate mixture into the DAB mixture.
4. Transfer the tissue back to the perforated wells and wash for 5 min in working PBS *without* azide. Repeat once.
5. Wash the tissue for 10 min in 2 % serum in working PBS *without* azide solution. Repeat once. Fill the non-perforated wells with ABC mixture.
6. Transfer the tissue to the non-perforated wells and incubate at room temperature (on a shaker plate) for 1 h.
7. After incubating in ABC, transfer the tissue back into perforated wells and wash for 5 min in working PBS *without* azide. Repeat once.
8. Wash the tissue for 10 min in Tris buffer. Repeat once. Filter the DAB solution during these washes.
9. Immediately before the DAB wash, add 30 % H_2O_2 (6.67 μL/200 mL solution). Place the perforated wells in the DAB wash and shake for 5–10 min. During this wash, watch the tissue to ensure it does not get too dark (which would result in too much background staining and hinder quantification).
10. Wash the tissue for 5 min in PBS *without* azide. Repeat four times.
11. Mount the tissue on gelatin-coated slides and allow to dry overnight.
12. On the next day, rinse the slides for 5 min each in 70, 95, and 100 % ethanol solutions. Then rinse for at least 5 min in clearing solution and coverslip. Clear the slides with cotton swabs and clearing solution until clean.

3.4 Imaging and Stereology

Unbiased stereology is a procedure used to quantify particles of interest in a tissue or sample [25]. In addition, stereology refers to the ability to make these measurements, including surface area, volume, number, and length, in three dimensions [25]. When using stereology to measure angiogenesis, it is imperative to utilize random sampling and imaging procedures that minimize the likelihood of bias (e.g., double counting). To do this, our laboratory has used a variation of the Disector Principle and point counting [25]. This section will describe these procedures in detail.

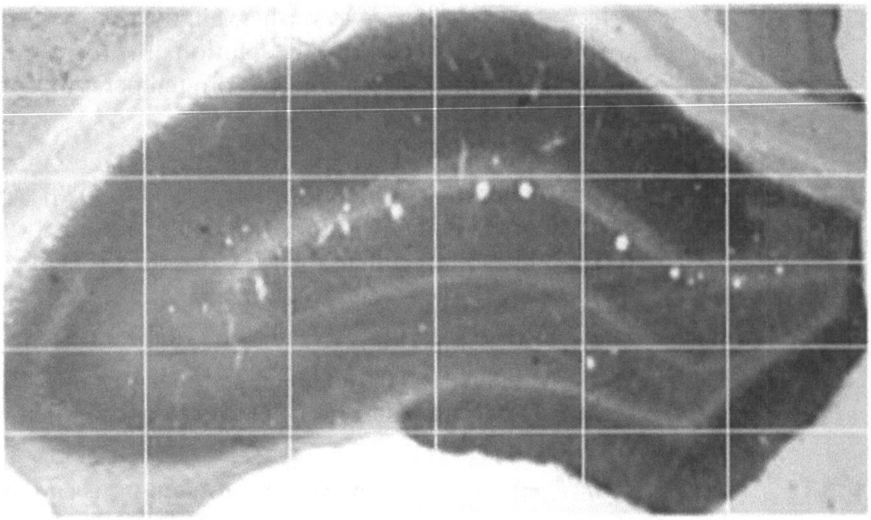

Fig. 5 The hippocampus with the imaging grid superimposed. Each *rectangle* represents one image. Once the entirety of the structure has been captured in nonoverlapping images, a random selection of images is chosen for quantification

3.4.1 Random Sampling

There are several ways to select an unbiased representative sample of tissue to analyze. According to Mouton [25], one way to accomplish this is to use a random number generator to determine the intervals at which tissue is sampled. We use a variation of this approach in our laboratory. We use ice cube trays to slice tissue serially into 6 wells, each slice going into the next well with the process repeated for the entirety of the structure(s) of interest such that several slices occupy each well of the ice cube tray. Before performing immunohistochemistry or staining the tissue, a subset of slices from each well is taken for further analysis. This method not only allows for a random selection of tissue but also ensures that each slice is the same distance apart from other slices in that well, giving a representative sample of tissue from the entire structure or brain region. Keeping track of how many slices are taken will allow for a volume measurement during analysis (see below).

Random sampling will also have to be done for images (imaging is described below). Once the entirety of the region of interest has been captured in non-overlapping images, a random number generator can be used to randomly select a subset of the images to analyze.

3.4.2 Imaging

In order to ensure that the analysis is unbiased, every part of the structure of interest should be captured in non-overlapping images. To do this, an invisible grid is superimposed over the structure such that several images are obtained for the structure in each tissue slice. Please *see* Fig. 5 for an illustration of this concept.

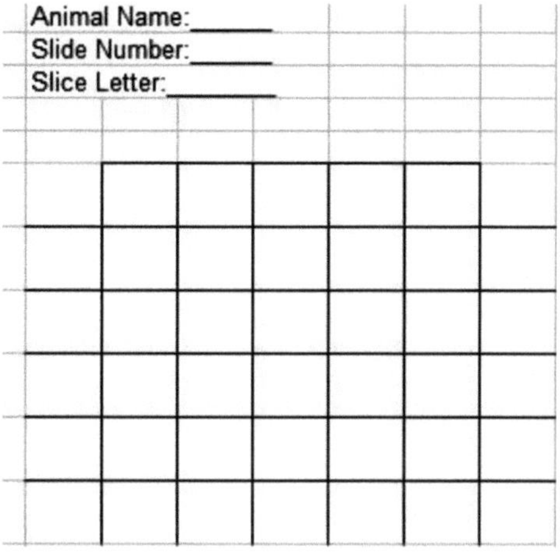

Fig. 6 A sample imaging data sheet

To generate the imaging grid, two landmarks of the structure of interest can be selected, and the triangulation point between those two landmarks can be used as the starting point for taking images. The process for imaging the hippocampus is described below, but can be adapted to any structure.

1. Set the microscope objective at the desired magnification.
2. Find the most lateral point of the region of interest (ROI). Record the location of this point with the Vernier scale that measures medial/lateral movement of the microscope stage.
3. Find the most ventral point of the ROI and record this number with the ruler that measures dorsal/ventral movement of the stage. *See* Fig. 6 for a sample data sheet for recording point locations and image numbers.
4. Calculate the additional points on the recording sheet. These calculations are made depending on the individual microscope (*see* **Note 10**). The triangulation point of these two locations is where the first image should be taken. If the ROI is not present at the triangulation point, simply skip the coordinates and continue onto the next ones.
5. After taking the first image, progress to the next point in the medial/lateral plane (staying at the same coordinate on the dorsal/ventral plane), and take the next image. Continue to progress along this plane until the ROI is no longer present.
6. After each row of images is taken, advance to the next dorsal/ventral coordinate and return to the original lateral/medial coordinate. Proceed to all coordinates on that row and take images. Repeat this process until the entire ROI is captured (*see* Fig. 7 for a completed imaging sheet).

Animal Name:_EX-10____
Slide Number:___1___
Slice Letter:___A____
Hemisphere:___Left____

6.6	x	x	x	x	x	x
7.1	x	16	17	18	x	x
7.6	11	12	13	14	15	x
8.1	5	6	7	8	9	10
8.5	x	x	1	2	3	4
	146.7	146.1	145.5	144.9	144.3	143.7

Fig. 7 A completed imaging data sheet. The *x* indicates that the region of interest is not visible within the image

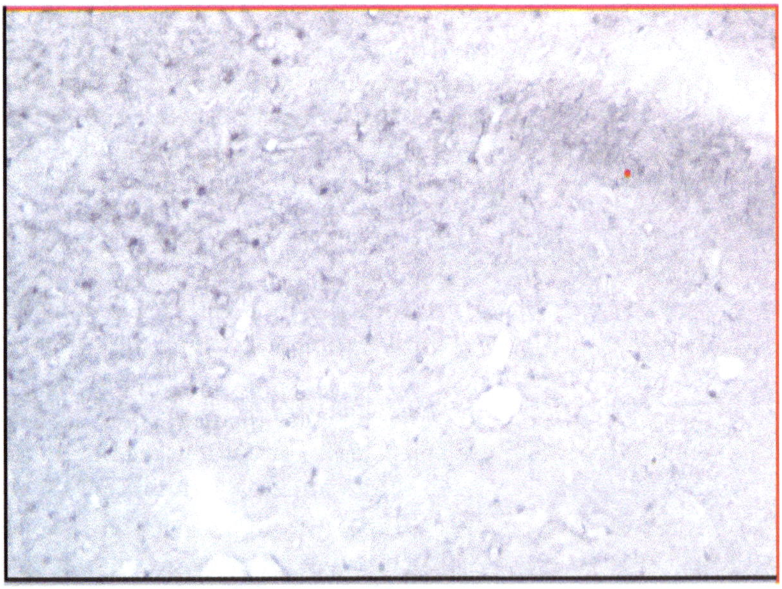

Fig. 8 Illustration of the Disector Principle (inclusion and exclusion borders). Particles or objects of interest that fall on the *black borders*, but not the *red borders*, should be counted

7. After all imaging has been completed, use a random number generator to select a subset of images for analysis and quantification. To avoid experimenter bias, all images should be coded so that during analysis the experimenter is unaware of the condition/group from which each particular image has been sampled.

3.4.3 Image Analysis

For quantification of angiogenesis in a region of interest, we use a combination of stereological techniques, namely, the Disector Principle and point counting [25]. The Disector Principle includes the use of inclusion and exclusion borders (*see* Fig. 8), which ensures that objects are not counted more than once if they happen to appear in more than one sampled image [25]. Point counting allows for the estimation of the area of the particles/objects of interest within a region of interest [25].

A photograph editing program, such as Adobe Photoshop, along with plug-ins that allow for point counting can be used to analyze the images (e.g., Image Processing Tool Kit—Reindeer Graphics, Inc.). Freeware such as ImageJ [26] can also be used for point grid analysis.

1. With an image open, determine the ROI within the image. In some cases the ROI may not occupy the entire image, or other structures may be present.
2. Superimpose the point grid atop the image and determine the number of points that fall within the region of interest.
3. Count how many points fall on or within blood vessels.
4. Once you have these two numbers for each image, compile and sum the number of points falling on or within blood vessels, and sum the number of points falling within the ROI. Do this for each animal.
5. To calculate the density of blood vessels in the region of interest, use the following formula:

   (Number or points falling on capillaries ÷ Number of points within the ROI) × 100.
6. In order to make a volume measurement, the area of each point must be known. ImageJ provides this information. According to Mouton [25], the volume of blood vessels in the structure of interested can be estimated by summing the points on or within blood vessels and multiplying that figure by the area per point (this must be corrected for the magnification by dividing the area of the points by the squared magnification).

4 Notes

1. When preparing stock paraformaldehyde solution, do not allow the temperature of the deionized water to exceed 65 °C. Additionally, when adding sodium hydroxide, wait a few minutes between drops; this will prevent addition of excessive sodium hydroxide and will avoid the necessity of making large adjustments to the pH of the solution.
2. Secondary antibody depends on the host of the primary antibody. For example, if the primary antibody is anti-RECA-1 produced in mouse, then the secondary antibody would have to be anti-mouse (this would be produced in a different host such as goat, horse, or sheep). The serum is chosen based on the host of the secondary antibody. For example, continuing

with the example above, the secondary antibody could be anti-mouse produced in horse. Thus horse serum would be used for all serum, blocking, and antibody solutions.

3. The dilution of the primary antibody and the concentration of the secondary antibody can be adjusted to allow for the best visualization of labeling. We suggest performing a dilution run that includes different dilutions within the range of dilutions recommended by the vendor and a few dilutions just outside of that range. For example, if the suggested dilution range is 1/1,000 to 1/2,000, an appropriate dilution run might include 1/500, 1/1,000, 1/2,000, and 1/3,000. These dilutions can also be run with various concentrations of secondary antibody (e.g., 0.5 or 1 %) to determine the best combination of the primary dilution and the secondary concentration which yields the most reliable labeling.
4. The DAB/ammonium nickel sulfate hexahydrate mixture can stir for several hours. The longer it stirs, the darker in color the mixture will get. Based on our observations, labeling is not affected by stirring for shorter or longer periods, but the experimenter should be sure to allow enough time for the DAB and nickel ammonium sulfate to go into solution and get mixed together and filtered.
5. All DAB solutions should be disposed of in proper hazardous materials collection containers. After the DAB wash and after direct disposal in the proper vessel, all dishware and other items should be washed thoroughly in a bleach solution and then rinsed thoroughly. All implements that have been used for DAB should be separated from other dishware and laboratory equipment and used only for the purpose of DAB handling. Also, when working with DAB, we advise double gloving. All disposable solid materials that have come into contact with DAB (e.g., weigh boats, filter paper, gloves) should also be deposited in the appropriate hazardous materials container.
6. During perfusion, the quantities of working PB and paraformaldehyde are flexible depending on the quality of the perfusion. More working PB can be added if the organs do not appear to be free of blood (i.e., if they are not white or tan in color). More or less paraformaldehyde can be used as well. If the limbs and organs feel stiff to the touch, then not all 400 mL of the paraformaldehyde needs to be used. Conversely, if the body does not feel stiff to the touch, more paraformaldehyde should be added. Experimenters should be sure to make some extra solution should more be required during the perfusion.
7. The quality of the stain and the length of time that tissue should remain in the stain are variable, depending on the age of the solution. Also, as the tissue proceeds through the rinses,

some staining will be lost. The abovementioned times are not absolute. If a lot of stain is being lost in the later alcohol washes, then the tissue should remain in those washes for shorter periods of time. It has been our experience that closely watching the tissue during these final rinses is imperative to ensure that the stain is not appreciably lost.

8. Immunohistochemistry is a multiday process, and we have found consistent results when the process is preformed over a period of 3 days. However, just as in determining the appropriate dilution for primary and secondary antibodies, the process can be adjusted to last more or less days. Some experimentation with methodology is encouraged to determine the procedure that maximizes the quality of the labeling.
9. We have found that shaking the tissue (for an hour or two) after placing it in primary antibody solution may help improve labeling.
10. Researchers will have to determine how far to move the stage in each direction to ensure that images do not overlap. As an example, at a magnification of 40×, our microscope requires movement of 0.6 mm on the lateral/medial plane and 0.5 mm on the dorsal/ventral plane to ensure that images do not overlap. Whether addition or subtraction is required depends on the hemisphere being imaged.

References

1. Hillman C, Motl R, Pontifex M et al (2006) Physical activity and cognitive function in a cross-section of younger and older community-dwelling individuals. Health Psychol 25(6): 678–687. doi:10.1037/0278-6133.25.6.678
2. Hillman CH, Snook CM, Jerome GJ (2003) Acute cardiovascular exercise and executive control function. Int J Psychophysiol 48:307–314. doi:10.1016/S0167 8760(03)00080-1
3. Winter B, Breitenstein C, Mooren F et al (2007) High impact running improves learning. Neurobiol Learn Mem 87:597–609. doi:10.1016/j.nlm.2006.11.003
4. Kim S, Ko I, Kim B et al (2010) Treadmill exercise prevents aging-induced failure of memory through an increase in neurogenesis and suppression of apoptosis in rat hippocampus. Exp Gerontol 45:357–365. doi:10.1016/j.exger.2010.02.005
5. Uysal N, Tugyan K, Kayatekin B et al (2005) The effects of regular exercise in adolescent period on hippocampal neuron density, apoptosis and spatial memory. Neurosci Lett 383: 241–245. doi:10.1016/j.neulet.2005.04.054
6. Vaynman S, Ying Z, Gomez-Pinilla F (2004) Hippocampal BDNF mediates the efficacy of exercise on synaptic plasticity and cognition. Eur J Neurosci 20:2580–2590. doi:10.1111/j.1460- 9568.2004.03720.x
7. Larson E, Wang L, Bowen J et al (2006) Exercise is associated with reduced risk for incident dementia among persons 65 years of age and older. Ann Intern Med 144:73–81
8. van Praag H, Shubert T, Zhao C et al (2005) Exercise enhances learning and hippocampal neurogenesis in aged mice. J Neurosci 25(38): 8680–8685. doi:10.1523/jneurosci.173105.2005
9. Devine J, Zafonte R (2009) Physical exercise and cognitive recovery in acquired brain injury: a review of the literature. PM R 1:560–575. doi:10.1016/j.pmrj.2009.03.015
10. Grealy M, Johnson DA, Rushton SK (1999) Improving cognitive function after brain injury: the use of exercise and virtual reality. Arch Phys Med Rehabil 80:661–667
11. Griesbach G, Hovda D, Gomez-Pinilla F (2009) Exercise-induced improvement in cognitive performance after traumatic brain-injury in rats is dependent upon BDNF activation. Brain Res 1288:105–115. doi:10.1016/j.brainres.2009.06.045

12. Kluding P, Tseng B, Billinger S (2011) Exercise and executive function in individuals with chronic stroke: a pilot study. J Neurol Phys Ther 35:11–17
13. Quaney B, Boyd L, McDowd J et al (2009) Aerobic exercise improves cognition and motor function poststroke. Neurorehabil Neural Repair 23:879–885
14. Black J, Isaacs K, Anderson B et al (1990) Learning causes synaptogenesis, whereas motor activity causes angiogenesis, in cerebellar cortex of adult rats. Proc Natl Acad Sci U S A 87:5568–5572
15. Ding Y, Li J, Zhao Y et al (2006) Cerebral angiogenesis and expression of angiogenic factors in aging rats after exercise. Curr Neurovasc Res 3:15–23
16. Isaacs K, Anderson B, Alcantara A et al (1992) Exercise and the brain: angiogenesis in the adult rat cerebellum after vigorous physical activity and motor skill learning. J Cereb Blood Flow Metab 12:110–119
17. Kleim JA, Cooper NR, Vandenberg PM (2002) Exercise induces angiogenesis but does not alter movement representations within rat motor cortex. Brain Res 934:1–6
18. Sikorski A, Hebert N, Swain R (2008) Conjugated linoleic acid (CLA) inhibits new vessel growth in the mammalian brain. Brain Res 1213:35–40. doi:10.1016/j.brainres.2008.01.096
19. Swain R, Harris A, Wiener E et al (2003) Prolonged exercise induced angiogenesis and increases cerebral blood volume in primary motor cortex of the rat. Neuroscience 117:1037–1046. doi:10.1016/S0306-4522(02)00664-4
20. van der Borght K, Kobor-Nyakas D, Klauke K et al (2009) Physical exercise leads to rapid adaptations in hippocampal vasculature: temporal dynamics and relationship to cell proliferation and neurogenesis. Hippocampus 19:928–936. doi:10.1002/hipo.20545
21. Kerr A, Steuer E, Pochtarev V et al (2010) Angiogenesis but not neurogenesis is critical for normal learning and memory acquisition. Neuroscience 171:214–226. doi:10.1016/j.neuroscience.2010.08.008
22. Thompson KJ, Bulinski SC, Powell SK, Sikorski AM, Swain RA (2000) Time-dependent expression of the tyrosine kinase receptors FLK-1 and FLT-1 in the cerebellum of the exercised rat. Society for Neuroscience Abstract, New Orleans
23. Bulinski SC, Thompson KJ, Powell SK, Sikorski AM, Swain RA (2000) Increased immunolabeling of Flk-1 receptors in primary motor cortex of the adult rat following exercise. Society for Neuroscience Abstract, New Orleans
24. Yu B, Yu C, Robertson RT (1994) Patterns of capillaries in developing cerebral and cerebellar cortices of rats. Acta Anat (Basel) 149:128–133
25. Mouton P (2002) Principles of unbiased stereology: an introduction for bioscientists. The Johns Hopkins University Press, Baltimore
26. Rasband WS ImageJ. U S National Institutes of Health, Bethesda, Maryland, USA http://imagej.nih.gov/ij/. Accessed 1997–2012

Part III

Methods of Examining Cerebral Angiogenesis

Chapter 14

Histological Assessment of Angiogenesis in the Hypoxic Central Nervous System

Moises Freitas-Andrade, Jacqueline Slinn, Claudie Charlebois, and Maria J. Moreno

Abstract

Angiogenesis, the sprouting of new capillaries from preexisting vessels, is an integral part of both normal development and numerous pathological conditions such as tumor growth, inflammation, and stroke. The development of angiogenesis assays has been critical in understanding this process in both the context of disease and normal physiology. With the growing availability of antibodies against angiogenic markers as well as advances in microscopy and imaging analysis software, a more comprehensive assessment of the angiogenesis process is beginning to take form (Milner et al., Stroke 39:191–197, 2008; Freitas-Andrade et al., J Cereb Blood Flow Metab 32:663–675, 2012; Li et al., Glia 58:1157–1167, 2010; Dore-Duffy and LaManna, Antioxid Redox Signal 9:1363–1371, 2007). This chapter describes an in vivo method of inducing brain angiogenesis in mice by chronic exposure to mild hypoxia. In addition, a detailed procedure of quantifying angiogenesis using multiple immunofluorescent labeling of mouse brain tissue sections is also presented.

Key words Cerebral angiogenesis, Hypoxia, Immunofluorescence, Brain endothelial cells, Astrocytes, Pericytes, Smooth muscle cells

1 Introduction

Angiogenesis is a highly complex multistep process involving the spatial and temporal orchestration of several pro- and anti-angiogenic factors. Oxygen plays a critical role in this process. Hypoxia promotes, directly and indirectly, the expression of several angiogenic factors and receptors in the endothelium and the surrounding hypoxic tissue. In the central nervous system (CNS), angiogenic factors activate cells of the neurovascular unit (NVU), consisting of endothelial cells and integral neighboring cells including pericytes, smooth muscle cells, astrocytes, and neurons [1, 2, 5, 6]. Together, these different cell types interact and are an essential part of the angiogenic process in the CNS. During angiogenesis, detachment of pericytes and smooth muscle cells destabilizes the

Richard Milner (ed.), *Cerebral Angiogenesis: Methods and Protocols*, Methods in Molecular Biology, vol. 1135, DOI 10.1007/978-1-4939-0320-7_14, © Springer Science+Business Media New York 2014

vasculature and promotes the breakdown of inter-endothelial tight junctions. Concomitantly, hypoxic neurons and particularly astrocytes secrete pro-angiogenic factors and chemoattractants that guide endothelial cells to the hypoxic area [3, 7]. In the final stages of angiogenesis, endothelial cells return to a quiescent state by perfusion of the new vessels. The recruitment of pericytes, astrocytes, and local neurons further stabilizes the nascent microvasculature.

Acute mild hypoxic (10 %) conditions, such as those encountered in high altitudes (~5,500 m), elicit systemic cardiovascular and respiratory adaptations, including increased cerebral blood flow (CBF) and ventilation to maintain oxygen delivery to the brain [8]. However, prolonged or chronic exposure to mild hypoxia results in a more complex mechanism of adaptation in the brain through the formation of new blood vessels and tissue remodeling [4].

This model of systemic (whole body) chronic mild hypoxia has been used experimentally to unravel the mechanism(s) through which hypoxia stimulates angiogenesis in the CNS [3, 4, 9]. LaManna et al. (1992) [10] have shown that rats exposed to 21 days of continuous hypoxia exhibit a significant increase in hematocrit (up to 71 %), cerebral blood flow (CBF), and brain microvessel density, suggesting stimulation of angiogenesis [10]. Under these conditions, astrocytes are also activated, proliferate, and reorganize their astrocytic end feet to match the sprouting microvessels [3].

Systemic chronic mild hypoxia in animals can be induced by two methods: (1) "normobaric hypoxia," decreasing the partial pressure of oxygen in a normobaric chamber [11], and (2) "hypobaric hypoxia," decreasing the overall barometric pressure in "Wright chambers," which are chambers for housing small animals capable of withstanding a partial vacuum [11]. Animals subjected to either of these two hypoxic methods have not shown significant differences in their mechanisms of adaptation to hypoxia [11].

In this chapter, we describe a method of inducing brain angiogenesis in mice using systemic chronic mild hypoxia ("normobaric hypoxia"). A comprehensive method of assessing cerebral vascular remodeling, using multiple immunofluorescent labeling and image analysis, is also presented. Two different imaging analysis approaches will be discussed; the first one focuses on evaluating individual images captured with either a fluorescence or a confocal laser scanning microscope, and the second one focuses on evaluating mosaic images composed of multiple z-stack stitched images captured with a confocal laser scanning microscope.

2 Materials

2.1 Components for the Mouse Model of Chronic Systemic Hypoxia

1. Female mice: at least four to five mice are recommended for each test group (*see* **Note 1**).
2. Normobaric hypoxic chamber, Forma Anaerobic System model 1024 (Fisher Scientific, Nepean, ON, Canada), attached

to a gas tank composed of 10 % O_2 and N_2 balance and fed into the chamber (*see* **Note 2**).

3. MI-730 Micro-Oxygen Electrode (Microelectrodes, INC. Bedford, USA) to measure oxygen levels in the chamber.
4. Mouse weighing scale.

2.2 Materials for Perfusion and Brain Isolation

1. Fisherbrand Heparinized Microhematocrit Capillary Tubes (Fisher Scientific, Nepean, ON, Canada).
2. Hemato-Seal capillary tube sealant.
3. Glass scale—50 mm.
4. Eppendorf tubes.
5. 10-cc syringe.
6. 18-gauge needle.
7. 10× phosphate-buffered saline (PBS), pH 7.4 diluted to 1× in Milli-Q water. For each mouse, 10 ml of PBS is used for perfusion.
8. Styrofoam container with lid (large enough to hold several mouse brains and 150 ml beaker) filled with dry ice.
9. Foil paper.
10. 150 ml beaker filled with 100 ml isopentane. Place the beaker, with isopentane, in the styrofoam container (**item 8**, above).
11. Halothane B.P.
12. Scissors and forceps.
13. Surgical tape.

2.3 Components for Sectioning

1. Cryostat.
2. Small paint brushes, to manipulate tissue sections.
3. Forceps.
4. Razor blades.
5. Superfrost Plus microscope slides.
6. Dry ice.
7. Microscope slide-box.
8. Tissue-Tek CRYO-OCT Compound kept at 4 °C (*see* **Note 3**).

2.4 Materials for Immunofluorescent Labeling

1. ImmEdge Hydrophobic Barrier Pen.
2. Glass (or plastic) staining dishes for 20 slides.
3. StainTray slide staining system (Sigma-Aldrich, Oakville, ON, Canada).
4. Permeabilizing solution: Dilute 0.25 % Triton-X-100 in PBS.
5. Blocking solution: Dilute 0.25 % Triton-X-100 and 10 % normal goat serum (NGS) in 1× PBS (*see* **Note 4**).

Table 1
Detailed information of primary antibodies used for staining of the neurovascular unit [2]. Protein concentrations are the final concentrations applied to tissue samples

Antibody	Concentration	Supplier	Catalog number	Species/isotype
Platelet endothelial cell adhesion molecule 1 (PECAM1)/CD31	5.0 μg/ml	BD Bioscience	557355	Rat IgG_{2a}, k
Vascular endothelial growth factor (VEGF)	5.0 μg/ml	R&D Systems	AF564	Polyclonal goat IgG
Nerve/glial antigen 2 (NG2) chondroitin sulfate proteoglycan	3.3 μg/ml	Millipore	AB5320	Polyclonal rabbit IgG
Desmin	2.35 μg/ml	DakoCytomation	M0760	Mouse IgG_1, k
α-smooth muscle actin	14.7 μg/ml	Sigma-Aldrich	A2547	Mouse IgG2a
Fibrinogen	25 μg/ml	DakoCytomation	F0111	Rabbit conjugated with FITC
Glial fibrillary acidic protein (GFAP)	5.8 μg/ml	DakoCytomation	Z0334	Polyclonal rabbit IgG

6. Primary antibody solution: Dilute the antibodies in 1× PBS and 5 % NGS. Details of the primary antibodies are described in Table 1 (*see* **Notes 5** and **6**).
7. Alexafluor®-conjugated secondary antibodies (Invitrogen, Burlington, ON, Canada) were used at 1:500 dilution in 1× PBS (*see* **Note** 7).
8. Coverslips 12 mm (Bellco Biotechnology, NJ, USA).
9. Hoechst 2 μg/ml in PBS (*see* **Note 8**).
10. Dako fluorescence mounting media (DakoCytomation, Burlington, ON, Canada).
11. Fluorescence microscope with a 10–60× objective, attached to a digital camera.
12. Confocal laser scanning microscope.
13. Image-Pro® Plus and Image-Pro Analyzer 7.0 (Olympus Canada Inc, Markham, Canada) (*see* **Note 9**).
14. FV10-ASW 3.1.01 (Olympus Canada Inc, Markham, Canada).

3 Methods

3.1 Induction of Cerebral Angiogenesis in Mice

1. Calibrate the oxygen probe and measure the percentage of oxygen in the hypoxic chamber. Purge the chamber with a gas mixture composed of 10 % O_2 and N_2 balance; the oxygen probe should register the same percentage of oxygen (10 % O_2) as the gas mixture (*see* **Note 10**).

2. Take note of the weight of each animal used in the experiment. Place four to five mice in a group in one mouse cage with plenty of food, water, and bedding (*see* **Note 1**). Ideally, the mice should be grouped by time point to facilitate removing the mice from the hypoxic chamber at each time point.
3. Place the mouse cages in the hypoxia chamber, one cage per time point (*see* **Note 11**). A similar number of control mice should also be kept in the same room under similar conditions except at normal oxygen levels (normoxia). Measure the oxygen levels in the chamber and manually purge the hypoxic chamber to reach 10 % oxygen according to the manufacturer's instructions (*see* **Note 12**).

3.2 Mouse Perfusion and Brain Isolation

1. At the end of each time point, set up the perfusion equipment in a well-ventilated area (fume hood) with a precision vaporizer for inhalant anesthetics (either isoflurane or halothane) for laboratory animals. The dissecting pan and tools (scissors and forceps) should all be placed in the fume hood as well as cold 1× PBS, enough volume to perfuse the mice (allow 10 ml/mouse).
2. Place 2 × 2 cm square foil paper (1 square foil paper per mouse) flat on dry ice. Insert the 150 ml beaker filled with isopentane in dry ice and keep the temperature of isopentane at −70 °C. Isopentane can be cooled by dropping small pieces of dry ice in solution.
3. Begin with the control mice (*see* **Note 13**). Record the weight of the animal. Deeply anesthetize the mouse with 4 % halothane B.P. under an oxygen flow rate of 2 L/min. Ensure the animal is in the surgical plane before making an incision, by testing whether the mouse reacts to a tail or toe pinch. Lie the mouse on its back and tape the limbs onto the dissecting pan.
4. Open the thoracic cavity, expose the heart, and visualize both the right and left external jugular veins. Insert an 18-gauge needle, attached to a syringe filled with 10 ml cold PBS, through the left ventricle. Hold the needle in place with one hand, while with the other hand, snip both jugular veins (*see* **Note 14**) and collect the blood with the Heparinized Microhematocrit Capillary Tubes. Cover both ends of the tubes with Hemato-Seal capillary tube sealant, and place the tubes in the tube rack.
5. Slowly perfuse the animal with 10 ml PBS until the fluid is clear (*see* **Note 15**).
6. After perfusion is complete, decapitate the animal and quickly remove the brain by cutting along the sagittal suture via the foramina of the skull. Place the brain in the 150 ml beaker filled with 100 ml cold (−70 °C) isopentane.
7. Once the brain is snap-frozen, place the brain on labeled foil paper on dry ice. Wrap the brains in the foil paper, then place

each brain in a small individually labeled plastic bag, and store at −80 °C for cryosectioning.

8. After all the control animals have been processed, measure the hematocrit. Centrifuge the capillary tubes at 10621 × *g* for 5 min (*see* **Note 16**). Using the glass scale, measure the length of the packed red cell volume, and divide by the total length of the blood sample. Then multiply this value by 100 to get the percent hematocrit for each mouse.
9. Repeat **steps 1** through **6** for the hypoxic group of mice.

3.3 Sectioning of Fresh Frozen Mouse Brain

1. Before sectioning, adjust the temperature in the cryostat to −21 °C (*see* **Note 17**). Make sure all necessary sectioning equipment is inside the cryostat: blades, paintbrushes, forceps, and specimen chucks.
2. Label the microscope slide-box appropriately and put the box in the styrofoam container filled with dry ice.
3. Transport brains on dry ice from −80 °C storage and place inside the cryostat (*see* **Note 18**). Allow the frozen brains to equilibrate to temperature inside the cryostat (45 min to 1 h) before sectioning.
4. Set the cryostat to cut 10 μm thick sections.
5. Cover the cryostat specimen chuck with a thin layer of OCT embedding compound and allow to freeze. Dispense another small amount of OCT onto the first frozen layer and immediately, with cold forceps, grasp the front end of the brain and immerse the posterior end of the brain (cerebellum and part of posterior cortex) into OCT embedding medium.
6. With the cold forceps, keep the brain in the proper orientation until the OCT around the cerebellum and posterior cortex of the brain has solidified completely. The brain should be securely fixed onto the specimen chuck and is now ready for sectioning (*see* **Note 19**).
7. Place the specimen chuck onto the microtome and orient the brain in relation to the cutting edge, so that it is perpendicular to the cryostat blade for coronal sectioning.
8. Collect the tissue sections by bringing the microscope slide close to the brain section (*see* **Note 20**). The brain section should "jump"/adhere onto the slide due to electrostatic forces on the surface of the slide (*see* **Note 21**). Because of the difference in temperature, between the slide and tissue section, as the section touches the slide, the section will flatten and stick to the slide.
9. Allow the glass slides containing brain sections to air-dry on a clean surface for 3–5 min, depending on relative humidity.

10. Position the microscope containing brain sections in the microscope slide-box in dry ice.
11. Once sectioning is complete, store the microscope slide-boxes at −80 °C, until immunofluorescence experiments.

3.4 Immunofluorescent Labeling of Fresh Frozen Mouse Brain Sections

1. Remove the sections from −80 °C storage and allow the sections to air-dry for 2 min.
2. Remove any excess water from the slide and circumscribe the tissue section with a hydrophobic barrier pen. Allow pen reagent to dry before immersing the slide into aqueous solution.
3. Place all the slides in a slide rack and fill all the slide holders with appropriate solutions (methanol and PBS).
4. Fix the brain sections by immersing the slides in ice-cold methanol for 10 min.
5. Wash the slide twice (2 min per wash) in PBS.
6. Remove each slide from the rack and apply approximately 50–100 μl of permeabilizing solution to each section and place the slides in the staining tray. Incubate the slides for 10 min at room temperature (RT).
7. Wash the slides three times (5 min per wash) in PBS.
8. Remove each slide from the rack and apply approximately 50–100 μl of blocking solution to each section, and place the slides in the staining tray. Incubate slides for 1 h at RT.
9. Carefully shake the blocking solution off the slide, and replace with primary antibody (Table 1) diluted in primary antibody solution. Incubate the slides for 1 h at RT or overnight at 4 °C in the staining tray (*see* **Note 22**).
10. Wash the slides three times (2 min per wash) in PBS.
11. Remove each slide from the rack and apply approximately 50–100 μl of appropriate secondary Alexafluor® antibodies at 1:500 dilution in PBS for 45 min to1 h at RT.
12. Wash the slides three times (2 min per wash) in PBS.
13. Repeat **steps 9–12** for the other set of antibodies (*see* **Note 23**).
14. After the final secondary incubation and final wash with PBS (**step 12**), stain the section with 50 μl of Hoechst (2 μg/ml) for 15 min.
15. Rinse the slides in PBS and dip the slides in Milli-Q water for 30 s.
16. Before the section dries completely, add 10 μl of Dako fluorescence mounting media, and place a glass coverslip over the section. Remove the excess Dako by pressing gently the coverslip with forceps.
17. Allow mounting media to harden at 4 °C overnight (*see* **Note 24**).

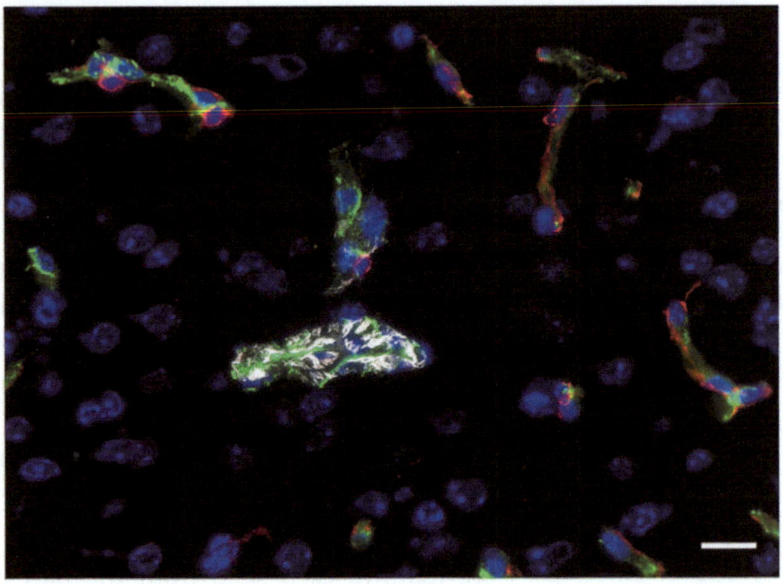

Fig. 1 (**a**) Triple-immunofluorescence staining performed on sections of cerebral cortex from hypoxic mice, using the endothelial marker CD31 (*green*), pericyte markers NG2 (*red*), and vascular smooth muscle cell (VSMC) marker α-SMA (*white*). Scale bar = 100 μm

18. Capture individual images of the stained tissue section using a fluorescence microscope with a 10–60× objective attached to a high-quality digital camera (Fig. 1) (*see* **Note 25**) or mosaic stitched images (Fig. 3) using a motorized confocal laser scanning microscope.

3.5 Quantitative Analysis of Individually Acquired Immunofluorescently Labeled Mouse Brain Section Images

1. Open the image analysis program Image-Pro® Plus (*see* **Note 9**).
2. Load the image to be analyzed (*see* **Note 25**). All channels will appear on separate windows. Each channel has to be manipulated separately.
3. From the main menu bar, click on "Enhance" and choose "Display Range" (display range window will open).
4. Adjust the contrast by moving the vertical lines on either side of the graph right or left in the display range window (*see* **Note 26**).
5. Click on "Process" on the main menu and choose "Filters."
6. Use the "Flatten" filter options to balance the image intensity.
7. Use default settings in Enhancement tab. Press apply.
8. On the main menu bar, choose "Measure" and then "count/size." Click close.
9. Select ranges in "count/size" window and adjust the upper and lower size limits that will be counted by the program (*see* **Note 27**).

10. In the "count/size" window menu, click on "Measure" and choose "Select Measurements."
11. Choose parameters to be measured: area, perimeter length, etc.
12. In the "count/size" window, click on "Options" and make the following changes: (1) clean borders, none; (2) outline style, outline; (3) label style, none (objects in the image window will not be numbered and easier to visualize); and finally (4) label color, green (or color of choice), then press OK.
13. In the count/size window, click on "count" and the program will automatically count all the objects in the image (Fig. 2).
14. Examine which objects are selected in the image; most objects should be outlined in green color (*see* **Note 28**).
15. To remove undesired objects from the "In range total count," right click on the objet, then choose count/size, then hide object.
16. It is possible to save the outline for future reference: in the count/size window, click on file then choose "save outlines." Give this file the same name as the quantified image. The image can be then be opened at a later time, and the objects selected can be examined by loading the saved outline file.
17. The raw and averaged data can be exported to an excel file from within Image-Pro® Plus. This is done by clicking on "Measure," in the main menu bar, and then "Data Collector." A new window will appear called "Data collector." Choose the parameters to be exported into the excel file, e.g., name of image, date, and count/size data. If multiple images are to be calculated together, data can be appended into one excel sheet.
18. The averaged as well as raw data from the calculated total area, length, number, etc., of immunopositive cells for a particular marker can be exported from the three sections per animal using Image-Pro® Plus. The average of each parameter for the four mice per group is then calculated (*see* **Note 29**).

3.6 Procedure for Capturing Mosaic Confocal Images

If a motorized confocal laser scanning microscope (CLSM) is available, a more comprehensive analysis of whole tissue section can be performed by combining z-stack image acquisition and stitching to create a confocal mosaic image that widens the field of view with high resolution (Fig. 3).

1. With computer, lasers, and microscope turned ON, open FV10-ASW 3.1.01 program.
2. Under Device Menu (DM), click open Microscope Controller, and then open Multi Area Time Lapse.
3. Select 10× objective from the Microscope Controller Window (MCW).
4. Place the microscope slide on the stage.

5. In order to focus through the eyepiece, select DAPI signal on MCW. Nuclei staining with DAPI or Hoechst is usually very intense and well defined, which facilitates a rapid focusing of the sample.
6. Select Course under the Jogsense Window under Focus Handle in MCW.
7. To switch to fluorescent mode, click EPI lamp button on Image Acquisition Control window.
8. Using the eyepiece, move the objective and bring the sample into focus.
9. To establish the limits of the mosaic image, under Multi Area Time Lapse Controller (MATLC) window, click icon box Mosaic Outline (MO).
10. Position the edge of tissue sample approximately in the middle of the field of view.
11. In MO box click on the first icon () on the left side to mark field of view as Point #1.
12. Repeat positioning the edge limits of the tissue section in the center and continue marking it with points until entire outline is formed creating your region of interest (ROI).
13. Select the appropriate objective for final image acquisition and refocus.
14. Turn off the EPI lamp whenever not viewing the sample to avoid bleaching.
15. Under the Image Acquisition IAC window, click on Dye List button.
16. Select the appropriate dyes to be imaged. Click Apply. Click close.
17. Make sure the Confocal Aperture (CA) is NOT on automatic. The on/off button for this is located directly under the CA gauge. For larger sections acquired using lower magnification objectives, more light is required. Set CA at 300 μm to begin. This aperture can be adjusted later if needed. If the CA is set too high, too much light is allowed to pass and can lead to aberrant signals in the form of wavy lines.
18. Go to Acquisition Setting, under Mode, select speed 2.0 μs/pixel (FAST), and place resolution in mid to low range. These settings are best for a quick evaluation of the acquired image avoiding bleaching. Change course focus to Fine.
19. Under IAC click XY Repeat for a live view of the sample. Using the fine focus, bring the sample into its sharpest view.
20. Optimize the live view by adjusting in order of priority: CA (for aperture openness), focus, Gain, laser levels, voltage levels,

and Offset. Press Ctrl H to place live view in Hi/Low mode for adjustments. Red indicates saturation of signal, and blue indicates background. Adjust to have minimal amount of both red and blue pixels. This will ensure setting fluorescence pixels to their brightest signal without being over saturated and setting the background to get a high-quality black.

21. Continue to adjust settings. Attempt to keep Gain at lowest level and only increase if laser and/or voltage levels are already set at a high level.
22. View different areas of the ROI to check for consistency in signal capture. Set Speed setting and Aspect Ratio in Acquisition Setting to determine final resolution of mosaic image.
23. Under MO click Apply. A list of images will be created under the Registration Point List (RPL) box.
24. Under the Options drop-down menu, click File and Folder and indicate desired destination for images.
25. Click Ready in RPL, and click Play.
26. When final image is captured, go to the Device drop-down menu and click Multi Area Time Lapse Viewer.
27. Under MATLV click Select the Action button. Click Stitch. In the Stitch View, click the Show button.
28. Click Save As. Choose the appropriate file type. It is best to choose OIB type for best quality.
29. Under Data Manager, take note of the size of the image (in micrometers) listed as Image Size (unit converted) if analysis is required.

3.7 Procedure for Adding Z Sectioning to Image Capture

1. The first step is to determine whether the section is well leveled by looking at each quadrant of the section. In quadrant 1, lower the objective until the signal just disappears (goes out of focus). Click Set 0 in the Acquisition box. Raise the objective bringing the section back into focus, and continue up until the section goes out of focus again. Observe and make note of the Set 0 number. The measurement will be in micrometers and indicates approximately how thick the section is. Repeat in each quadrant, but do not click the Set 0 button again. If each quadrant has similar measurements, this indicates the section is level. If not, the section is tilted. On a removable stage holder, a piece of masking tape may be used to boost up a low corner. Attempt to make the section as level as possible. Repeat the quadrant measurements to verify.
2. Go to the quadrant whose section is lowest and reset the Set 0 button at the bottom of the section. Go to the quadrant whose section is the highest, and raise the objective until the section just goes out of focus. Make note of this number and round off

to the nearest whole micrometer. For example, if the tissue is 14.5 μm thick, round it off to 15 μm.

3. Choose how many z sections are to be obtained. These are called Steps (measured in μm).
4. Under the IAC click the Depth button located under the XY button to the immediate right of XY Repeat. Click the XY button to obtain a sample image. When the Z stack is completed, click the Series Done button that appears over the Depth button. In the 2D view window, drop the menu down from the Projection/Topography mode icon box on the far right. Click the Intensity projection over the Z axis button. This will create a single image from the individual slices. Note: Because each slice is added together, the intensity is also added. The new single slice may be too bright. Lower laser levels accordingly so that final image is appropriate.
5. Make final adjustments to Speed and Aspect Ratio, and continue with remaining steps of creating the Registration Point List, file destination, etc.
6. When the final image is captured, go to the Device menu and click MATLV. Click Select Action, click Projection Image, and click Projection Image for all Points. This last step compresses each z stack into one image.
7. Click Image Type. Click Stitch. Click Show and Save As.

3.8 Quantitative Analysis of Mosaic Confocal Images

1. Open Image-Pro Analyzer 7.0 (*see* **Note 9**)
2. Open OIB image to be analyzed. Make sure that the image is split into single channel windows in addition to composite image.
3. Select the single channel image to be analyzed.
4. Under the Edit drop-down menu, select Convert To/Grey Scale 8. Input Range of 0–4,095 will be converted to Output Range of 0–255 (8 bits per pixel) meaning instead of having 4,095 shades of grey, you will work with 255. Click Convert. This conversion is necessary to reduce the resolution size of the image for the purpose of analysis. Mosaic images can be quite large. The computer system capacity may not be always adequate to handle such large size images.
5. Under Measure drop-down menu, click Calibration/Spatial Calibration Wizard.
6. In Spatial Calibration window, select appropriate units to match OIB file units, e.g., μM.
7. Click Next. Click Draw Reference Line. Stretch Reference Line to each edge of image. In the scaling window, type in the micrometer measurement obtained from the OIB file. Click OK, click Finish.

8. Using the converted image, click the Measure drop-down menu and make sure Manual is highlighted. Click Count. Then, click Select Ranges. In this window adjust the range so that the signal/objects you are interested in will be measured. Signals in the lower range will be dimmer, and signals closer to 255 will be the brightest. Selecting a range further away from zero will help to remove background signal (the dimmest pixels). If background levels are too high, objects of interest will bleed into one another so that two brighter objects may be counted as one.
9. The counting of objects can be refined by the application of a size parameter. This can be done through selecting the Measure drop-down window in the Count/Size window. Click Select Measurements. Highlight Area in the Filter range dialog box. Below, enter the range of the size to be counted. Refining the range of area will help to eliminate objects that are too small or large which may be artifacts. Other parameters can be measured, e.g., the sum of the intensity of the signal.
10. Continue to make adjustments to area size and intensity to optimize object count to achieve the best representation of signal selection.
11. Useful tools include Zoom In, Pan Image Tool, and Pan all Images Simultaneously, Irregular AOI (if wanting to measure an area within an area).
12. Area of the entire tissue section can be done through the Measure drop-down menu. Select Measurements. Under Features click on the Create Polygon Feature icon. Draw a line around the tissue section. Double click to close outline.
13. Statistics of the data can be viewed by selecting Statistics under the View drop-down menu. Data includes, sum of the area of all objects, its standard deviation, etc. Data can be exported to an Excel file.
14. Save all images as necessary.

4 Notes

1. While male mice can be used, female mice are less likely to fight if housed together. This is critical if space within the chamber is limited to only a few rodent cages. All animals must be of a similar age.
2. Different types of hypoxic chambers are available [10, 12]. However, it is critical to confirm hypoxia by measuring oxygen levels in the chamber during the entire length of the experiment. We used a highly sensitive oxygen probe for our experiments; however, other less expensive probes are available (Vernier, Beaverton, OR, USA).

3. Fisher Scientific Canada has since replaced this brand with the following product: Andwin Scientific Tissue-Tek* CRYO-OCT (Cat. No.: 14-373-65 Andwin Scientific No.:4583).
4. We find that it is best to prepare this solution fresh each time.
5. Pericytes are mural cells that belong to the same cell lineage as vascular smooth muscle cells [13]. Depending on the size of the vessel (intermediate size to small microvessels), transitional forms of pericytes and smooth muscle cells express common markers, such as α-smooth muscle actin (α-SMA), desmin, and nerve/glial antigen 2 (NG2) chondroitin sulfate proteoglycan, among others. These markers also vary depending on the organ, type of vessel, and developmental stages, which makes the identification of these cells very challenging [13–15]. The use of multiple markers to clearly identify smooth muscle cells and particularly pericytes is necessary (Fig. 1).
6. The fibrinogen antibody indicated in Table 1 recognizes both native form and active fragments D and E. The active fragments of fibrinogen have been shown to disrupt the endothelial barrier and induce microvascular leakage [16].
7. In our laboratory, we aliquot 20 μl of the secondary antibodies into 0.5 ml Eppendorf tubes and store at −20 °C. After thawing the tube for the immunofluorescence experiment, we then store the antibodies at 4 °C. Invitrogen has an extensive list of high-quality secondary antibodies. When selecting secondary antibodies, the following parameters should be considered: reactivity, host, isotype, and conjugate. The choice of conjugate is dependent on the type of fluorescence filter cube installed in the microscope. In our experiments we primarily used Alexa Fluor 488, 568, and 647 simultaneously along with DAPI or Hoechst.
8. We have used DAPI as well in our staining protocol.
9. Image-Pro® Plus and Image-Pro Analyzer 7.0 are very powerful image analysis software; however, other software such as imageJ can also perform similar quantitative image analysis.
10. If using the hypoxic chamber for the first time, it would be a sensible precaution to measure oxygen levels for at least a week before the start of any experiments. This will confirm whether hypoxia is maintained in the chamber over a prolonged period of time.
11. The Forma Anaerobic System model 1024 comes with an interchange compartment that permits a fast and simple transfer from normoxic to hypoxic conditions and vice versa. The interchange should not be used to introduce the mice into the actual chamber. The vacuum created in the interchange is stressful to the mice. Mice should be placed in the already hypoxic chamber directly, and the whole chamber should then

be re-purged to 10 % oxygen manually, according to the manufacturer's instructions.

12. Mouse cages will have to be replaced with new cages, fresh food, bedding, and water, every few days. If the hypoxic chamber has a glove box setup with an interchange compartment, new cages can be introduced in the chamber without removing the mice from hypoxia. This procedure is not possible for some hypoxic chamber models. However, hypoxic mice can be exposed to normoxia temporarily during cage cleaning, without affecting the experimental outcome [17].
13. Always start with control group first, to ensure perfusion is successful and that there are no problems in the procedure, before taking the corresponding hypoxic animals out of the chamber.
14. Only the brain will be perfused with PBS when the right and left external jugular veins are cut. To perfuse the whole animal, the right atrium should be snipped instead of the jugular veins.
15. In our original description of this method [2], we used cold heparinized saline to perfuse the mice. Subsequently, we have used cold PBS instead with the same results.
16. Hematocrit capillary tubes are placed inside Eppendorf tubes for centrifugation. However, we have constructed metal adapters that can be fitted in the Eppendorf tubes, which then secures the capillary tubes in place. Alternatively, small amounts of cotton can also be placed in the Eppendorf tubes; this would also protect and secure the capillary tubes from damage during centrifugation.
17. While −21 °C is an ideal starting temperature for brain sectioning, the temperature may have to be adjusted during the procedure, in order to optimize the conditions for cutting. The quality of sections is highly dependent on humidity levels, sharpness and angle of the blade, as well as several other factors that are beyond the scope of this chapter. However, several good troubleshooting tips can be found in the following websites: http://www.ihcworld.com/royellis/problems/mainpage.htm and http://pathologyinnovations.com/frozen_section_technique.htm
18. It is a sensible precaution to section all the brains from the same group, before introducing brains from different experimental groups into the cryostat.
19. We do not cover the brain sample completely with OCT embedding medium. OCT is only used to fix the brain onto the specimen chuck for sectioning. In this way, the brain can be completely visualized and easily oriented in relation to the cutting edge of the cryostat blade. Moreover, we have found that, in some cases, OCT can distort the brain tissue during sectioning.

20. In our original study [2], the entire frontal brain section of the mouse (Bregma 2.80 mm to Bregma 1.18 mm) was sectioned [2]. We collected one brain section per microscope slide. However, depending on the type of experiments, a maximum of three sections can easily fit on the slide for immunofluorescence.
21. Occasionally, brain sections will not "jump"/adhere onto the microscope slide. If this occurs, brain sections can be carefully picked up, with small paintbrushes, and placed on the frozen microscope slide, inside the cryostat. The tissue section can then "melt" and stick on the microscope slide by warming the backside of the slide with index finger.
22. It is important to determine the optimal labeling conditions and ensure no cross reactivity for each individual antibody. Proper positive (replacing the primary antibodies with the same concentration of isotype matched IgG) and negative (omission of primary antibodies and incubating tissue with only secondary antibodies) controls must be performed.
23. In triple antibody labeling experiments, we normally incubate primary antibodies one at a time followed by the secondary antibody (after each primary incubation). However, incubating simultaneously with multiple primary antibodies at a time, followed by simultaneous incubation with appropriate secondary antibodies, is also possible.
24. After fluorescence mounting media has hardened, sealing edge of the coverslips with clear nail polish can further protect section.
25. In our original study [2], we used four mice per genotype (knockout and wild type), treatment condition (normoxia and hypoxia), and time point (7, 14, and 21 days). Three brain sections from the frontal cortex (Bregma 2.80 mm, Bregma 1.70 mm, and Bregma 1.18 mm) were used per animal. Four images per section were acquired, two from the right and two from the left side of the cerebral cortex (12 images/animal) (Fig. 2).
26. When using high-quality antibodies that are optimized for immunofluorescent conditions, very little adjustment of the contrast will be required. Furthermore, when antibodies perform well, the high contrast will also improve quantification of vessels by the image analysis program.
27. The upper and lower limit values should be consistent for all the images. We set the lower limits to 10 μm^2 (Fig. 2).
28. If small or large objects (artifacts) are selected by the program and are obviously not vessels, these artifacts can be excluded. In the count/size window, click on "Measure" and then

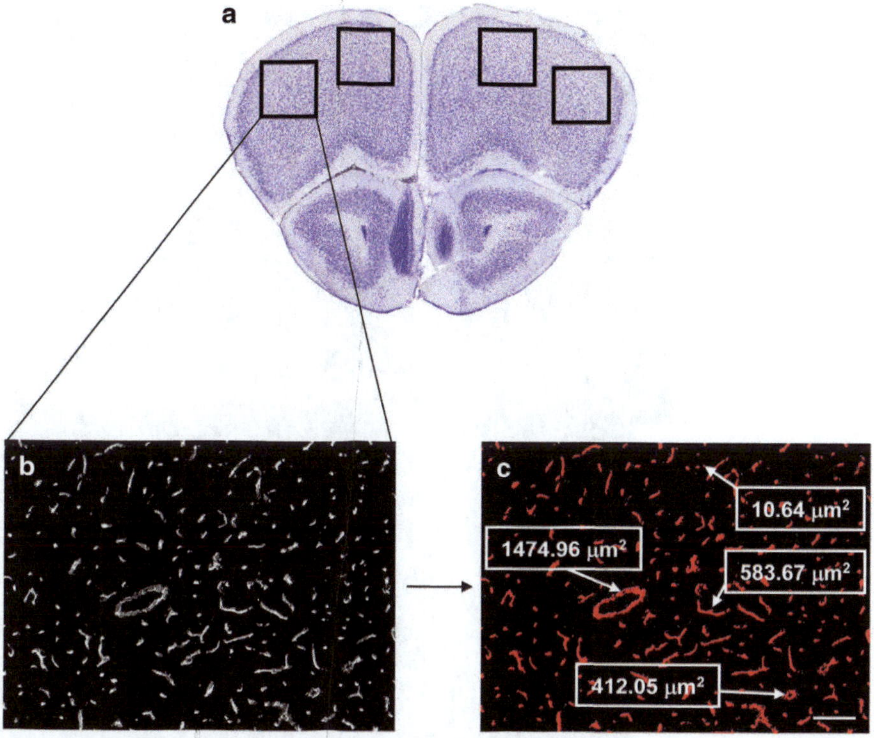

Fig. 2 Example of the analysis performed by the image analysis program, Image-Pro® Plus (**a–c**). (**a**) Representative image of one of the three sections used from the frontal cortex (Bregma 2.80 mm); the *outlined boxes* represent the four 10× images taken from the cerebral cortex. (**b**) A grey scale representation of one of the 10× images immunostained with the CD31 vessel marker and analyzed by the image analysis software. (**c**) An example of the "intensity range selection" analysis performed by Image-Pro® Plus; the areas in *red* delineate the surface of the CD31+ vessels measured by the software. Some examples of microvessel size (area) are denoted in the figure; the largest object is 1,474.96 μm² and the smallest object is 10.64 μm². Scale bar = 100 μm

choose "Select Measurements." In the "filter ranges" window, highlight "Area" and this will activate the "Edit Range"; click on it then press OK. Move the vertical lines on either side of the graph left or right to deselect objects that are not vessels. However, keep the lower limits consistent to 10 μm². In the "Select Measurements" window, click "Measure" and a new count will appear with the excluded objects.

29. Occasionally, artifacts may be present and will be counted as objects, e.g., fibers or dust. These can be removed manually by zooming in on the object enough to click on it so that it appears in the object attributes window also found under the View drop-down menu. Click on the hide button and the object will be removed from the total count.

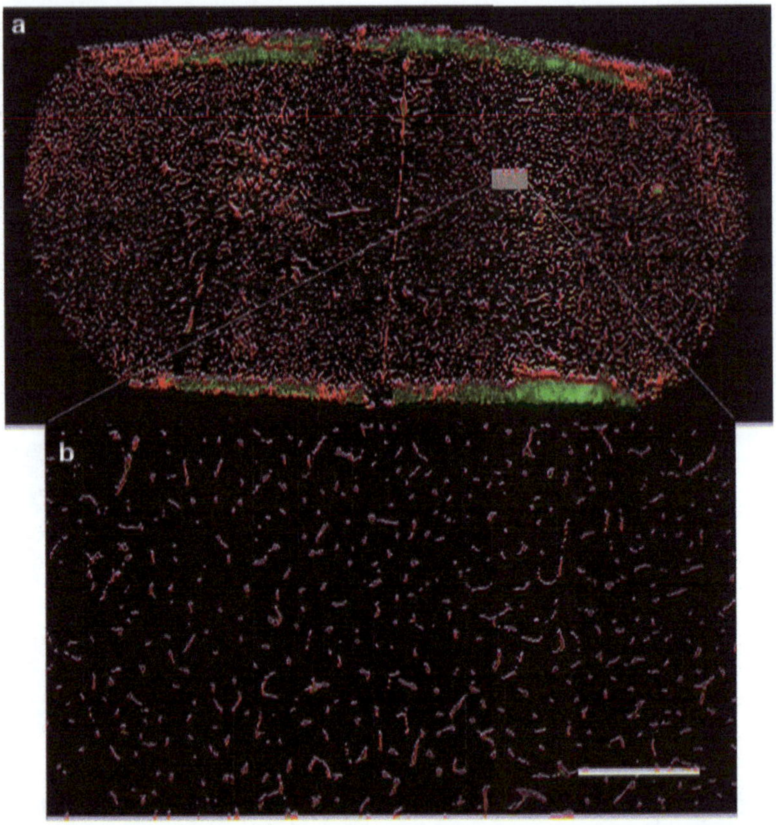

Fig. 3 (**a**) Mosaic image of a 10 μm brain section stained with CD31 analyzed by Image-Pro software. Vessels (*in green*) are outlined (*in red*) as single objects for the purpose of counting and total area quantification. Care must be given to obtaining a slide free from artifacts such as folding of tissue as observed in the upper and lower edges of this tissue section. (**b**) Magnification of one of the individual images that comprise the mosaic image. Scale bar: 100 μm

Acknowledgements

This work was supported by a grant from Heart and Stroke Foundation of Ontario (HSFO) and a scholarship (M.F.A.) from HSFO/Canadian Institute of Health Research (CIHR)/Canadian Stroke Network (CSN)/Astra Zeneca.

References

1. Milner R, Hung S, Wang X, Berg GI, Spatz M, del Zoppo GJ (2008) Responses of endothelial cell and astrocyte matrix-integrin receptors to ischemia mimic those observed in the neurovascular unit. Stroke 39:191–197
2. Freitas-Andrade M, Carmeliet P, Charlebois C, Stanimirovic DB, Moreno MJ (2012) PlGF knockout delays brain vessel growth and maturation upon systemic hypoxic challenge. J Cereb Blood Flow Metab 32:663–675

3. Li L, Welser JV, Dore-Duffy P, del Zoppo GJ, Lamanna JC, Milner R (2010) In the hypoxic central nervous system, endothelial cell proliferation is followed by astrocyte activation, proliferation, and increased expression of the alpha 6 beta 4 integrin and dystroglycan. Glia 58:1157–1167
4. Dore-Duffy P, LaManna JC (2007) Physiologic angiodynamics in the brain. Antioxid Redox Signal 9:1363–1371
5. Freitas-Andrade M, Carmeliet P, Stanimirovic DB, Moreno M (2008) VEGFR-2-mediated increased proliferation and survival in response to oxygen and glucose deprivation in PlGF knockout astrocytes. J Neurochem 107: 756–767
6. Stanimirovic DB, Friedman A (2012) Pathophysiology of the neurovascular unit: disease cause or consequence? J Cereb Blood Flow Metab 32:1207–1221
7. Beck H, Acker T, Puschel AW, Fujisawa H, Carmeliet P, Plate KH (2002) Cell type-specific expression of neuropilins in an MCA-occlusion model in mice suggests a potential role in post-ischemic brain remodeling. J Neuropathol Exp Neurol 61:339–350
8. LaManna JC, Chavez JC, Pichiule P (2004) Structural and functional adaptation to hypoxia in the rat brain. J Exp Biol 207:3163–3169
9. Ratan RR, Siddiq A, Smirnova N, Karpisheva K, Haskew-Layton R, McConoughey S, Langley B, Estevez A, Huerta PT, Volpe B, Roy S, Sen CK, Gazaryan I, Cho S, Fink M, LaManna J (2007) Harnessing hypoxic adaptation to prevent, treat, and repair stroke. J Mol Med (Berl) 85:1331–1338
10. LaManna JC, Vendel LM, Farrell RM (1992) Brain adaptation to chronic hypobaric hypoxia in rats. J Appl Physiol 72:2238–2243
11. Xu K, Lamanna JC (2006) Chronic hypoxia and the cerebral circulation. J Appl Physiol 100:725–730
12. Al Ahmad A, Gassmann M, Ogunshola OO (2009) Maintaining blood-brain barrier integrity: pericytes perform better than astrocytes during prolonged oxygen deprivation. J Cell Physiol 218:612–622
13. Armulik A, Abramsson A, Betsholtz C (2005) Endothelial/pericyte interactions. Circ Res 97:512–523
14. Diaz-Flores L, Gutierrez R, Madrid JF, Varela H, Valladares F, Acosta E, Martin-Vasallo P, Diaz-Flores L Jr (2009) Pericytes. Morphofunction, interactions and pathology in a quiescent and activated mesenchymal cell niche. Histol Histopathol 24:909–969
15. Krueger M, Bechmann I (2010) CNS pericytes: concepts, misconceptions, and a way out. Glia 58:1–10
16. Guo M, Daines D, Tang J, Shen Q, Perrin RM, Takada Y, Yuan SY, Wu MH (2009) Fibrinogen-gamma C-terminal fragments induce endothelial barrier dysfunction and microvascular leak via integrin-mediated and RhoA-dependent mechanism. Arterioscler Thromb Vasc Biol 29:394–400
17. Ward NL, Moore E, Noon K, Spassil N, Keenan E, Ivanco TL, LaManna JC (2007) Cerebral angiogenic factors, angiogenesis, and physiological response to chronic hypoxia differ among four commonly used mouse strains. J Appl Physiol 102:1927–1935

Chapter 15

Examining Vascular Remodeling in the Hypoxic Central Nervous System

Amin Boroujerdi, Jennifer V. Welser-Alves, and Richard Milner

Abstract

The goal of this chapter is to highlight techniques used to determine the role of molecular mechanisms involved in remodeling of cerebral blood vessels. Enhanced vascularization in the central nervous system (CNS) is seen in many diseases including stroke, cancer, and multiple sclerosis (MS). However, despite the prevalence of this phenomenon in these different pathological conditions, the exact nature of how it occurs still remains unclear. To better understand the process of cerebrovascular remodeling, we use the chronic hypoxia model, in which a vigorous and robust angiogenic remodeling response takes place. In this model, mice are placed in a hypoxic chamber (8 % O_2 for up to 14 days), which results in strong vascular remodeling and increased vessel density within the CNS. Using an immunofluorescent (IF)-based approach, different aspects of this vascular remodeling response can be examined. By employing this method, we have shown that chronic mild hypoxia triggers both angiogenic (capillary sprouting) and arteriogenic (widening of arterial vessels) responses. Furthermore, we have used this system to define both the expression pattern and potential role of candidate adhesion molecules in this vascular remodeling process. Thus, the techniques described in this chapter can be used to define the importance of different molecular mechanisms in vascular remodeling in the CNS.

Key words Vascular remodeling, Hypoxia, Central nervous system (CNS), Angiogenesis, Arteriogenesis, Extracellular matrix (ECM) proteins, Fibronectin, Integrin, Endothelial cell, Proliferation

1 Introduction

Vascular remodeling is a physiological response that occurs in various conditions such as embryonic development, female reproductive cycles, tissue regeneration, and repair [1]. In addition to its normal physiological role, in the CNS, vascular remodeling also occurs in many disease conditions including ischemic stroke, multiple sclerosis, cancer, and Alzheimer's disease [2–4]. However, much is still unknown about the exact mechanisms by which vascular remodeling occurs in these various diseases and how vascular remodeling affects the progression of these conditions.

Following chronic mild hypoxia (14 days at 8 % O_2), an increase in blood vessel density is observed in the CNS [5, 6]. This increase

Richard Milner (ed.), *Cerebral Angiogenesis: Methods and Protocols*, Methods in Molecular Biology, vol. 1135,
DOI 10.1007/978-1-4939-0320-7_15, © Springer Science+Business Media New York 2014

in vascularity is one of the homeostatic mechanisms that acts to maintain CNS oxygenation in the face of reduced levels of oxygen availability. Many studies have used the chronic mild hypoxia model to better understand the mechanisms involved in vascular remodeling. A variety of IF markers have been used in this model to identify the events and important factors associated with vascular remodeling. For instance, the number of endothelial cells (identified by an anti-CD31 antibody) that express the cell proliferation marker Ki67 strongly increases after 4 days of mild hypoxia [5]. This suggests that endothelial cell proliferation is an important and early part of the angiogenic response to chronic mild hypoxia. Size distribution analysis of blood vessels has shown that in addition to increased numbers of smaller vessels during hypoxic-induced vascular remodeling, counts of middle and large size vessels also increase, suggesting an active arteriogenic response [6]. This was confirmed by the finding that the number of vessels staining positive for alpha smooth muscle actin (α-SMA), an indicator of arterial vessels, also increased during 14 days of hypoxic exposure [6]. These results signify that cerebral hypoxia not only induces angiogenesis as previously described but also strongly promotes an arterogenic remodeling response.

In terms of molecular mechanisms, vascular remodeling in this system is associated with altered expression of different β1 integrin receptors present on blood vessels. Studies have shown a switch occurring from high $\alpha5\beta1$/low $\alpha6\beta1$ during angiogenesis to low $\alpha5\beta1$/high $\alpha6\beta1$ integrin in mature vessels [6, 7]. Concomitant with changes in α5β1 integrin, the extracellular matrix (ECM) protein ligand fibronectin showed parallel changes in expression level during the remodeling response [6, 8, 9]. These findings are consistent with previous studies defining roles for integrins and ECM proteins at different stages of the remodeling process, including cell proliferation, cell migration, growth, and differentiation of blood vessels [9–11]. Based on these combined studies, it is clear that the combination of the chronic mild hypoxia model and IF techniques offers a powerful approach to better understand the mechanisms underlying cerebrovascular remodeling.

2 Materials

2.1 Chronic Hypoxia System

1. OxyCycler hypoxic chamber and control system (Biospherix, Redfield, NY).
2. Compressed gas source of oxygen.
3. Compressed gas source of nitrogen.
4. Compressed mixed gas source of carbon dioxide (1 %)/ oxygen (99 %).

5. Three gas regulators, one for each compressed gas source.
6. Thermometer (for calibration).
7. 1/4″ tubing to connect the compressed supply gas source to the units.

2.2 Tissue Preparation for Freezing Mice Brains

1. Dissecting instruments including scissors and forceps.
2. Euthanizing agent (Nembutal).
3. Perfusion apparatus (mini-pump variable flow; Fisher Scientific).
4. Cold phosphate buffer saline (PBS).
5. Tissue-Tek OCT embedding compounds.
6. Tissue-Tek Cryomolds.

2.3 Tissue Storage and Sectioning

1. Tissue-Tek OCT embedding compounds.
2. Feather low profile disposable microtome blades. Check for most suitable blades to be used, as this may vary with the type of cryostat.
3. Microscope slides.
4. Cryostat-microtome capable of cutting 10 μm thick sections.

2.4 Immunofluorescence

1. Humidified incubation chamber.
2. Slide racks.
3. Washing containers.
4. Acetone/methanol mix (50:50). Store at −20 °C.
5. PBS stored at room temperature.
6. PAP pen for immunostaining.
7. DAKO antibody diluent.
8. Primary antibodies: hamster anti-CD31 (1:100) (Abcam, Cambridge, MA); rabbit anti-Ki67 (1:1500) (Vector Laboratories, Burlingame, CA), rat anti-CD105 (1:100), and rat anti-integrin α5, α6 subunit (1:100) (BD Pharmingen, La Jolla, CA); and rabbit anti-fibronectin (1:400), rabbit anti-laminin (1:500), and rabbit anti-alpha smooth muscle actin-Cy3 (1:500) (Sigma, St. Louis, MO).
9. Secondary antibodies: DyLight 594 and 488 anti-Armenian hamster (1:300) (Biolegend, San Diego, CA), Cy3 anti-rabbit and anti-rat (1:200) (Jackson ImmunoResearch, Baltimore, PA), and Fluor 488 anti-rabbit alexa (1:500) (Invitrogen).
10. Mounting medium: Aqua-Poly/Mount (Polyscience).
11. Microscope cover glass (24 × 50) #1(0.13–0.17 mm).

2.5 Analysis of Markers of Vascular Remodeling

Volocity software (PerkinElmer) (*see* **Note 1**).

3 Methods

3.1 Chronic Hypoxia Model

1. Calibrate the OxyCycler hypoxic chamber. Please refer to the manual for full calibration instructions.
2. Place the mouse cages in the hypoxia chamber.
3. Fully open the main knob on the nitrogen tank, and increase the regulator to 10 psi flow rate.
4. Repeat for the oxygen tank.
5. Start the hypoxia program and monitor the chamber to ensure that the oxygen level in the chamber reaches 8 %.
6. Monitor the mouse cages and the level of the tanks daily (*see* **Note 2**).
7. Remove the mouse cages after different periods of hypoxia (up to 14 days).

3.2 Tissue Preparation: Freezing Mice Brains

1. Euthanize the mice by administering 1 ml of Nembutal at a dose of 5 mg/ml via intraperitoneal injection (*see* **Note 3**).
2. While the mouse is heavily sedated, open the thoracic cavity and expose the heart. Make a nick in the right atrium with scissors, and perform perfusion by injecting a PBS into the left ventricle using the mini pump. We infuse at least 10 ml PBS per mouse to ensure a thorough perfusion to remove as much blood as possible (*see* **Note 4**).
3. Following perfusion, remove the head and spray with 70 % alcohol to damp the fur.
4. Cut through skin down the midline and reflect back.
5. Remove the top of the skull without touching the brain.
6. Carefully cut through the dura mater along the length of the brain and reflect back.
7. Working forward from the back, gently separate the brain from the floor of the skull using the scalpel to cut through the nerve roots.
8. When the brain has been freed, gently place it on a Petri dish or parafilm. Remove any debris and bisect the brain into the right and left lobes, i.e., make a vertical cut.
9. Place each lobe into a Cryomold with the freshly cut side face down, and fill the mold with tissue-Tek OCT and store at −80 °C.

3.3 Tissue Storage and Sectioning

1. For cutting brain tissue, set the chamber temperature to −19 °C. This temperature may have to be varied slightly for optimal cutting.
2. Transfer the brain tissue from the −80 °C freezer to the cryostat chamber, and allow the temperature to equilibrate for 30–45 min prior to cutting.
3. Mount a disposable blade in the cryostat, being careful of the sharp edges. Allow the blade to cool before sectioning.
4. Adjust the blade angle, usually between 0 and 5 degrees. This angle is variable.
5. Put a small amount of OCT on a chilled specimen disc, and press the brain tissue into the OCT before it freezes. More OCT can be added later to ensure specimen is firmly mounted onto the specimen disc.
6. Ensure the handwheel on the side of the cryostat is locked, so that the specimen holder is in the upper position and place specimen disc into the holder.
7. Adjust the cutting advance to 30 μm and advance/retreat the brain tissue block until it just touches the blade.
8. Trim the tissue block until you have a complete section through the tissue by rotating the main cutting handle clockwise. At this point, reduce the cutting advance to 10 μm.
9. Lower the anti-roll plate and turn the main cutting handle to produce a section. The speed of the cutting process depends on the type of tissue to be cut. Generally, the harder the tissue, the faster the section is cut (*see* **Note 5**).
10. Mount the sections onto a glass slide and place the slides in a slide box in the cryostat chamber. When the box is full, transfer to the −80 °C freezer and store until the slides are needed.
11. When sectioning is complete, dispose of the blade in a sharps container, and remove the specimen from the specimen disc. Cover the cut surface of the specimen with OCT to protect the surface, and store at −80 °C until further sections are required.
12. With a cold brush, sweep all the tissue shaving into a BioHazard bag and dispose.

3.4 Immunofluorescence

1. Remove the slides from the −80 °C freezer and air-dry for 15 min.
2. With the PAP pen, draw a water repellent circle around the brain sections, leaving at least 5 mm from the edge of the section and allow to dry.
3. Load the slides into a rack and immerse them into staining dishes containing −20 °C acetone/methanol mix (50:50) for 3 min to fix.

4. Remove the racks containing the slides, place in a new staining dish, and rapidly flood with PBS, taking care not to damage the section. Do not allow the section to dry out.
5. Wash 3× 5 min with PBS, each time replacing with fresh PBS.
6. Lay the slides horizontal in the humidified incubation chamber. Add the primary antibody/antibodies (either single or mix of antibodies) to the sections. Dilute the antibodies in DAKO antibody diluent. Ensure that the sections are covered and the antibody solution is retained within the water repellent circle. Incubate at 4 °C overnight or for 2 h at 37 °C.
7. Discard the primary antibody and wash the section 3× 5 min each with PBS.
8. Perform all subsequent steps under dimmed room lights.
9. Dilute the corresponding fluorescent-conjugated secondary in DAKO antibody diluent. Ensure that the sections are covered and that the antibody solution is retained within the water repellent circle. Place the slides in the incubation chamber, and store at room temperature for 2 h or incubate at 37 °C for half an hour (*see* **Note 6**).
10. Discard the secondary antibody/antibodies and wash the sections 3× 5 min each with PBS.
11. Mount the sections in Aqua polymount. Allow the medium to come to room temperature before use. Place a couple of small drops of the mounting medium onto the glass slide. Gently place the coverslip onto the glass slide. Leave for 10 s to allow the weight of the microslide to spread the mounting medium over the section (*see* **Note 7**).
12. Allow the polymount to dry and set before viewing the slides (generally the next day). Slides can be viewed immediately or stored at 4 °C.
13. Examine the staining under a fluorescent microscope with excitation at 488 (green emission) and 570 (red emission). Examples of staining of frozen brain sections for the markers CD31 (endothelial-specific marker) and Ki67 (marker of cell proliferation) are shown in Fig. 1. Staining for fibronectin and the α5 integrin can be seen in Fig. 2.
14. Take three images at 20× for each section per mouse.

3.5 Analysis of Specific Vascular Markers Using the Volocity Software

1. Using the Volocity software, create a new file and import the images taken with the microscope.
2. Analysis of blood vessel markers can be performed by generating a protocol in the software with certain parameters. Specific parameters of each marker can be analyzed, including total counts, sum area, expression levels, and variation in vessel size.

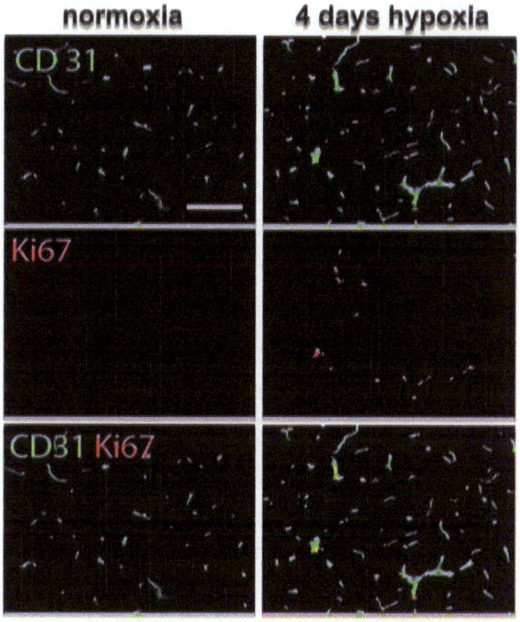

Fig. 1 Evaluation of endothelial cell proliferation in the hypoxic CNS. Dual-IF for CD31 (endothelial-specific marker) and Ki67 (marker of cell proliferation) was performed on frozen brain sections taken from mice under normoxic (control) or hypoxic conditions. Scale bar = 50 μm. Note that 4 days of hypoxia promoted a marked endothelial cell proliferation response

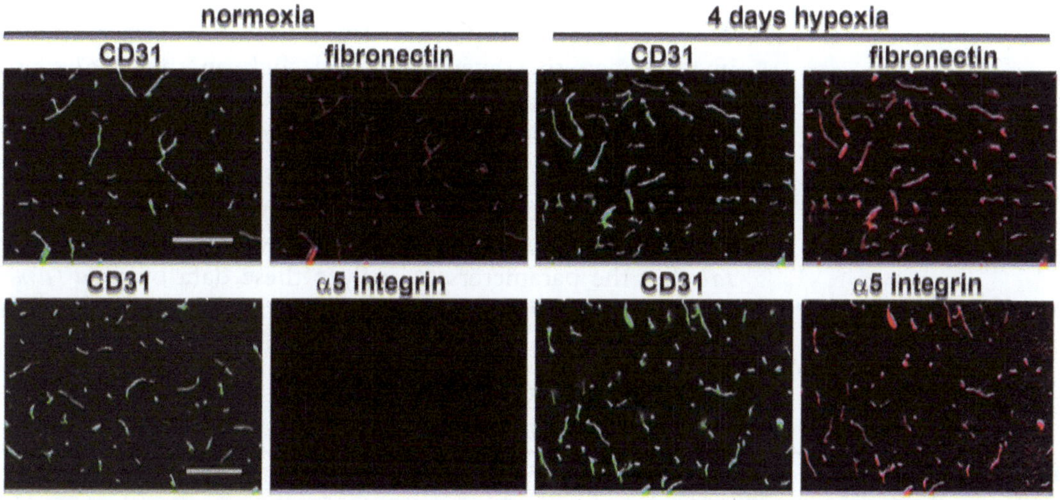

Fig. 2 Hypoxic induction of fibronectin and the α5β1 integrin on cerebral blood vessels. Dual-IF for CD31 (endothelial-specific marker) and either the fibronectin or the fibronectin receptor, α5β1 integrin, was performed on frozen brain sections taken from mice under normoxic (control) or hypoxic conditions. Scale bar = 50 μm. Note that 4 days of hypoxia promoted a marked increase in endothelial expression of fibronectin and the α5β1 integrin

3. Click on the measurement tab above the image and drag the first parameter task, and find objects using *standard deviation (SD) intensity* into the measurement list. Adjust the SD intensity accordingly only allowing the marker staining to be highlighted by the software (*see* **Note 8**).
4. Drag the second parameter task, and find objects using *excluding objects by size* into the measurement list. Adjust accordingly to exclude objects that are too small to be vessels (*see* **Note 9**).
5. Drag the third parameter task, and find objects using *close objects* into the measurement list. Adjust accordingly the iteration number allowing fragment vessels to be counted as an individual vessel (*see* **Note 10**).
6. By clicking on the *measurements tab* and turning on the *update feedback*, this will generate a measurement table.
7. The table can be saved by clicking on the *make measurement item* under the *measurement tabs.*
8. To determine the total count of vessels in your image, open the saved measurement table and click on the *analysis tab* followed by *analyze this data tab.* Set the parameters to analyze these data by ID, summarize by *count*, and organize the data by *type.*
9. To determine the total sum area of vessels in your image, open the saved measurement table and click on the *analysis tab* followed by *analyze this data tab.* Set the parameters to analyze these data by *area* (μm^2), summarize by *sum*, and organize the data by *type.*
10. To determine the expression levels of vessel markers in your image, open the saved table and click on the *analysis tab* followed by analyze this *data tab.* Set the parameters to analyze these data by *mean*, summarize by *mean*, and organize the data by *type.*
11. To determine variation in the size of vessels, open the saved table and click on the *analysis tab* followed by *analyze this data tab.* Set the parameters to analyze these data by *area* (μm^2), summarize by sum, and organize the data by *type.* Click on the *organization tab* in the data analysis window. Check the group rows into bins of size in increments of *10.*

4 Notes

1. The Volocity Demo software is available at the PerkinElmer webpage. In addition Image J can also be used for those familiar with the program.
2. When conducting hypoxia experiments, it is good practice to monitor the health of the mice daily by looking through the

transparent window each day and, on alternate days, opening the chamber briefly to inspect the health of the mice and replace the bedding and replenish water and food supplies.

3. An alternative method of inducing deep surgical plane anesthesia is to administer an intraperitoneal injection of 100 mg/kg ketamine and 10 mg/kg xylazine.
4. A thorough perfusion is required to remove as much blood as possible. This makes the IF staining a lot cleaner and reduces the possibility of blood components contributing significantly to the IF signal. This is especially important for our studies of vascular remodeling as we are focused specifically on protein expression on vascular cells, e.g., endothelial cells, smooth muscle cells, and pericytes.
5. If the sections are not cutting well, adjust the temperature of the cryostat chamber. Generally, if sections appear "shattered," the temperature is too low and so the chamber temperature should be warmed 1–2 °C. If the sections stick to the cryostat base plate as they are cut, the temperature is too warm and so the cryostat chamber should be chilled 1–2 °C. After temperature adjustment, allow at least 5 min for the tissue to equilibrate. If tissue sections show scores or lines, ensure no debris is attached to the anti-roll plate or the underside of the blade. Check the blade for nicks on cutting edge, and if present, replace the blade.
6. You can obtain better specific staining by centrifuging the diluted secondary antibodies before covering the sections (5 min at 15,000 rpm (9,800 × *g*) in a benchtop microfuge).
7. Air bubbles are undesirable in the mounting medium. Slow careful lowering of the slide onto the coverslip and mounting medium reduces their appearance. In some cases, trapped air bubbles may be eased out of the section by gentle pressure applied to the coverslip by use of a blunt pencil. Take care not to crack the coverslip or damage the tissue section.
8. A value of 2 standard deviations in the lower limit and 0 in the upper limit is generally effective in highlighting the various blood vessel markers.
9. With images taken at 20×, excluding objects less than 25 μm^2 generally prevents the software from highlighting nonspecific staining.
10. For the close object parameter, a value of 5 iterations is generally used. This value allows highlighted images 5 pixels apart to be counted as a single vessel as opposed to artificially representing an individual vessel, which would artificially skew the total count number.

Acknowledgments

This work was supported by the American Heart Association (Western State Affiliate) Postdoctoral Fellowship (A.B.) and by the National Multiple Sclerosis Society, Harry Weaver Neuroscience Scholar Award (R.M.), and Postdoctoral Fellowship (J.V.W.).

References

1. Forget MA, Desrosiers RR, Beliveau R (1999) Physiological roles of matrix metalloproteinase: implications for tumor growth and metastasis. Can J Physiol Pharmacol 77(7): 465–480
2. Kirk S, Frank AJ, Karlik S (2004) Angiogenesis in multiple sclerosis: is it good, bad or an epiphenomenon? J Neurol Sci 217(2):125–130
3. Polverini PJ (1995) The pathophysiology of angiogenesis. Crit Rev Oral Biol Med 6(3): 230–247
4. Winkler JD, Jackson JR (2000) Chronic inflammation and angiogenesis. In: Rubanyi GM (ed) Angiogenesis in health and disease: basic mechanisms and clinical applications. Marcel Dekker, New York, pp 407–416
5. Li L, Welser JV, Dore-Duffy P, del Zoppo GJ, LaManna JC, Milner R (2010) In the hypoxic central nervous system, endothelial cell proliferation is followed by astrocytes activation, proliferation, and increased expression of the α6β4 integrin and dystroglycan. Glia 58(10): 1157–1167
6. Boroujerdi A, Welser-Alves JV, Tigges U, Milner R (2012) Chronic cerebral hypoxia promotes arteriogenic remodeling events that can be identified by reducing endoglin (CD105) expression and a switch in β1 integrins. J Cereb Blood Flow Metab 32:1820–1830
7. Milner R, Campbell IL (2002) Developmental regulation of β1 integrins during angiogenesis in the central nervous system. Mol Cell Neurosci 20:616–626
8. Milner R, Hung S, Erokwu B, Dore-Duffy P, LaManna JC, del Zoppo GJ (2008) Increased expression of fibronectin and the alpha5 beta 1 integrin in angiogenic cerebral blood vessels of mice subject to hypobaric hypoxia. Mol Cell Neurosci 38(1):43–52
9. Li L, Welser-Alves JV, van der Flier A, Boroujerdi A, Hynes RO, Milner R (2012) An angiogenic role for the α5β1 integrin in promoting endothelial cell proliferation during cerebral hypoxia. Exp Neurol 237(1):46–54
10. Brooks PC, Klemke RL, Schon S, Lewis JM, Schwartz MA, Cheresh DA (1997) Insulin-like growth factor receptors cooperates with integrin alpha v beta 5 to promote tumor cell dissemination in vivo. J Clin Invest 99:1390–1398
11. Lorger M, Krueger JS, O'Neal M, Staflin K, Felding-Habermann B (2009) Activation of tumor cell integrin αvβ3 controls angiogenesis and metastatic growth in the brain. Proc Natl Acad Sci U S A 106(26):10666–10671

Chapter 16

Analysis of Cerebral Angiogenesis in Human Glioblastomas

Michel Mittelbronn, Peter Baumgarten, Patrick N. Harter, and Karl H. Plate

Abstract

The formation of new blood vessels is a major hallmark in the process of malignant transformation in human glioblastomas. In diffusely infiltrating gliomas, enhanced angiogenesis is associated with decreased patient survival rates and therefore serves as a central diagnostic criterion according to the WHO (World Health Organization) classification of tumors of the central nervous system (CNS). However, the assessment of what a newly built blood vessel really is and how the extent of glioma-associated angiogenesis can be estimated in vivo is often a highly subjective procedure with imprecise criteria depending on the experience of the neuropathologist. The increased interest in translational medicine and anti-angiogenic treatment strategies implies that basic researchers in glioma angiogenesis are frequently asked to validate their findings in patient material to provide evidence for potential clinical relevance of their results. Therefore, more precise methods and measurement techniques are needed to objectively measure the extent of angiogenesis in human glioblastoma samples. The present synopsis provides an overview about morphological methods to assess the formation of new blood vessels by quantitative imaging using histological and immunohistochemical marker profiles.

Key words Human glioblastoma, Angiogenesis, Immunohistochemistry, Endothelial cells, Mural cells, Pericytes, VEGF

1 Introduction

Despite promising innovations in brain tumor treatment, the median survival in unselected patient cohorts suffering from the most frequent and malignant primary brain tumor, the glioblastoma WHO grade IV, is still under 1 year [1]. Besides the highly proliferative and migratory potential, angiogenic processes as adaptive regulatory responses to hypoxia are necessary for continuous supply of oxygen and nutrition during malignant progression in glioblastomas. A large variety of proangiogenic molecules, of which vascular endothelial growth factor (VEGF) is the most prominent and potent representative, have been shown to be released by glioma cells, thereby contributing to the formation of new blood vessels (*for review see* ref. 2). Although the formation of new blood

Richard Milner (ed.), *Cerebral Angiogenesis: Methods and Protocols*, Methods in Molecular Biology, vol. 1135, DOI 10.1007/978-1-4939-0320-7_16, © Springer Science+Business Media New York 2014

vessels is considered as a negative prognostic factor in glioblastomas in general, it became more widely accepted that the angiogenic pattern itself is an independent prognostic factor in multivariate survival analysis [3]. Patients exhibiting an increase in the amount of regularly formed microvessels ("classic" vascular pattern) show significantly longer survival times as compared to the group with prominent vascular abnormalities in shape, size, and complexity of blood vessels resulting in glomeruloid, garland-like, or clustered bizarre vascular formations [3]. Thus, not only the number of vessels but especially the vascular morphology seems to be associated with tumor malignancy. However, the interobserver agreement concerning the degree of angiogenesis in glioblastoma is often very poor even among experienced neuropathologists, which is most likely related to the fact that different stages of angiogenesis are only poorly defined in the classical pathological diagnostic setting [4]. Therefore, there is a strong need for standardization of glioblastoma vessel morphology which has been tried to be resolved by the introduction of histologic morphometry. Distinct parameters such as the total microvascular area (TVA), the microvascular density (MVD), the proliferating capillary index (PCI), or the microvessel pericyte coverage index (MPI) had been introduced to evaluate the extent of ongoing angiogenesis in glioblastomas in a quantitative manner [5]. In addition, to objectify and quantify angiogenesis in glioblastoma, it is also important to have knowledge about the degree of vascularization in different areas of the human CNS under nonneoplastic conditions. The present synopsis provides an overview about the currently used histomorphological tools and methods to assess angiogenesis in human glioblastomas in vivo.

2 Materials

2.1 Fixatives (See Note 1)

The composition of the fixation solution (for 1 L) should be as follows:

1. 100 ml commercially available formalin (aqueous solution of 37–40 % formaldehyde).
2. 900 ml distilled (or tap) water.
3. Disodium hydrogen phosphate, anhydrous: Na_2HPO_4 (6.5 g/L).
4. Sodium dihydrogen phosphate, monohydrate: $NaH_2PO_4{\cdot}H_2O$ (4.0 g/L).
5. pH (at 20 °C): 7.0 ± 0.2.

2.2 Reagents for Gomori's Silver Impregnation for Reticulin Fibers (Type III Collagen) (See Note 2)

1. Distilled water/tap water.
2. 0.5 % potassium permanganate.
3. Potassium metabisulfite.
4. Freshly prepared 2 % ferric ammonium sulfate.
5. Ammoniacal silver nitrate solution.

6. 20 % unbuffered formalin.
7. 0.1 % gold chloride.
8. 2 % potassium metabisulfite.
9. 1 % sodium thiosulfate.
10. Nuclear fast red aluminum sulfate (0.1 g nuclear fast red dye in 100 ml 5 % aqueous aluminum sulfate, boil, cool down, and then filter, possibly adding one to two drops of glacial acetic acid).
11. Ethanol.
12. Xylene.

2.3 Immunohistochemical Analysis of Glioblastoma-Associated Angiogenic Factors

Angiogenic processes can be assessed by immunohistochemical preparations using antibodies specifically directed against distinct epitopes of molecules involved in angiogenesis. However, recently a tremendous change in morphological studies has taken place mainly due to the introduction of automated staining systems (e.g., DAKO, Menarini Diagnostics, Roche Diagnostics, Thermo Fisher Scientific) which provide highly standardized and reproducible results with a much higher sensitivity and specificity than usually reached before by using manual methods. Therefore, instead of providing staining protocols for immunohistochemical or immunofluorescent preparations in this paragraph, we would rather give an overview about the most reliable markers adding helpful information about distinct molecules or cell types involved in angiogenesis. Since many trustworthy companies currently sell well-established antibodies for immunohistochemical analyses, and the choice often depends on the species in which the antibodies were generated (e.g., for double/triple stainings), here we will not recommend any specific providers.

2.3.1 Antibodies for the Assessment of Endothelial Cells

1. CD31/PECAM-1 (*see* **Note 3**).
2. CD34 (*see* **Note 4**).
3. von Willebrand factor (*see* **Note 5**).
4. Tyrosine kinase with immunoglobulin-like and EGF-like domains 2 (Tie-2) (*see* **Note 6**).
5. Vascular endothelial growth factor receptor 1–3 (VEGFR1–3) (*see* **Note 7**).
6. CD105 (endoglin) (*see* **Note 8**).

2.3.2 Antibodies for the Assessment of Mural Cells (See **Note 9**)

1. Alpha-smooth muscle actin (α-SMA) (*see* **Note 10**).
2. CD248 (*see* **Note 11**).
3. Platelet-derived growth factor receptor-β (PDGFR-β) (*see* **Note 12**).

2.3.3 Antibodies for the Assessment of Basement Membrane Constituents

1. Collagen IV (*see* **Notes 13** and **14**).

2.3.4 Antibodies for the Assessment of Proliferative/Mitotic Activity (see ***Note 15****)*

1. Ki-67 (MIB-1) (*see* **Note 16**).
2. pHH3 (phospho-histone H3) (*see* **Note 17**).

3 Methods

3.1 Tissue Fixation and Preparation

1. Fix the tissues for 24–48 h in fixative as described above.
2. Embed the tissues in paraffin blocks according to routine protocols.
3. Sections of a constant thickness of approximately 3–5 μm should be prepared using a microtome.
4. Place the slices on glass slides (e.g., SuperFrost glass slides allowing strong electrostatic attraction of the tissue specimens).
5. After deparaffinization steps, tissue sections can be processed in further histological or immunohistochemical analyses.

3.2 Gomori's Silver Impregnation for Reticulin Fibers (Type III Collagen)

The reticulin silver stain (Fig. 1) should be performed as follows (*according to* ref. 6). *Do not use metal instruments!*

1. Deparaffinize the tissue slices and rehydrate the samples in descending series of ethanol to distilled water.
2. Oxidation in 0.5 % potassium permanganate for 2–3 min (potassium permanganate has to be purple; the solution is unfeasible if the color is brownish).
3. Rinse the brain slices in running tap water for 5 min.
4. Incubate the brain slices for 2 min in potassium metabisulfite.
5. Rinse the slices in running tap water for 5–10 min.
6. Sensitize the slices in freshly prepared 2 % ferric ammonium sulfate for 1–2 min.
7. Rinse in running tap water for 5 min followed by two washing steps in distilled water for 2 min each time.
8. Incubate the sections in ammoniacal silver nitrate solution for 1–2 min.
9. Rinse the sections quickly in distilled water for 5–10 s (1–3 times); this step is important for the silver impregnation (if rinse cycles are too short, impregnation is too compact; if rinse cycles are too long, vice versa).
10. Reduce the slides in 20 % unbuffered formalin for 5 min (sections should turn dark brown to black immediately after the formalin solution is applied).
11. Rinse the sections in running tap water for 5 min.
12. Tone the slides in 0.1 % gold chloride for at least 10 min.

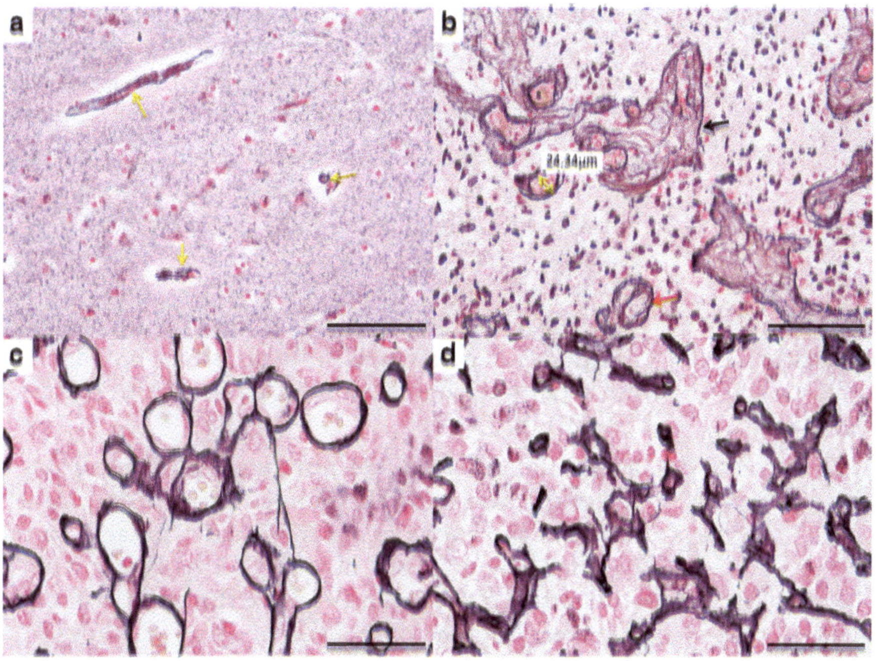

Fig. 1 Gomori's silver impregnation method for reticulin (collagen type III) fibers. (**a**) Normal human white matter is depicted showing small capillaries (*yellow arrows*), all exhibiting an internal diameter of 7–8 μm. (**b**–**d**) Different examples of human glioblastomas are presented showing (**b**) increasing internal diameters of blood vessels (*yellow arrow*) as well as bizarre newly formed blood vessels with glomeruloid (*red arrow*) or garland-like (*black arrow*) morphology, (**c**) clusters of proliferating enlarged microvessels, and (**d**) a classical form of increased microvascular density mainly consisting of small to moderately enlarged capillaries (scale bar = 100 μm in **a**, **b** and 50 μm in **c**, **d**)

13. Rinse in running tap water for 2 min.
14. Reduce the sections in 2 % potassium metabisulfite for 1–2 min.
15. Rinse in running tap water for 2 min.
16. Fix the sections in 1 % sodium thiosulfate for 1 min (not longer since fine reticulin fibers would be bleached).
17. Rinse the sections in running tap water for 10–15 min.
18. Counterstain the nuclei for 5–10 min in nuclear fast red aluminum sulfate.
19. Dehydrate the sections in ascending series of ethanol and xylene, before mounting.

3.3 Assessment of the Microvascular Density (MVD)

The assessment of all subsequent angiogenic parameters can be performed with a standard microscope and corresponding image analysis software with counting and measurement functions (*see* **Notes 18** and **19**).

1. For the analysis of the microvascular density, the "hot-spot" method is frequently used [7], although its diagnostic and

prognostic usefulness for glioblastoma patients is still a matter of debate [4]. The area showing the highest MVD should be determined at low magnification and subsequently analyzed for the amount of microvessels at 200× magnification.

2. Each microvessel as determined by reticulin or immunohistochemical staining for endothelial cells (which can be distinguished from the surrounding glioblastoma cells) should be counted.

3.4 Assessment of the Microvascular Area (MVA) (See Note 20)

1. The MVA is calculated by calculating the total vessel area using an image analysis software in a predefined area (e.g., hot spot) and magnification (total vessel area is indicated in μm or mm).
2. Most studies use CD34 immunohistochemistry as a tool for the calculation of the MVA in gliomas. It remains to be determined if the encouraging MVA data can also be reproduced by using immunohistochemistry for basement membrane constituents (e.g., collagen IV) since this approach might increase the overall outline of the MVA.

3.5 Assessment of the Proliferating Capillary Index (PCI)

1. In contrast to the "static" parameters (MVD, MVA, MID, MPI, and vessel morphology), the proliferating capillary index (PCI) provides an instantaneous assessment of ongoing vascular proliferation which might be a very helpful index in the evaluation of treatment effects in resections of glioblastoma recurrences.
2. For this purpose, it is important to co-label vascular structures (e.g., CD31) with antibodies indicating proliferative or mitotic activity (e.g., Ki-67 or pHH3). The PCI is calculated by the ratio of microvessels with proliferating endothelial cells and the total number of microvessels.
3. As increased data point to an involvement of mural cells in the vascular proliferation of glioblastomas, we recommend to expand the PCI to the subendothelial vascular compartments using basement membrane constituents (e.g., collagen IV) to define the external border of the glioblastoma vasculature.

3.6 Assessment of the Mean Intercapillary Distance (MID) (See Note 21)

1. The MID can be determined by measuring the distances between a microvessel and its neighboring ones (as determined by reticulin or immunohistochemical stainings for endothelial cells).
2. MID is significantly higher in normal human CNS (approximately 120 μm) as compared to glioblastoma (approximately 80 μm) tissue [8].

3.7 Assessment of the Microvascular Pericyte Coverage Index (MPI) (See Note 22)

1. The MPI reflects the fraction of blood vessels associated with mural cells.
2. This can be easily assessed by calculating the ratio of mural cells detected by immunohistochemical stainings for α-SMA-positive and the number of blood vessels as detected by CD31- or CD34-positive endothelial cells (e.g., an MPI of 30 % indicates that 30 % of all detected blood vessels are associated with α-SMA-positive mural cells).
3. In glioblastoma samples, the MPI index has been determined to be low as 10–20 % indicating that most tumor vessels are not surrounded by mural cells [5].

3.8 Assessment of Morphological Abnormalities (Shape, Size, Complexity)

1. Several studies indicated that MVD might not be predictive for glioblastoma patient survival, and its assessment shows high inter-rater variability [4].
2. In contrast, morphological abnormalities in the glioma vasculature including vessels with glomeruloid and garland-like morphology were significantly associated with negative patient survival (Figs. 1 and 2).
3. Although morphological assessment of the glioblastoma vasculature is more difficult than some of the aforementioned unbiased methods and requires good neuropathological knowledge, it seems to be an indispensable component in the evaluation of angiogenesis in glioblastomas.
4. Classical histomorphological staining methods such as reticulin silver impregnation (Fig. 1) might be of advantage since they are easy to perform, provide the full outline of the glioblastoma vasculature, and do not depend on more complex immunohistochemical settings which often differ between laboratories. However, more objective criteria are needed for the pure morphological assessment of glioblastoma vasculature to facilitate meaningful comparison between different studies.

4 Notes

1. Before assessing angiogenesis in glioblastoma samples in vivo, it is important to choose an appropriate fixation method to avoid artifacts and get the best morphological results. While in basic research tissue is often cryoembedded, it is well recognized that this method is often associated with a lot of artifacts. Due to its chemical characteristics, water forms hydrogen bonds upon freezing and expands while being cooled down, leading to ice crystals. In thinner sections, these ice crystals often

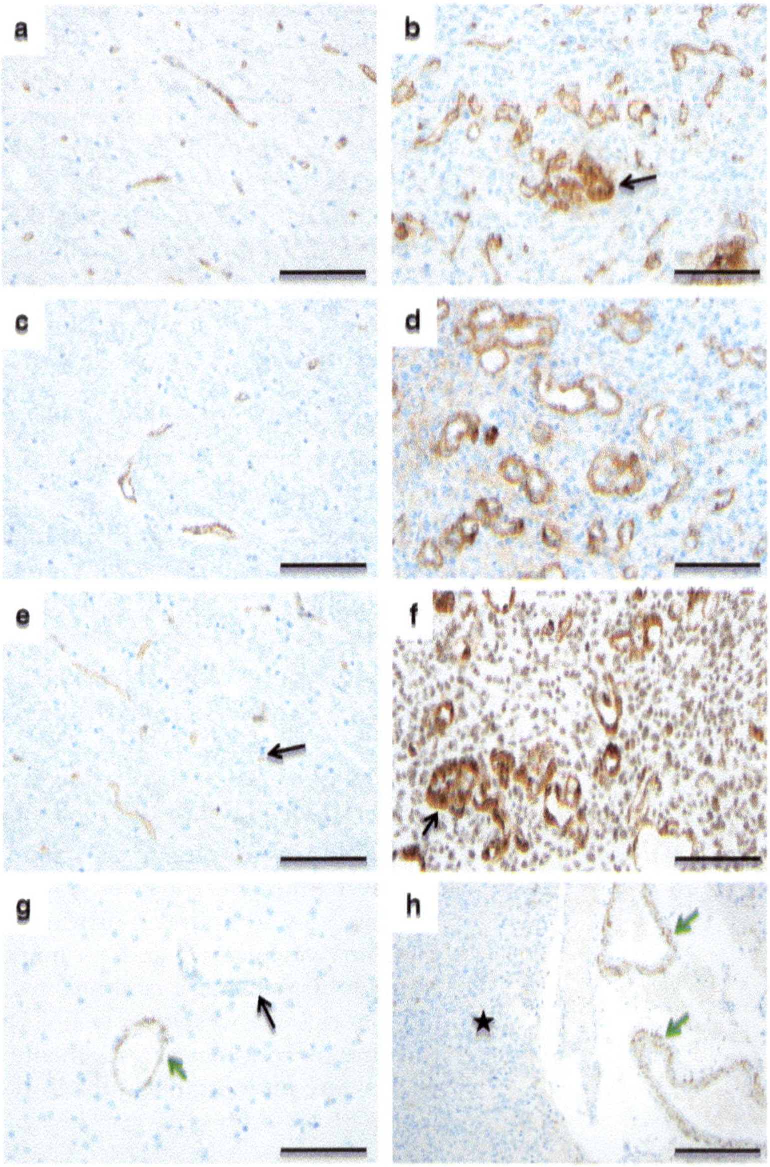

Fig. 2 Immunohistochemical analysis of vascular markers of human specimens from normal cortex (**a**, **c**, **e**, **g**) and glioblastoma (**b**, **d**, **f**, **h**). Immunohistochemistry for CD31 (**a**, **b**) and CD34 (**c**, **d**) reveals a marked increase in the microvascular density (MVD) in glioblastoma tissue as compared to the normal white matter (scale bar = 100 μm). Anti-αSMA stainings (**e**, **f**) show only a thin layer of mural cells in the normal white matter (*black arrow* in **e**), while in glioblastoma samples much thicker and multilayered (*black arrow* in **f**) mural cells are observed. (**g**, **h**) Desmin immunoreactivity is mainly observed in larger vessels (*green arrows*) in both normal brain and glioblastoma samples, most probably restricted to smooth muscle cells but not reacting with pericytes covering smaller vessels (*black arrow* in **g** or *asterisk* in **h**)

become visible as empty spaces within the tissue. In contrast to water, fatty tissue only freezes well at lower temperatures, but then becomes harder. This feature makes it very difficult to get well-preserved tissue sections from fat-rich tissues. Unfortunately, CNS tissue, in which glioblastoma diffusely infiltrates, is very rich in lipids due to an abundant amount of myelin, thereby CNS tissue often exhibits prominent morphological artifacts upon cryoembedding. To provide best morphology for the assessment of angiogenesis in glioblastomas, we recommend fixation in formaldehyde-based solutions since they also preserve lipid-rich tissues. In contrast to the expansion artifacts created by cryoembedding, formaldehyde-based fixations more likely lead to shrinkage artifacts. However, the shrinkage can be minimized if the formaldehyde concentration is lowered and buffer is added and if fixation time ranges from 24 to 48 h and the fixation temperature does not exceed room temperature [9]. Formaldehyde-based solutions only penetrate the tissue very slowly reaching approximately 0.5–1 mm/h. If native glioblastoma tissue is available, macroscopic sectioning in pieces of about 3–5 mm thickness prior to fixation enables a more equal fixation. Longer fixation times exceeding weeks or months lead to an increased cross-linking of tissue proteins, making harsher pretreatments necessary for immunohistochemical investigations.

2. During the angiogenic process, both endothelial and mural cells induce remodeling of the extracellular matrix (ECM) by secreting enzymes that digest the basement membrane [10]. The maturation of newly formed capillaries is strongly organized and characterized by sequentially deposited laminins, fibronectins, and different collagens. For instance, type I and type III collagens are absent during vascular sprouting. However, they accumulate in the perivascular space following microvessel maturation, thereby restoring the perivascular ECM network of the microvessels [10]. Although immunohistochemical and immunofluorescent analyses might be useful to assess molecular involvement in angiogenic ECM remodeling, degree of vascular sprouting, or stage of vessel maturation, results may strongly differ between institutions and researchers as a result of different pretreatment methods or techniques of preparation and analysis of antibody-based detection methods. Therefore, an alternative fast and simple method to gain first insights of the vascular network of normal brain and glioblastoma samples is the Gomori's silver impregnation method, which reveals reticulin fibers. These fibers mainly consist of type III collagen, thereby not only indicating preexisting capillaries and blood vessels but also providing an overview about the extent of mature microvessels. An additional advantage is

the fact that type III collagen is almost exclusively expressed in the basement membrane of blood vessels in the human CNS (Fig. 1). However, one drawback of this technique is that infrequently desmoplastic reactions or independent sarcomatous components occur in glioblastomas which also positively react in the reticulin silver staining method. Therefore in these cases, only vessel-associated staining should be taken into account for the assessment of angiogenesis in the normal CNS and in glioblastomas.

3. CD31 which is also known as platelet endothelial cell adhesion molecule-1 (PECAM-1) is constitutively expressed by normal (Fig. 2a) and intra-tumoral endothelial cells (Fig. 2b); however, some expression is also found on platelets and subsets of myeloid and lymphoid cells. CD31 might therefore serve as a reference marker for the general amount of endothelial cells although, in contrast to other endothelial cell markers, CD31 expression levels in human glioblastomas are not associated with patient survival [11].
4. The cell adhesion molecule CD34 is also used to assess the content of blood vessels in glioblastomas since it is also constitutively expressed by endothelial cells (Fig. 2c, d) [12]. However, in contrast to CD31, the evaluation of CD34-positive cells in glioblastomas has to be performed with caution since over 50 % of all glioblastomas also express CD34 on neoplastic glial cells [13]. CD34 was also detected on angiogenic tip cells in human vascular endothelial cell cultures; however, the tip-stalk cell concept of sprouting angiogenesis remains to be corroborated in human glioblastomas [14]. As CD34 is also strongly expressed on normal endothelial cells in vivo, it is not a useful tool to identify actively ongoing angiogenic processes.
5. The procoagulant protein von Willebrand factor (vWF) is synthesized by all endothelial cells throughout the body. Most pioneering studies investigating angiogenesis in human gliomas therefore used vWF as the reference for the total amount of endothelial cells [15, 16]. Despite the more frequent use of antibodies against CD31 (or CD34), vWF-based analyses of angiogenesis are still a reliable method in use [17].
6. The endothelial angiopoietin (Ang) receptor Tie-2 is a key player in regulating angiogenesis, vascular maturation, and permeability depending on the ratio of its major ligands Ang-1 and Ang-2 that act in an antagonistic manner [18]. Being already present at low levels in the normal human brain endothelium, Tie-2 is strongly induced upon angiogenic processes in malignant gliomas [19]. Therefore, Tie-2 is a useful tool to directly distinguish preexisting quiescent blood vessels from vessels in a proangiogenic state within human glioblastoma samples.

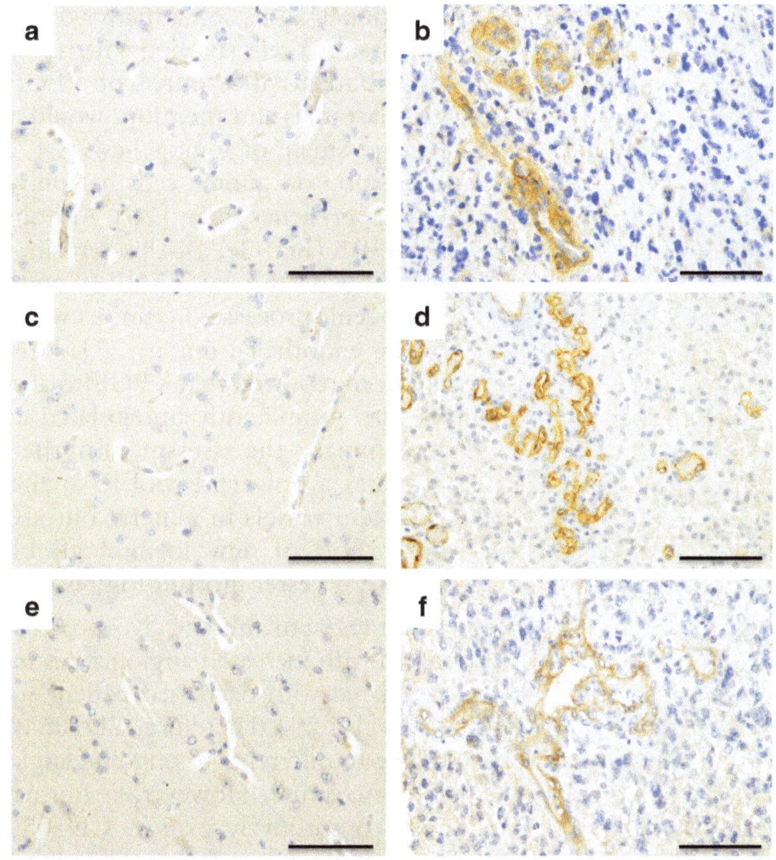

Fig. 3 Immunohistochemical analysis of VEGFR1 (**a**, **b**), VEGFR2 (**c**, **d**), and VEGFR3 (**e**, **f**) showing negative results in all normal CNS specimens (**a**, **c**, **e**) but a strong endothelial upregulation in glioblastoma samples (**b**, **d**, **f**) (scale bar = 100 μm)

At least in animal models, a small population of myeloid cells and pericyte precursor cells also express Tie-2, but it remains to be established to which extent this might also be the case for human tissue and therefore could be of importance for the assessment of angiogenesis in human glioblastomas [20].

7. Besides the use of markers that are constitutively expressed on endothelial cells, assessment of factors which are exclusively upregulated only on endothelia in malignant gliomas is of particular importance. Being absent in the normal human brain (Fig. 3a, c), VEGFR1 and VEGFR2 are expressed in high-grade gliomas, most prominently in glioblastomas (Fig. 3b, d) [21, 22]. The assessment of VEGFR1 has to be performed with caution, since a soluble VEGFR1 variant exists [23]. Thus, apart from the VEGFR1 staining signal, histological blood vessel morphology should be unequivocally identified in the analysis of VEGFR1-related angiogenesis in human glioblastomas.

Additionally, it is still a matter of debate if and to what extent glioblastoma cells and bone marrow-derived myeloid cells might contribute to the source of VEGFR1 expression in human glioblastomas and therefore would also have to be excluded in the assessment of angiogenesis [24, 25]. In contrast, VEGFR2 expression on glioma cells has only been demonstrated by in vitro approaches so far [26]. Although less data is available for VEGFR3 (Fig. 3e, f), this tyrosine kinase receptor, which is mainly activated by VEGF-C and VEGF-D, is involved in angiogenic processes in tumors where it maintains the integrity of the endothelial lining [27]. However, in glioblastomas, it has been reported that VEGFR3 and its major ligand VEGF-C are also prominently upregulated in perinecrotic areas [28]. In summary, the assessment of the VEGF receptors not only provides a powerful tool to analyze the amount expressed by blood vessels in general but also might help to quantify the extent of new formed glioblastoma-associated blood vessels, if vessel morphology is carefully taken into account.

8. The integral membrane glycoprotein CD105 (endoglin) is associated with increased angiogenesis in human glioblastomas and associated with decreased patient survival if expressed at high levels [11, 29]. Therefore, CD105 constitutes a promising candidate to differentiate ongoing angiogenesis from the preexisting vasculature. However, as normal human brain endothelial cells constitutively express CD105, it does not discriminate between quiescent and proliferating vessels [30].

9. Previously misinterpreted as endothelial proliferation, a paradigmatic change occurred after the discovery that pericytes and/or smooth muscle cells also proliferate and therefore contribute to microvascular proliferation in glioblastoma [31]. Since it has been recently shown that the density of mural cells within gliomas is also a significant prognostic factor in gliomas, there is also a need for the assessment of mural cells and their proliferative index in the setting of analyzing angiogenic processes [32].

10. Within the tumor vasculature, it is difficult to clearly differentiate between pericytes and smooth muscle cells by conventional histological or immunohistochemical methods. The ultrastructural analyses using electron microscopy may allow for a better differentiation; however, this method is time-consuming and does not offer the possibility to analyze larger tumor areas. Therefore, the term "mural cell" is currently used to describe the fraction of subendothelial cells within the basement membrane. The most frequently used and reliable marker for the detection of mural cells is alpha-smooth muscle actin [33]. A great advantage of the use of α-SMA is its lack of expression within normal CNS tissue apart from mural cells (Fig. 2e).

Proliferating mural cells in glioblastomas can easily be detected using anti-α-SMA immunohistochemistry (Fig. 2f). In contrast, other markers which have been presented as markers of mural cells such as desmin (Fig. 2g, h) are not reliably detectable on mural cells of smaller capillaries. As already mentioned for reticulin silver staining and CD34 immunohistochemistry, anti-α-SMA staining in human glioblastoma has to be evaluated with caution and morphologically attributed to blood vessels since glioma cells might also express α-SMA upon sarcomatous transformation or within a desmoplastic reaction [34].

11. More recently, the search for other mural cell markers which might more specifically distinguish between normal pericytes and glioblastoma-associated mural cells revealed CD248 (endosialin) as being absent from the normal CNS vasculature but highly expressed in glioblastomas [35].
12. The discovery that PDGFR-β-deficient mice are unable to develop a normal pericytic coverage paved the way for PDGFR-β as a reliable marker for pericytes [36]. In human high-grade gliomas, the recruitment of smooth muscle cells/pericytes to newly formed blood vessels is increased, which is reflected by higher expression of PDGFR-β [32]. PDGFR-β is most strongly expressed within vascular proliferations in glioblastomas [37]. Of note, apart from smooth muscle cells and pericytes, neoplastic glial cells also strongly express PDGFR-β in high-grade gliomas and therefore have to be excluded by means of morphology or with the use of double-labeling experiments in the assessment of angiogenic processes in human glioblastomas [38].
13. In larger glomeruloid or garland-like vascular proliferations in glioblastomas, it might be difficult to analyze total vascular area only with the use of markers for endothelial or mural cells. For a better assessment of the outline of blood vessels in normal or neoplastic CNS tissues, molecules of the basement membrane can be investigated including laminins, collagens, as well as expression of the cellular extracellular matrix (ECM) protein receptors, integrins.
14. Very frequently, collagen IV expression is assessed to define the basement membrane of blood vessels in gliomas [39]. The use of collagen IV as a basement membrane marker has the advantage that in contrast to some other constituents of the basement membrane, it is virtually absent in the normal neuropil and in glioblastoma cells. In contrast to type III collagens/reticulin, collagen IV is mainly expressed in the lamina densa of the basal lamina which serves as a bridge between basal and reticular laminae.
15. An unequivocal criterion for acute ongoing vascular proliferation is the detection of mitotic activity in endothelial or mural cells.

Since the analysis of mitotic activity based on classical hematoxylin and eosin (HE) staining is highly subjective and not easy to perform for histologically unexperienced researchers, the use of immunohistochemical proliferation markers helps to objectify the extent of proliferation in vascular cells. It is most practical to assess vascular proliferative activity in double staining using a proliferation marker in combination with a marker for the basement membrane (e.g., collagen IV) to comprise all cells of the vasculature in the analysis [39].

16. Ki-67 is probably one of the most frequently assessed markers in pathology, and the corresponding antibody MIB-1 is eternalized in the term "MIB-1 index" which is often used to describe the percentage of proliferating cells. Ki-67 is absent from cells in the G0-phase of the cell cycle; however, it is expressed in all other phases including G1, S, and G2 and during mitosis.
17. Specific phosphorylation of histone H3 occurs during mitosis or meiosis which is followed by dephosphorylation upon exit from the mitotic cycle [40]. In addition to the general proliferation rate as assessed by KI-67 analysis, pHH3 may provide further information about the fraction of cells entering the mitotic phase.
18. Before assessing the angiogenic changes in human glioblastomas, it is important to have knowledge about vascular patterns in the normal human CNS. First, capillary density in normal human brain areas shows strong regional variations, reaching twice as high levels in gray as compared to white matter areas [41]. While in normal adult white matter, capillaries only cover about 5–7 % of the total brain area, 13–15 % of the gray matter is covered by CNS microvessels. In addition, vascular density decreases with age [42]. It is therefore important to have clinical and neuroradiological information about the location of the tumor material to inform an accurate estimation of the degree of ongoing angiogenesis in glioblastoma tissue.
19. This method provides a reproducible tool to assess MVD, for example, in glioblastomas that were treated with different approaches or in comparison to normal CNS tissue samples from a similar brain region. Since the method of Weidner et al. was developed for breast carcinoma samples, the vascular heterogeneity in glioblastomas might be much more pronounced not least due to their diffuse growth pattern with constantly varying cell density. It might be worth increasing the number of visual fields (e.g., 10 fields at 200× magnification) in the analysis of glioblastoma samples to gain a more complete overview about the extent of ongoing angiogenesis rather than pursuing the "hot-spot" method which might be suitable for more homogenously vascularized neoplasms.

20. Although MVD is routinely used to determine glioblastoma angiogenesis, it recently turned out that microvascular area (MVA) provides a higher prognostic value for glioblastoma patients [43]. This might be related to the fact that MVD does not take into account the heterogeneity in vascular sizes, thereby potentially underestimating the negative prognostic impact of bizarre glomeruloid or garland-like vessels. In addition to its superior prognostic value compared to MVD, MVA also correlated well with relative cerebral blood volume (rCBV) as assessed by neuroradiological MR imaging.
21. Hlatky et al. criticize that MVD and MID do not reflect the angiogenic activity since MID only indicates the balance of pro- and anti-angiogenic factors within a microregion [44]. Since the delivery of nutrients and oxygen to glioblastoma cells is essential for their survival, the number of tumor cells that can be supplied between capillaries is limited, leading to necrotic transformation when tumor cell demand exceeds vascular supply. In a prostate carcinoma model, a sharp border between viable and necrotic tumor cells was observed at a distance of 110 μm from the next tumor microvessel [44]. There might be strong inter- and intraindividual differences in the perivascular distance being sufficient for supply of glioblastoma cells since cell density is very heterogenous and glioma cells might quickly perform a metabolic switch from an invasive to an angiogenic phenotype or vice versa [45].
22. These results were interpreted as a maturation deficit of glioblastoma-associated microvessels. Since anti-angiogenic treatment strategies might also interfere with pericyte functions, the MPI might constitute an additional tool to assess angiogenic processes in glioblastomas.

Acknowledgment

The authors would like to thank Maika Dunst, Tatjana Starzetz, and Cornelia Zachskorn for the histological preparations.

References

1. Johnson DR, O'Neill BP (2012) Glioblastoma survival in the United States before and during the temozolomide era. J Neurooncol 107(2): 359–364
2. Plate KH, Scholz A, Dumont DJ (2012) Tumor angiogenesis and anti-angiogenic therapy in malignant gliomas revisited. Acta Neuropathol 124(6):763–775
3. Birner P, Piribauer M, Fischer I, Gatterbauer B, Marosi C, Ambros PF, Ambros IM, Bredel M, Oberhuber G, Rössler K, Budka H, Harris AL, Hainfellner JA (2003) Vascular patterns in glioblastoma influence clinical outcome and associate with variable expression of angiogenic proteins: evidence for distinct angiogenic subtypes. Brain Pathol 13(2):133–143
4. Preusser M, Heinzl H, Gelpi E, Schonegger K, Haberler C, Birner P, Marosi C, Hegi M, Gorlia T, Hainfellner JA, European Organization for Research and Treatment of

Cancer Brain Tumor Group (2006) Histopathologic assessment of hot-spot microvessel density and vascular patterns in glioblastoma: poor observer agreement limits clinical utility as prognostic factors: a translational research project of the European Organization for Research and Treatment of Cancer Brain Tumor Group. Cancer 107(1):162–170

5. Eberhard A, Kahlert S, Goede V, Hemmerlein B, Plate KH, Augustin HG (2000) Heterogeneity of angiogenesis and blood vessel maturation in human tumors: implications for antiangiogenic tumor therapies. Cancer Res 60(5):1388–1393
6. Mulisch M, Welsch U (2010) Romeis. Mikroskopische technik, 18th edn. Spektrum, Heidelberg
7. Weidner N, Semple JP, Welch WR, Folkman J (1991) Tumor angiogenesis and metastasis—correlation in invasive breast carcinoma. N Engl J Med 324(1):1–8
8. Yoshii Y, Sugiyama K (1998) Intercapillary distance in the proliferating area of human glioma. Cancer Res 48(10):2938–2941
9. Fox CH, Johnson FB, Whiting J, Roller PP (1985) Formaldehyde fixation. J Histochem Cytochem 33(8):845–853
10. Nicosia RF, Madri JA (1987) The microvascular extracellular matrix. Developmental changes during angiogenesis in the aortic ring-plasma clot model. Am J Pathol 128:78–90
11. Yao Y, Kubota T, Takeuchi H, Sato K (2005) Prognostic significance of microvessel density determined by an anti-CD105/endoglin monoclonal antibody in astrocytic tumors: comparison with an anti-CD31 monoclonal antibody. Neuropathology 25(3):201–206
12. Sie M, Wagemakers M, Molema G, Mooij JJ, de Bont ES, den Dunnen WF (2009) The angiopoietin 1/angiopoietin 2 balance as a prognostic marker in primary glioblastoma multiforme. J Neurosurg 110(1):147–155
13. Galloway M (2010) CD34 expression in glioblastoma and giant cell glioblastoma. Clin Neuropathol 29(2):89–93
14. Siemerink MJ, Klaassen I, Vogels IM, Griffioen AW, Van Noorden CJ, Schlingemann RO (2012) CD34 marks angiogenic tip cells in human vascular endothelial cell cultures. Angiogenesis 15(1):151–163
15. Abdulrauf SI, Edvardsen K, Ho KL, Yang XY, Rock JP, Rosenblum ML (1998) Vascular endothelial growth factor expression and vascular density as prognostic markers of survival in patients with low-grade astrocytoma. J Neurosurg 88(3):513–520
16. Leon SP, Folkerth RD, Black PM (1996) Microvessel density is a prognostic indicator for patients with astroglial brain tumors. Cancer 77(2):362–372
17. Essock-Burns E, Phillips JJ, Molinaro AM, Lupo JM, Cha S, Chang SM, Nelson SJ (2013) Comparison of DSC-MRI post-processing techniques in predicting microvascular histopathology in patients newly diagnosed with GBM. J Magn Reson Imaging 38(2):388–400. doi:10.1002/jmri.23982
18. Scharpfenecker M, Fiedler U, Reiss Y, Augustin HG (2005) The Tie-2 ligand angiopoietin-2 destabilizes quiescent endothelium through an internal autocrine loop mechanism. J Cell Sci 118(Pt 4):771–780
19. Stratmann A, Risau W, Plate KH (1998) Cell type-specific expression of angiopoietin-1 and angiopoietin-2 suggests a role in glioblastoma angiogenesis. Am J Pathol 153(5):1459–1466
20. De Palma M, Venneri MA, Galli R, Sergi Sergi L, Politi LS, Sampaolesi M, Naldini L (2005) Tie2 identifies a hematopoietic lineage of proangiogenic monocytes required for tumor vessel formation and a mesenchymal population of pericyte progenitors. Cancer Cell 8(3):211–226
21. Chan AS, Leung SY, Wong MP, Yuen ST, Cheung N, Fan YW, Chung LP (1998) Expression of vascular endothelial growth factor and its receptors in the anaplastic progression of astrocytoma, oligodendroglioma, and ependymoma. Am J Surg Pathol 22(7):816–826
22. Plate KH, Breier G, Weich HA, Mennel HD, Risau W (1994) Vascular endothelial growth factor and glioma angiogenesis: coordinate induction of VEGF receptors, distribution of VEGF protein and possible in vivo regulatory mechanisms. Int J Cancer 59(4):520–529
23. Lamszus K, Ulbricht U, Matschke J, Brockmann MA, Fillbrandt R, Westphal M (2003) Levels of soluble vascular endothelial growth factor (VEGF) receptor 1 in astrocytic tumors and its relation to malignancy, vascularity, and VEGF-A. Clin Cancer Res 9(4):1399–1405
24. de Groot JF, Piao Y, Tran H, Gilbert M, Wu HK, Liu J, Bekele BN, Cloughesy T, Mehta M, Robins HI, Lassman A, DeAngelis L, Camphausen K, Chen A, Yung WK, Prados M, Wen PY, Heymach JV (2011) Myeloid biomarkers associated with glioblastoma response to anti-VEGF therapy with aflibercept. Clin Cancer Res 17(14):4872–4881
25. Steiner HH, Karcher S, Mueller MM, Nalbantis E, Kunze S, Herold-Mende CJ (2004) Autocrine pathways of the vascular endothelial growth factor (VEGF) in glioblastoma multiforme: clinical relevance of radiation-induced increase of VEGF levels. J Neurooncol 66(1–2):129–138
26. Knizetova P, Ehrmann J, Hlobilkova A, Vancova I, Kalita O, Kolar Z, Bartek J (2008)

Autocrine regulation of glioblastoma cell cycle progression, viability and radioresistance through the VEGF-VEGFR2 (KDR) interplay. Cell Cycle 7(16):2553–2561
27. Kubo H, Fujiwara T, Jussila L, Hashi H, Ogawa M, Shimizu K, Awane M, Sakai Y, Takabayashi A, Alitalo K, Yamaoka Y, Nishikawa SI (2000) Involvement of vascular endothelial growth factor receptor-3 in maintenance of integrity of endothelial cell lining during tumor angiogenesis. Blood 96(2):546–553
28. Jenny B, Harrison JA, Baetens D, Tille JC, Burkhardt K, Mottaz H, Kiss JZ, Dietrich PY, De Tribolet N, Pizzolato GP, Pepper MS (2006) Expression and localization of VEGF-C and VEGFR-3 in glioblastomas and haemangioblastomas. J Pathol 209(1):34–43
29. Sica G, Lama G, Anile C, Geloso MC, La Torre G, De Bonis P, Maira G, Lauriola L, Jhanwar-Uniyal M, Mangiola A (2011) Assessment of angiogenesis by CD105 and nestin expression in peritumor tissue of glioblastoma. Int J Oncol 38(1):41–49
30. Holley JE, Newcombe J, Whatmore JL, Gutowski NJ (2010) Increased blood vessel density and endothelial cell proliferation in multiple sclerosis cerebral white matter. Neurosci Lett 470(1):65–70
31. Wesseling P, Schlingemann RO, Rietveld FJ, Link M, Burger PC, Ruiter DJ (1995) Early and extensive contribution of pericytes/vascular smooth muscle cells to microvascular proliferation in glioblastoma multiforme: an immuno-light and immuno-electron microscopic study. J Neuropathol Exp Neurol 54(3):304–310
32. Sato S, Sato Y, Hatakeyama K, Marutsuka K, Yamashita A, Takeshima H, Asada Y (2011) Quantitative analysis of vessels with smooth muscle layer in astrocytic tumors: correlation with histological grade and prognostic significance. Histol Histopathol 26(4):497–504
33. Takeuchi H, Hashimoto N, Kitai R, Kubota T, Kikuta K (2010) Proliferation of vascular smooth muscle cells in glioblastoma multiforme. J Neurosurg 113(2):218–224
34. Ang LC, Perry JR, Bilbao JM, Ozane W, Peschke E, Young B, Nelson N (1996) Postirradiated and nonirradiated gliosarcoma: immunophenotypical profile. Can J Neurol Sci 23(4):251–256
35. Simonavicius N, Robertson D, Bax DA, Jones C, Huijbers IJ, Isacke CM (2008) Endosialin (CD248) is a marker of tumor-associated pericytes in high-grade glioma. Mod Pathol 21(3):308–315
36. Lindahl P, Johansson BR, Levéen P, Betsholtz C (1997) Pericyte loss and microaneurysm formation in PDGF-B-deficient mice. Science 277(5323):242–245
37. Plate KH, Breier G, Farrell CL, Risau W (1992) Platelet-derived growth factor receptor-beta is induced during tumor development and upregulated during tumor progression in endothelial cells in human gliomas. Lab Invest 67(4):529–534
38. Haberler C, Gelpi E, Marosi C, Rössler K, Birner P, Budka H, Hainfellner JA (2006) Immunohistochemical analysis of platelet-derived growth factor receptor-alpha, -beta, c-kit, c-abl, and arg proteins in glioblastoma: possible implications for patient selection for imatinib mesylate therapy. J Neurooncol 76(2):105–109
39. Wesseling P, van der Laak JA, Link M, Teepen HL, Ruiter DJ (1998) Quantitative analysis of microvascular changes in diffuse astrocytic neoplasms with increasing grade of malignancy. Hum Pathol 29(4):352–358
40. Nowak SJ, Corces VG (2004) Phosphorylation of histone H3: a balancing act between chromosome condensation and transcriptional activation. Trends Genet 20(4):214–220
41. Hunter JM, Kwan J, Malek-Ahmadi M, Maarouf CL, Kokjohn TA, Belden C, Sabbagh MN, Beach TG, Roher AE (2012) Morphological and pathological evolution of the brain microcirculation in aging and Alzheimer's disease. PLoS One 7(5):e36893
42. Brown WR, Moody DM, Thore CR, Challa VR, Anstrom JA (2007) Vascular dementia in leukoaraiosis may be a consequence of capillary loss not only in the lesions, but in normal-appearing white matter and cortex as well. J Neurol Sci 257(1–2):62–66
43. Hu LS, Eschbacher JM, Dueck AC, Heiserman JE, Liu S, Karis JP, Smith KA, Shapiro WR, Pinnaduwage DS, Coons SW, Nakaji P, Debbins J, Feuerstein BG, Baxter LC (2012) Correlations between perfusion MR imaging cerebral blood volume, microvessel quantification, and clinical outcome using stereotactic analysis in recurrent high-grade glioma. AJNR Am J Neuroradiol 33(1):69–76
44. Hlatky L, Hahnfeldt P, Folkman J (2002) Clinical application of antiangiogenic therapy: microvessel density, what it does and doesn't tell us. J Natl Cancer Inst 94(12):883–893
45. Rajcevic U, Petersen K, Knol JC, Loos M, Bougnaud S, Klychnikov O, Li KW, Pham TV, Wang J, Miletic H, Peng Z, Bjerkvig R, Jimenez CR, Niclou SP (2009) iTRAQ-based proteomics profiling reveals increased metabolic activity and cellular cross-talk in angiogenic compared with invasive glioblastoma phenotype. Mol Cell Proteomics 8(11):2595–2612

Chapter 17

Quantitative Cerebral Blood Flow Measurements Using MRI

Eric R. Muir, Lora Talley Watts, Yash Vardhan Tiwari, Andrew Bresnen, Qiang Shen, and Timothy Q. Duong

Abstract

Magnetic resonance imaging can be utilized as a quantitative and noninvasive method to image cerebral blood flow. The two most common techniques used to detect cerebral blood flow are dynamic susceptibility contrast (DSC) perfusion MRI and arterial spin labeling perfusion MRI. Herein we describe the use of these two techniques to measure cerebral blood flow in rodents, including methods, analysis, and important considerations when utilizing these techniques.

Key words Cerebral blood flow, Dynamic susceptibility contrast, Arterial spin labeling, Magnetic resonance imaging

1 Introduction

Magnetic resonance imaging (MRI) can be used to noninvasively image cerebral blood flow (CBF) to study normal physiology and pathophysiology with high temporal and spatial resolution. CBF MRI has been used to investigate vascular remodeling in brain disorders, including stroke and cancer [1].

There are two common techniques to measure CBF based on MRI: (1) dynamic susceptibility contrast (DSC) perfusion MRI [2, 3], which involves administration of an exogenous intravascular contrast agent, or (2) arterial spin-labeling (ASL) perfusion MRI which noninvasively and magnetically labels the endogenous water in blood [4]. DSC works by imaging the dynamic passage of a bolus of contrast agent. The area under the signal ($\Delta R2^*$) versus time curve approximates cerebral blood volume. The mean transit time of the bolus can also be derived from the signal versus time curves and CBF is determined using cerebral blood volume divided by mean transit time. DSC is efficient but some subjects with renal disease develop side effects (such as nephrogenic systemic fibrosis) and DSC is incompatible with dynamic functional MRI due to the

Richard Milner (ed.), *Cerebral Angiogenesis: Methods and Protocols*, Methods in Molecular Biology, vol. 1135,
DOI 10.1007/978-1-4939-0320-7_17,

long half-life of the contrast agent allowing only one measurement per bolus injection.

Arterial spin labeling (ASL) noninvasively utilizes arterial blood water as an endogenous tracer by magnetically labeling, either by inverting or saturating, the hydrogen spins of the incoming blood (i.e., in the neck area). When the magnetically labeled blood water flows into the region of interest (i.e., the brain), the mixing of blood and brain water changes the water T1 of the brain. By measuring the labeling efficiency (equivalent of an arterial input function), different mathematical models can be used to calculate CBF. ASL is totally noninvasive, and the magnetically labeled water has a short half-life (approximately blood spin-lattice relaxation time constant T_1) making it possible to perform multiple repeated measurements, which can be used to augment spatial resolution and/or signal-to-noise ratio. Both DSC and ASL CBF techniques have been used in humans. This chapter describes the MRI procedures to image CBF using the DSC and ASL techniques in rodents.

2 Materials

2.1 Animals and Agents

1. Rats (200–250 g).
2. Anesthetics (isoflurane, urethane, or pentobarbital, etc.; *see* **Note 1**).
3. Common surgical tools and supplies.
4. Catheters and PE-50 tubing for contrast agent injection outside the scanner.
5. Contrast agent for DSC method (but not ASL): Magnevist (gadolinium diethyltriaminepentaacetic acid Gd-DTPA) or ProHance (gadolinium 10-(2-hydroxy-propyl)-1,4,7,10-tetraazacyclododecane-1,4,7-triacetic acid).

2.2 MRI

1. Bruker 7 Tesla scanner (Billerica, MA).
2. 40-G/cm BGA12 gradient insert (ID = 12 cm, 120-μs rise time).
3. Animal holder equipped with ear and tooth bars.
4. Radiofrequency (RF) transmitter/receiver coil for brain imaging.
5. RF transmitter coil for arterial spin labeling at the neck position for ASL (but not DSC option) (*see* **Note 2**).
6. Switch box to actively detune RF coils for ASL (but not DSC option) (*see* **Note 3**).

2.3 Peripheral MRI-Compatible Monitoring Equipment and Animal Support

1. Oximetry (heart rate, arterial oxygen saturation)—(MouseOx, Starr Life Sciences).
2. Noninvasive respiration monitoring via force transducer—(MR-compatible small animal monitoring & gating system, SA Instruments).

3. Circulating warm water bath.
4. Temperature feedback regulator (Digi-sense, Cole Palmer) and temperature sensor.
5. Anesthetic delivery system, such as vaporizer, if needed.
6. Standard infusion pump for DSC option.

3 Methods

3.1 Animal Preparation

1. Place a rat in an anesthesia box with 4–5 % isoflurane in air continuously flowing to the box until the animal is sufficiently anesthetized. Check the paw reflexes by pinching the paws and looking for reflex withdrawal to ensure adequate anesthesia.
2. Maintain the anesthesia level using 1.5 % isoflurane mixed with house air delivered by nose cone placed over the animal's nose. Different anesthetics can be used if preferred, but be aware that this will affect CBF (*see* **Note 1**).
3. Place the rat into the head holder with ear bars and tooth bar to immobilize the skull and minimize animal motion throughout the scanning procedures.
4. The rat is then moved to the scanner holder which should have a temperature-controlled water pad to maintain the rat's body temperature at 37 ± 0.5 °C. A rectal temperature probe should be inserted using lubrication to prevent damage to the internal organs (*see* **Note 4**).
5. Place the force transducer to monitor respiration underneath the rat. Connect the MouseOx sensor to one of the hind paws and secure with tape. The MouseOx system will enable the user to monitor the rat's physiology throughout the imaging procedures.

3.2 MRI (See Fig. 1)

1. For ASL MRI, the animal should be positioned so that the labeling coil is at the neck, to allow labeling of blood flowing through the carotid arteries (*see* **Note 5**).

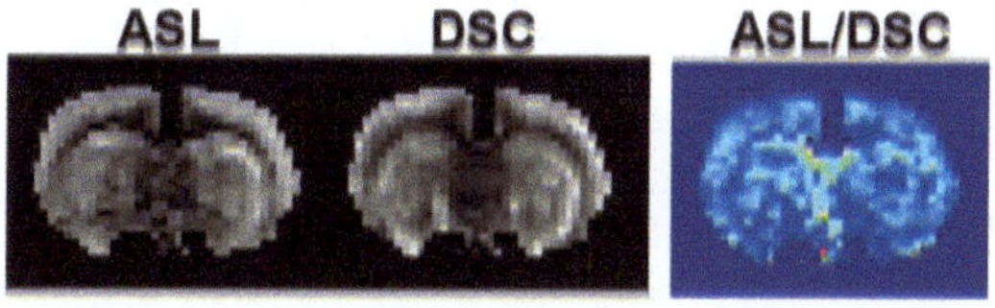

Fig. 1 Examples of CBF maps (in ml/g/min) from both ASL and DSC methods in the same rat. The ratio of CBF from ASL/DSC is also shown. Some brain structures should be able to be distinguished in the CBF maps, such as the corpus callosum with low blood flow. Figure adapted from Tanaka et al. [12]

2. Position the RF coil as close to the center of the region of interest as possible. Secure the RF coil with tape (*see* **Note 6**).
3. Tune and match the RF coil by adjusting the 1H resonance frequency (300 MHz at 7 T) and impedance.
4. Use a position scan to position the area of interest in the *x*, *y*, and *z* directions to ensure the area of interest is centered in the scanner. Increase the field of view (FOV) if needed.
5. Run auto-shim or manual shim as needed to improve the image quality (*see* **Note 7**).
6. Calibrate the RF pulses for given pulse shapes and durations. This can be set up to occur automatically.
7. Perform a pilot scan using a 2D gradient-echo FLASH or RARE sequence (10–30 s). Based on the pilot scan, plan five to eight 1.5 mm coronal slices to cover the region of interest in the brain.
8. T2-weighted images are acquired using the RARE pulse sequence (echo time per echo = 6.5 ms) with two different effective echo times (52 and 104 ms), echo train length = 16, and 16 signal averages. Typical parameters are data matrix = 64 × 64, FOV = 2.56 cm × 2.56 cm, eight 1.5-mm coronal slices, spectral width = 50 kHz, TR = 2–3 s, 90° flip angle with pulse shape Gaussian or Sinc3, and pulse duration 1–2 ms.
9. For ASL, single-shot, gradient-echo, echo-planar-imaging (EPI) acquisition is used. Paired images are acquired alternately—one with arterial spin labeling (labeled image) and the other without (control). Typical MR parameters are data matrix = 64 × 64, FOV = 2.56 cm × 2.56 cm, eight 1.5-mm coronal slices, TE = 12 ms, and TR = 3 s (90° flip angle). Continuous arterial spin labeling (cASL) is employed with separate imaging and labeling coils, using a –2.7 s square radio frequency pulse for the labeling coil in the presence of a 1.0 G/cm gradient along the flow direction such that the condition of adiabatic inversion is satisfied. The sign of the frequency offset of the label pulse is switched for non-labeled images. The number of averages is typically 40 or more, depending on the required SNR (*see* **Note 5**). Please see the following references for more details [5–11].
10. For DSC, single-shot, gradient-echo, echo-planar-imaging (EPI) acquisition is used with matrix = 64 × 64, FOV = 2.56 cm × 2.56 cm, three to five 1.5-mm slices, TE = 20 ms, TR = 0.333 s, and 22° flip angle. Preload the i.v. line with 0.2 mmol/kg of gadolinium contrast agent in a volume of 0.15–0.2 ml (typically 3 ft of PE-50 tubing will hold such a volume). Start the DSC acquisition consisting of a scanning period of 1 min. Twenty seconds into the acquisition, deliver the contrast agent in a single bolus and flush with saline. Continue the DSC acquisition for another 40 s (*see* **Note 8**).

3.3 Animal Recovery

1. At the conclusion of imaging, remove the rat from the scanner. Discontinue the anesthesia once you have removed the rat from the holder for the scanner. Place the rat on a heating pad on the low setting until it is fully awake and ambulatory.
2. Once the rat is awake and ambulatory, it can then be returned to its home cage for recovery.
3. The animal should be monitored throughout the day for signs of distress.

3.4 Image Analysis

1. Image analysis can be done using custom written codes or various freely available MRI analysis tools. Image calculation can be done using custom written codes or software available online such as ASLtbx (University of Pennsylvania) or DSCoMAN (Duke) [9, 13]. If needed, image co-registration can be performed using software such as Statistical Parametric Mapping (SPM, University College London). Image display can be done with software such as SPM or STIMULATE (University of Minnesota). Several other free software programs are also available to process and display MRI images (*see* **Note 9**).
2. For ASL acquisitions, CBF images (S_{CBF}) with intensity in units of ml/g/min are calculated [10, 11, 14] pixel by pixel using

$$S_{CBF} = \frac{\lambda / T_1 \cdot (S_c - S_L)}{(S_L + (2\alpha - 1)S_c)},$$

where S_C and S_L are signal intensities of the control and labeled images, respectively. λ, the brain-blood partition coefficient of water, is ~0.9 ml/g for averaged gray and white matter [15]. T_1 is the longitudinal relaxation time of the brain, which is about 1.6–1.8 s at 7 T [16]. α is the labeling efficiency which has been measured to be 0.75–0.9 in animal models [5, 7] (*see* **Notes 2** and **9**).
3. For DSC-CBF calculation, the change in the transverse relaxation rate ($\Delta R2^*$) is calculated using

$$\Delta R2 * (t) = \frac{-\ln(S(t) / S_0)}{TE},$$

where $S(t)$ is the signal intensity at time t, S_0 is the pre-contrast baseline signal intensity, and TE is the pulse sequence echo time. Hemodynamic parameters can be generated by deconvolving the change in tissue concentration over the first pass of contrast agent with an arterial input function using singular value decomposition [2, 3]. The area under the curve of $\Delta R2^*$ versus time gives the cerebral blood volume (CBV). The mean transit time (MTT) can be determined from the shape of the curve, and CBF is calculated as CBV/MTT (*see* **Note 10**).

4 Notes

1. Different anesthetic agents have different effects on cerebral blood flow values. Comparisons of CBF across different studies need to be made with caution.
2. For ASL we described the use of a continuous ASL sequence with separate labeling and imaging coils. Many variations of ASL exist and different scanners may have different sequences available or may have hardware limitations preventing some methods. All methods give reasonably similar results with some advantages and disadvantages. For more information, *see* [17, 18].
3. For ASL, the imaging coil needs to be detuned while the labeling coil is transmitting to avoid artifacts and inaccurate CBF values.
4. Stable maintenance of animal physiology under anesthesia is important since CBF is affected by physiological parameters, such as blood gases and temperature. Maintain the animal's body temperature within 37 °C ± 0.5 °C, as body temperature decreases under anesthesia. A circulating temperature-controlled heat pad is ideal for this. Additionally, the rodent's physiology should be monitored continuously using a MouseOx system while under anesthesia.
5. Variability in the common carotid and vertebral arteries across animals can lead to quantification errors with ASL. Proper calibration of RF power and efficient labeling must be ensured to avoid quantification errors. This can be done by acquiring multiple short CBF scans with arrayed RF labeling power to find the optimal power to provide the maximum CBF. However, too high of labeling power can begin to directly saturate the brain and will be seen as very bright areas in the CBF maps in the posterior brain.
6. For surface coils, avoid pressing the coil too hard on the animal's head as it would increase "loading," which decreases SNR.
7. High-order localized shimming on the brain can improve image quality and reduce distortion in EPI images. However, for ASL, we have found that this can cause poor shim quality outside of the brain leading to poor labeling efficiency at the neck, so first-order global shimming may be preferable.
8. Injection of the contrast agent must be done as a bolus (as fast as possible but without damaging the vessels) through the tail vein or femoral vein.
9. Note that there are multiple equation models to calculate CBF via ASL and DSC approaches. We presented here common models for CBF calculation. The accuracy of CBF measurements by MRI is still an active area of research.
10. Magnetic resonance angiography and venography may also be relevant to readers interested in investigating angiogenesis.

Acknowledgements

This work is supported in part by the National Institute of Neurological Disorders and Stroke, the National Institute of Health (R01 NS045879), and the Established Investigator Award (EIA 0940104N) from the American Heart Association.

References

1. Brown GG, Clark C, Liu TT (2007) Measurement of cerebral perfusion with arterial spin labeling: part 2. Applications. J Int Neuropsychol Soc 13:526–538
2. Ostergaard L, Sorensen AG, Kwong KK, Weisskoff RM, Gyldensted C, Rosen BR (1996) High resolution measurement of cerebral blood flow using intravascular tracer bolus passages. Part II: experimental comparison and preliminary results. Magn Reson Med 36:726–736
3. Ostergaard L, Weisskoff RM, Chesler DA, Gyldensted C, Rosen BR (1996) High resolution measurement of cerebral blood flow using intravascular tracer bolus passages. Part I: mathematical approach and statistical analysis. Magn Reson Med 36:715–725
4. Williams DS, Detre JA, Leigh JS, Koretsky AP (1992) Magnetic resonance imaging of perfusion using spin inversion of arterial water. Proc Natl Acad Sci U S A 89:212–216
5. Shen Q, Fisher M, Sotak CH, Duong TQ (2004) Effects of reperfusion on ADC and CBF pixel-by-pixel dynamics in stroke: characterizing tissue fates using quantitative diffusion and perfusion imaging. J Cereb Blood Flow Metab 24:280–290
6. Shen Q, Huang S, Du F, Duong TQ (2011) Probing ischemic tissue fate with BOLD fMRI of brief oxygen challenge. Brain Res 1425:132–141
7. Shen Q, Meng X, Fisher M, Sotak CH, Duong TQ (2003) Pixel-by-pixel spatiotemporal progression of focal ischemia derived using quantitative perfusion and diffusion imaging. J Cereb Blood Flow Metab 23:1479–1488
8. Muir ER, Shen Q, Duong TQ (2008) Cerebral blood flow MRI in mice using the cardiac-spin-labeling technique. Magn Reson Med 60:744–748
9. Shen Q, Ren H, Cheng H, Fisher M, Duong TQ (2005) Functional, perfusion and diffusion MRI of acute focal ischemic brain injury. J Cereb Blood Flow Metab 25:1265–1279
10. Sicard K, Shen Q, Brevard ME, Sullivan R, Ferris CF, King JA, Duong TQ (2003) Regional cerebral blood flow and BOLD responses in conscious and anesthetized rats under basal and hypercapnic conditions: implications for functional MRI studies. J Cereb Blood Flow Metab 23:472–481
11. Sicard KM, Duong TQ (2005) Effects of hypoxia, hyperoxia and hypercapnia on baseline and stimulus-evoked BOLD, CBF and CMRO2 in spontaneously breathing animals. Neuroimage 25:850–858
12. Tanaka Y, Nagaoka T, Nair G, Ohno K, Duong TQ (2011) Arterial spin labeling and dynamic susceptibility contrast CBF MRI in postischemic hyperperfusion, hypercapnia, and after mannitol injection. J Cereb Blood Flow Metab 13:1403–1411
13. Liu ZM, Schmidt KF, Sicard KM, Duong TQ (2004) Imaging oxygen consumption in forepaw somatosensory stimulation in rats under isoflurane anesthesia. Magn Reson Med 52:277–285
14. Duong TQ, Silva AC, Lee SP, Kim SG (2000) Functional MRI of calcium-dependent synaptic activity: cross correlation with CBF and BOLD measurements. Magn Reson Med 43:383–392
15. Herscovitch P, Raichle ME (1985) What is the correct value for the brain-blood partition coefficient for water? J Cereb Blood Flow Metab 5:65–69
16. Barbier EL, Liu L, Grillon E, Payen JF, Lebas JF, Segebarth C, Remy C (2005) Focal brain ischemia in rat: acute changes in brain tissue T1 reflect acute increase in brain tissue water content. NMR Biomed 18:499–506
17. Barbier EL, Lamalle L et al (2001) Methodology of brain perfusion imaging. J Magn Reson Imaging 13:496–520
18. Wu WC, St Lawrence KS et al (2010) Quantification issues in arterial spin labeling perfusion magnetic resonance imaging. Top Magn Reson Imaging 21(2):65–73

Chapter 18

Fluorescent Angiogenesis Models Using Gelfoam® Implanted in Transgenic Mice Expressing Fluorescent Proteins

Robert M. Hoffman

Abstract

Fidler's group described an in vivo angiogenesis assay utilizing Gelfoam® sponges impregnated with agarose and proangiogenic factors. Vessels were detected by staining with fluorescent antibodies against CD31. We showed that Gelfoam® implanted in transgenic mice expressing the nestin promoter-driven green fluorescent protein (ND-GFP mice) was rapidly vascularized with ND-GFP-expressing nascent blood vessels. Angiogenesis in the Gelfoam® was quantified by measuring the total length of ND-GFP-expressing nascent blood vessels in a skin flap by in vivo fluorescence microscopy imaging. The ND-GFP-expressing nascent blood vessels formed a network on the surface of the basic fibroblast growth factor (bFGF)-treated Gelfoam®. We then developed a color-coded imaging model that can visualize the interaction between α_v integrin linked to green fluorescent protein (GFP) in osteosarcoma cells and blood vessels in Gelfoam® vascularized after implantation in red fluorescent protein (RFP) transgenic nude mice. The implanted Gelfoam® became highly vascularized with RFP-expressing vessels in 14 days. 143B osteosarcoma cells expressing α_v integrin-GFP were injected into the Gelfoam® after transplantation of Gelfoam®. After cancer cell injection, cancer cells interacting with blood vessels were observed in the Gelfoam® by color-coded confocal microscopy through the skin flap window. We developed another color-coded Gelfoam®-based imaging model that can visualize the anastomosis between blood vessels. RFP-expressing vessels in vascularized Gelfoam®, previously transplanted into RFP transgenic mice, were re-transplanted into ND-GFP mice. Skin flaps were made and anastomosis between the GFP-expressing nascent blood vessels of ND-GFP transgenic nude mice and RFP blood vessels in the transplanted Gelfoam® could be imaged. Our results demonstrate that the Gelfoam® in vivo angiogenesis model in combination with fluorescent protein labeling of blood vessels is a powerful system for use in the discovery and evaluation of agents influencing vascularization.

Key words Gelfoam®, Angiogenesis, Vascularization, Anastomosis, Green fluorescent protein, Red fluorescent protein, Nestin, Transgenic, Nude mice, Implantation, Skin flap, Imaging

1 Introduction

Angiogenesis occurs in many normal situations, including embryonic development, tissue regeneration, the menstrual cycle, and also pathologically in cancer, proliferative retinopathy, rheumatoid

Richard Milner (ed.), *Cerebral Angiogenesis: Methods and Protocols*, Methods in Molecular Biology, vol. 1135, DOI 10.1007/978-1-4939-0320-7_18, © Springer Science+Business Media New York 2014

arthritis, and other diseases [1, 2]. Previous angiogenesis assays have used the chorioallantoic membrane assay [3, 4], the monkey iris neovascularization model [5], the disc angiogenesis assay [6], and the cornea [7–12].

Fidler's group developed an in vivo angiogenesis assay utilizing Gelfoam® sponges impregnated with 0.4 % agarose and specific growth factors. Basic fibroblast growth factor (bFGF), vascular epidermal growth factor (VEGF), tumor growth factor-alpha (TGF-α), and endothelial growth factor (EGF) stimulated angiogenesis in the implanted Gelfoam® in nude mice. Vessels growing in the Gelfoam® were detected after fixation, sectioning, and staining with fluorescent antibodies against CD31 [13].

We have developed a fluorescent Gelfoam® angiogenesis model by using transgenic mice in which the nestin promoter drives green fluorescent protein (ND-GFP). In this mouse, ND-GFP is expressed only in nascent blood vessels, which enables nascent blood vessels to be visualized by their GFP expression.

We then developed an imageable in vivo angiogenesis assay after transplantation of Gelfoam® into ND-GFP mice. We demonstrated that a Gelfoam® was rapidly vascularized with GFP-expressing vessels in the presence of an angiogenesis stimulator. Anti-angiogenesis agents inhibited this process. Thus, this rapid and simple new in vivo assay can rapidly identify angiogenic stimulators and inhibitors [14].

For example, Gelfoam® treated with bFGF and transplanted into the subcutis on both flanks of the ND-GFP transgenic mice led to rapid vascularization with ND-GFP-expressing blood vessels. This was analyzed by making skin flaps at day 7 after transplantation of Gelfoam®. Angiogenesis was quantified by measuring the length of ND-GFP-expressing nascent blood vessels in the Gelfoam® in the skin flap by in vivo fluorescence microscopy imaging (Fig. 1). The vessels on the surface were counted under fluorescence microscopy. Implanted Gelfoam® that was not treated with bFGF was not vascularized [14]. Co-localization of ND-GFP fluorescence and CD31 in frozen sections of the vascularized Gelfoam® was detected by immunohistochemistry (IHC) [14].

In addition to investigating the influence of pro-angiogenic factors, the effect of anti-angiogenic factors can also be examined in this system. For instance, ND-GFP mice that received intraperitoneal injections of doxorubicin (5 μg/g) at days 0, 1, and 2 after transplantation of Gelfoam®, with or without bFGF, had fewer ND-GFP-expressing nascent blood vessels than control NaCl-treated mice [14].

In another study, we implanted Gelfoam® into red fluorescent protein (RFP)-expressing transgenic nude mice followed by implantation of α_v integrin-GFP-expressing human 143B osteosarcoma cells in order to visualize the behavior of αv integrin in cancer cells interacting with blood vessels, using color-coded imaging [15].

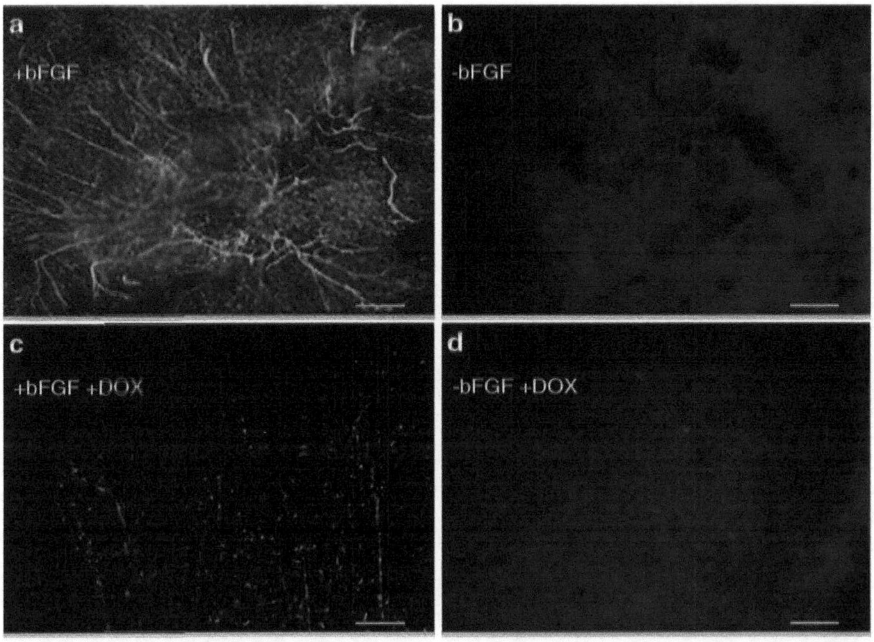

Fig. 1 Angiogenesis of implanted Gelfoam® with green fluorescent protein (GFP)-expressing vessels. Nestin-driven GFP (ND-GFP) mice were given daily intraperitoneal (ip) injections of 0.9 % NaCl solution at day 0, 1, and 2 after transplantation of Gelfoam® with or without basic fibroblast growth factor (bFGF). (**a**) At day 7 after transplantation of Gelfoam® with bFGF, ND-GFP-expressing nascent blood vessels formed a network on the surface of Gelfoam® in the skin flap. The ND-GFP-expressing nascent blood vessels had many branches that were connected to each other. (**b**) At day 7 after transplantation of Gelfoam®, without bFGF, into the subcutis of ND-GFP transgenic mice, the ND-GFP-expressing nascent blood vessels were very sparse. (**c**, **d**) The Gelfoam®-transplanted ND-GFP mice were treated with 5 μg/g doxorubicin (DOX) at day 0, 1, and 2 after transplantation. Doxorubicin significantly decreased the blood vessel density in the (**c**) presence or (**d**) absence of bFGF. Bar = 500 μm [14]

To stimulate Gelfoam® angiogenesis with RFP-expressing blood vessels, Gelfoam® was treated with 300 ng bFGF and then implanted into the subcutis on both flanks of RFP nude mice. Seven days after transplantation, skin flaps were made and observed by confocal microscopy. Vessels had not yet appeared in the Gelfoam®. Fourteen days after transplantation, RFP-expressing vessel structures were observed on the surface of the Gelfoam®, and 21 days after transplantation, RFP-expressing vessels were observed to be penetrating deeply in the Gelfoam®. Twenty-eight days after transplantation, RFP-expressing vessels were thick and long and had invaded deeply into the Gelfoam® [15]. At this time-point, frozen sections of Gelfoam® in the subcutis of RFP transgenic nude mice were made and showed that RFP-expressing blood vessels were growing into the Gelfoam®. Immunohistochemical (IHC) staining showed that CD31 and RFP fluorescence co-localized in the RFP-expressing blood vessels in the Gelfoam® [15].

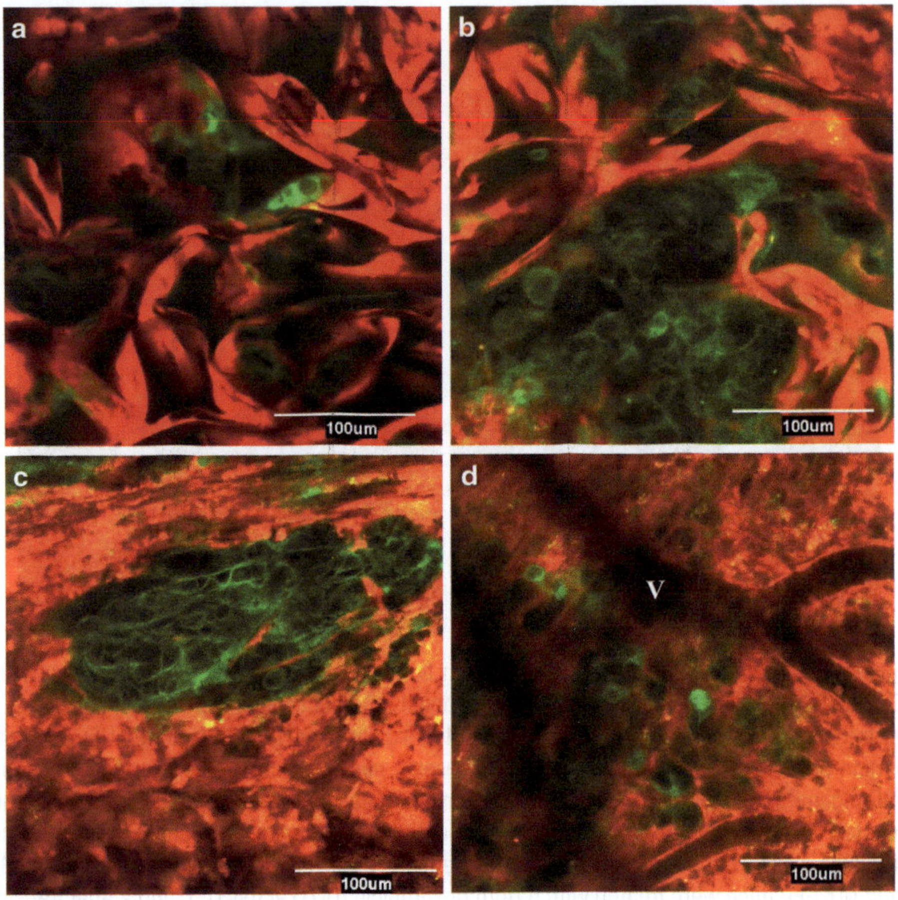

Fig. 2 In vivo visualization of the interaction between 143B cells expressing α_v integrin-GFP and RFP vessels in Gelfoam®. Skin flaps were made 7 days after transplantation of 143B cancer cells expressing α_v integrin-GFP, into the Gelfoam®, previously transplanted in RFP transgenic nude mice. (**a**) Seven days after cell injection, GFP-expressing cancer cells were observed in the Gelfoam®. Bar = 100 μm. (**b**) Fourteen days after injection, GFP-expressing cancer cells were found proliferating in the Gelfoam®. Bar = 100 μm. (**c**) Twenty-one days after injection, α_v integrin-GFP-expressing cancer cells were found to be proliferating in the Gelfoam®. α_v Integrin-GFP was strongly expressed around the surface of the cells. Bar = 100 μm. (**d**) Twenty-eight days after injection, cancer cells strongly expressed α_v integrin-GFP around a small RFP-expressing vessel (V). Bar = 100 μm [15]

Seven days after 143B cell injection, GFP-expressing cancer cells were observed in the Gelfoam® in the RFP nude mice. Twenty-eight days after injection, cancer cells strongly expressing α_v integrin-GFP were observed around a small vessel, suggesting that the pattern of expression of α_v integrin-GFP in the cells is associated with their interaction with blood vessels [15] (Fig. 2).

A fluorescent Gelfoam® vessel anastomosis imaging model was then developed. We generated Gelfoam® containing RFP-expressing vessels by implantation of Gelfoam® to RFP mice as described above. After the Gelfoam® was vascularized with RFP vessels, it was transplanted this into ND-FP transgenic nude mice in order to observe vessel anastomosis using color-coded imaging [16].

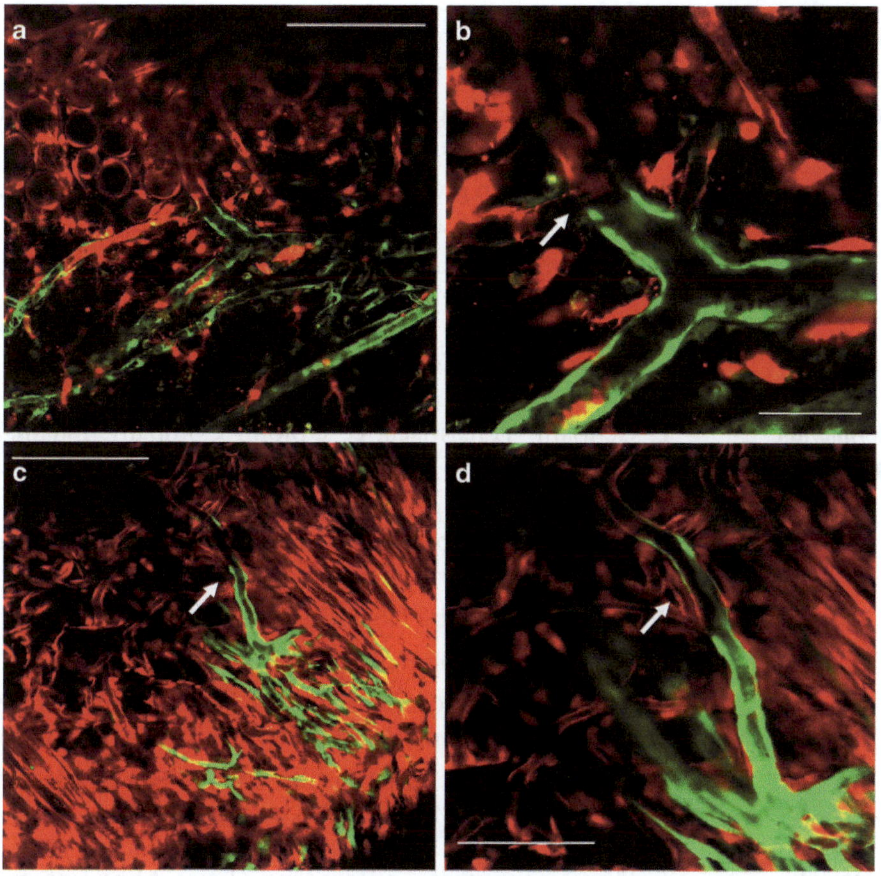

Fig. 3 In vivo visualization of anastomosis between red fluorescent protein (RFP) vessels in Gelfoam® and host vessels in nestin-driven green fluorescent protein (ND-GFP) transgenic nude mice. Skin flaps were made after re-transplantation of vascularized Gelfoam®, previously transplanted in RFP nude mice. (**a**) Fourteen days after re-transplantation, RFP-expressing vessels were observed and anastomosed to ND-GFP-expressing blood vessels (*arrow*). Bar = 200 μm. (**b**) At higher magnification (×60), vessel anastomosis was imaged (*arrow*). Bar = 30 μm. (**c**) Twenty-one days after re-transplantation, RFP-expressing vessels and ND-GFP-expressing blood vessels were anastomosed (*arrow*). Bar = 200 μm. (**d**) At higher magnification (×40), a hybrid vessel with mixed expression of RFP and ND-GFP was imaged (*arrow*). Bar = 100 μm [16]

RFP-expressing blood vessels anastomosed to host ND-GFP-expressing blood vessels in the Gelfoam® by 14 days after re-transplantation. Twenty-one days after re-transplantation, RFP vessels and ND-GFP vessels were further fused, suggesting that Gelfoam® transplanted with blood vessels could fuse with host blood vessels and enhance blood flow to organs in need of revascularization (Fig. 3).

These studies demonstrate that Gelfoam® is a powerful tool for observing angiogenesis in vivo and the interactions between vessels. This will allow further understanding of the mechanisms underlying tumor and normal angiogenesis, which can then be exploited for vessel replacement in regenerative medicine as well as for targeting tumor angiogenesis [16].

2 Materials

2.1 Transgenic Fluorescent Mice

1. Red fluorescent protein (RFP) transgenic mice: females weighing 25–30 g (AntiCancer, Inc., San Diego, CA).
2. Nestin-driven green fluorescent protein (ND-GFP) nude mice, 25–30 g (AntiCancer, Inc., San Diego, CA).
3. All animal studies were conducted in accordance with principles and procedures outlined in the National Research Council's Guide for the Care and Use of Laboratory Animals under PHS Assurance Number A3873-01 [15].

2.2 Experimental Materials

1. Gelfoam® (Pharmacia & Upjohn Company, Kalamazoo, MI, USA) (5 × 5 mm).
2. Basic fibroblast growth factor (bFGF; Chemicon, Temecula, CA, USA). Gelfoam® was impregnated with 300 ng basic fibroblast growth factor in 75 μL RPMI 1640 medium.
3. Anti-CD31 MAb (1:50; Chemicon).
4. Substrate-chromogen: 3,3′-diaminobenzidine.
5. 143B human osteosarcoma cells (5×10^5) expressing α_v integrin-GFP (AntiCancer Inc., San Diego, CA, USA) [17].
6. 0.5 ml 28G latex-free insulin syringe (TYCO Health Group LP, Mansfield, MA, USA).

2.3 Equipment

1. Fluorescent stereo microscope (Leica Microsystems GmbH, Wetzlar, Germany). Selective excitation of GFP was produced through a D425/60 band-pass filter and a 470 DCXR dichroic mirror. Emitted fluorescence was collected through a long-pass filter (GG475; Chroma Technology, Brattleboro, VT, USA) on a Hamamatsu C5810 3-chip cooled color charge-coupled device (CCD) camera (Hamamatsu Photonics, Bridgewater, NJ, USA).
2. Image pro plus 3.1 software (Media Cybernetics, Silver Spring, MD, USA).
3. Fluoview FV1000 confocal laser scanning microscope (Olympus Corp., Tokyo, Japan).
4. FV10-ASW Fluoview software (Olympus) and ImageJ (NIH, Bethesda, MD, USA) images were not modified beyond the standard adjustment of intensity levels [15, 18].
5. XLUMPLFL 20× (0.95 numerical aperture) water-immersion objective for imaging. GFP was excited at 488 nm, and RFP was excited at 559 nm with an argon laser.

3 Methods

3.1 Implantation of Gelfoam®

1. Anesthetize the ND-GFP transgenic nude mice [19] or RFP transgenic nude mice [20], 6–8 weeks old, with tribromoethanol (*see* **Note 1**).
2. Transplant the bFGF-treated Gelfoam® into the subcutis on both flanks of the ND-GFP or RFP transgenic mice (*see* **Note 2**).
3. Raise skin flaps at day 7 after transplantation of Gelfoam® under anesthesia (*see* **Note 3**).
4. Quantify angiogenesis by measuring length of ND-GFP-expressing nascent blood vessels in the Gelfoam® in the skin flap by in vivo fluorescence microscopy imaging. In addition, count the vessels on the surface using fluorescence microscopy.

3.2 Reimplantation of Gelfoam®

1. Anesthetize 4-week-old ND-GFP transgenic nude mice [14] with a ketamine mixture as described above.
2. Remove Gelfoam® previously implanted in the RFP transgenic nude mice 14 days after implantation and reimplant into the subcutis on the flank of ND-GFP transgenic nude mice.
3. Observe the Gelfoam® via a skin flap with a confocal laser-scanning microscope 14 and 21 days after reimplantation [16].

3.3 Immuno-fluorescence Staining of Blood Vessels (See Notes 4 and 5)

1. Detect co-localization of ND-GFP or RFP fluorescence and CD31 in frozen sections of vascularized Gelfoam® using standard immunohistochemical techniques and using an anti-rat immunoglobulin horseradish peroxidase detection kit.
2. Express the experimental data as the mean ± sd. Perform statistical analysis using the two-tailed Student's *t*-test [14, 15].

3.4 Making Skin Flaps (See Notes 6 and 7)

1. Anesthetize RFP transgenic nude mice with ketamine mixture via s.c. injection.
2. Make an arc-shaped incision in the abdominal skin from the axillary to the inguinal region.
3. Separate the subcutaneous connective tissue to free the skin flap without injuring the vessel.
4. Lay the mice flat and spread the skin flap and fix on a flat stand [14].
5. Directly image the Gelfoam® with a FV1000 confocal microscope.
6. Close the skin with a 6-0 nylon suture [15, 21].

3.5 Imaging of 143B Osteosarcoma Cells Expressing α_v Integrin-GFP Interacting with RFP Vessels in Gelfoam® (See Notes 8–10)

1. Inject the 143B α_v integrin-GFP cells into the Gelfoam® previously implanted in transgenic RFP nude mice with a 0.5 ml 28G latex-free insulin syringe.
2. Raise skin flaps on days 7, 14, 21, and 28 after injection on the inside surface of the Gelfoam® and directly image.
3. Close the skin with a 6-0 suture [15, 21].

4 Notes

1. The GFP and RFP nude mice appear to have a life span similar to that of non-GFP nude mice, such that long-term angiogenesis can be carried out [7].
2. Gelfoam® is well suited for angiogenesis and more superior to Matrigel™.
3. Skin flaps provide higher resolution imaging than transplantable "windows" [22, 23]. In vivo color-coded imaging of blood vessels and cancer cells is facilitated by using the Olympus IV100 Intravital Laser Scanning Microscope or Olympus FV1000 confocal microscope [14].
4. It is important to minimize autofluorescence of tissue and body fluids by using proper filters. Excitation filters should have a narrow band as close to 490 nm as possible to specifically excite GFP whose peak is distinct from autofluorescence of the skin, tissues, and fluid of the animal. In addition, proper band-pass emission filters should be used with a cutoff of approximately 515 nm (Chroma Technology) [11–13].
5. Bleeding should be avoided at the surgical site, as hemoglobin will absorb the incident excitation light [11–13].
6. When doing a skin flap or open-biopsy procedures, it is essential to hydrate the animal by spraying normal saline on the open tissue [11–13].
7. When doing repeat procedures such as a skin flap, open biopsy, or other invasive procedures, it is critical to maintain a properly sterile operation field [11–13]. Skin flaps can be repeated 2–3 times per week.
8. The luciferase (Luc) reporter technique requires that animals are anaesthetized and restrained so that sufficient photons can be collected overtime to construct a pseudo-image. Furthermore, this process must be carried out in an almost light-free environment and animals must be injected at each imaging session with the luciferin substrate, which has to reach every tumor cell to be useful. Detection of Luc-labeled cells in vivo is very low resolution and cellular imaging is not possible and the clearance of the luciferin results in an unstable signal.

These limitations preclude studies that would be perturbed by anesthesia, restraint, or substrate injection and also make high-throughput screening infeasible [24].

9. In comparison with the luciferase reporter, GFP has a much stronger signal and can therefore be used to image unrestrained animals—irradiation with non-damaging blue light is the only step needed. Real images can be captured using fairly simple apparatus and there is no need for total darkness. The fluorescence intensity of GFP is very strong and individual cells or even subcellular structures can be imaged in the animal [25].
10. Be careful of misconceptions in the literature stating the "limitations" of in vivo imaging with fluorescent proteins. For example, the following was just published "Whole animal fluorescence imaging in vivo suffers from low signal-to-noise as a result of background autofluorescence, modeling-dependent photon quantification, photobleaching, low tissue penetration, and low resolution" [26]. Despite enormous evidence to the contrary, as outlined above, this type of misinformation on limitations of fluorescent protein imaging continues to be published even in 2013.

References

1. Folkman J (1995) Angiogenesis in cancer, vascular, rheumatoid and other disease. Nat Med 1:27–31
2. Folkman J (1990) What is the evidence that tumors are angiogenesis dependent? J Natl Cancer Inst 82:4–6
3. Auerbach R, Kubai L, Knighton D, Folkman J (1974) A simple procedure for long term cultivation of chicken embryos. Dev Biol 41:391–394
4. Crum R, Szabo S, Folkman J (1985) A new class of steroids inhibits angiogenesis in the presence of heparin or a heparin fragment. Science 230:1375–1378
5. Miller JW, Stinson WG, Folkman J (1993) Regression of experimental iris neovascularization with systemic alpha-interferon. Ophthalmology 100:9–14
6. Passaniti A, Taylor RM, Pili R, Guo Y, Long PV, Haney JA, Pauly RR, Grant DS, Martin GR (1992) A simple, quantitative method for assessing angiogenesis and antiangiogenic agents using reconstituted basement membrane, heparin, and fibroblast growth factor. Lab Invest 67:519–528
7. Alessandri G, Raju K, Gullino PM (1983) Mobilization of capillary endothelium in vitro induced by effectors of angiogenesis in vivo. Cancer Res 43:1790–1797
8. Deutsch TA, Hughes WF (1979) Suppressive effects of indomethacin on thermally, induced neovascularization of rabbit corneas. Am J Ophthalmol 87:536–540
9. Korey M, Peyman GA, Berkowitz R (1977) The effect of hypertonic ointments on corneal alkali burns. Ann Ophthalmol 9:1383–1387
10. Mahoney JM, Waterbury LD (1985) Drug effects on the neovascularization response to silver nitrate cauterization of the rat cornea. Curr Eye Res 4:531–535
11. Li WW, Grayson G, Folkman J, D'Amore PA (1991) Sustained-release endotoxin. A model for inducing corneal neovascularization. Invest Ophthalmol Vis Sci 32:2906–2911
12. Epstein RJ, Hendricks RL, Stulting RD (1990) Interleukin-2 induces corneal neovascularization in A/J mice. Cornea 9:318–323
13. McCarty MF, Baker CH, Bucana CD, Fidler IJ (2002) Quantitative and qualitative in vivo angiogenesis assay. Int J Oncol 21:5–10
14. Amoh Y, Li L, Katsuoka K, Bouvet M, Hoffman RM (2007) GFP-expressing vascularization of Gelfoam® as a rapid in vivo assay of angiogenesis stimulators and inhibitors. Biotechniques 42:294–298
15. Uehara F, Tome Y, Yano S, Miwa S, Mii S, Hiroshima Y, Bouvet M, Maehara H, Kanaya F, Hoffman RM (2013) A color-coded imaging

model of the interaction of αv integrin-GFP expressed in osteosarcoma cells and RFP expressing blood vessels in Gelfoam® vascularized in vivo. Anticancer Res 33:1361–1366
16. Uehara F, Tome Y, Reynoso J, Mii S, Yano S, Miwa S, Bouvet M, Maehara H, Kanaya F, Hoffman RM (2013) Color-coded imaging of spontaneous vessel anastomosis *in vivo*. Anticancer Res 33(8):3041–3045
17. Tome Y, Sugimoto N, Yano S, Momiyama M, Mii S, Maehara H, Bouvet M, Tsuchiya H, Kanaya F, Hoffman RM (2013) Real-time imaging of αv integrin molecular dynamics in osteosarcoma cells *in vitro* and *in vivo*. Anticancer Res 33(8):3021–3025
18. Uchugonova A, Zhao M, Weinigel M, Zhang Y, Bouvet M, Hoffman RM, Koenig K (2013) Multiphoton tomography visualizes collagen fibers in the tumor microenvironment that maintain cancer-cell anchorage and shape. J Cell Biochem 114:99–102
19. Amoh Y, Yang M, Li L, Reynoso J, Bouvet M, Moossa AR, Katsuoka K, Hoffman RM (2005) Nestin-linked green fluorescent protein transgenic nude mouse for imaging human tumor angiogenesis. Cancer Res 65:5352–5357
20. Yang M, Reynoso J, Bouvet M, Hoffman RM (2009) A transgenic red fluorescent protein-expressing nude mouse for color-coded imaging of the tumor microenvironment. J Cell Biochem 106:279–284
21. Yang M, Baranov E, Wang J-W, Jiang P, Wang X, Sun F-X, Bouvet M, Moossa AR, Penman S, Hoffman RM (2002) Direct external imaging of nascent cancer, tumor progression, angiogenesis, and metastasis on internal organs in the fluorescent orthotopic model. Proc Natl Acad Sci U S A 99:3824–3829
22. Jain RK, Munn LL, Fukumura D (2002) Dissecting tumour pathophysiology using intravital microscopy. Nat Rev Cancer 2:266–276
23. Ritsma L, Steller EJ, Beerling E, Loomans CJ, Zomer A, Gerlach C, Vrisekoop N, Seinstra D, van Gurp L, Schäfer R, Raats DA, de Graaff A, Schumacher TN, de Koning EJ, Rinkes IH, Kranenburg O, van Rheenen J (2012) Intravital microscopy through an abdominal imaging window reveals a pre-micrometastasis stage during liver metastasis. Sci Transl Med 4:158ra145
24. Hoffman RM (2005) The multiple uses of fluorescent proteins to visualize cancer in vivo. Nat Rev Cancer 5:796–806
25. Hoffman RM (ed) (2012) In vivo cellular imaging using fluorescent proteins: methods and protocols. In: Walker JM (series ed) Methods in molecular biology, vol 872. Humana Press, New York
26. Kocher B, Piwnica-Worms D (2013) Illuminating cancer systems with genetically engineered mouse models and coupled luciferase reporters in vivo. Cancer Discov 3:616–629

Chapter 19

Laser Speckle Contrast Imaging to Measure Changes in Cerebral Blood Flow

Ian R. Winship

Abstract

Laser speckle contrast imaging (LSCI) is a powerful tool capable of acquiring detailed maps of blood flow in arteries and veins on the cortical surface. Based on the blurring of laser speckle patterns by the motion of blood cells, LSCI can be combined with a variety of optical imaging preparations to acquire high-spatiotemporal resolution images of blood flow, and track changes in blood flow over time, using relatively simple instrumentation. Here, we describe methods for LSCI of cerebral blood flow via a thin skull imaging preparation in mice or rats. This preparation allows precise semiquantitative mapping of changes in blood flow over time using straightforward surgical protocols and equipment.

Key words Laser speckle contrast imaging, Arterioles, Veins, Cerebral blood flow

1 Introduction

A number of direct and indirect methods can be used to image cerebral blood flow in humans and animal models. Techniques such as magnetic resonance (MR) perfusion imaging, computed tomography (CT) perfusion, xenon-enhanced CT, single-photon emission CT, and positron-enhanced tomography can be used to measure cerebral blood flow in humans but do not permit individual resolution of small vessels. Direct visualization of individual vessels can be obtained using cerebral digital angiography, CT angiography, MR angiography, and transcranial Doppler sonography [1, 2]. However, access to these modalities is limited by their requirement for rare and expensive instrumentation. Moreover, their utility for use in experimental animal models is reduced in many cases by their limited spatial and temporal resolution.

Laser speckle contrast imaging (LSCI) is a technique for optically recording blood flow that can be used to acquire high-resolution, semiquantitative maps of blood flow in vasculature on the cortical surface in animal models. Importantly, LSCI permits wide-field mapping of blood flow on the cortical surface with

Richard Milner (ed.), *Cerebral Angiogenesis: Methods and Protocols*, Methods in Molecular Biology, vol. 1135,
DOI 10.1007/978-1-4939-0320-7_19, © Springer Science+Business Media New York 2014

resolution to individual vessels and high temporal resolution [3–9]. Requiring relatively inexpensive instrumentation, LSCI has been shown effective in mapping changes in blood flow in visually identified blood vessels, such as those that occur downstream of an occlusion (Fig. 1a, b) [4, 10–14]. LSCI of cerebral blood flow can be performed through the intact skull or through a thinned skull preparation (described below) or craniotomy preparation (with skull removed).

LSCI maps of blood flow are based on blurring of a characteristic laser speckle pattern produced by illumination of the brain surface (or skull) with coherent laser light. Importantly, the speckle pattern is dynamic and blurred by the motion of particles (such as blood cells) on or below the illuminated surface. By analyzing fluctuations in the speckle pattern in a particular image, speckle contrast values that provide a measure of blood flow can be calculated. Analysis for laser speckle involves determining the speckle contrast factor (K). K is a measure of the local spatial contrast of the laser speckle pattern and is defined as the ratio of the standard deviation to the mean intensity ($K = \sigma_s / I$) in a small region of the speckle image (typically 5×5 or 7×7 pixels). Speckle contrast and motion of the scattering particles are inversely related, with K values near 1 suggesting no blood flow in that region and values closer to zero reflecting greater blood flow. Plotting K values allows for maps for maps of blood flow that can reveal changes in the pattern of blood flow over time, including enhanced collateral perfusion, reperfusion, or angiogenesis (Fig. 1) [3, 10, 14–18].

While LSCI permits sensitive mapping of blood flow in surface vessels, the precise quantification of blood flow velocity is difficult because the exact relationship between speckle contrast and blood flow velocity remains undefined [7, 8]. Moreover, LSCI is susceptible to artifacts that can result from differences in the optical characteristics of the imaging window or changing parameters in the image acquisition system. Because these artifacts can vary between animals and imaging sessions [3, 6, 14, 15], care must be taken when comparing between animals and interpreting repeated imaging data. LSCI is best restricted to describing relative changes in the pattern of blood flow within animals rather than quantifying blood flow velocity, and careful control of imaging parameters is necessary to allow comparison between imaging sessions [3, 15]. These considerations are particularly important in the case of repeated imaging with long intervals and/or movement of the animal between sessions (*see* **Note 1**). Despite these caveats, LSCI is a powerful tool for mapping changes in blood flow on the cortical surface in animal models. Here, we describe materials and methods for LSCI through a thin skull preparation in rodents.

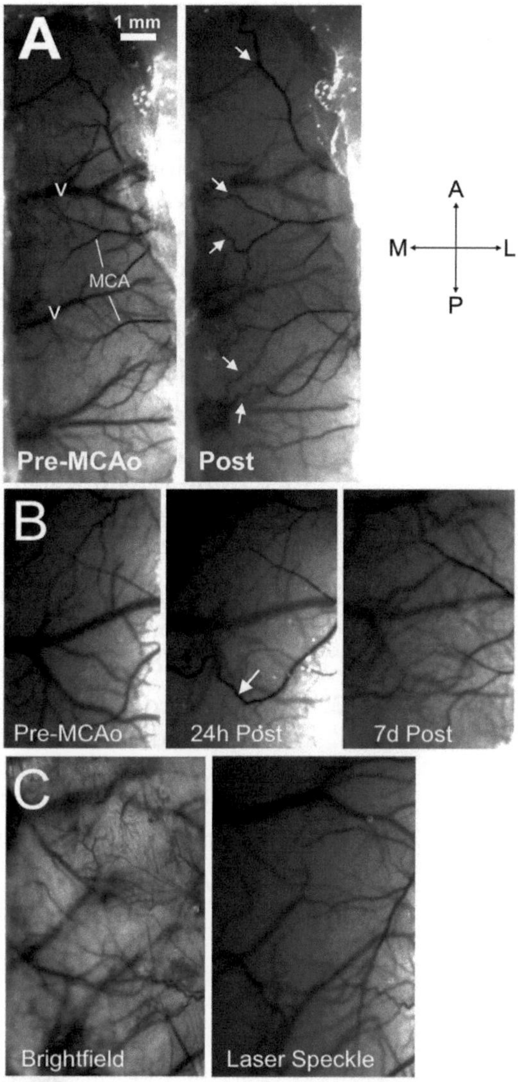

Fig. 1 Laser speckle contrast imaging (LSCI) of dynamic blood flow. (**a**) LSCI provides high-resolution maps of blood flow in surface veins (V) and distal segments of the middle cerebral artery (MCA) in the sensorimotor cortex before (*left panel*) and after (*right panel*) middle cerebral artery occlusion (MCAo). Note anastomatic connections (*arrows, right panel*) between distal segments of the anterior and middle cerebral artery that appear after occlusion. (**b**) LSCI can be performed through a thin skull imaging preparation that is stable over several days to weeks. Images show maps of speckle contrast before MCAo, 24 h after MCAo, and 7 days after MCAo. Note anastomatic connections that are open 24 h after stroke (*large arrow*) and the possible formation of new smaller vessels at 7 days post-MCAo. (**c**) Thinned skull preparations do not require the skull to be completely transparent (i.e., extremely thin) as laser speckle imaging can be performed through a partially thinned and smoothed skull. Compare brightfield (*left*) and speckle contrast (*right*) images of the same imaging preparation. (**a**) is modified from Armitage et al. [10]

2 Materials

2.1 Imaging Apparatus

LSCI is most frequently performed using a video "macroscope" (Fig. 2a) [19–22] or through an upright microscope frame and low-magnification objective. The key instrumentation required for LSCI are a video camera with optics and image acquisition software, a laser diode, and equipment to stabilize the animal during imaging. Notably, high dynamic range cooled cameras are not required for LSCI and inexpensive cameras have reliably been used to map blood flow with LSCI [8]. The equipment listed below from our systems provide examples of compatible equipment, but other configurations have been described (e.g., *see* ref. 22):

1. *CCD camera and image acquisition software*: EPIX PIXCI EL1 software-controlled frame-grabber (Epix, Inc.) coupled with a DALSA 1M60 Pantera CCD camera (DS-21-01M60, Teledyne DALSA Inc.).
2. *Laser diode module*: 785 nm laser diode module with adjustable power output (1–20 mW) and a 2.4 × 3.4 mm beam diameter (LDM785, Thorlabs, Inc.) (*see* **Note 2**).
3. *Optics.* Macroscopic optics can be created using two face-to-face mounted video lenses (Fig. 2a, *see* **Note 3**) [19, 20, 23] or a macro zoom lens (e.g., Edmund Scientific, NT58-240 [22]). Microscope objectives with magnification between 1.5× and 4× are effective for LSCI on an upright microscope frame [12, 21].
4. *Animal holder and stage.* Anesthetized rodents must be secure and immobile during LSCI to reduce artifact. Our lab uses

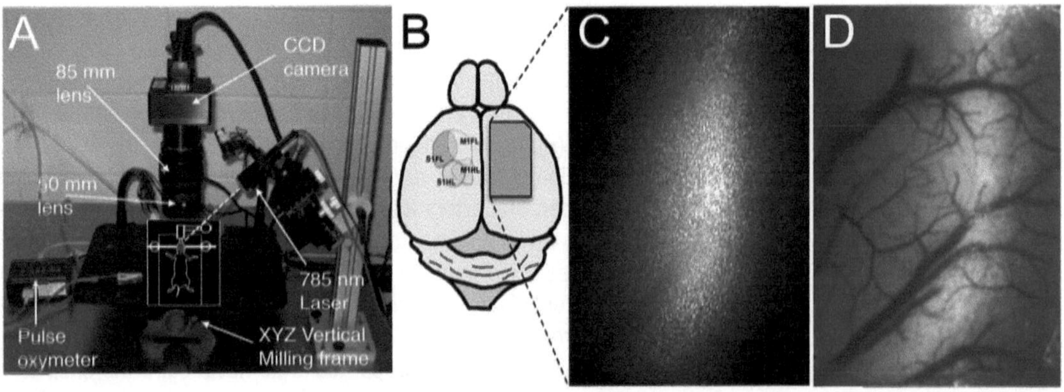

Fig. 2 (**a**) Video macroscope with tandem lens configuration for laser speckle contrast imaging (LSCI). (**b**) Approximate location of imaging windows used to assess changes in collateral blood flow following middle cerebral artery occlusion. (**c**) A raw speckle image. (**d**) Map of speckle contrast derived from raw speckle image in (**c**). (**b**) is modified from Armitage et al. [10]

custom animal plates that integrate with standard stereotaxic equipment (*see* **Note 4**) mounted on an XYZ milling frame (*see* Fig. 2a, Sherline Products Inc.) for the macroscope or onto a Sutter MPC-78 large moving stage platform (Sutter Instruments Company) for imaging on the upright microscope.

2.2 Surgical Supplies

1. DC temperature regulator, heated pad, and rectal probe (FHC, Inc.).
2. Isoflurane (or other anesthetic; *see* **Note 5**).
3. Anesthetic work station including precision vaporizer, induction chamber, and stereotax-compatible masks/bite adapters for rats or mice.
4. Nitrous oxide.
5. Oxygen.
6. Eye ointment (Liposic ophthalmic gel, Bausch and Lomb).
7. Dental drill and drill bits (0.7–1.3 mm diameter).
8. Sterile surgical instruments, including scalpels, fine forceps, hemostats, and bulldog serrefines.
9. Rounded scalpel.
10. Sterile Gelfoam cotton tip applicators and gauze.
11. Pulse oximeter (MouseOx, Starr Life Sciences, Inc.).
12. Betadine surgical scrub and solution.
13. Syringes (1 ml) and needles (27 g).
14. Small animal hair clippers.
15. 70 % ethanol.
16. Hot bead sterilizer (for surgical instruments).
17. Silk sutures (5.0, Ethicon).
18. Bupivicaine (5 mg/ml in sterile saline).
19. Buprenorphine or other analgesic.
20. HEPES-buffered artificial cerebrospinal fluid (ACSF) (*see* **Note 6**).
21. Low-melt agarose (Sigma Type IIIA, #A9793).
22. Dental cement.

3 Methods

3.1 Surgical Procedures

1. Animal experiments must be performed in accordance with institutional and governmental authorities. All procedures herein are within the guidelines of the Canadian Council on Animal Care and were approved by the Animal Care and Use Committee for Health Sciences at the University of Alberta.

2. Rats or mice are allowed to habituate to handling and the animal housing facility for at least 1 week prior to surgery.
3. Prior to surgery, the surgical suite is sterilized. Sterilize all surgical instruments using hot bead sterilizer or autoclave.
4. Anesthetize the rats with 4.0 % isoflurane (70:30, O_2:N_2O) in an induction chamber. Carefully monitor the surgical depth. Absence of the toe pinch reflex signifies the surgical plane. Once surgical depth is achieved, transfer the animal to the stereotaxic apparatus or animal plate with ear bars and bite adapter and switch to anesthetic delivery via a mask.
5. If desired, inject 1 mg/kg of dexamethasone (subcutaneous, s.c., in the scruff of the neck) to reduce cerebral swelling and inflammation and reduce airway secretions during surgery (*see* **Note 7**).
6. Maintain the body temperature at 36–37 °C using an electric heating pad coupled to a DC temperature regulator with feedback from a rectal probe. Heart rate, breathing rate, oxygen saturation, pulse distension, and breath distension should be monitored via pulse oxymetry to confirm that physiological parameters remain stable during surgery and imaging. Respiration and flesh color (footpads) should also be constantly monitored to evaluate animal health.
7. Place eye ointment onto the eyes to prevent desiccation. Repeat every 30 min during surgery and imaging.
8. Shave the skin overlying the desired imaging location. To sterilize the planned incision site, apply Betadine surgical scrub, lather, and leave for 5 min before wiping away with moist gauze. Next, apply Betadine povidone/iodine topical anesthetic solution and allow to dry.
9. Inject 0.01 ml (mice) or 0.1 ml (rats) of the local anesthetic bupivicaine (5 mg/ml in saline, s.c.) at two sites along the incision path (0.005 or 0.05 ml per site). Allow 10 min for bupivicaine to take effect.
10. Make a 1–2 cm longitudinal incision along the midline of the skull. Reflect skin laterally to expose the skull, and hold in place with bulldog serrefines. Remove the periosteum with fine forceps. Clean the skull surface with sterile gauze and cotton tip applicators until dry and clear of blood (*see* **Note 8**).
11. Define the borders of the area of skull over the cortex to be thinned using a permanent marker (*see* **Note 9**).
12. Start thinning the skull over the region of interest with a precision dental drill and sterilized drill bit (be sure to clean with soap and water after surgery and sterilize with hot bead sterilizer prior to surgery). To avoid producing excess heat and potentially damaging the brain, continually move the drill bit around

the cranial window and drill in a stochastic manner. Frequently flush the exposure with sterile HEPES-buffered ACSF and clear with sterile gauze or Gelfoam. Allow to dry and repeat the thinning process with the drill (*see* **Notes 10** and **11**).

13. When the bone has been thinned to approximately one-half to one-third of its original thickness, use a rounded scalpel for the final stages of thinning. Gradual thinning and smoothing is achieved by gentle scraping, angling the scalpel away from the direction of motion. The rounded scalpel can be used on wet or dry skull, changing direction frequently. Be sure to thin the exterior edges of the window as the skull in the center of the window tends to thin most quickly. The skull is sufficiently thin for LSCI when blood vessels on the cortical surface of the brain are clearly visible when the skull is dry and well defined when the skull is wet (Fig. 1c).
14. Once the skull thinning is complete, place a drop of 1.3 % low-melt agarose (at 37–38 °C) dissolved in HEPES-buffered ACSF on the surface of the thinned skull and immediately seal with a coverslip. After 5 min, the coverslip can be fixed in place with small drips of cyanoacrylate glue around the exterior of the coverslip, being sure to keep the imaging surface of the coverslip clean.
15. If desired, apply dental cement around the circumference of the coverslip to provide a protective border and seal the imaging preparation (*see* **Note 12**).
16. Securely suture the skin around the dental cement.
17. At this point, the animal is ready for imaging. If imaging is to be performed, continue to image acquisition (*see* section below).
18. If the animal is to be returned to cage and recovered from anesthetic, clean the incised skin with warm water and Betadine solution. Remove from the anesthetic.
19. After surgery, keep the animal warm and closely monitor for breathing or motor difficulties. When the animal has regained consciousness, administer 0.03 mg/kg Buprenorphine (s.c.) or other approved analgesics or local anesthetics as necessary. Provide the animals with free access to a prerecorded amount of food or water. Track the food consumption, body weight, and clinical signs consistent with institutional policies during the recovery period.

3.2 Image Acquisition

1. Ensure the animal is in surgical anesthesia and monitor physiological parameters as described previously.
2. Transfer the animal plate to the imaging apparatus, ensuring that the head is securely fixed and the coverslip is clean.

A cotton tip applicator dampened with 70 % ethanol can be used to gently clean the coverslip if required.

3. Position the imaging window under the lens or objective and then focus the image on the cortical surface vasculature under brightfield illumination (Fig. 1c, *see* **Note 13**).
4. Turn on the laser diode module and position the beam over the thinned skull in the region of interest (Fig. 2c). Set the exposure time to 5–15 ms. Some reports indicate that 5 ms exposure time is optimal for LSCI of cerebral blood flow (*see* **Note 14**).
5. For optimal contrast, laser speckles should be sampled at Nyquist criterion [**24, 25**]. That is, the pixel size should be set to half of speckle diameter or smaller. Minimum speckle diameter can be estimated a priori using the formula

$$S = 2.44\lambda(1 + M)N$$

where λ is the illumination wavelength, M is the imaging system magnification, and N is the lens f number [**24**].

6. Acquire 100–400 images of the speckle pattern at 10–20 Hz. Raw speckle images can be saved as a single tiff file with a sequence of images.

3.3 Image Processing and Analysis

1. Analysis of the spatial speckle contrast values in laser speckle images is easily performed using ImageJ software (National Institute of Mental Health, Bethesda, MD; *see* **Note 15**).
2. Average the 100–400 images in the stack to create an average of the raw speckle images.
3. To create individual images of variance, each speckle image in the original stack of raw speckle images should then be processed with ImageJ's two-dimensional variance filter (5×5 or 7×7 radius). Average all the variance images and create a single 32-bit image of the standard deviation by taking the square root of the averaged variance image. This image reflects the average standard deviation of the laser speckle over the acquisition period (*see* **Note 16**).
4. Divide the mean standard deviation image by the mean of the raw speckle images to create an image of the speckle contrast, K ($K = \sigma_s / I$) (Fig. 1). These "maps" of speckle contrast illustrate blood flow with resolution to individual cortical surface vessels.
5. Measuring the speckle contrast in defined regions of interest within the lumen of surface veins or arteries or in parenchymal tissue provides an estimate of relative blood flow over time. While speckle contrast is not a quantitative measure of blood

flow, *changes* in speckle contrast are inversely related to *changes* in blood flow. A more accurate representation of blood flow can be achieved by calculating correlation times (τ_c) that are approximately inversely proportional to blood perfusion. The relationship between speckle contrast, K, and τ_c is given by

$$K = \left[\frac{\tau_c}{2T} \{1 - \exp(-2T / \tau_c)\} \right]^{1/2}$$

where T is the exposure time of the camera. However, the demonstration that correlation times are inversely proportional to blood flow is rooted in a number of simplifying assumptions based on literature from other imaging modalities and may not hold under all imaging and blood flow conditions [8]. Focusing on within animal comparisons can thereby reduce some of the uncertainty about the semiquantitative nature of LSCI (*see* **Note 17**).

4 Notes

1. Speckle contrast is susceptible to artifacts that may result from changing optical parameters (e.g., changing quality or clarity of optical window, skull regrowth, intensity or position of laser), so care must be taken to ensure that all optical parameters (including exposure duration and aperture settings) are kept consistent between imaging sessions. These concerns are amplified in chronic experiments where controlling changes in optical parameters are made more difficult by movement of the animal and long durations between imaging sessions.
2. Wavelength is not critical as laser wavelengths beyond 600 nm are not well absorbed, and laser diode modules are widely available at several wavelengths in the red and infrared spectrum. If a larger beam diameter is desired for even illumination of large imaging windows, the laser diode module can be coupled to a beam expander (e.g., BE02M-B, Thorlabs, Inc.) or diverging lens.
3. Figure 2a shows the tandem lens macroscope configuration. In this case, an 85 mm lens is coupled with a 50 mm lens to give 1.7× magnification. These designs permit close working distance imaging with a sharp depth of focus and light collection performance well above most commercial microscope objectives of comparable magnification [19, 20].
4. Harrison, Sigler, and Murphy [23] provide a comprehensive description of the plates adapted for use on our laser speckle contrast imaging systems (*see* Fig. 3 of ref. 23). Plates for mouse imaging are described, with rat plates including a

posterior extension to support larger body size. For LSCI, rats and mice are fitted with ear bars and a bite adapter (with anesthetic delivery) (David Kopf Instruments) to secure the head.

5. Choice of anesthetic will depend on experimental design (e.g., acute imaging vs. repeated imaging with recovery from anesthetic between sessions). Isoflurane anesthesia is frequently used as it is an effective surgical anesthetic and doses can be closely regulated. However, isoflurane is vasodilatory and impairs vasoreactivity, limitations that must be evaluated in light of experimental goals.

6. HEPES-buffered ACSF:

   Dissolve the following in 500 ml of distilled H_2O.

NaCl	3.94 g
KCl	0.20 g
$MgCl_2 \times 6H_2O$	0.102 g
$CaCl_2 \times 2H_2O$	0.132 g
NaHEPES	0.651 g

Adjust the pH to 7.4 and sterilize by vacuum filtration (0.02 μm pore).

7. Dexamethasone is often administered during craniotomy preparations to reduce brain swelling. While not required during thin skull surgeries, dexamethasone (1 mg/kg, s.c.) can be injected after induction to reduce inflammation and decrease airway secretions during surgery if anti-inflammatory actions are not a confound to experimental outcomes.

8. Skull thinning with the dental drill is best achieved with a dry, clear skull surface. Bleeding from incisions should be cleared away. Bleeding may also occur from the skull surface or from the skull during thinning. In most cases, bleeding can be stopped with gentle bone thinning with the dental drill or rounded scalpel. Sterile saline and Gelfoam are effective for stopping small bleeds, and saline will soften the bone to facilitate gentle scraping with a rounded scalpel.

9. Skull landmarks and stereotaxic coordinates can be used to define the region of interest, whose borders can then be marked with permanent marker prior to thinning. For example, to evaluate middle cerebral artery flow to forelimb and hindlimb sensorimotor cortex in rats, a region of skull ±3 mm anterior/posterior of Bregma and 1–5 mm lateral from the midline can be thinned (Fig. 2b) [26].

10. When thinning with the dental drill, the skull should be dry to reduce the risk of skull puncture. During scraping of the skull with the rounded scalpel, moistening the skull with sterile

saline or artificial cerebrospinal fluid will facilitate smoothing of the skull surface. In either case, use extreme caution, as puncturing the skull generally requires termination of the experiment.

11. Mix HEPES-buffered ACSF and Type III-A low-melt agarose and heat to approximately 60–70 °C in the water bath. Once the agarose solution is completely clear, with no undissolved agarose, cool to approximately 38 °C and apply a small quantity to the thinned skull.
12. This imaging preparation (agarose-sealed with a coverslip and dental cement border) is stable over days (e.g., Fig. 1b), but other options are available for acute or chronic imaging schedules. If imaging will consist of baseline imaging then a second session much later, sealing with dental cement is not necessary. After baseline imaging, the coverslip and agarose can be removed and the wound sutured. Re-thin and smooth the thin skull window and reapply coverslip at the next imaging session. For chronic, repeated imaging, a chronic craniotomy preparation is frequently used, though this approach requires careful surgery [10]. Similarly, a polished and reinforced thin skull preparation [27] allows repeated imaging over weeks to months without additional surgery. In this preparation, the thinned skull is polished; then the coverslip is fixed in place with cyanoacrylate cement. *See* Drew et al. [27] for detailed methods.
13. LED illumination at 532 nm provides good contrast for vasculature (Fig. 1c) but can also accentuate any blood on or below the skull surface. The wavelength of illumination is not critical; the purpose is merely to focus the optics on the surface of the brain. Image acquisition parameters (e.g., exposure time) do not have to be consistent with speckle imaging.
14. The speckle contrast-to-noise ratio is minimal for exposure times of 5 ms, suggesting this is the optimal exposure for LSCI, though other values are also reported in the literature [28]. More recently, multi-exposure speckle imaging (MESI) has been shown to more accurately measure blood flow, though this requires more sophisticated instrumentation [3, 14].
15. Methods herein describe basic analysis of *spatial* contrast during LSCI. Other approaches requiring more intensive calculations have also been reported. Of note, analysis of temporal contrast (or mixed spatiotemporal analyses) is reported to be less susceptible to artifacts from static scatters and may improve quantitative accuracy [29–32].
16. Averaging more raw speckle files reduces noise and improves image quality. However, if temporal changes in flow are of interest during the acquisition period, individual speckle

images can be analyzed by dividing each individual standard deviation image by the mean intensity.

17. LSCI provides wide-field maps of blood flow with excellent spatial and temporal resolution. However, due to the uncertainties over the precise relation between speckle contrast and blood flow velocity, it is considered semiquantitative. True quantitative measures of blood flow velocity or flux and resolution below the surface vasculature can be attained using in vivo two-photon microscopy [33].

Acknowledgements

I.W. is supported by Alberta Innovates Health Solutions, the Heart and Stroke Foundation of Canada, and the Canadian Institutes of Health Research. Thanks to Gomathi Ramakrishnan for assistance with the manuscript figures. Figures 1a and 2b are modified from Armitage et al. [10].

References

1. Liebeskind DS (2003) Collateral circulation. Stroke 34:2279–2284
2. Shuaib A, Butcher K, Mohammad AA, Saqqur M, Liebeskind DS (2011) Collateral blood vessels in acute ischaemic stroke: a potential therapeutic target. Lancet Neurol 10:909–921
3. Parthasarathy AB, Tom WJ, Gopal A, Zhang X, Dunn AK (2008) Robust flow measurement with multi-exposure speckle imaging. Opt Express 16:1975–1989
4. Strong AJ, Bezzina EL, Anderson PJ, Boutelle MG, Hopwood SE, Dunn AK (2006) Evaluation of laser speckle flowmetry for imaging cortical perfusion in experimental stroke studies: quantitation of perfusion and detection of peri-infarct depolarisations. J Cereb Blood Flow Metab 26:645–653
5. Tom WJ, Ponticorvo A, Dunn AK (2008) Efficient processing of laser speckle contrast images. IEEE Trans Med Imaging 27:1728–1738
6. Boas DA, Dunn AK (2010) Laser speckle contrast imaging in biomedical optics. J Biomed Opt 15:011109
7. Duncan DD, Kirkpatrick SJ (2008) Can laser speckle flowmetry be made a quantitative tool? J Opt Soc Am A Opt Image Sci Vis 25:2088–2094
8. Dunn AK (2012) Laser speckle contrast imaging of cerebral blood flow. Ann Biomed Eng 40:367–377
9. Dunn AK, Devor A, Dale AM, Boas DA (2005) Spatial extent of oxygen metabolism and hemodynamic changes during functional activation of the rat somatosensory cortex. Neuroimage 27:279–290
10. Armitage GA, Todd KG, Shuaib A, Winship IR (2010) Laser speckle contrast imaging of collateral blood flow during acute ischemic stroke. J Cereb Blood Flow Metab 30:1432–1436
11. Shin HK, Nishimura M, Jones PB, Ay H, Boas DA, Moskowitz MA, Ayata C (2008) Mild induced hypertension improves blood flow and oxygen metabolism in transient focal cerebral ischemia. Stroke 39:1548–1555
12. Zhang S, Murphy TH (2007) Imaging the impact of cortical microcirculation on synaptic structure and sensory-evoked hemodynamic responses in vivo. PLoS Biol 5:e119
13. Li P, Murphy TH (2008) Two-photon imaging during prolonged middle cerebral artery occlusion in mice reveals recovery of dendritic structure after reperfusion. J Neurosci 28:11970–11979
14. Parthasarathy AB, Kazmi SM, Dunn AK (2010) Quantitative imaging of ischemic

stroke through thinned skull in mice with Multi Exposure Speckle Imaging. Biomed Opt Express 1:246–259

15. Ayata C, Dunn AK, Gursoy-OZdemir Y, Huang Z, Boas DA, Moskowitz MA (2004) Laser speckle flowmetry for the study of cerebrovascular physiology in normal and ischemic mouse cortex. J Cereb Blood Flow Metab 24:744–755
16. Rege A, Seifert AC, Schlattman D, Ouyang Y, Li KW, Basaldella L, Brem H, Tyler BM, Thakor NV (2012) Longitudinal in vivo monitoring of rodent glioma models through thinned skull using laser speckle contrast imaging. J Biomed Opt 17:126017
17. Rege A, Thakor NV, Rhie K, Pathak AP (2012) In vivo laser speckle imaging reveals microvascular remodeling and hemodynamic changes during wound healing angiogenesis. Angiogenesis 15:87–98
18. Rege A, Murari K, Seifert A, Pathak AP, Thakor NV (2011) Multiexposure laser speckle contrast imaging of the angiogenic microenvironment. J Biomed Opt 16:056006
19. Bonhoeffer T, Grinvald A (1996) In: Toga AW, Mazziotta JC (eds) Brain mapping: the methods. Academic, San Diego, pp 55–97
20. Ratzlaff EH, Grinvald A (1991) A tandem-lens epifluorescence macroscope: hundred-fold brightness advantage for wide-field imaging. J Neurosci Methods 36:127–137
21. Sigler A, Goroshkov A, Murphy TH (2008) Hardware and methodology for targeting single brain arterioles for photothrombotic stroke on an upright microscope. J Neurosci Methods 170:35–44
22. Ponticorvo A, Dunn AK (2010) How to build a Laser Speckle Contrast Imaging (LSCI) system to monitor blood flow. J Vis Exp (45). pii: 2004. doi: 10.3791/2004
23. Harrison TC, Sigler A, Murphy TH (2009) Simple and cost-effective hardware and software for functional brain mapping using intrinsic optical signal imaging. J Neurosci Methods 182:211–218
24. Kirkpatrick SJ, Duncan DD, Wells-Gray EM (2008) Detrimental effects of speckle-pixel size matching in laser speckle contrast imaging. Opt Lett 33:2886–2888
25. Thompson O, Andrews M, Hirst E (2011) Correction for spatial averaging in laser speckle contrast analysis. Biomed Opt Express 2:1021–1029
26. Paxinos G, Watson C (2007) The rat brain in stereotaxic coordinates, 6th edn. Academic, San Diego
27. Drew PJ, Shih AY, Driscoll JD, Knutsen PM, Blinder P, Davalos D, Akassoglou K, Tsai PS, Kleinfeld D (2010) Chronic optical access through a polished and reinforced thinned skull. Nat Methods 7:981–984
28. Yuan S, Devor A, Boas DA, Dunn AK (2005) Determination of optimal exposure time for imaging of blood flow changes with laser speckle contrast imaging. Appl Opt 44:1823–1830
29. Duncan DD, Kirkpatrick SJ, Wang RK (2008) Statistics of local speckle contrast. J Opt Soc Am A Opt Image Sci Vis 25:9–15
30. Kirkpatrick SJ, Duncan DD, Wang RK, Hinds MT (2007) Quantitative temporal speckle contrast imaging for tissue mechanics. J Opt Soc Am A Opt Image Sci Vis 24:3728–3734
31. Li P, Ni S, Zhang L, Zeng S, Luo Q (2006) Imaging cerebral blood flow through the intact rat skull with temporal laser speckle imaging. Opt Lett 31:1824–1826
32. Qiu J, Li P, Luo W, Wang J, Zhang H, Luo Q (2010) Spatiotemporal laser speckle contrast analysis for blood flow imaging with maximized speckle contrast. J Biomed Opt 15:016003
33. Shih AY, Driscoll JD, Drew PJ, Nishimura N, Schaffer CB, Kleinfeld D (2012) Two-photon microscopy as a tool to study blood flow and neurovascular coupling in the rodent brain. J Cereb Blood Flow Metab 32(7):1277–1309

Chapter 20

Laser Doppler Flowmetry to Measure Changes in Cerebral Blood Flow

Brad A. Sutherland, Tamer Rabie, and Alastair M. Buchan

Abstract

Laser Doppler flowmetry (LDF) is a method by which relative cerebral blood flow (CBF) of the cortex can be measured. Although the method is easy to employ, LDF only measures relative CBF, while absolute CBF cannot be quantified. LDF is useful for investigating CBF changes in a number of different applications including neurovascular and stroke research. This chapter will prepare the reader for rodent experiments using LDF with two preparations. The closed skull preparation can be used to monitor CBF with an intact skull, but in adult rats, thinning of the skull is required to obtain an accurate cortical CBF signal. The open skull preparation requires a craniotomy to expose the surface of the brain and the LDF probe is held close to the surface to measure cerebral perfusion.

Key words Laser Doppler flowmetry, Cerebral blood flow, Rodent, Relative changes, Probe, Perfusion

1 Introduction

When studying neurovascular pathophysiology, methods for measuring cerebral blood flow (CBF) are required. These methods should be reliable and noninvasive and provide high temporal and spatial resolution. Traditionally, autoradiography was used to assess CBF using the tracer molecule [^{14}C] iodoantipyrine [1]. Although this method has high spatial resolution, it does not provide any temporal information since collection of data requires removal of the brain. More recently, optical methods have been developed which circumvent this problem.

Laser Doppler flowmetry (LDF) is one of those recently developed methods. In this method, a laser beam is emitted by a fiber-optic probe at a position over the cerebral cortex where CBF measurement is sought. When the light hits moving cells, mostly erythrocytes within vessels, the wavelength is changed (Doppler shift), whereas stationary cells do not cause such a change. The magnitude of the Doppler shift is proportional to the number and velocity of the moving erythrocytes. Thus, when the backscattered

Richard Milner (ed.), *Cerebral Angiogenesis: Methods and Protocols*, Methods in Molecular Biology, vol. 1135, DOI 10.1007/978-1-4939-0320-7_20, © Springer Science+Business Media New York 2014

light is detected by a receiving optic fiber probe, the signal is analyzed and the flow velocity can be determined [2, 3]. In addition to measuring CBF [4], this method has been used to measure blood flow in different organs such as the kidney, gastric mucosa, or sciatic nerve [5–7]. Although this method provides an instantaneous measurement of CBF (high temporal resolution), no spatial information can be obtained as the measurement is limited to the site where the emitting and receiving probe is placed. In addition, only relative changes in CBF can be measured whereas absolute quantification of CBF in mL/100 g/min is not possible [4]. Other methods based on similar principles such as laser speckle contrast imaging can be used to assess both spatial and temporal changes in CBF and this has been described in Chapter 20. LDF probes that measure spatial as well as temporal changes in CBF are being developed, such as the PeriScan PIM 3 system (Perimed, Järfälla, Sweden), but the protocols in this chapter will only refer to the single-point LDF probes. The penetration depth of LDF probes is critical for experimental success and can be influenced by a number of factors, including the distance between emitting and receiving fibers, emission wavelength, and the tissue itself [8].

Laser Doppler flowmetry has a wide range of applications for scientific research, particularly in physiological and pathophysiological conditions where changes in CBF are required to be monitored. Ischemic stroke is defined as an occlusion of a vessel leading to a reduction in cerebral tissue perfusion, and LDF has been used in animal models of ischemic stroke. Routine assessment of CBF is required following middle cerebral artery occlusion (MCAO) to ensure that an occlusion has been made [9]. LDF can also monitor the extent of reperfusion once the occlusion has been removed [9]. Models of cortical spreading depression (CSD) where dramatic changes in cortical CBF occur leading to neurovascular dysfunction are another application where LDF could be used to assess these CBF changes [10]. Blood flow in the brain is controlled by signals sent from both astrocytes and neurons to the vasculature [11] and so LDF measurements are useful for neurovascular coupling experiments such as whisker or forepaw stimulation models [12]. However, it is important to note that these experiments will only be assessing relative changes in CBF, and so in order to compare between results of different experiments, the LDF probe must be calibrated with specially designed motility standards or measure absolute CBF using another method. The protocols described below will prepare the reader for the use of LDF in a closed skull or open skull preparation that could potentially be used for MCAO, CSD, neurovascular coupling, or other studies investigating changes in CBF.

2 Materials

2.1 Laser Doppler Flowmetry

1. LDF system (*see* **Note 1**).
2. Computer and software to record CBF measurements (*see* **Note 2**).
3. LDF probe (*see* **Note 3**).
4. LDF probe holder (use manufacturer-specific probe holders).
5. Rod that fits the LDF probe holder (this will be manufacturer specific as well).
6. Superglue and accelerant such as Loctite Tak Pak kit (RS Components, Corby, UK).
7. Small vessel or beaker for the accelerant.
8. Pipette tip to aspirate the accelerant.
9. Micromanipulator (World Precision Instruments, Sarasota, FL, USA).

2.2 Animals, Anesthesia, and Preparation for Surgery

1. Male Wistar rats 250–300 g from your local laboratory animal supplier (*see* **Note 4**).
2. Scales.
3. Anesthesia induction chamber.
4. Isoflurane delivery unit with flowmeters that can deliver isoflurane carried in other gases.
5. Isoflurane.
6. 100 % oxygen medical grade.
7. 100 % nitrous oxide medical grade.
8. Shaver.
9. 70 % alcohol.
10. Any chlorhexidine-based antiseptic such as hibitane.
11. Cotton swabs (sterilized).
12. Stereotaxic frame (optional) (David Kopf Instruments, Tujunga, CA, USA).
13. Ruler (optional—only to identify the region of interest relative to bregma if the stereotaxic frame is not used).
14. Marker pen.

2.3 Surgical Tools

1. Surgical scalpel (sterilized). Surgical scissors (sterilized) (Fine Science Tools).
2. Retractor (sterilized).
3. Curved forceps (sterilized).
4. A microdrill with a small drill bit (e.g., 1.2 mm diameter) (e.g., Ideal Microdrill, Cellpoint Scientific, Gaithersburg, MD, USA).

5. Probing tool to lift the bone flap of the craniotomy (sterilized).
6. Needle (25G or smaller), curved or bent.

2.4 Open Skull Preparation

1. Artificial cerebrospinal fluid (aCSF): 120 mM NaCl, 2.8 mM KCl, 22 mM $NaHCO_3$, 1.45 mM $CaCl_2$, 1 mM Na_2HPO_4, 0.876 mM $MgCl_2$, and 2.55 mM d-glucose (recipe based on [13]). To make 1 L of aCSF, measure out 7.013 g NaCl, 0.209 g KCl, 1.848 g $NaHCO_3$, 0.142 g Na_2HPO_4, 0.083 g $MgCl_2$, and 0.459 g d-glucose. Add 500 mL distilled water and dissolve. Once all powders have dissolved, top up the solution to 1 L with distilled water. Just before use, bubble the solution with 5 % CO_2 and 95 % O_2 and add 0.161 g $CaCl_2$ and keep bubbling for 10–15 min (*see* **Note 5**).
2. 5 % CO_2/95 % O_2 gas.

3 Methods

Two preparation methods will be described to measure CBF with LDF, a closed skull preparation and an open skull preparation.

3.1 Closed Skull Preparation

1. Make sure the LDF system is turned on and connected to any relevant software on a computer to measure CBF.
2. Weigh the rat using scales.
3. Anaesthetize the rat by placing it in an induction chamber linked to an isoflurane delivery unit (*see* **Note 6**) and deliver 4 % isoflurane in 30 % oxygen and 70 % nitrous oxide at a total flow rate of 1 L/min (*see* **Note 7**).
4. Once the rat has been fully anesthetized as assessed by lack of righting reflex (*see* **Note 8**), shave the top of the head quickly and put the rat in a stereotaxic frame (*see* **Note 9**) and maintain anesthesia at 2 % isoflurane delivered in 30 % oxygen and 70 % nitrous oxide.
5. Once the rat is stable within the stereotaxic frame, sterilize and disinfect the shaved area of the head with 70 % alcohol and a chlorhexidine-based antiseptic.
6. Make an incision using a scalpel or surgical scissors through the skin along the midline from between the ears to between the eyes (Fig. 1a).
7. Using blunt dissection with scissors, open the area underneath the skin to each side of the skull above the temporalis muscles (*see* **Note 10**).
8. Use a retractor to retract the skin on either side of the incision to keep the area above the skull open.

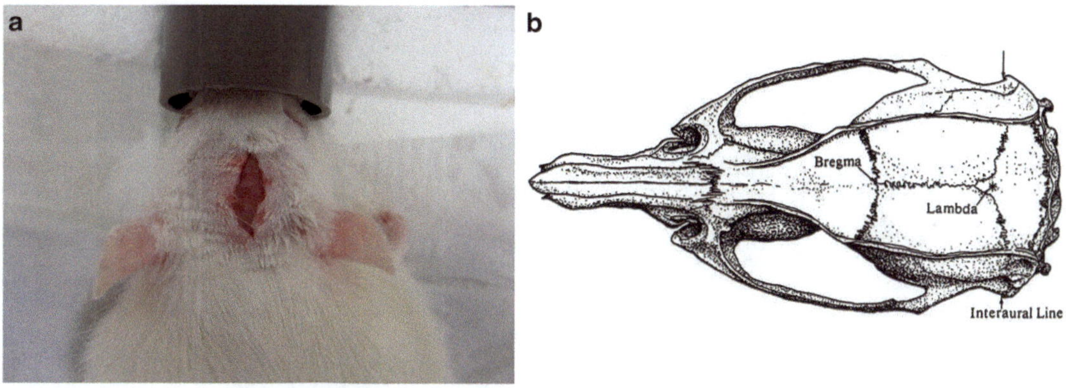

Fig. 1 Making the incision and identifying the bregma in the rat. (**a**) Using a scalpel or surgical scissors, the first incision should be made through the skin of the back of the head down the midline between the ears and eyes. This will allow access to the skull. This procedure was approved by the Home Office (UK) and University of Oxford Animal Ethics Committee. (**b**) Locate the bregma which is at the intersection of the two rostral sutures and the midline. Use the bregma as the starting point to determine the coordinates at which CBF is going to be measured with an LDF probe. (**b**) This image was reproduced from [14] with permission from Elsevier Academic Press

9. Using the blunt side of a scalpel blade, carefully scrape off the connective tissue to expose the skull. Clean the skull with cotton swabs.
10. Identify the bregma (Fig. 1b) and use a ruler or the stereotaxic frame to identify the area of interest where CBF will be measured and mark this with a marker pen.
11. Use a microdrill to thin the skull over the area of interest relative to the bregma until only a small translucent sheet of bone remains (*see* **Note 11**). Ensure that the size of the drilled area is large enough to fit the probe and its holder (if using).
12. Prepare the accelerant by spraying it into a small vessel or beaker and use a pipette tip to aspirate a small amount into the tip (*see* **Note 12**).
13. Now apply the LDF probe to the thin skull. The process of applying the LDF probe depends on the type of probe that is being used. The two types of probes that are mostly used to measure CBF in a thin skull preparation are a needle probe (*see* **Note 13** and Fig. 2a) (*see* Subheading 3.1, **step 13a**) or a disposable probe attached to an adapter (*see* **Note 14** and Fig. 2b) (*see* Subheading 3.1, **step 13b**):
    (a) If using a needle probe, a probe holder will need to be used (Fig. 2c). To ensure that your thin skull placement of LDF probe will provide a good signal, place the needle probe in the probe holder and hold it on the thin skull and read the CBF output. If the signal is good, continue; if

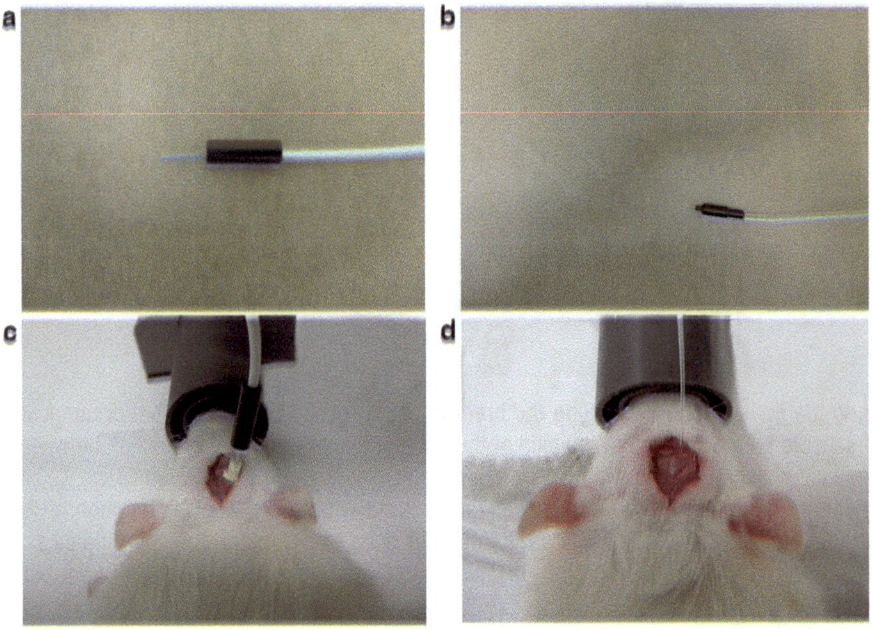

Fig. 2 Types of LDF probes for a closed skull preparation. (**a**) The needle probe is a small straight probe used in both closed skull and open skull preparations. (**b**) The disposable probe with adapter is a versatile flexible probe that is used for closed skull preparations only. (**c**) The needle probe can be directly attached to the rat skull perpendicularly through a probe holder. (**d**) The tip of the disposable probe can be directly glued to the rat skull perpendicularly and this can be cut and reused at the end of the experiment. The LDF probes pictured in this figure are courtesy of Oxford Optronix Ltd (Oxford, UK). These procedures were approved by the Home Office (UK) and University of Oxford Animal Ethics Committee

not, then improve the thin skull preparation until there is an adequate CBF signal from the LDF probe (*see* **Note 15**). Remove the holder and take out the probe. Use a rod or something similar to put in the probe holder (*see* **Note 16**). Apply a small amount of superglue around the holder's rim and gently lower the holder onto the thin skull. Apply a small amount of accelerant from the pipette tip onto the rim of the holder to set the glue and secure the attachment (*see* **Note 17**). Once the holder is fixed, secure the attachment further by adding some more glue and accelerant around the outside of the holder. Once the glue has set, remove the rod or similar if using and insert the probe into the secure holder so the tip of the probe touches the thin skull (*see* **Note 18** and Fig. 2c).

(b) If using a disposable probe attached to an adapter, then a probe holder is not required. To ensure that your thin skull placement will provide a good signal, attach the disposable probe to the adapter (Fig. 2b) and hold it on the thin skull and read the CBF output. If the signal is good, continue; if not, then improve the thin skull preparation until there is an adequate CBF signal from the LDF probe (*see* **Note 15**). Apply a drop of superglue over the thin skull area where you wish to place the probe. Place the disposable probe tip into the glue so it touches the skull and apply some accelerant to aid the glue setting process (*see* **Note 19**). Once the glue has set, the disposable probe should now remain in place (Fig. 2d).

14. Now that the LDF probe is in place, begin recording CBF using any software coupled to the LDF system or using the output screen on the LDF system itself.
15. Carry out the proposed experiment investigating changes in CBF by recording from the LDF probe continuously.

3.2 Open Skull Preparation

1. Follow **steps 1–10** in Subheading 3.1 as per the closed skull preparation protocol.
2. Use a drill to thin the skull over the area where the craniotomy will be. Drill around the perimeter of the area (all 4 edges) so only a thin sheet of bone remains on the outer edge of the area of interest (*see* **Note 20** and Fig. 3a).
3. Using a probing tool, lift under one corner of the thin skull and detach the edges of the thin skull from the thick skull by running the probing tool along three edges (*see* **Note 21**). Lift the detached flap of skull to expose the cortex underneath.
4. Remove the dura. Use a curved needle to nick the dura and make a small hole. Then use a small hook tool to pull the dura away to the edge of the craniotomy and expose the brain (*see* **Note 22**).

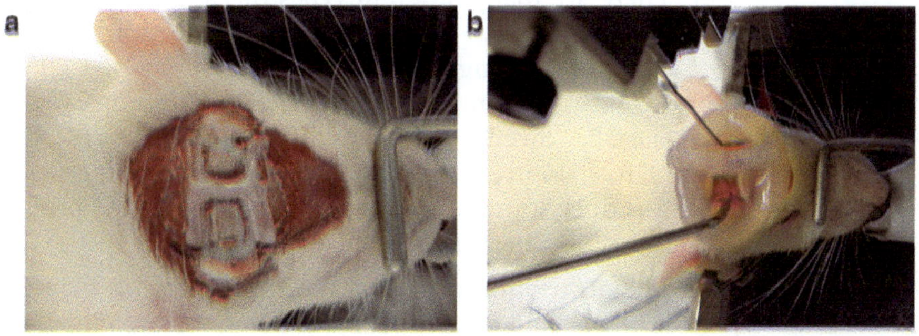

Fig. 3 Using an LDF probe for an open skull preparation. (**a**) To drill a craniotomy over a large area of the skull, the perimeter of the area is drilled until a thin sheet of bone runs along the four edges of the area. The rat pictured is undergoing two craniotomies. (**b**) The skull and dura have been removed, and an agarose well has been formed over the craniotomy to keep the aCSF over the exposed brain. A stainless steel LDF probe has been placed over the exposed brain without touching it. In this rat, the second craniotomy has a stimulating probe inserted into the brain. The LDF probe pictured in this figure is courtesy of Perimed (Järfälla, Sweden). These procedures were approved by the Home Office (UK) and University of Oxford Animal Ethics Committee

5. Make a well by heating 1 % agarose to a smooth gel and mould it around the craniotomy. Wait for the agarose to set, and superfuse aCSF over the exposed brain to keep the brain under physiological conditions (*see* **Note 23**).
6. Place a stainless steel or needle LDF probe in a micromanipulator. Move the probe with a micromanipulator to the exact position required over the cortex leaving only a few millimeters between the tip of the probe and the surface of the brain (*see* **Note 24** and Fig. 3b).
7. Now that the LDF probe is in place, begin recording CBF using any software coupled to the LDF system or using the output screen on the LDF system itself.
8. Carry out the proposed experiment investigating changes in CBF by recording from the LDF probe continuously.

4 Notes

1. There are many different types of LDF systems and the abilities of each system depend on the manufacturer. Generally, the same principles are used and the specifications are similar but each system operates slightly differently. Some examples are Perimed (Järfälla, Sweden), Moor Instruments (Axminster, UK), and Oxford Optronix (Oxford, UK).
2. Each LDF system will require certain software specifications in order to obtain accurate quantitative CBF data, and this can usually be purchased from the manufacturer of that particular LDF system. Therefore, it is important to note the software requirements when purchasing an LDF system.
3. There are many different types of LDF probes which all have their advantages and disadvantages depending on what application it is being used for. Manufacturers of LDF systems all have their own LDF probes. Some examples are Perimed (Järfälla, Sweden), Moor Instruments (Axminster, UK), and Oxford Optronix (Oxford, UK). Subheading 3.1, **step 13** describes some of the different types of LDF probes that can be used to measure CBF.
4. The description of these experiments is with male Wistar rats, but they can equally be done with other weights and strains of rat or also in mice. It is important to maintain consistency of the types of animals used for all experiments, i.e., weight, sex, and strain.
5. Do not add the $CaCl_2$ to the aCSF without bubbling with 5 % CO_2 95 % O_2 because it will precipitate with the $NaHCO_3$ making the solution unusable. Therefore, ensure the solution

is adequately bubbled once the $CaCl_2$ has been added immediately before use.

6. Depending on the type of anesthesia being used, the method of inducing anesthesia will be different. The above method describes how to induce anesthesia in a chamber that will be filled with 4 % isoflurane. There are other types of anesthetics that can be used to induce anesthesia including injectable anesthetics, such as ketamine and xylazine, and urethane.
7. Our lab uses 4 % isoflurane delivered with 30 % oxygen and 70 % nitrous oxide. Nitrogen can be used instead of nitrous oxide, while 100 % oxygen can also be used. The important thing is to maintain consistency between experiments so comparisons can be made.
8. Induction of anesthesia normally takes 3–5 min. To ensure the rat has no righting reflex, tip the induction chamber on an angle to see if the rat is responsive to the change in orientation of the chamber. If the rat does not get to its feet, it has no righting reflex.
9. A stereotaxic frame can be used for these experiments, but other animal holders can be used to similar effect. The head should remain still to ensure that there are no movement artefacts while measuring CBF with laser Doppler. For instance, if you plan to carry out another procedure such as middle cerebral artery occlusion where you would need to turn the animal over, a stereotaxic frame is not required. Ensure that the ear bars of the stereotaxic frame are in the ear canal of the rat, because if they are not, the head will not be stable in the frame.
10. This blunt dissection will allow easier retraction and a greater field of view.
11. For a rat, use the drill to thin the skull that is intended to be imaged; otherwise the rat's skull is too thick to obtain an accurate LDF signal of the brain. Drilling should cease when a thin translucent sheet of bone over the area of interest can be seen, and blood vessels on the surface of the brain should be visible through the thin skull. The thin skull should be flat to improve the quality of the LDF signal. The skull is susceptible to bleeding while drilling. If the skull does bleed, apply pressure with a swab and clean with saline. Also, a few drops of saline should be added to the skull while drilling so the cortex below the skull is not damaged due to friction-induced heat. If using mice or very young rats, the skull will be thin enough to image blood flow and so thinning the skull is not required.
12. The pipette tip, through capillary action, will take up the liquid accelerant, but if more is needed, a pipette or syringe can be attached to the tip.

13. The needle LDF probe (Fig. 2a) can come in a number of variants depending on the manufacturer, but essentially, it is a small straight probe that can fit in a probe holder that is attached to the skull (Fig. 2c), or it can be used with a micromanipulator over an exposed brain for exact positioning. The tip of this probe emits a laser that can be received by a receiver within the probe. These probes are versatile and can be used for a number of applications.
14. The disposable probe with adapter is versatile and flexible, and most manufacturing companies offer this variant of the LDF probe. The adapter cable runs from the LDF system and a disposable probe can be connected to the adapter (Fig. 2b). The tip of the disposable probe can be glued on the skull for recording (Fig. 2d), and at the end of the experiment, the tip of the disposable probe can be cut off and the probe reused. The Doppler shift signal emitted from the cortex is transmitted through the disposable probe and adapter to the LDF system. This type of probe is most useful when the animal needs to be moved after probe placement such as for MCAO procedures.
15. It is important for any experiment when measuring CBF that an adequate signal is achieved that can accurately measure any changes in CBF (e.g., if CBF is going to be increased experimentally, the CBF signal must not be too high to begin with). Therefore, prior to gluing the LDF probe to the skull, it is important that the site where the probe will be attached has an adequate CBF signal. One problem associated with the thin skull approach is that the CBF measured will be sourced from both the cortex and the skull itself. Therefore, if the skull is not thin enough where the CBF will be measured, the ability to detect CBF changes in the cortex will be reduced. To improve this, the skull must be as thin as possible but as flat as possible to maximize the signal from the cortex through the skull.
16. The superglue used to glue the probe holder to the skull can damage the LDF probe and will prevent accurate CBF quantification. To prevent glue moving up the middle of the holder and damaging the LDF probe, use a rod or something similar. Do this by inserting the rod into the holder as it is glued in place on the skull. Alternatively, if there is no rod, use a needle attached to a syringe to remove any residual glue after the holder has been placed before sticking in the needle probe.
17. When placing the holder, use forceps to keep the holder still on the flat thin skull while the accelerant sets the glue.
18. When the probe is inserted into the probe holder, it is important that the tip of the probe touches the skull gently and does not apply any pressure on the skull and therefore on the brain as this can adversely affect CBF.

19. To keep the probe in place while an accelerant is used to speed up the setting of the glue, the probe needs to be held securely with forceps or similar. If the probe is not staying in place once the glue has set, add some more glue and accelerant to strengthen the attachment of the probe to the skull.
20. When drilling the skull to make a craniotomy, it is important that the skull and brain suffer as minimal damage as possible. The most efficient way to drill a large area of the skull for a craniotomy is not to drill the whole area but to drill the perimeter of the area that will be taken off (Fig. 3a). This ensures that the central area of the skull, and subsequently the brain underneath, is not affected by the drilling procedure. Use the drill to thin and soften the skull around all four edges of the perimeter, until there is only a thin translucent sheet of bone that runs along all four edges. Once this is done, and the pial vessels can be clearly seen, the skull is ready to be lifted off.
21. To obtain a clean craniotomy, the four edges of the skull to be taken off must have been drilled so only a thin sheet of bone remains (*see* **Note 20**). Use a probing tool to carefully break through one corner of the thin skull. Once the small opening has been made, carefully run the probing tool along three of the four edges of the thin skull so the skull becomes detached. Then the skull can easily be lifted with forceps and removed.
22. Removing the dura is optional as it will not affect the CBF signal. It is less invasive to leave the dura on, and so if you only want to measure CBF and not do anything else with the exposed brain, then there is no need to remove the dura. However, if the experiment involves superfusing drugs topically over the cortex, then the dura will need to be removed so the brain can be accessed. Removing the dura is one of the trickiest parts of this procedure. When using the bent needle, ensure it is parallel to the surface of the brain so that when the dura is nicked, the needle does not stab the brain. When using the hook, gently lift the dura and pull it away from the center to the edge of the craniotomy. Sometimes, there can be a lot of bleeding from the surface vessels of the cortex. It is important to minimize this as much as possible by using saline and applying gentle pressure with a cotton swab.
23. Artificial cerebrospinal fluid must be used if the brain is exposed, to keep the brain close to physiological conditions. aCSF must be periodically changed so that the brain can receive fresh nutrients.
24. Be careful when lowering the probe towards the exposed brain that the tip of the probe does not touch the brain, as this can damage the brain and adversely affect the CBF measurements gathered.

References

1. Sakurada O, Kennedy C, Jehle J, Brown JD, Carbin GL, Sokoloff L (1978) Measurement of local cerebral blood flow with iodo [14c] antipyrine. Am J Physiol 234:H59–H66
2. Bonner R, Nossal R (1981) Model for laser doppler measurements of blood flow in tissue. Appl Opt 20:2097–2107
3. Stern MD (1975) In vivo evaluation of microcirculation by coherent light scattering. Nature 254:56–58
4. Dirnagl U, Kaplan B, Jacewicz M, Pulsinelli W (1989) Continuous measurement of cerebral cortical blood flow by laser-doppler flowmetry in a rat stroke model. J Cereb Blood Flow Metab 9:589–596
5. Kiel JW, Riedel GL, DiResta GR, Shepherd AP (1985) Gastric mucosal blood flow measured by laser-doppler velocimetry. Am J Physiol 249:G539–G545
6. Stern MD, Bowen PD, Parma R, Osgood RW, Bowman RL, Stein JH (1979) Measurement of renal cortical and medullary blood flow by laser-doppler spectroscopy in the rat. Am J Physiol 236:F80–F87
7. Sundqvist T, Oberg PA, Rapoport SI (1985) Blood flow in rat sciatic nerve during hypotension. Exp Neurol 90:139–148
8. Jakobsson A, Nilsson GE (1993) Prediction of sampling depth and photon pathlength in laser doppler flowmetry. Med Biol Eng Comput 31:301–307
9. Chen RL, Nagel S, Papadakis M, Bishop T, Pollard P, Ratcliffe PJ et al (2012) Roles of individual prolyl-4-hydroxylase isoforms in the first 24 hours following transient focal cerebral ischaemia: insights from genetically modified mice. J Physiol 590:4079–4091
10. Piilgaard H, Lauritzen M (2009) Persistent increase in oxygen consumption and impaired neurovascular coupling after spreading depression in rat neocortex. J Cereb Blood Flow Metab 29:1517–1527
11. Attwell D, Buchan AM, Charpak S, Lauritzen M, Macvicar BA, Newman EA (2010) Glial and neuronal control of brain blood flow. Nature 468:232–243
12. Lecrux C, Toussay X, Kocharyan A, Fernandes P, Neupane S, Levesque M et al (2011) Pyramidal neurons are "neurogenic hubs" in the neurovascular coupling response to whisker stimulation. J Neurosci 31:9836–9847
13. Mathiesen C, Caesar K, Thomsen K, Hoogland TM, Witgen BM, Brazhe A et al (2011) Activity-dependent increases in local oxygen consumption correlate with postsynaptic currents in the mouse cerebellum in vivo. J Neurosci 31:18327–18337
14. Paxinos G, Watson C (1997) The rat brain in stereotaxic coordinates, 3rd edn. Academic, San Diego, CA

Part IV

Determining the Role of Candidate Genes in Cerebral Angiogenesis

Chapter 21

Defining the Role of HIF and Its Downstream Mediators in Hypoxic-Induced Cerebral Angiogenesis

Xiaoyan Sun, Constantinos P. Tsipis, Girriso F. Benderro, Kui Xu, and Joseph C. LaManna

Abstract

Now that some of the basic mechanisms that underlie hypoxia-induced cerebral angiogenesis have been described, it has become clear that the hypoxia-inducible transcription factors, HIF-1 and HIF-2, play an important role in the process by causing the upregulation of vascular endothelial growth factor (VEGF). The heterogeneity of the brain parenchyma means that further progress in understanding capillary pathophysiology requires techniques that allow determination of the roles of individual components of the neurovascular unit. Multi-stain fluorescence co-localization techniques provide one such approach.

Key words Hypoxia, Hypoxic adaptation, Brain, Angiogenesis, Immunohistochemistry, HIF, VEGF, GLUT-1, S100, MAP2, NeuN, COX-2, Neurovascular unit, Angioplasticity

1 Introduction

There are at least two major contributing molecular pathways to hypoxia-induced cerebral angiogenesis. These are as follows: (1) hypoxia-inducible factor (HIF-1, -2) accumulation resulting in upregulation of vascular endothelial growth factor (VEGF) and (2) upregulation of endothelial cyclooxygenase-2 (COX-2), subsequent to production of prostaglandin-E2 and autocrine release of angiopoeitin-2 (Ang-2) [1]. Previous studies have used Western blot analyses to study HIF and its downstream gene products to investigate these phenomena. For example, the data shown in Fig. 1 clearly demonstrate an increased accumulation of VEGF protein in brain cortical tissue after 4 and 7 days of exposure to 8 % O_2/balance N_2 (ambient pO_2~60 mmHg) in 2–3-month-old C57BL/6J mice [2]. Nevertheless, results generated from tissue homogenates mask the individual roles of neurons, astrocytes, endothelial cells, and pericytes—the principle components of the neurovascular unit [3].

Richard Milner (ed.), *Cerebral Angiogenesis: Methods and Protocols*, Methods in Molecular Biology, vol. 1135, DOI 10.1007/978-1-4939-0320-7_21, © Springer Science+Business Media New York 2014

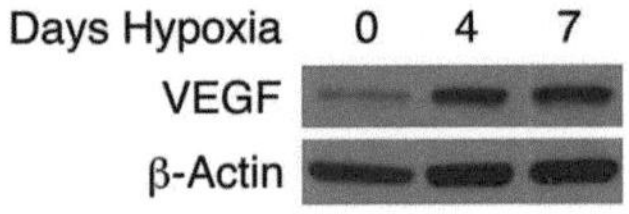

Fig. 1 Western blot analysis of VEGF in whole cell lysates. VEGF is expressed at baseline under normoxic conditions and upregulated in the cerebral cortex of C57BL/6J mice after 4 and 7 days of hypoxia

One approach to identifying the individual contributions of the neurovascular unit components is to use double and triple immunohistochemistry. This method uses cell-type-specific markers in concert with specific antibodies to localize the molecule of interest, in this case VEGF. For this purpose, the cell-/organelle-specific markers are GLUT-1, S100, MAP2, NeuN, and DAPI. GLUT-1 identifies the glucose transporter associated with endothelial cells that make up the blood–brain aqueous barrier [4]. GLUT-1 is also associated with glial cells and the chorid plexus cells, but at a much lower concentration [5]. With this stain, functional cerebral capillaries are easy to identify and quantify (refer to Chapter 6). The brain-specific acidic protein, S-100, is an excellent stain specific for astrocytes [6]. MAP2 (microtubule associated protein-2) antibodies identify processes of differentiated neurons through reaction with high molecular weight neurotubules [7]. NeuN (neuronal nuclei) antibody, as the name implies, stains the nuclei of neurons [8]. These immunohistochemical stains can be used in co-localization studies to begin the process of defining the individual roles of the cellular components of the neurovascular unit in cerebral angioplasticity [9].

Because the process of hypoxia-induced angiogenesis occurs over several weeks [10], it was important to choose hypoxic exposure durations that would be meaningful. Based on previous studies in the mouse, 4 days and 7 days of exposure were chosen to be the most significant physiologically [2]: Four days, because this represents a period of maximal VEGF as indicated by Western blot analyses, and 7 days because this is the time when the first capillaries are seen sprouting [11].

Figure 2 shows the results. There is little or no co-localization of VEGF in capillary endothelial cells until 7 days. Baseline VEGF is associated predominantly in neurons and some astrocytes. Neuronal processes have significant amounts of VEGF at 4 days and somewhat less at 7 days, at which time VEGF has disappeared from the neuronal cell bodies. The astrocytes appear to contain maximum VEGF at 4 days.

When we add to this picture the finding that the COX-2 signaling pathway is upregulated significantly in endothelial cells after 7 days of hypoxic exposure (Fig. 3), a pattern begins to emerge that can be described by the cartoon shown in Fig. 4. Much work

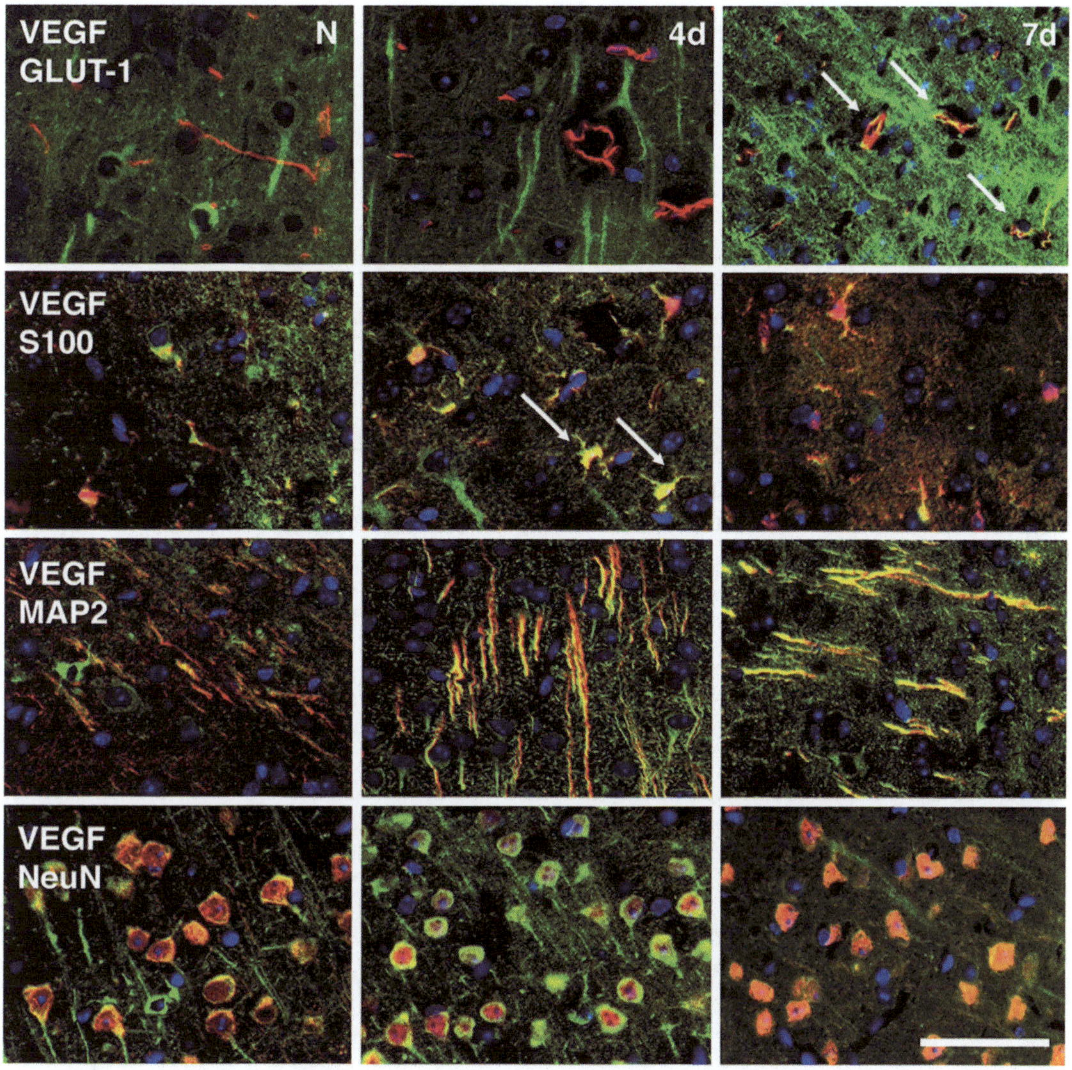

Fig. 2 Co-localization of VEGF (*green*) in endothelial cells (GLUT-1), astrocytes (S100), and neurons (MAP2, NeuN) under normoxic (N), 4, and 7 days of hypoxia. Scale bar; 50 μm

remains to be done to allow the elucidation of the regulatory mechanisms of cerebral angiogenesis, but it is clear that multi-stain histochemistry will remain an important and useful tool.

2 Materials

All solutions must be prepared using ultrapure water (purified to 18 MΩ cm sensitivity at 25 °C) using reagent-grade chemicals at room temperature. 1× PBS (phosphate-buffered saline) and 4 % PFA (paraformaldehyde) should be stored at 4 °C after preparation until use. Solutions can be stored at room temperature unless otherwise noted.

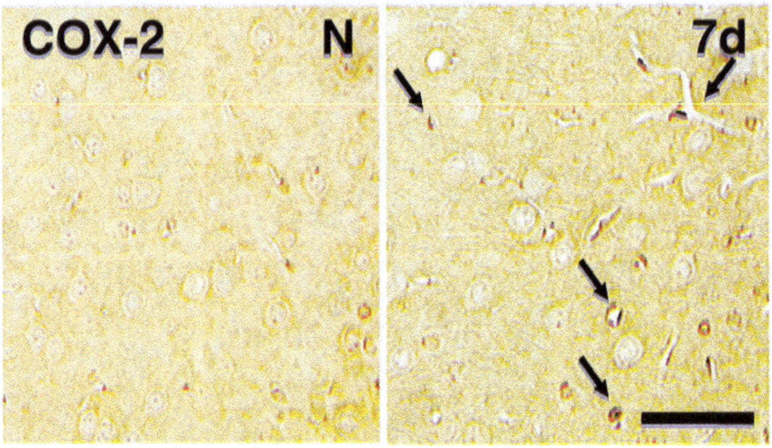

Fig. 3 COX-2 expression in the mouse cerebral cortex under normoxic (N) and 7 days of hypoxia. COX-2 is induced in both endothelial cells and neurons at 7 days of hypoxia; *arrows* indicate COX-2 positive endothelial cells. Scale bar; 50 μm

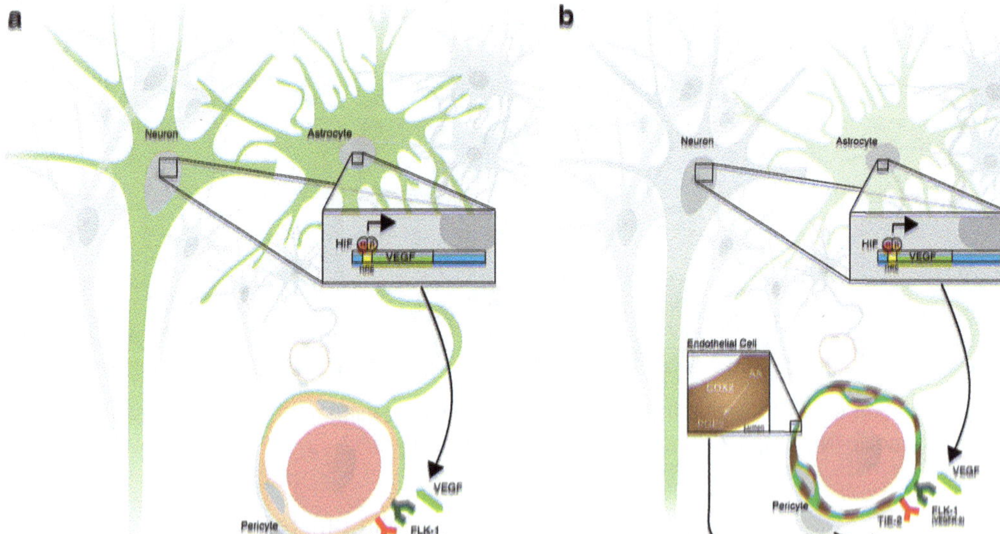

Fig. 4 Scheme showing the hypothesized roles of the individual components of the neurovascular unit in hypoxia-induced angiogenesis. (**a**) Normoxic baseline. (**b**) Activated neurovascular unit at 7 days of hypoxia. Abbreviations in the figure: *HIF* hypoxia-inducible factor, *HRE* hypoxia response element, *VEGF* vascular endothelial growth factor, *FLK-1 or VEGFR-2* fetal liver kinase 1 or vascular endothelial growth factor receptor 2, *AA* arachidonic acid, *COX-2* cyclooxygenase 2, *PGE-2* prostaglandin E2, *ANG-1* angiopoietin 1, *ANG-2* angiopoietin 2, and *TIE-2* endothelium-specific tyrosine kinase receptor 2

2.1 Hypoxia Chamber

1. Modified Wright Chamber for hypobaric hypoxic exposure (*see* Chapter 6 for more details).

2.2 Perfusion and Fixation Solutions

1. 1× Phosphate-buffered saline (PBS) with heparin: 0.01 M PBS, pH 7.4. Add 1 mL of heparin sodium to 1 L of 0.01 M PBS. Store at 4 °C.
2. 4 % Paraformaldehyde in PBS: Heat 100 mL of 0.01 M PBS, pH 7.4 in the fume hood to 60 °C. Weigh 4.0 g of paraformaldehyde and transfer to PBS. Allow the solution to completely dissolve and adjust pH to 7.4 adjust if necessary. Bring to room temperature and filter the solution through a 0.45 μm filter. Store at 4 °C.
3. Millonig's buffer (Leica Biosystems).

2.3 Immunohistochemistry Solutions

Immunohistochemistry solutions are to be prepared at the time of staining procedure and cannot be made and stored ahead of time unless noted otherwise.

1. Tris-buffered saline (1× TBS): 25 mM Tris, 150 mM NaCl, 2 mM KCl, pH 7.4. Weigh 8 g NaCl, 0.2 g KCl, and 3 g Tris and transfer to 800 mL of ultrapure water. Allow the solution to completely dissolve and titrate with 12 N HCl to pH 7.4. Adjust the final volume to 1 L and store at room temperature.
2. 1× TBST wash buffer: Add 1 mL of Tween-20® to 1 L of 1× TBS buffer and mix vigorously on a stir plate.
3. Antigen retrieval buffer: 10 mM sodium citrate, pH 6.0. Weigh 2.941 g of sodium citrate and transfer to 1 L of ultrapure water. Allow the solution to completely dissolve and titrate with 12 N HCl to pH 6.0. Add 0.5 mL Tween-20® and allow to mix entirely before storage.
4. Blocking solution: 1× TBST, pH 7.4, 10 % normal serum. Store on ice for the duration of preparation.

2.4 Antibodies and Conjugates

1. VEGF Antibody, 1:100 dilution (Novus Biologicals, CO, USA).
2. S100 Antibody, 1:200 dilution (Abcam, CA, USA).
3. Glut-1 Antibody, C-20, 1:200 dilution (Santa Cruz Biotechnology, TX, USA).
4. MAP2 Antibody, 1:200 dilution (Millipore, CA, USA).
5. NeuN Antibody, 1:200 dilution (Millipore, CA, USA).
6. COX-2 Antibody, 1:50 dilution (Cayman Chemical, MI, USA).
7. Biotinylated anti-rabbit IgG (H+L) 1:200 secondary antibody dilution for anti-COX-2.
8. Horseradish peroxidase (HRP) affinity purified secondary antibodies: Rabbit anti-goat IgG-HRP, Rabbit anti-mouse IgG-HRP, Goat anti-rabbit IgG HRP (Invitrogen, NY, USA). All dilutions are 1:500.

9. Alexa Fluor-labeled secondary antigen solutions: Alexa 488 conjugated anti-rabbit IgG (H+L), Alexa 569 conjugated rabbit anti-mouse IgG (H+L), Alexa 569 conjugated rabbit anti-goat IgG (H+L) (Invitrogen, NY, USA). All solutions are stock 2 μg/mL concentration.
10. VECTASHELD® Mounting Medium with DAPI (Vector Laboratories, CA, USA).

3 Methods

3.1 Exposure to Hypoxia

1. Mouse of appropriate age are placed in the hypobaric chamber (*see* Chapter 6 for more details).
2. Individual animals are weighed and maintained for food and water on a regular basis and clean bedding is provided every day or two, necessitating return to normobaric conditions for up to 1 h.

3.2 Perfusion and Fixation

1. Body weight: Remove mice from chamber and record body weight in grams.
2. Deeply anesthetize mice with isofluorane and create a midsagittal thoracic incision exposing the heart and carefully clip the right atrium. Collect blood with a capillary tube for hematocrit (HCT) determination.
3. Insert a butterfly needle into the apex of the heart and perfuse transcardially with ~30 mL ice-cold 1× PBS with heparin placed at about 100 cm above heart level, and allow for perfusion to continue for at least 5 min, or until the liver is blanched.
4. Close the PBS line and start the 4 % PFA flow immediately through the butterfly needle and allow fixation to occur for 5 min, again, keeping the flow rate low enough to ensure the structural integrity of the cerebral vasculature.
5. Decapitate at the level of the neck and carefully remove the skin and underlying cranium with sharp scissors by making an incision down the midline and carefully peeling back the cranium using forceps. Remove the brain by carefully lifting it from the base of the skull.
6. Place the brain in 4 % PFA solution for a 24-h postfixation period at 4 °C. Wash the brain with 1×PBS and transfer the brain to storage solution (*see* **Note 1**).

3.3 Tissue Processing

1. Place the brain in a labeled tissue-processing cassette and wash under cold running tap water for 1 h.
2. Remove the cassettes from running water and place in 80 % ethanol overnight (~14 h) at 4 °C.

3. Transfer the cassettes to 95 % ethanol for 90 min at room temperature and repeat this step three times with fresh 95 % ethanol for a total of 4.5 h.
4. Transfer the cassettes to 100 % ethanol for 90 min at room temperature and repeat this step three times with fresh 100 % ethanol for a total of 4.5 h.
5. Transfer the cassettes to xylene for 90 min at room temperature and repeat this step three times with fresh xylene for a total of 4.5 h.
6. Remove the cassettes and transfer to molten paraffin and store overnight at 60 °C.
7. Transfer the cassettes to a vacuum oven and apply pressure for 2 h at 80 °C.
8. Remove the brains from the tissue-processing cassettes and embed them in paraffin using an embedding mold and store at 4 °C overnight to allow paraffin to harden. Tissue blocks can be stored at room temperature forever.

3.4 Sectioning

1. Cut 5 μm coronal sections of cerebral cortex between Bregma +1.2 mm and 0.7 mm.
2. Using a hot water bath set to a temperature of 42 °C, mount each section onto Superfrost Plus™ slides.
3. Allow the slides to dry on the slide rack overnight. Slides can then be stored in a slide box.

3.5 Immunohistochemistry

All slides should be dried on a slide warmer overnight at 40–45 °C prior to starting staining preparation. All steps are to be performed at room temperature unless otherwise noted:

1. Deparaffinize: Place the slides in 100 % xylene for 5 min. Repeat this step twice for a total of 15 min in three separate xylene baths (*see* **Note 2**).
2. Rehydrate with graded alcohols: Transfer the slides to 100 % ethanol for 3 min. Repeat this step for a total of 6 min in two separate 100 % ethanol baths. Transfer the slides to 95 % ethanol for 3 min. Repeat this step for a total of 6 min in two separate 95 % ethanol baths. Transfer the slides to 70 % ethanol for 3 min, followed by 50 % ethanol for 3 min.
3. Transfer the slides to double distilled (dd) H_2O for 3 min. Repeat this step twice for a total of 9 min in three separate washes.
4. Preheat the antigen retrieval solution in a Coplin jar (Fisher Scientific) to 95 °C in a water bath. Transfer the slides into 95 °C antigen retrieval for 30 min. Allow the slides to reach room temperature (~20 min) before continuing the procedure.

5. Wash: Transfer the slides to ddH_2O for 5 min and repeat this step one more time for a total of 10 min in two separate washes.
6. Wash: Transfer the slides to 1× TBST buffer for 5 min and repeat this step one more time for a total of 10 min in two separate washes.
7. Incubate the slides with 100–300 μL of 10 % blocking buffer for 60 min at room temperature inside a humid chamber (*see* **Note 3**).
8. Prepare the first primary antibody solution by adding the appropriate antibody and dilution (*see* Subheading 2.4) to the 10 % blocking buffer.
9. Drain off blocking buffer from the slides and apply 100–300 μL of the first primary antibody solutions overnight at 4 °C in the humid chamber (*see* **Note 4**).
10. Wash: Transfer the slides to 1× TBST buffer for 5 min and repeat this step two more times for a total of 15 min in three separate washes.
11. Prepare the HRP-labeled secondary antibody solution by adding the appropriate antibody and dilution (*see* Subheading 2.4) to 10 % blocking buffer.
12. Incubate the slides with 100–300 μL of the HRP-labeled secondary antibody solution for 60 min at room temperature (*see* **Note 5**).
13. Wash: Transfer the slides to 1× TBST buffer for 5 min and repeat this step two more times for a total of 15 min in three separate washes.
14. Apply 100–300 μL of the Alexa Fluor® (Invitrogen) labeled antibody solution to the slides and immediately shield from light. Allow the incubation to proceed for 30 min at room temperature inside a dark humid chamber (*see* **Note 6**).
15. Wash: Transfer the slides to 1× TBST buffer for 5 min and repeat this step two more times for a total of 15 min in three separate washes (*see* **Note 7**).
16. After washing, repeat **steps 6–14** using the second target antibody to achieve double staining. The repeated steps should be performed while again minimizing slide exposure to light (*see* **Note 8**).
17. Coverslip the slides with VECTASHELD® Mounting Medium with DAPI. Store at 4 °C in the dark if the slides are not to be viewed immediately (*see* **Note 9**).
18. Slides can be stored at 4 °C in the dark for up to 10 weeks; any period longer than that requires storage at −20 °C (*see* **Note 10**).
19. Capture digital photographs at 400× magnification of the cerebral cortex with a high-density digital camera connected to a fluorescence microscope (*see* **Note 11**).

4 Notes

1. The fixed mouse brain can be stored in Millonig's buffer (Leica Biosystem). Storage in 4 % PFA solution should only be temporary; Millonig's buffer can help to preserve tissue for extended periods of time without degrading tissue immunoreactivity.
2. Xylene is a hazardous chemical and requires clearance from the institutional safety office for use. All steps requiring xylene must be performed under a chemical fume hood with proper ventilation.
3. The humid chamber prevents slides from drying out and can be constructed using a layer of damp paper towels at the bottom of a container with a sealable lid.
4. Alternatively, primary antibody incubation can be done at room temperature for 1 h.
5. Remove as much of the previous solution as possible to prevent uneven staining and further dilution of the current solution. Routine titration of primary antibodies is done to find the lowest dilution that elicits the strongest contrast for optimal signal.
6. All Alexa Fluor® conjugates (Invitrogen) are supplied at a stock concentration of 2 μg/mL. To optimize the signal, it is important to use predefined positive and negative controls for contrast adjustment. FITC-conjugated antibodies can be substituted in this step as well.
7. It is critical to keep slides in the dark during this step by shielding the wash chamber with tinfoil.
8. Ensure that slides do not dry out throughout this lengthy procedure. Covering the humid chamber during subsequent incubations with thin plastic film can prevent slides from drying. After the first Alexa Fluor® antibody has been applied, it is imperative to limit light exposure as much as possible to preserve the fluorescent signal strength.
9. DAPI is a useful counterstain to indicate nuclei in immunofluorescence when used with AlexaFluor®-labeled green or red markers.
10. Sections are rehydrated during this procedure and require a temperature of −20 °C for long-term storage.
11. Same-field fluorescent images from separate channels were captured at a magnification of 400× with an Aquos EXi digital camera (QImaging, Surrey, BC, CA) connected to an Eclipse E600 microscope (Nikon Instruments Inc., Melville, NY, USA) and subsequently merged with Image J (NIH, Bethesda, MD, USA).
12. COX-2 immunohistochemistry is accomplished using the standard technique as detailed in Chapter 6.

References

1. Dore-Duffy P, LaManna JC (2007) Physiologic angiodynamics in the brain. Antioxid Redox Signal 9(9):1363–1371
2. Benderro GF, Lamanna JC (2011) Hypoxia-induced angiogenesis is delayed in aging mouse brain. Brain Res 1389:50–60
3. Iadecola C (2004) Neurovascular regulation in the normal brain and in Alzheimer's disease. Nat Rev Neurosci 5(5):347–360
4. Harik SI et al (1990) Immunocytochemical localization of the erythroid glucose transporter: abundance in tissues with barrier functions. J Neurosci 10:3862–3872
5. Simpson IA et al (2007) Supply and demand in cerebral energy metabolism: the role of nutrient transporters. J Cereb Blood Flow Metab 27(11):1766–1791
6. Matus A, Mughal S (1975) Immunohistochemical localisation of S-100 protein in brain. Nature 258(5537):746–748
7. Izant JG, McIntosh JR (1980) Microtubule-associated proteins: a monoclonal antibody to MAP2 binds to differentiated neurons. Proc Natl Acad Sci USA 77(8):4741–4745
8. Mullen RJ et al (1992) NeuN, a neuronal specific nuclear protein in vertebrates. Development 116(1):201–211
9. LaManna JC (2012) Angioplasticity and cerebrovascular remodeling. Adv Exp Med Biol 737:13–17
10. Harik N et al (1996) Time-course and reversibility of the hypoxia-induced alterations in cerebral vascularity and cerebral capillary glucose transporter density. Brain Res 737(1–2):335–338
11. Mironov V et al (1994) Architectural alterations in rat cerebral microvessels after hypobaric hypoxia. Brain Res 660:73–80

Chapter 22

Inducible Gene Deletion in Glial Cells to Study Angiogenesis in the Central Nervous System

Hye Shin Lee and Joseph H. McCarty

Abstract

Most organs and tissues of the vertebrate body harbor elaborate network of blood vessels with diverse functions that are determined, in part, by cues within the local environment (Warren and Iruela-Arispe, Curr Opin Hematol 17:213–218, 2010). How vascular endothelial cells decipher these cues to promote normal blood vessel development and physiology remains largely uncharacterized. In this review, we will focus on genetic strategies to analyze glial regulation of blood vessel growth and sprouting within the microenvironment of the retina, a component of the central nervous system (CNS) that contains a complex web of blood vessels with many unique features, including a blood-retinal barrier (Abbott et al., Nat Rev Neurosci 7:41–53, 2006). Blood vessels promote retinal development and homeostasis and alterations in vascular functions can lead to various developmental and adult-onset retinal pathologies (Fruttiger, Angiogenesis 10:77–88, 2007). How glial cells control retinal endothelial cell growth and sprouting remains largely uncharacterized. We will detail methodologies involving inducible Cre-lox technologies to acutely ablate genes of interest in CNS glial cells. These methods allow for precise spatial and temporal regulation of gene expression to study how glial cells in the retinal microenvironment control angiogenesis and blood-retinal barrier development.

Key words Itgb8, Retina, Astrocyte, Inducible Cre, GFAP-CreERT2, Tamoxifen, Retinal angiogenesis, Neurovascular

1 Introduction

The vertebrate retina has a planar cytoarchitecture that enables high-resolution analysis of the vascular cytoarchitecture [4, 5]. In mice the retina is vascularized during neonatal development when angiogenic blood vessels emanating from the optic nerve proliferate and sprout along a preformed network of glial fibrillary-associated protein (GFAP)-expressing astroglial cells to form a primary vascular plexus [6]. After approximately 10 postnatal days, a second wave of angiogenesis occurs when blood vessels within the primary plexus invade radially along Muller glial cell processes and vascularize the inner and outer neuronal layers of the retina. Alterations in glial-endothelial cell communication can lead to neonatal and

Richard Milner (ed.), *Cerebral Angiogenesis: Methods and Protocols*, Methods in Molecular Biology, vol. 1135, DOI 10.1007/978-1-4939-0320-7_22, © Springer Science+Business Media New York 2014

adult-onset CNS pathologies [7, 8]. For example, heritable mutations in the NORRIN gene lead to Norrie disease, which is typified by blindness due to retinal hypovascularity during development [9]. Norrin also regulates blood vessel sprouting under pathological conditions including retinopathy of prematurity, a retinal vascular disease that commonly afflicts premature infants and is typified by shifts in oxygen levels [10]. Elevated expression of VEGF-A as a result of acute hypoxia has been reported in retinopathy of prematurity [11, 12]. Since pathways involved in physiological retinal angiogenesis and blood-retinal barrier development are altered in Norrie disease and retinopathy of prematurity, identifying normal mechanisms of glial regulation of the retinal vasculature will yield new insights into retinal vascular deficits.

Retinal glial cells communicate with blood vessels via secreted growth factors and ECM proteins [13]. Integrins are cell surface receptors for extracellular matrix (ECM) ligands, which play crucial roles in vascular development and homeostasis. αvβ8 integrin and its ECM protein ligands, the latent transforming growth factor βs (TGFβs), play particularly important roles in glial regulation of brain and retinal blood vessel development. During mouse embryogenesis, genetic ablation of αv or β8 integrin expression leads to abnormal CNS angiogenesis and intracerebral hemorrhage [14–16]. These defects are due to impaired integrin adhesion to latent TGFβs, since TGFβ1 and TGFβ3 double mutant mice develop nearly identical phenotypes as αv and β8 gene knockouts [17]. This same glial-regulated signaling axis also plays an important role in promoting normal retinal vascular development [18–20]. Interestingly, similar pathologies have been reported in mice deficient for components of the Wnt signaling pathway, suggesting functional cross talk with integrins and TGFβs [21–24].

One important methodology that has enabled analysis of cell-cell and cell-ECM communication between glia and endothelial cells in the retina involves Cre-lox [25]. In the case of integrin αvβ8, ablation of either integrin gene in the neuroepithelium via Cre expression under the control of the minimal Nestin promoter (Nestin-Cre) [26] reproduces whole-body knockout phenotypes, whereas endothelial cell-specific knockouts develop a normal vasculature [16, 27]. Similarly, selective ablation of Tgfbr2 or Alk5 in the vascular endothelium, but not neuroepithelium, results in defective angiogenesis and hemorrhage in the brain [28] and retina [18–20]. Taken together, Cre-lox-based strategies are powerful tools for identifying molecular mechanisms that regulate retinal vascular development, which may provide fundamental insights into pathological angiogenesis in the CNS.

2 Materials

2.1 Experimental Mice

1. GFAP-CreERT2 [29].
2. Integrin β8+/− [14].
3. Integrin β8flox/flox [16].
4. R26-loxSTOPlox-EYFP (Jackson Laboratory, stock #006148).

2.2 Tamoxifen Injection

1. Tamoxifen stock solution (10 mg/ml): Dissolve 10 mg of tamoxifen (Sigma) in 250 μl of 100 % ethanol by vortexing. Add 750 μl of sunflower seed oil and vortex vigorously to mix. Protect from light and store at 4 °C. This stock solution can be used for up to 3 days.
2. 1 ml syringe.
3. 30G needle.

2.3 Genotyping

1. Ear tag and ear tag applicator (National Band and Tag Co., Newport, KY).
2. STE buffer: 10 mM Tris (pH 8.5), 5 mM EDTA, 0.2 % SDS, 200 mM NaCl, and 0.1 mg/ml DNAase-free Proteinase K (USB Scientific, Cleveland, OH).
3. TE: 10 mM Tris (pH 8.0), 1 mM EDTA.
4. MangoMix (Bioline, Taunton, MA).
5. Synthetic DNA primers.
6. Agarose.
7. 50× TAE: Add 242 g Tris-base, 57.1 ml glacial acetic acid, and 100 ml of 0.5 M EDTA and sterile water up to 1 L.
8. Quanti-Marker 1 kb DNA ladder (ISC BioExpress, Kaysville, UT).

2.4 Flat Mount Retinal Preparations

1. Dissection tools.
2. 4 % paraformaldehyde (PFA) in PBS: To make 500 ml, weigh 20 g paraformaldehyde and add in 400 ml distilled water, add 25 μl of 10 N NaOH, and heat at 60 °C for 15 min until PFA powder is completely dissolved. Add 50 ml 10×PBS and adjust volume to 500 ml with dH_2O. Filter using filter paper, and aliquot into 50 ml tubes and store at −20 °C.
3. Blocking solution: 1 % BSA with 0.5 % Triton X-100 in PBS.
4. Washing solution: 0.5 % Triton X-100 in PBS.
5. Primary antibodies: rabbit polyclonal antibody for GFP (Abcam), mouse monoclonal antibody for GFAP (Millipore, Billerica, MA), and rat anti-mouse CD31 antibody (BD Pharmingen).

6. Secondary antibodies: Goat anti-rabbit Alexa Fluor 488, goat anti-mouse Alexa Fluor 594, and goat anti-rat Alexa Fluor594 (Invitrogen).
7. VECTASHIELD mounting medium for fluorescence (Vector Laboratories, Burlingame, CA).

2.5 Preparation of Frozen Sections for Immuno-fluorescence Staining

1. 30 % sucrose/PBS solution: Dissolve 30 g of sucrose in 50 ml sterile water and add 10 ml 10×PBS. Adjust final volume to 100 ml with sterile water.
2. Tissue-Tek OCT compound and cryomold (Sakura Finetek USA, Torrance, CA).
3. Microslides.
4. Mini Pap Pen.
5. Blocking solution: 10 % normal goat serum in PBS.
6. Washing solution: 0.1 % Tween-20 in PBS.

3 Methods

Targeted gene ablation is a powerful tool to analyze in vivo functions for signaling pathways in development and physiology. One common method involves the use of Cre-lox recombination strategies that involve Cre recombinase regulated by cell type-specific promoters together with target genes flanked by loxP recombination sequences [30]. Several transgenic and knock-in strains have been developed that express Cre in CNS astrocytes. These include GFAP-Cre [31, 32], Nestin-Cre [26], S100β-Cre [33, 34], and BLBP-Cre [35]. Since these gene promoters are also active in embryonic neural stem and progenitor cells, one limitation in using these lines is that other lineages such as neurons and oligodendrocyte precursors also display Cre activation, which limits analyses of gene functions selectively in astrocytes [36].

One strategy to control temporal and spatial patterns of gene ablation is to employ inducible Cre variants. CreERT2 consists of Cre conjugated with a mutated estrogen receptor (ERT2) that binds poorly to endogenous estrogen but with high affinity to the estrogen derivative tamoxifen [37, 38]. CreERT2 is sequestered in the cytoplasm bound to chaperones until tamoxifen binds to ERT2. Upon binding to tamoxifen, the protein structure of CreERT2 is altered, allowing translocation to the nucleus and recombination of floxed alleles. Since the retinal vasculature develops postnatally, the inducible Cre strategy is an attractive approach to study detailed gene functions in retinal glial cells and blood vessels. By injecting tamoxifen into mice at varying postnatal ages, one can also dissect the knockout effects in primary angiogenic sprouting, secondary vascular plexus formation, vessel maturation, and blood-retinal barrier development and homeostasis [5]. Various CreERT2 lines have been established to allow conditional gene deletion in retinal endothelial cells such as Cdh5-CreERT2 [19] and PDGFB-CreERT2 [39]. To study the influence of astrocyte-expressed genes in retinal angiogenesis, GFAP-CreERT2 transgenic mice are useful because retinal astrocytes display strong GFAP expression, while endothelial cells sprout across the astrocyte network. In this review, we will introduce methods for generating the GFAP-CreERT2 strain [29] that harbors a EYFP reporter and floxed integrin β8 as an example of targeted gene ablation. This protocol will also review methods for administrating tamoxifen into neonatal mice as well as other general methods for analyzing glial control of retinal angiogenesis.

3.1 Mouse Breeding

While the GFAP and GLAST promoters restrict gene expression to many CNS astroglial cells and neural progenitors, in GFAP-CreERT2 and GLAST-CreERT2 strains, tamoxifen exposure results in mosaic patterns of CreERT2 activation [29, 40]. Such a mosaic Cre activation pattern has the advantage of producing less severe overall defects but a more focal phenotype. However, it is difficult to distinguish which cells have CreERT2 activation and normal gene expression because this largely relies on the specificity of antibodies to target proteins. In this protocol we detail a method for generating a double transgenic mouse line harboring floxed β8 integrin alleles and a Cre-inducible EYFP reporter [41] to identify cells with tamoxifen-dependent Cre activation (*see* Fig. 1):

1. Breed a GFAP-CreERT2 heterozygote male (GFAP-CreERT2+/−) with a β8 heterozygote female (β8+/−) to generate GFAP-CreERT2+/−;β8+/− male.
2. Determine the genotype of the F1 progeny by genotyping for either GFAP-CreERT2 or β8 integrin. According to standard Mendelian ratio, approximately 25 % of the F1 progeny should be GFAP-CreERT2+/−;β8+/− (*see* **Note 1**).
3. Breed homozygous β8flox/flox females with homozygous R26-loxSTOPlox-EYFP males. All F1 progeny will be heterozygous for both β8flox or R26-loxSTOPlox-EYFP alleles (β8f/+;EYFP+/−).
4. Interbreed the F1 progeny to generate homozygous β8flox/flox;R26-loxSTOPlox-EYFP. Determine genotype of the F2 progeny, approximately 25 % of them will be β8f/f;EYFP+/+.
5. Breed a GFAP-CreERT2+/−;β8+/− male with β8f/f;EYFP+/+ female. There will be about 25 % Cre+/−;β8f/+;EYFP+/−, 25 % Cre+/−;β8f/−;EYFP+/−, 25 % +/+;β8f/+;EYFP+/−, and 25 % +/+;β8f/−;EYFP+/− in the litter (*see* **Note 2**).

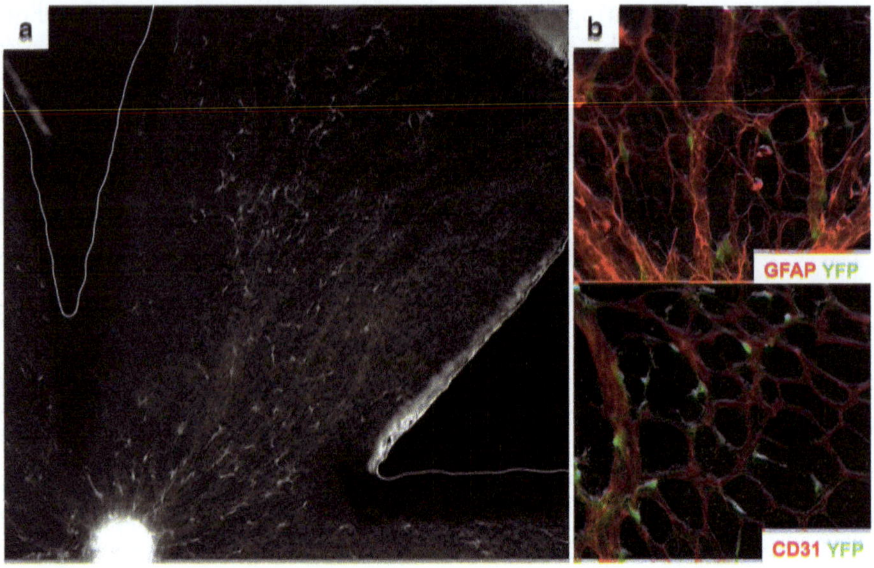

Fig. 1 Tamoxifen-inducible activation of CreERT2 in retinal astrocytes. (**a**) Flat mount retinal staining of P6 GFAP-CreERT2;R26-EYFP mouse injected with tamoxifen from P1 to P3. Retinal tissue was immunofluorescently labeled with anti-GFP to visualize CreERT2 activity in astrocytes. (**b**) P6 GFAP-CreERT2;R26-EYFP mouse retina immunofluorescently labeled with anti-GFAP (*red*) to visualize astrocytes and anti-GFP (*green*) to visualize cells with CreERT2 activity (*upper panel*). P6 GFAP-CreERT2;R26-EYFP mouse retina immunofluorescently labeled with anti-CD31 (*red*) to visualize endothelial cells and anti-GFP (*green*) to visualize cells with tamoxifen-dependent CreERT2 activity (*lower panel*)

3.2 Intragastric Tamoxifen Injection

1. On the day of injection, prepare 1 mg/ml tamoxifen solution by diluting 100 μl of 10 mg/ml stock solution to 900 μl sunflower seed oil. Mix well by vortexing and protect the solution from light. Use the 1 mg/ml solution within 1 day after preparation.
2. Aspirate 700 μl of 1 mg/ml tamoxifen solution with a 1 ml syringe; remove all the air bubbles before placing needle. Place on the 30G needle and push the plunger up to remove air in the needle.
3. Hold the tail and hind limbs with the left index and middle finger and the shoulders with thumb and ring finger. Hold the pup tightly so it doesn't move while injecting. Take care to not overly extend the torso since this may cause leakage during or after injection.
4. Inject 100 μl of tamoxifen solution directly into the stomach. In the case of neonates, the stomach is visible by the presence of milk. Pierce about half of the needle toward the stomach and slowly push the plunger. Support the syringe with left thumb so that the needle does not move while injecting. After injecting 100 μl, leave the needle inside for 1–2 s and slowly pull it out (*see* **Note 3**).
5. Inject the same dose of tamoxifen to all pups in the litter, once a day for 3 consecutive days (*see* **Note 4**).

3.3 Mouse Genotyping

For analysis of multiple litters, PCR-based genotyping can be performed before sacrifice using genomic DNA isolated from the tail or other tissues. For example, ear tagging with clips and genotyping can be performed with live mice at P12 or older, although for animals younger than P12, the ear tag tends to detach. Alternatively, mice can be genotyped postmortem, for example, after cardiac perfusion to preserve brain morphology:

1. Place an ear tag into the ear tag applicator. Grab the mouse with one hand and clamp the ear tag into one ear. Snip a piece of tissue from the other ear and insert it into a 1.5 ml microcentrifuge tube pre-labeled with the same number as the ear tag.
2. Add 500 μl of STE to each tube and incubate the ear snip in STE at 55 °C in a heating chamber with constant rotation for 12–18 h.
3. Vortex well until the tissue mass dissolves. Centrifuge for 10 min at 13,000 rpm.
4. Prepare new microcentrifuge tubes, label the number of each ear tag, and add 500 μl of 100 % ethanol into each tube. Transfer the supernatant containing STE with digested tissue to the tube with 100 % ethanol. Mix well by invert mixing up to 20 times until the DNA precipitates are visible.
5. Prepare 200 μl of TE in new tubes labeled with ear tag numbers. Swirl the genomic DNA precipitates with a 2 0 μl tip and place it in TE solution.
6. Incubate the solution at 37 °C for 1 h with shaking at 1,400 rpm.
7. Prepare PCR reaction tubes: Take 1 μl of genomic DNA and add 24 μl of PCR master mix as shown in Table 1. Run the PCR program as detailed in Table 2. Primers for identifying select genotypes are listed in Table 3.
8. After the PCR reactions are completed, load 20 μl of each PCR sample into a 1.5 % agarose gel. Separate the PCR products by applying 120 V for 30–60 min using the DNA ladder as a reference.

Table 1
PCR reagent mixtures for each genotype. Listed are the various amounts of each reagent required to perform PCR-based genotyping for the various gene knockout and transgenic mice

	GFAP-CreERT2	β8 flox	β8+/−	R26-EYFP
2× Mango Taq Mix	12.5	12.5	12.5	12.5
Primers (50 μM)	0.5	0.26	0.26	0.26
50 % Glycerol	5	0	0	0
Nuclease free H_2O	Adjust to 25 μl	Adjust to 25 μl	Adjust to 25 μl	Adjust to 25 μl

Table 2
Summary of PCR cycles optimized for each genotype. Shown are details for the number of PCR cycles, the duration for each cycle, and the expected PCR product sizes related to genotyping the various gene knockout and transgenic mice

	GFAP-CreERT2	β8 flox	β8+/−	R26-EYFP
Denaturation (1 cycle)	95 °C for 5 min	95 °C for 5 min	95 °C for 5 min	94 °C for 3 min
Amplification (30 cycles)	95 °C for 30 s 56 °C for 30 s 72 °C for 1 min	95 °C for 45 s 58 °C for 30 s 72 °C for 1 min	95 °C for 45 s 58 °C for 30 s 72 °C for 1 min	94 °C for 30 s 58 °C for 1 min 72 °C for 1 min
Extension (1 cycle)	72 °C for 5 min	72 °C for 10 min	72 °C for 10 min	72 °C for 2 min
Expected product size	~300 bp	WT ~250 bp Floxed ~370 bp	WT ~300 bp Null ~500 bp	WT ~500 bp mut ~300 bp

Table 3
Primer sequences to detect transgenic or knockout alleles. The nucleotide sequences are provided for the forward and reverse DNA primers used to genotype knockout and transgenic mice described in this review

PCR		Primers
GFAP-CreERT2	Cre-F Cre-R	5′-ACCAGCCAGCTATCAACTC-3′ 5′-TATACGCGTGCTAGCGAAGATCTC CATCTTCCAGCAG-3′
EYFP	oIMR4982 oIMR8545 oIMR8546	5′-AAGACCGCGAAGAGTTTGTC-3′ 5′-AAAGTCGCTCTGAGTTGTTAT-3′ 5′-GGAGCGGGAGAAATGGATATG-3′
β8+/−	β8 wt-F β8 wt-R β8 null-F β8 null-R	5′-ATTATCTGGTTGATGTGTCAGC-3′ 5′-AGAGAGGAACAAATATCCTTCCC-3′ 5′-AGAGGCCACTTGTGTAGCGCCAAG-3′ 5′-GGAGGCATACAGTCTAAATTGT-3′
β8 flox	β8 flox-F β8 flox-R	5′-GAGATGCAAGAGTGTTTACC-3′ 5′-CACTTTAGTATGCTAATGATGG-3′

3.4 Flat Mount Retinal Staining

One advantage to using the retina as a model to study angiogenesis is the ability to visualize the primary and secondary vascular networks using endothelial cell-specific markers and confocal microscopy. Flat mount retinal preparation is a common method to visualize the primary and secondary vascular plexi. In combination with inducible knockout strategy in astrocytes, this protocol can be applied to analyze localized effects of astrocytes on endothelial tip cell growth, vessel maturation, and homeostasis. Below we provide protocols of cardiac perfusion of fixative as well as flat mount retinal staining. Cardiac perfusion is recommended in order to minimize background fluorescent signal and preserve brain and retinal morphologies; however, if the mice are sacrificed before P12, cardiac perfusion is technically challenging so skip to **steps 4–6**:

1. Anesthetize the mice by injecting 20 mg/g body weight Avertin solution intraperitoneally. Wait until they are deeply anesthetized (*see* **Note 5**).
2. Position the mouse on its back on a Styrofoam bed covered with a paper towel. Fix four limbs with pins.
3. Moisten the fur with 70 % ethanol and expose the body cavity using scissors until the diaphragm is exposed.
4. Snip the diaphragm with sharp scissors and cut both sides of the ribs so you can access the beating heart.
5. Using blunt forceps, grasp the heart carefully and insert the needle of syringe filled with 4 % PFA into the left ventricle. Cut the right atrium. Apply mild pressure to the plunger so that the fixative goes into the heart. If the perfusion is working well, the color of the liver should become pale within a minute. Perfuse 10–15 ml of fixative for one adult mouse (*see* **Notes 6** and 7).
6. To remove the eyes, make incisions surrounding each eye and open the eyelid with forceps. Place the curved forceps carefully below the eye and remove the entire eye with the residual optic nerve attached.
7. Place the eye in 4 % PFA/PBS for 16 h at 4 °C with gentle shaking.
8. Wash with PBS and transfer the eyes to a 10 cm dish containing 20 ml of PBS.
9. Dissect the retina from the eye under a dissection microscope. Grasp the optic nerve with forceps and make an incision at the center of the cornea using curved scissors. From the hole of the corneal center, make four incisions toward four directions. Peel out the sclera and carefully remove the lens, iris, and ciliary body. Take care to not burst the lens. It is also important not to damage the retina while removing adjacent tissues.
10. Using a transfer pipette, transfer the retinal cup into a single well of a 24-well plate containing cold PBS. Keep the plate on ice until all the retinas in the experimental group are dissected.
11. Remove the PBS and add 1 ml of blocking solution into each well. Incubate overnight with gentle shaking.
12. Prepare appropriate titers of primary antibodies diluted in blocking solution. For example, for double staining of YFP cells and astrocytes, dilute rabbit polyclonal anti-GFP antibody

and mouse monoclonal anti-GFAP antibody at 1:500. Alternatively, to double label YFP-expressing cells and endothelial cells, add rat anti-CD31 antibody at 1:100 dilution. Replace the blocking solution with primary antibody solution and incubate overnight at 4 °C with gentle shaking. Include isotype-matched IgGs as a negative control.

13. Wash five times with 0.5 % TritonX-100 in PBS at room temperature. Allow 1 h for each with gentle shaking.
14. Dilute fluorescence-labeled secondary antibodies in blocking solution. For double staining of YFP cells and astrocytes or blood vessels, mix goat anti-rabbit IgG conjugated to Alexa Fluor 488 and goat anti-mouse or rat IgG conjugated to Alexa 594 at 1:750, respectively. Incubate overnight at 4 °C with gentle shaking. Cover the plate with aluminum foil to protect from the light (*see* **Note 8**).
15. Wash five times with 0.5 % Triton X-100 in PBS at room temperature, 1 h for each wash with shaking.
16. For flat mounting, transfer the retina into a 10 cm dish containing PBS. Under the dissection microscope, make four incisions from the edge of the cup toward the optic nerve head. This will allow flattening of the cup.
17. Carefully place the retina on a microslide using a transfer pipette as inner surface facing up. Remove the remaining solution completely with a 200 μl pipette, and the retina should be flattened on the glass slide. Put a drop of fluorescence mounting medium on the center of the cover slide and cover the microslide as the mounting medium fills the tissue.
18. Capture images using a confocal fluorescent microscope (*see* **Note 9**).

3.5 Immunofluorescence Staining of Retinal Sections

Although retinal flat mount staining is a useful method to analyze primary vascular networks, there are limitations to analysis of the secondary plexus in deeper retinal layers. This issue can be addressed by longitudinally sectioning through the retina. Below we will provide a general method for preparation of frozen retinal sections and an example of immunofluorescent staining of these sections:

1. Perfuse mice with 4 % PFA and postfix the eyeballs in 4 % PFA/PBS for 16 h at 4 °C.
2. Wash off the fixative with cold PBS. Dehydrate the tissue by serial sucrose gradient. Replace PBS with 10 % sucrose/PBS solution and incubate for 2 h at 4 °C with gentle rotation. Increase the sucrose gradient by replacing 10–20 % and then 20–30 % sucrose/PBS and incubate for 16–24 h at 4 °C for each step.
3. Using a dissection microscope, remove the cornea and the lens.

4. Fill the cryomold with the OCT compound in. Pick up the retina using curved forceps and put in the cryomold. Place the retina in the right orientation so that cross sections of retinal layers are obtained.
5. Place the cryomold containing the OCT-embedded retina on a flat surface of dry ice until the OCT compound is frozen. Store the frozen block in a –80 °C freezer.
6. Slice the frozen block using a cryostat at 7 μm thickness and collect sections on microslides. Store the frozen sections in a –80 °C freezer until they are used for staining.
7. Remove the OCT compound by immersing slides in PBS, three times for 5 min each.
8. Wipe away the excess PBS solution from the slides and draw a border around the tissue using a Pap pen. Fill inside of the border with PBS (*see* **Note 10**).
9. Remove the PBS using a vacuum aspirator unit and add the blocking solution. Incubate for 1 h at room temperature.
10. Mix the appropriate titer of primary antibodies in blocking solution. In case of double staining of YFP and astrocytes, add rabbit polyclonal anti-GFP antibody and mouse monoclonal anti-GFAP antibody at 1:500 dilution. When staining vascular endothelial cells, use the rat anti-CD31 antibody at 1:100. Incubate overnight at 4 °C. Include control staining of another section with the same amount of isotype-matched IgG.
11. Wash three times with the washing solution, 15 min for each wash at room temperature.
12. Dilute the fluorescence-conjugated secondary antibodies in blocking solution at 1:750. Incubate for 2 h at room temperature. Protect from light exposure.
13. Wash three times with washing solution, 15 min for each wash at room temperature.
14. For mounting, remove the remaining solution on the slides. Drop 20 μl of fluorescence mounting medium on the tissue and carefully cover with cover slides.
15. Collect the images using a fluorescent microscope.

4 Notes

1. High levels of Cre recombinase have been reported to induce toxicity in developing mice [42, 43]. To minimize the toxic effect generated by Cre itself, we recommend maintaining the Cre mouse lines as heterozygotes and including Cre positive control for analysis.

2. Since β8+/− mice do not display any defect when compared to β8+/+ mice, we set up breeding to generate Cre+/−;β8flox/− as mutants in order to achieve efficient Cre-mediated gene deletion.
3. For consistent results, it is important to deliver the same dose of tamoxifen to every pup. There may be a drop of solution at the injection point due to leakage from the needle. If so, it is best to exclude the pup for future analysis.
4. Some of the reported CreERT2 mouse lines display basal levels of Cre activity. To test this possibility, we suggest using another litter injected with the same volume of sunflower seed oil to exclude any nonspecific effects due to basal (tamoxifen-independent) Cre levels or caused by carrier injection.
5. Optimally, the heart should still be beating when perfusing fixative. Monitor the depth of anesthesia by checking reflex response by firmly pinching a hind paw.
6. It is important to avoid high pressure levels when delivering the fixative, as this may cause rupture of microvessels and tissue damage. One alternative is to use a perfusion pump with fixed flow (3–4 ml/min).
7. PFA is a suspected carcinogen that requires careful handling. To avoid inhalation and eye contact, we suggest manipulating the powdered and liquid forms and performing perfusion steps in a ventilated chemical hood.
8. The cells with Cre activation induced by tamoxifen injection will also activate the R26 promoter leading to YFP expression. However, the YFP signal will fade upon fixation in 4 % PFA/PBS. Since the anti-GFP antibody cross-reacts with YFP, staining with anti-GFP antibody will strengthen the fluorescence of YFP-expressing cells. Make sure to use green fluorescence-labeled secondary antibody for anti-GFP since low levels of YFP fluorescence persist even after fixation.
9. The form of fluorescence dye conjugated to secondary antibody should be determined based on the filters installed in the microscope of use.
10. It is important not to dry tissues during staining steps. Drying tissue can cause high fluorescence background. We suggest using a staining tray that can maintain humidity during staining by placing water-absorbed paper towels on the bottom of the tray.

References

1. Warren CM, Iruela-Arispe ML (2010) Signaling circuitry in vascular morphogenesis. Curr Opin Hematol 17:213–218
2. Abbott NJ, Ronnback L, Hansson E (2006) Astrocyte-endothelial interactions at the blood-brain barrier. Nat Rev Neurosci 7: 41–53
3. Fruttiger M (2007) Development of the retinal vasculature. Angiogenesis 10:77–88
4. Gerhardt H, Golding M, Fruttiger M, Ruhrberg C, Lundkvist A, Abramsson A, Jeltsch M, Mitchell C, Alitalo K, Shima D, Betsholtz C (2003) VEGF guides angiogenic sprouting utilizing endothelial tip cell filopodia. J Cell Biol 161:1163–1177
5. Pitulescu ME, Schmidt I, Benedito R, Adams RH (2010) Inducible gene targeting in the neonatal vasculature and analysis of retinal angiogenesis in mice. Nat Protoc 5: 1518–1534
6. Kubota Y, Suda T (2009) Feedback mechanism between blood vessels and astrocytes in retinal vascular development. Trends Cardiovasc Med 19:38–43
7. Hawkins BT, Davis TP (2005) The blood-brain barrier/neurovascular unit in health and disease. Pharmacol Rev 57:173–185
8. Tam SJ, Watts RJ (2010) Connecting vascular and nervous system development: angiogenesis and the blood-brain barrier. Annu Rev Neurosci 33:379–408
9. Clevers H (2004) Wnt signaling: Ig-norrin the dogma. Curr Biol 14:R436–R437
10. Chen J, Smith LE (2007) Retinopathy of prematurity. Angiogenesis 10:133–140
11. Benjamin LE, Hemo I, Keshet E (1998) A plasticity window for blood vessel remodelling is defined by pericyte coverage of the preformed endothelial network and is regulated by PDGF-B and VEGF. Development 125:1591–1598
12. Scott A, Powner MB, Gandhi P, Clarkin C, Gutmann DH, Johnson RS, Ferrara N, Fruttiger M (2010) Astrocyte-derived vascular endothelial growth factor stabilizes vessels in the developing retinal vasculature. PLoS One 5:e11863
13. Engelhardt B, Sorokin L (2009) The blood-brain and the blood-cerebrospinal fluid barriers: function and dysfunction. Semin Immunopathol 31:497–511
14. Zhu J, Motejlek K, Wang D, Zang K, Schmidt A, Reichardt LF (2002) beta8 integrins are required for vascular morphogenesis in mouse embryos. Development 129:2891–2903
15. McCarty JH, Monahan-Earley RA, Brown LF, Keller M, Gerhardt H, Rubin K, Shani M, Dvorak HF, Wolburg H, Bader BL, Dvorak AM, Hynes RO (2002) Defective associations between blood vessels and brain parenchyma lead to cerebral hemorrhage in mice lacking alphav integrins. Mol Cell Biol 22:7667–7677
16. Proctor JM, Zang K, Wang D, Wang R, Reichardt LF (2005) Vascular development of the brain requires beta8 integrin expression in the neuroepithelium. J Neurosci 25: 9940–9948
17. Mu Z, Yang Z, Yu D, Zhao Z, Munger JS (2008) TGFbeta1 and TGFbeta3 are partially redundant effectors in brain vascular morphogenesis. Mech Dev 125:508–516
18. Hirota S, Liu Q, Lee HS, Hossain MG, Lacy-Hulbert A, McCarty JH (2011) The astrocyte-expressed integrin alphavbeta8 governs blood vessel sprouting in the developing retina. Development 138:5157–5166
19. Allinson KR, Lee HS, Fruttiger M, McCarty J, Arthur HM (2012) Endothelial expression of TGFbeta type II receptor is required to maintain vascular integrity during postnatal development of the central nervous system. PLoS One 7:e39336
20. Arnold TD, Ferrero GM, Qiu H, Phan IT, Akhurst RJ, Huang EJ, Reichardt LF (2012) Defective retinal vascular endothelial cell development as a consequence of impaired integrin alphaVbeta8-mediated activation of transforming growth factor-beta. J Neurosci 32:1197–1206
21. Daneman R, Agalliu D, Zhou L, Kuhnert F, Kuo CJ, Barres BA (2009) Wnt/beta-catenin signaling is required for CNS, but not non-CNS, angiogenesis. Proc Natl Acad Sci U S A 106:641–646
22. Liebner S, Corada M, Bangsow T, Babbage J, Taddei A, Czupalla CJ, Reis M, Felici A, Wolburg H, Fruttiger M, Taketo MM, von Melchner H, Plate KH, Gerhardt H, Dejana E (2008) Wnt/beta-catenin signaling controls development of the blood-brain barrier. J Cell Biol 183:409–417
23. Stenman JM, Rajagopal J, Carroll TJ, Ishibashi M, McMahon J, McMahon AP (2008) Canonical Wnt signaling regulates organ-specific assembly and differentiation of CNS vasculature. Science 322:1247–1250
24. Wang Y, Rattner A, Zhou Y, Williams J, Smallwood PM, Nathans J (2012) Norrin/Frizzled4 signaling in retinal vascular development and blood brain barrier plasticity. Cell 151:1332–1344
25. Mortensen R (2006) Overview of gene targeting by homologous recombination. Curr Protoc Mol Biol Chapter 23:Unit 23.21

26. Tronche F, Kellendonk C, Kretz O, Gass P, Anlag K, Orban PC, Bock R, Klein R, Schutz G (1999) Disruption of the glucocorticoid receptor gene in the nervous system results in reduced anxiety. Nat Genet 23:99–103
27. McCarty JH, Lacy-Hulbert A, Charest A, Bronson RT, Crowley D, Housman D, Savill J, Roes J, Hynes RO (2005) Selective ablation of alphav integrins in the central nervous system leads to cerebral hemorrhage, seizures, axonal degeneration and premature death. Development 132:165–176
28. Nguyen HL, Lee YJ, Shin J, Lee E, Park SO, McCarty JH, Oh SP (2011) TGF-beta signaling in endothelial cells, but not neuroepithelial cells, is essential for cerebral vascular development. Lab Invest 91:1554–1563
29. Hirrlinger PG, Scheller A, Braun C, Hirrlinger J, Kirchhoff F (2006) Temporal control of gene recombination in astrocytes by transgenic expression of the tamoxifen-inducible DNA recombinase variant CreERT2. Glia 54:11–20
30. Mortensen R (2007) Overview of gene targeting by homologous recombination. Curr Protoc Neurosci Chapter 4, Unit 4.29
31. Matthias K, Kirchhoff F, Seifert G, Huttmann K, Matyash M, Kettenmann H, Steinhauser C (2003) Segregated expression of AMPA-type glutamate receptors and glutamate transporters defines distinct astrocyte populations in the mouse hippocampus. J Neurosci 23:1750–1758
32. Nolte C, Matyash M, Pivneva T, Schipke CG, Ohlemeyer C, Hanisch UK, Kirchhoff F, Kettenmann H (2001) GFAP promoter-controlled EGFP-expressing transgenic mice: a tool to visualize astrocytes and astrogliosis in living brain tissue. Glia 33:72–86
33. Vives V, Alonso G, Solal AC, Joubert D, Legraverend C (2003) Visualization of S100B-positive neurons and glia in the central nervous system of EGFP transgenic mice. J Comp Neurol 457:404–419
34. Zuo Y, Lubischer JL, Kang H, Tian L, Mikesh M, Marks A, Scofield VL, Maika S, Newman C, Krieg P, Thompson WJ (2004) Fluorescent proteins expressed in mouse transgenic lines mark subsets of glia, neurons, macrophages, and dendritic cells for vital examination. J Neurosci 24:10999–11009
35. Anthony TE, Klein C, Fishell G, Heintz N (2004) Radial glia serve as neuronal progenitors in all regions of the central nervous system. Neuron 41:881–890
36. Mori T, Buffo A, Gotz M (2005) The novel roles of glial cells revisited: the contribution of radial glia and astrocytes to neurogenesis. Curr Top Dev Biol 69:67–99
37. Feil R, Brocard J, Mascrez B, LeMeur M, Metzger D, Chambon P (1996) Ligand-activated site-specific recombination in mice. Proc Natl Acad Sci U S A 93:10887–10890
38. Indra AK, Warot X, Brocard J, Bornert JM, Xiao JH, Chambon P, Metzger D (1999) Temporally-controlled site-specific mutagenesis in the basal layer of the epidermis: comparison of the recombinase activity of the tamoxifen-inducible Cre-ER(T) and Cre-ER(T2) recombinases. Nucleic Acids Res 27:4324–4327
39. Claxton S, Kostourou V, Jadeja S, Chambon P, Hodivala-Dilke K, Fruttiger M (2008) Efficient, inducible Cre-recombinase activation in vascular endothelium. Genesis 46:74–80
40. Mori T, Tanaka K, Buffo A, Wurst W, Kuhn R, Gotz M (2006) Inducible gene deletion in astroglia and radial glia—a valuable tool for functional and lineage analysis. Glia 54:21–34
41. Srinivas S, Watanabe T, Lin CS, William CM, Tanabe Y, Jessell TM, Costantini F (2001) Cre reporter strains produced by targeted insertion of EYFP and ECFP into the ROSA26 locus. BMC Dev Biol 1:4
42. Forni PE, Scuoppo C, Imayoshi I, Taulli R, Dastru W, Sala V, Betz UA, Muzzi P, Martinuzzi D, Vercelli AE, Kageyama R, Ponzetto C (2006) High levels of Cre expression in neuronal progenitors cause defects in brain development leading to microencephaly and hydrocephaly. J Neurosci 26:9593–9602
43. Naiche LA, Papaioannou VE (2007) Cre activity causes widespread apoptosis and lethal anemia during embryonic development. Genesis 45:768–775

Chapter 23

Bone Marrow Chimera Experiments to Determine the Contribution of Hematopoietic Stem Cells to Cerebral Angiogenesis

Marcia Regina Machein and Karl H. Plate

Abstract

The generation of bone marrow chimera in mice is a valuable tool to study a variety of cellular processes. Donor bone marrow cells expressing reporter genes have been used to study the process of cell differentiation and the mechanisms involved in bone marrow cell recruitment. Bone marrow cells bearing genetic manipulation have been used in bone marrow chimeras to elucidate the role of molecules in different physiological and pathological settings. Since in the normal adult brain angiogenesis does not occur, models of brain injury like ischemia and tumor growth have been used to study the contribution of bone marrow-derived cells to the cerebral vasculature. This chapter describes the procedures to perform bone marrow transplantation in order to study the contribution of bone marrow-derived cells to vascularization in an orthotopic glioma model.

Key words Bone marrow transplantation, Bone marrow chimera, Reporter genes, Angiogenesis

1 Introduction

During embryonic development mesodermal cells differentiate to form hemangioblasts, a common precursor of hematopoietic stem cells and endothelial progenitor cells (also denominated angioblasts). The vascular plexus arises from the angioblasts. This process is known as vasculogenesis [1]. In adult organisms new blood vessels arise mainly from sprouting of endothelial cells from the preexisting vessels, a process known as angiogenesis. For many years, it was assumed that postnatal neovascularization derives exclusively from the angiogenic process. In 1997 Asahara et al. [2] described for the first time the isolation of putative precursor cells from the peripheral blood with capacity to differentiate into endothelial cells in vitro and in vivo. This finding challenged the paradigm that vasculogenesis is restricted to embryogenesis. This first description began to direct the attention to bone marrow-derived endothelial progenitors as a source of vascular cells for cell-based

Richard Milner (ed.), *Cerebral Angiogenesis: Methods and Protocols*, Methods in Molecular Biology, vol. 1135,
DOI 10.1007/978-1-4939-0320-7_23, © Springer Science+Business Media New York 2014

anti- and proangiogenic therapies [3]. Shortly thereafter, the field of endothelial progenitor cells generated great interest among the scientific community. Since there was no marker available to phenotypically characterize the subpopulation of hematopoietic stem cells with endothelial progenitor ability, murine bone marrow studies have been used to study the process of postnatal vasculogenesis [4, 5].

Over the past years there has been growing discussion in the literature regarding endothelial progenitor cells [6, 7]. Several reports described the ability of endothelial progenitor cells to form blood vessels in vitro and in vivo. Conversely, growing experimental data suggested that the contribution of bone marrow cells to the vascular wall is minor, if any [8, 9]. However, now it has become more widely accepted that bone marrow-derived cells contribute to angiogenesis by a paracrine mechanism, namely, the secretion of proangiogenic molecules such as VEGF and metalloproteases, but not via transdifferentiation into endothelial cells and their subsequent integration into the vessel wall [10]. Studies of bone marrow cells lacking or overexpressing such molecules provide a tool to better understand the interaction of bone marrow cells and vascular growth in different physiological and pathological settings.

Bone marrow transplantation can be classified into short-term and long-term reconstitution. Short-term reconstitution occurs through myeloid and lymphoid precursors within 6–8 weeks after transplantation. The long-term repopulation is derived from true pluripotent hematopoietic stem cells. In short-term transplant models, rescue of the bone marrow might be due to committed hematopoietic progenitors, which do not have the ability to transdifferentiate into other cell lineages. In long-term bone marrow chimeras (>3 months), the reconstitution of the bone marrow by hematopoietic stem donor cells enables a durable and stable hematopoiesis for life [11].

In this chapter, we discuss the basic protocols used to establish bone marrow chimeras in mice and the subsequently morphological studies of the vascular network in a cerebral angiogenesis model (orthotopic glioma) (Fig. 1).

2 Materials

2.1 Mice

We used donor strain mice that carry ubiquitous expression of a reporter gene [the most useful for confocal immunofluorescence is green fluorescent protein (GFP) (C57Bl6-Tg (AVCTB-EGFP) 1 Osb/J)] for tracking bone marrow cells in cerebral tissue. We have also used mice with LacZ as reporter gene (Tg(Rosa 26)26 Sor) and mice with the expression of a reporter gene (GFP or LacZ) under the control of a tissue-specific promoter (FVB/N-Tg

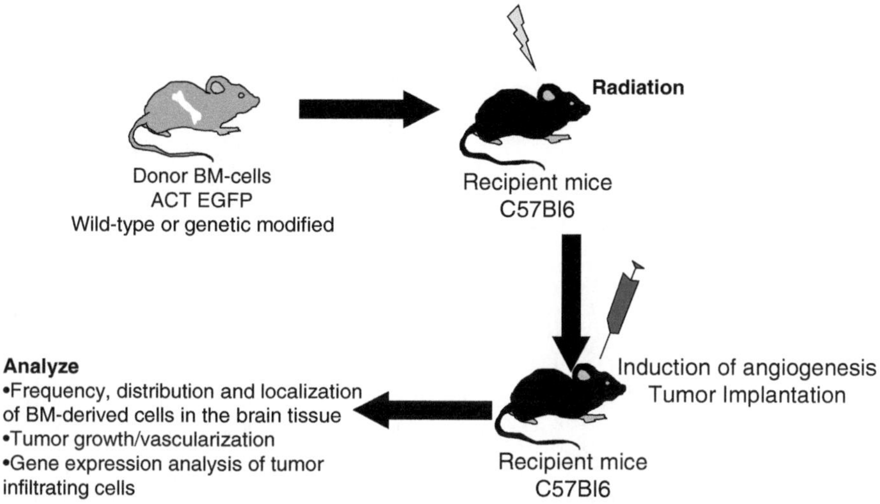

Fig. 1 Experimental scheme. Recipient mice are lethally irradiated and transplanted with unfractionated bone marrow cells from transgenic mice expressing ubiquitous GFP. After 6 weeks (short-term transplantation) mice are injected with syngeneic glioma cells to induce cerebral angiogenesis in a tumor model. This is followed by analysis of cell recruitment, vessel morphology, gene expression, and tumor growth and survival

(TIE2-lacZ)182Sato/J or Flk1-LacZ transgenic mice). All these mouse strains can be obtained from Jackson Laboratories. Mice should be 6–8 weeks old since younger mice have a higher number of hematopoietic stem cells.

2.2 Preparation of Bone Marrow Cells

1. PBS.
2. Dissection tools.
3. Petri dishes.
4. 22-gauge needle.
5. Sterile pads.
6. 50-mL conical tubes.
7. Cell counter or hemocytometer.
8. Microscope.
9. Red blood cell lysis buffer (eBioscience).
10. 40-μm cell strainer.

2.3 Irradiation Procedure

1. Irradiation box (Fig. 2).
2. Cobalt irradiation source.

2.4 Bone Marrow Reconstitution

1. 25-gauge needle.
2. Heat source.
3. Tube mice holder for intravenous injection.
4. Rat anti-mouse CD45 PE-labelled antibody (eBioscience or BD PharMingen).

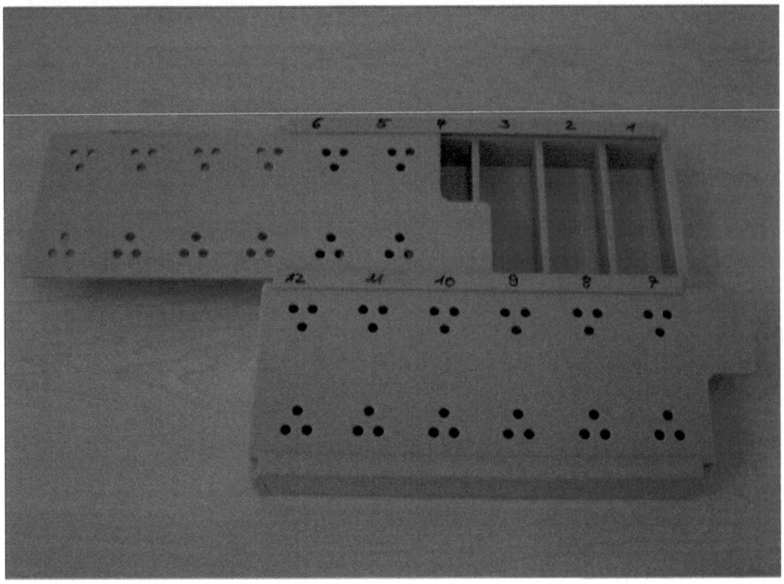

Fig. 2 Rectangular irradiation box. This box holds 12 mice for whole-body irradiation

5. FACS Calibur (Becton Dickinson).
6. Gentamicin sulfate (Refobacin).

2.5 Tumor Cell Preparation

1. DMEM with 10 % fetal bovine serum.
2. 0.02 % trypsin-EDTA solution.
3. DMEM.

2.6 Stereotactic Procedures

1. Small Animal Stereotaxic Instrument.
2. Mouse adaptor for inhalation narcosis (Harvard apparatus).
3. Isoflurane.
4. Scissors and forceps.
5. Monofilament 4-0 nylon.
6. Cotton bud.
7. 70 % ETHOH.
8. Saline.
9. Eye ointment.
10. Drill and drill bit.
11. Hamilton syringe.

2.7 Terminal Perfusion and Fixation

1. Pentobarbital.
2. Ketamine.

3. 4 % paraformaldehyde (PFA) dissolved in PBS. Dissolve 4 g of paraformaldehyde in 90 mL double distilled water. (Warning: work in the fume hood. PFA is toxic and should not be inhaled.) Heat the solution to ~60 °C. Add 2 N NaOH until solution becomes clear. Add 10 mL 10× PBS. Adjust pH to 7.4 with 1 N HCL.
4. Perfusion device.
5. Surgical instruments for mice perfusion and brain removal.

2.8 Vibratome Slicing

1. 5 % low melting agarose dissolved in PBS. Weigh 0.5 g low melting agarose. Add 10 mL PBS. Heat the solution to ~60 °C. Keep it at 4 °C for storage.
2. Cryomolds.
3. Leica Vibratome microtome.
4. Mounting medium Evanol: add 2.4 g of Mowiol to 6 g glycerol and mix well. Add 6 mL of H_2O. Stir for several hours at room temperature. Add 12 mL of 0.2 M Tris-Cl (pH 8.5) and heat to 53 °C until the solution has been dissolved. Do not heat above 60 °C. Centrifuge the mixture at 5,000 × g for 20 min. For fluorescence detection, add 2.5 % DABCO. Aliquot and store at −20 °C.

2.9 Free-Floating Immunofluorescence

All primary antibodies are diluted in 1 % bovine serum albumin (BSA) in PBS containing 0.1 % Triton.

2.9.1 Equipment

1. 24-well dishes.
2. Mounting medium Evanol.

2.9.2 Antibodies

1. Specific markers of endothelial cells [12]:
   - von Willebrand factor (vWF), localization: Cytoplasmatic rabbit antihuman vWF (Dako). Dilution, 1:200.
   - CD31 (PECAM), localization: Interendothelial cell contacts rat anti-mouse CD31 (BD Biosciences). Dilution, 1:100.
   - CD144 (VE-cadherin), localization: Interendothelial cell contacts, rat anti-mouse CD144 (BD Biosciences). Dilution, 1:100.
2. Markers of pericytes [13]:
   - Desmin (mouse anti-desmin, Dako Cytomation). Dilution, 1:100.
   - PDGFR-β (rat anti-mouse CD104b, eBioscience). Dilution, 1:200.
   - NG2 (rabbit anti-NG2 chondroitin sulfate proteoglycan, Millipore). Dilution, 1:200 [14, 15]. *See* **Note 7**.

3. Specific markers of smooth muscle cells:
   - Alpha-smooth muscle actin (mouse anti-SMA Sigma). Dilution, 1:100.
4. Other markers to phenotypically characterize the GFP-expressing bone marrow cells:
   - CD45 (pan leukocyte marker) rat anti-mouse CD45 (Southern Biotech). Dilution, 1:100.
   - F4/80 (microglia, macrophage marker) rat anti-mouse F4/80 (Serotec). Dilution, 1:200.
5. Secondary fluorescent-labelled antibodies diluted in 1 % BSA in PBS containing 0.1 % Triton.

3 Methods

3.1 Bone Marrow Transplantation (BMT)

3.1.1 Mice

For donor mice, use C57Bl6-Tg (AVCTB-EGFP) 1 Osb/J, which carries the enhanced green fluorescence protein (EGFP) driven by the universal chicken β-actin promoter and cytomegalovirus intermediate enhancer.

Use a syngeneic or congenic bone marrow transplantation approach. The term congenic refers to the strains, which differ genotypically only in one locus of particular genetic region. Syngeneic animals are genetically identical. In syngeneic or congenic transplants, significant immune rejection is not expected. Use C57Bl/6J mice (age 8–12 weeks) as recipient. AVCTB-EGFP mice can be backcrossed with other transgenic mice bearing specific gene deletions to study the roles of specific target genes in different cellular process in vivo involving bone marrow cells [16] (*see* **Note 1**).

3.1.2 Lethal Irradiation

Successful engraftment of the donor bone marrow requires depletion of host progenitor cells. For this purpose we use whole-body gamma irradiation using a cobalt source. We irradiate the mice in a self-constructed box in which the mouse is not able to turn, avoiding the need of anesthesia, which renders minimal distress to the animals (Fig. 2). This chamber can be used to irradiate 12 mice at once. The time of irradiation depends on the radioisotope decay. We used a single dose of 9 Gy. However, radiosensitivity is dependent on the background strain (Balb/c mice are more susceptive to irradiation) [17] (*see* **Note 2**).

3.1.3 Bone Marrow Cell Preparation

The femurs, tibias, and fibulas retrieved from one adult donor animal will yield around 5×10^7 total bone marrow cells. For one recipient mouse, at least 5×10^6 donor bone marrow cells should be injected.

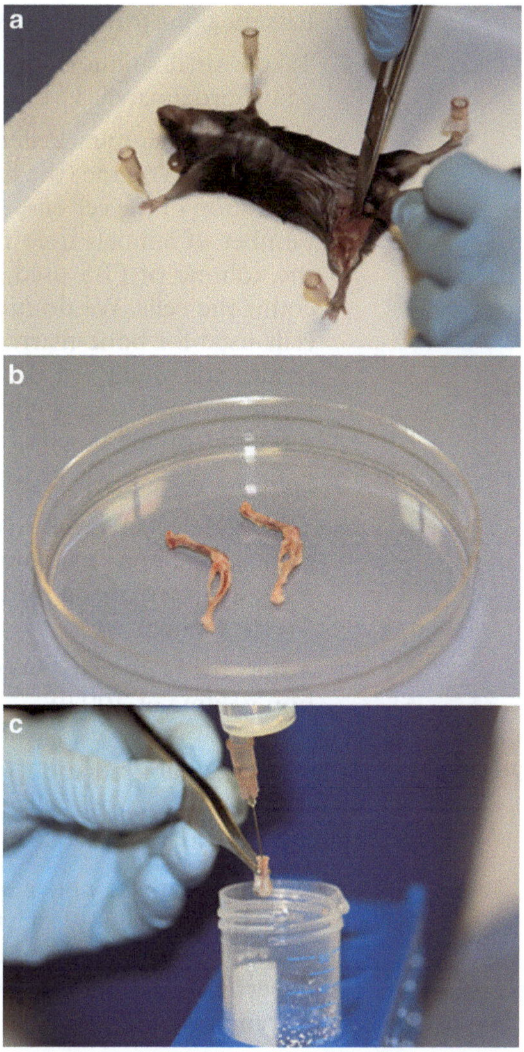

Fig. 3 Preparation of the bone marrow cells for BMT. (**a**) Remove the femur, fibulas, and tibias from donor mice. (**b**) Bones with soft tissues removed. (**c**) Flushing the bone marrow from the bones

1. Euthanize the donor animal.
2. Disinfect the mice with 70 % ethanol.
3. Remove the femurs, fibulas, and tibias using sharp dissection. If necessary, remove the pelvic bones. This will increase the number of bone marrow cells (Fig. 3a).
4. Remove as much muscle and connective tissue from the bones as possible, so that the bone marrow can be easily identified (Fig. 3b).
5. Clip the distal ends of each bone. Using a 22-G needle, flush the bone marrow with PBS from both ends (Fig. 3c).

6. Pass the bone marrow cell suspension through a 40-μm strainer.
7. Take a small aliquot (approx 500 μL) of cell suspension and add approximately 1 volume of red blood cell lysis buffer.
8. Incubate for 5 min. Pellet the cells and resuspend them in an appropriate volume (the final volume is dependent on the concentration of the cell suspension, which in turn depends on the number of animals used for extracting the bone marrow and the volume of PBS used to flush the bones completely) and count the cells. We do not perform red blood cell lysis in the cells used for bone marrow reconstitution because this might reduce the vitality of the cells. If a cell counter is used, red blood cell lysis might not be necessary since the erythrocytes can be identified based of their small size.
9. Spin the bone marrow cells for 10 min at 1,200 rpm.
10. The number of bone marrow cells required to successfully reconstitute bone marrow of recipient mice varies from 5×10^6 to 2×10^7 marrow-derived cells per recipient. We use routinely 5×10^6/animal.
11. Adjust the bone marrow cell density to have a maximum injection volume of 200 μL.

3.1.4 Bone Marrow Cell Injection

1 day after lethal irradiation, recipient mice should receive 5×10^6 unsorted bone marrow cells intravenously. The injection volume should not be larger than 200 μL.

1. Warm the mouse under a heat lamp but be sure not to overheat the animal (39 °C for no longer than 10 min). Mice should be monitored and checked for signs of dehydration and hyperthermia (rapid respiration, decreased activity, and red extremities).
2. Mice tails have two lateral veins and a single artery in the ventral side and a dorsal vein in the dorsal part of the tail. Place the animal in a holder device and stretch the tail with the hand that will not be holding the syringe.
3. Using 25-gauge needles, inject at the distal part of the lateral tail vein. Insert the need in flat angle, almost parallel to the tail. Do not aspirate. Inject the fluid slowly (Fig. 4).
4. Remove the needle and apply slight pressure to the puncture site until bleeding stops. If the injection is successful, the vein will blanch and there is no resistance. If swelling occurs at the site the injection or resistance is felt upon injection, remove the needle and reinsert above the insertion site. Puncture of the lateral vein is more difficult in black mice (*see* **Note 3**).

3.1.5 Housing of Mice After BMT

1. After BMT, mice should be kept in a ventilated-rack cage system. Gentamicin sulfate (Refobacin) should be given in the drinking water (0.2 mg/mL) to prevent infection.

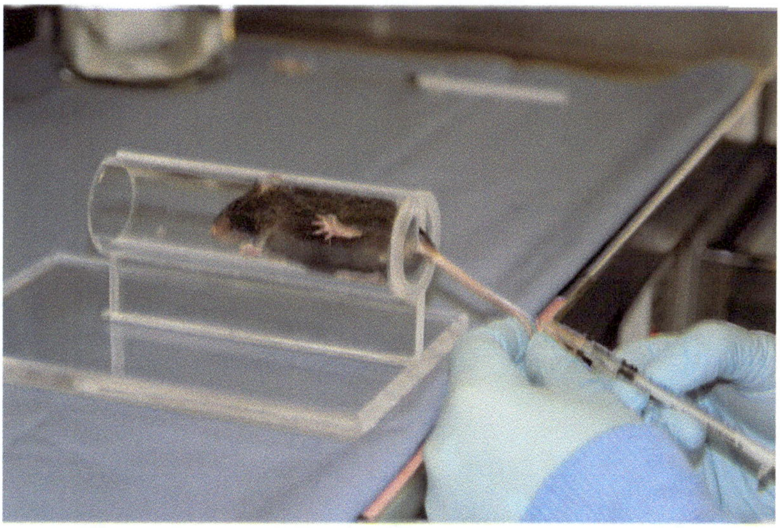

Fig. 4 Injection of the bone marrow cell suspension into the mouse lateral vein

2. If reconstitution of bone marrow with donor cells fails, mice will usually became ill and die in the first 2 weeks after irradiation. Animals should be monitored daily. Severely affected animals display weight loss, rough coat, and inactivity. They should be humanely euthanized.

3.1.6 Analysis of BM Reconstitution

6–8 weeks after grafting, reconstitution should be assessed by FACS analysis of peripheral blood.

1. Collect 100 μL peripheral blood from the tail vein and place in a tube which contains heparin (1 U/100 μL blood). Mix completely. Do not use citrate as this will adversely affect the labelled antibodies.
2. Stain the blood cells 1 μL with phycoerythrin (PE)-labelled CD45 or other hematopoietic lineage-specific antibodies. Incubate for 30 min on ice. After lysis of erythrocytes and washings, cell suspensions should be analyzed on a FACS Calibur. Analyze the lysed cells for GFP and PE (*see* **Note 4**).

3.2 Orthotopic Tumor Implantation

Implantation of GL261 cells is a syngeneic glioma model in C57Bl6. Because this model does not require an immunodeficient host, it resembles most closely the human disease.

3.2.1 Tumor Cell Preparation

The GL261 cell line is cultivated in DMEM containing 10 % serum.

1. Trypsinize GL261 cells and centrifuge at 800 rpm for 5 min.
2. Resuspend the cells in serum-free DMEM and count the cells; 10^5 cells should be adjusted to a volume 4–7 μL.

3.2.2 Stereotactic Tumor Implantation

1. Place the animals in the induction chamber and induce anesthesia (set the vaporizer to 4 % isoflurane).
2. Shave the head and immobilize the mouse in the stereotactic apparatus, first in the nose cone, and then fix the animal through the ear bars. Reduce the anesthesia to 2 % isoflurane. Cover the eyes with eye ointment in order to prevent corneal dehydration (isoflurane inhibits the blinking reflex). Inject 0.1 mg/kg buprenorphine subcutaneously.
3. Wipe the mouse's head with a sterile cotton swab dipped in 70 % ETHOH. Make a 1.5-cm sagittal incision in the mouse's scalp (begin the incision behind the eyes). Use a scissor to make the incision. The use of a scalpel can displace the mice from the device. Use a cotton swab to expose the skull surface. Localize the bregma and lambda.
4. Carefully tap the cell suspension. Avoid cell clumps. Draw cells up into the Hamilton syringe. Place the Hamilton syringe in the stereotactic device. Drill a 2-mm burr hole 2.2 mm laterally and 5.2 mm anterior of the lambda suture.
5. Introduce the needle to a deep of 4 mm and then retract to 3 mm to make a pocket where the cells can be injected. Slowly inject the cells over at least 3 min. Wait a further 5 min and slowly retrieve the needle over a period of at least 5 min (*see* **Note 5**). Apply bone wax to cover the burr hole and suture the scalp.
6. Animals should be monitored daily. Injection and growth of tumor cells can easily be evaluated by bioluminescence monitoring (*see* **Note 6**). Mice will develop sites of intracranial tumor around day 21 postimplantation. Typical symptoms are inactivity, weight loss, and rough coat.

3.3 Animal Perfusion and Slice Preparation

3.3.1 Perfusion

The usual method of whole-body perfusion is via the transcardial route. This is achieved through the infusion of PBS (to wash out the blood) followed by injection of 4 % PFA in PBS into the left ventricle (*see* **Note 7**).

1. Animals should be euthanized by an overdose of pentobarbital and ketamine. When the mice show no noxious reflex, secure the mice lying on their back on a Styrofoam board with small needles. Work in the fume hood.
2. Open the abdomen, open the diaphragm, and visualize the heart. Insert a 24-G needle attached to the infusion system into the left ventricle. Immediately cut the right atrium. Slowly perfuse the body with PBS until the fluid exiting the right atrium is clear. Switch to 4 % PFA. We inject PBS and fixative solution under gravity for 10–15 min (around 20 mL PFA).
3. Decapitate the animal and remove the brain.
4. Place the brains in 4 % PFA and gently rock overnight at 4 °C.
5. Wash in PBS 5 min, three times.

3.3.2 Slice Preparation

1. Embed the brains to perform coronal sections in 5 % low melting agarose in cryomolds. Keep cryomolds cold on ice.
2. Take the agarose block out of the cryomold. Cut out the excess of agarose. The pieces of agarose can be reused (microwave the pieces, but do not boil them).
3. Fill the tray with ice-cold PBS. Fix the agarose block with the brain in the specimen disc with superglue, in order to obtain coronal sections.
4. Cut at lower speed (between 4 and 6) and high vibration (7–8) sections of 50-μm thickness.
5. Collect the sections in PBS in 24-well plates.

3.4 Free-Floating Immunofluorescence

1. Select the slices to be stained and place them in a 24-well plate. Incubate overnight in PBS containing 1 % BSA with 0.3 % Triton (volume should be at least 250 μL/well to ensure floating of the slice).
2. Incubate the slices with the primary antibody in PBS with 1 % BSA and 0.1 % Triton overnight. To improve the penetration of the antibody, microwave the slices in ice for 15 min at 90 W (*see* **Notes 8** and **9**).
3. Wash the slices three times in PBS with 0.3 % Triton for 30 min each wash.
4. Incubate with the appropriate secondary antibody overnight. After three washes in PBS, wash in water and then mount the slices on a glass slide using a fine paintbrush. Cover with a cover slip and Evanol.
5. Analyze the sections using confocal laser microscopy (Fig. 5).

4 Notes

1. We have also performed allogenic BMT in genetically different mice strains (donor B6/129 in recipient FVB, donor FVB in C57/Bl6). When we used allogenic transplantation, we observed in a few animals graft-versus-host disease (GVHD) in general, several months after BMT. In contrast, we did not observe graft-versus-host reaction in congenic bone marrow transplantation (using Bl6 as donor and recipient).
2. Irradiation of the central nervous system (CNS) induces several changes in cerebral tissue inducing the upregulation of several chemokines with myeloattractant activity. Therefore, some protocols advocate the use of head protection, which allows irradiation excluding the brain [18].
3. Optimal technical procedures are crucial for successful donor's engraftment. Increased recipient mortality mostly results from

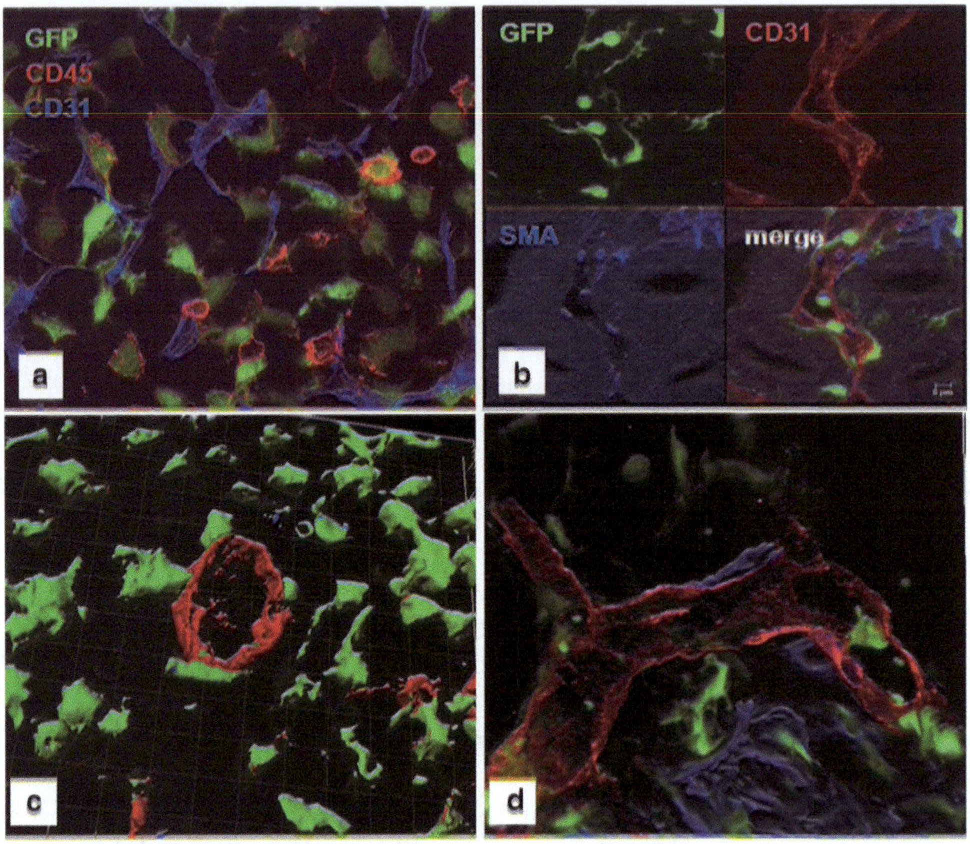

Fig. 5 Representative fluorescent staining and confocal analysis of brain vessels. (**a**) GFP+ derived bone marrow cells stained for CD45 in tumor tissue. Endothelial cells were stained for CD31. (**b**) GFP+ bone marrow cells in relation to a vessel stained with CD31 (endothelial cells) and SMA (mural cells). (**c, d**) After Z-stack acquisition, files were used for 3-D reconstruction using Imaris software. Staining of blood vessels in an intracerebral glioma using VE-cadherin (*red*). Bone marrow-derived cells are GFP positive (*green*)

error during the injection of the bone marrow cells; therefore, this procedure must be performed by trained personnel. In some cases we successfully transplanted recipient mice by injection of 10^7 cells intraperitoneally. Injection in the tail of BL6 is more difficult than in nude or FVB mice (because of the pigmentation of the tail). Be careful not to inject into the tail artery. This might cause sudden death of the animal.

4. When we euthanize our bone marrow chimeric mice, before perfusion with 4 % PFA, we also take the bone marrow to evaluate bone marrow reconstitution by FACS analysis using the following antibodies: cKit, Mac-1, B220, and GR-1.

5. For successful intracerebral tumor implantation, it is necessary to slowly inject the cells in a small volume. After injection it is necessary to wait at least 5 min before slowly retrieving the Hamilton needle.

6. Appropriate injection of tumor cells can be easily evaluated by bioluminescence monitoring. For this purpose a commercial available GL261-Luc 2 BiowareUltra (Perkin Elmer) is available.
7. Fixation of the tissues by PFA perfusion in vivo enhanced the detection of GFP.
8. For free-floating immunofluorescence of brain sections with intracranial tumors, we mostly used PDGFR-β as a marker for pericytes, since NG2 is also expressed in GL261 cells. To avoid nonspecific staining when using mouse antibodies in mouse tissue, monoclonal mouse antibodies should be labelled with Alexa fluor antibody labelling kit (Life Technologies).
9. When we used β-galactosidase as a reporter gene, we never succeeded in performing immunofluorescence for colocalization studies. At least in our hands, anti-β-galactosidase antibodies do not work well.

Acknowledgments

Our work is supported by the Deutsche Krebshilfe (Project Number 109410) and Deutsche Forschungsgemeinschaft (SFB/TR23 "Vascular differentiation and remodelling," Project C4).

References

1. Choi K, Kennedy M, Kazarov A, Papadimitriou JC, Keller G (1998) A common precursor for hematopoietic and endothelial cells. Development 125:725–732
2. Asahara T, Murohara T, Sullivan A, Silver M, van der Zee R, Li T, Witzenbichler B, Schatteman G, Isner JM (1997) Isolation of putative progenitor endothelial cells for angiogenesis. Science 275:964–967
3. Schatteman GC, Dunnwald M, Jiao C (2007) Biology of bone marrow-derived endothelial cell precursors. Am J Physiol Heart Circ Physiol 292:H1–H18
4. Asahara T, Masuda H, Takahashi T, Kalka C, Pastore C, Silver M, Kearne M, Magner M, Isner JM (1999) Bone marrow origin of endothelial progenitor cells responsible for postnatal vasculogenesis in physiological and pathological neovascularization. Circ Res 85:221–228
5. Murayama T, Asahara T (2002) Bone marrow-derived endothelial progenitor cells for vascular regeneration. Curr Opin Mol Ther 4:395–402
6. Urbich C, Dimmeler S (2004) Endothelial progenitor cells functional characterization. Trends Cardiovasc Med 14:318–322
7. Urbich C, Dimmeler S (2004) Endothelial progenitor cells, characterization and role in vascular biology. Circ Res 95:343–353
8. Machein MR, Renninger S, Lima-Hahn E, Plate KH (2003) Minor contribution of bone marrow-derived endothelial progenitors to the vascularization of murine gliomas. Brain Pathol 13:582–597
9. De Palma M, Venneri MA, Roca C, Naldini L (2003) Targeting exogenous genes to tumor angiogenesis by transplantation of genetically modified hematopoietic stem cells. Nat Med 9:789–795
10. Kozin SV, Duda DG, Munn LL, Jain RK (2011) Is vasculogenesis crucial for the regrowth of irradiated tumours? Nat Rev Cancer 11:532
11. Zhong JF, Zhao Y, Sutton S, Su A, Zhan Y, Zhu L, Yan C, Gallaher T, Johnston PB, Anderson WF, Cooke MP (2005) Gene expression profile of murine long-term reconstituting vs. short-term reconstituting hematopoietic stem cells. Proc Natl Acad Sci U S A 102:2448–2453
12. Machein MR, Knedla A, Knoth R, Wagner S, Neuschl E, Plate KH (2004) Angiopoietin-1 promotes tumor angiogenesis in a rat glioma model. Am J Pathol 165:1557–1570

13. Nasarre P, Thomas M, Kruse K, Helfrich I, Wolter V, Deppermann C, Schadendorf D, Thurston G, Fiedler U, Augustin HG (2009) Host-derived angiopoietin-2 affects early stages of tumor development and vessel maturation but is dispensable for later stages of tumor growth. Cancer Res 69:1324–1333
14. Ozerdem U, Alitalo K, Salven P, Li A (2005) Contribution of bone marrow-derived pericyte precursor cells to corneal vasculogenesis. Invest Ophthalmol Vis Sci 46:3502–3506
15. Rajantie I, Ilmonen M, Alminaite A, Ozerdem U, Alitalo K, Salven P (2004) Adult bone marrow-derived cells recruited during angiogenesis comprise precursors for periendothelial vascular mural cells. Blood 104:2084–2086
16. Kerber M, Reiss Y, Wickersheim A, Jugold M, Kiessling F, Heil M, Tchaikovski V, Waltenberger J, Shibuya M, Plate KH, Machein MR (2008) Flt-1 signaling in macrophages promotes glioma growth in vivo. Cancer Res 68:7342–7351
17. Duran-Struuck R, Dysko RC (2009) Principles of bone marrow transplantation (BMT), providing optimal veterinary and husbandry care to irradiated mice in BMT studies. J Am Assoc Lab Anim Sci 48:11–22
18. Mildner A, Schmidt H, Nitsche M, Merkler D, Hanisch UK, Mack M, Heikenwalder M, Bruck W, Priller J, Prinz M (2007) Microglia in the adult brain arise from Ly-6ChiCCR2+ monocytes only under defined host conditions. Nat Neurosci 10:1544–1553

Chapter 24

Novel Methods for Accurate Identification, Isolation, and Genomic Analysis of Symptomatic Microenvironments in Atherosclerotic Arteries

Mark Slevin, Maribel Baldellou, Elspeth Hill, Yvonne Alexander, Garry McDowell, Christopher Murgatroyd, Michael Carroll, Hans Degens, Jerzy Krupinski, Norma Rovira, Mohammad Chowdhury, Ferdinand Serracino-Inglott, and Lina Badimon

Abstract

A challenge facing surgeons is identification and selection of patients for carotid endarterectomy or coronary artery bypass/surgical intervention. While some patients with atherosclerosis develop unstable plaques liable to undergo thrombosis, others form more stable plaques and are asymptomatic. Identification of the cellular signaling mechanisms associated with production of the inflammatory, hemorrhagic lesions of mature heterogenic plaques will help significantly in our understanding of the differences in microenvironment associated with development of regions susceptible to rupture and thrombosis and may help to predict the risk of plaque rupture and guide surgical intervention to patients who will most benefit. Here, we demonstrate detailed and novel methodologies for successful and, more importantly, accurate and reproducible extraction, sampling, and analysis of micro-regions in stable and unstable coronary/carotid arteries. This information can be applied to samples from other origins and so should be useful for scientists working with micro-isolation techniques in all fields of biomedical science.

Key words Laser capture, Atherosclerotic plaques, Microfluidity cards, PCR

1 Introduction

Vascular/inflammatory-based diseases are an important and frequent cause of death and disability in the world today. For example, stroke and heart disease account for approximately 11 % of all deaths in the European Union (EU) [1]. Those who survive often have significant long-term morbidity. In the EU, cardio-/cerebrovascular disease (CVD) is the number one cause of death, and the total number of stroke deaths is estimated at 1.24 million/annum (with 16 million first-ever strokes/annum worldwide) [2]. 49 % percent of all European deaths are caused by CVD, and 2.76

Richard Milner (ed.), *Cerebral Angiogenesis: Methods and Protocols*, Methods in Molecular Biology, vol. 1135, DOI 10.1007/978-1-4939-0320-7_24,

million of these are from ischemic heart disease. The EU economic burden of stroke is estimated at approximately £169 billion annually, both through direct costs to the NHS and through disability, and this accounts for approximately 4 % of the gross national income [1]. Therefore, new technologies able to provide deeper insights into the mechanisms through which the process of atherosclerosis occurs, leading to unstable plaque formation and thrombosis, are vital in order to identify new therapeutic targets to slow down or stop progression of this disease.

If we could determine which patients with atherosclerosis are most at risk of stroke or myocardial infarction, we could not only begin to develop targeted therapy but also target intervention to those who will benefit from it. For example, it is recognized that neovessel formation and associated inflammation play an important role in plaque stability [3]. However, at present, the precise molecular mechanisms leading to induction and proliferation of these processes within actively developing plaques are not fully understood, nor how exactly this affects plaque stability. Hence, detailed study of the microenvironments of unstable plaques and of susceptible regions at different stages of development in the appropriate animal models, for instance, following isolation using laser-capture microdissection (LCM), could reveal the key switches from which we could formulate anti-atheroma therapies.

Such technology is an implicit requirement since we know that regions of stable and unstable microvessels coexist within a single carotid plaque [4]. Vulnerable plaques contain areas of instability and are microvessel-rich, associated with inflammatory infiltration, which are structurally weak and poorly invested with smooth muscle cells. These are therefore leaky and prone to hemorrhage [5]. Adjacent areas may appear more stable and contain patent functional vessels with pericyte coatings and much lower levels of inflammation [6]. Hence, it is important to understand why these two local regions have developed differently. Therefore, a detailed analysis of the mechanisms of creation of these potentially thrombotic "hot spots" is essential. An important question to answer is how does the plaque microenvironment contribute to the development of unstable intimal lesions prone to thrombosis?

The technology described in this paper is potentially very exciting. It has opened up the possibility of looking at arterial and plaque tissue more closely, examining specific microenvironments within, and isolating individual tissue components and cells for transcriptomal analysis. This strategy involving selection of areas of interest from complex tissue using LCM has been in use now for over a decade for the side-by-side comparison of nano-quantities of cells in a variety of diseases. During this time, technological developments have been abundant and have reduced some of the limitations clearly apparent in early versions of the systems. However, step-by-step analysis of this technology reveals that the production of accurate, reproducible, and meaningful data from LCM-dissected samples is not as simple as was first anticipated. Therefore, here we have established an optimal series of protocols that have enabled us to successfully compare paired plaque/adventitial vascular microenvironments for the first time producing reliable and verifiable data.

2 Materials

2.1 Immunohistochemistry and Analysis of Target Areas

1. SuperFrost® slides.
2. Membrane slides (Leica Microsystems, Milton Keynes, UK).
3. ABC kit (Vector Laboratories, Peterborough, UK).
4. Acustain Harris Hematoxylin.
5. 30, 50, 75, and 100 % ethanol.
6. Xylene.
7. VectaMount permanent mounting media (Vector Laboratories, California, USA) (*see* **Notes 1** and **2**).

2.2 Software for Image Processing and Mapping

1. High-speed Marlin F-146C camera (Allied Vision Technologies, Stadtroda, Germany).
2. Mirax Histology System (Zeiss, Jena Germany/3DHistech, Budapest, Hungary).
3. 3DHistech's Mirax Viewer software.
4. Adobe Photoshop software.

2.3 LCM and cDNA Preparation Extraction and Analysis

1. RLT buffer (RNA lysis buffer supplied by Qiagen) and methodology/materials associated with RNA extraction as part of the Qiagen RNeasy kit for pico/nano cDNA extraction (Qiagen, Crawley, Sussex, UK).
2. Buffer RW (RNA wash buffer supplied by Qiagen); RNase-free water; buffer RPE (for washing membrane bound RNA-part of the Qiagen kit); collection tubes; RNeasy mini/micro spin columns.
3. Other requirements: β-mercaptoethanol, RNase-free pipette tips, microcentrifuge, 70 % ethanol, vortex mixer.
4. RNA picochips (Agilent 2100) and Bioanalyzer (e.g., Agilent Technologies 2100; Agilent, Berkshire, UK).
5. For processing, RNase Zap; nuclease-free water; RNA ladder and other supplied kit components from the nano-chip assay kit.
6. Applied Biosystems amplification kit (PN 4384557B).
7. PALM system (Zeiss) from PEN membrane slides.

2.4 Selection of Housekeeping Genes and Low-Density Microarray Analysis

1. TaqMan housekeeping gene expression arrays from Life Technologies containing all the components required to produce quantitative real-time PCR data.
2. 96-well plate format, endogenous controls.

For selection process *see* Subheading 3.7.

2.5 Materials for RNA-cDNA Pre-amplification

1. TaqMan PreAmp Master Mix Kit (PN 4384267) containing all components required including PreAmp Master Mix (2×).
2. Gene Expression Master Mix.
3. Optical 96-well reaction plates.

2.6 Microarray Analysis

1. TaqMan microfluidity card reagent kit and nuclease-free water.
2. High-capacity cDNA reverse transcription kit.
3. High-capacity RNA-cDNA Master Mix.
4. High-capacity RNA-cDNA kit.
5. TaqMan array custom microfluidity cards.
6. TaqMan gene set.
7. TaqMan gene signature array.

3 Methods

Methodological details of extraction of RNA, preparation of cDNA, pre-amplification, and microarray preparations have not been described in detail, as they are available in booklet form within the appropriate listed kits. All solutions should be prepared in RNase-free conditions at 4 °C and utilized in sterile RNase-free hoods.

3.1 Dissection of Tissue and Preparation of Sections

1. Surgically dissect appropriate material at sites of plaque development.
2. Dissect into fragments for OCT embedding in a sterile mRNase-free environment in a class II cabinet and keep the tissue on a bed of cardice.
3. Prepare cross-sectional pieces of tissue around 3–5 mm thick and embed directly in OCT.
4. Cryostat serial sections (5–9 μL thick), in RNase-free conditions, onto SuperFrost poly-l-lysine-coated slides (for IHC) and laser-capture (membrane-coated; PEN-polyethylene naphthalate) slides (Arcturus system or Leica laser-capture systems).
5. Place serial sections, alternatively, on normal and LCM slides to ensure minimal chances of changes in tissue morphology between sections examined by immunohistochemistry (IHC) and sampling of adjacent material (*see* **Note 3**).

3.2 Selection and Processing of Arterial Samples

1. Obtain the coronary/carotid artery samples from transplant patients immediately following the operation and snap freeze.
2. Select regions of tissue with the aim to compare plaque-bearing regions with evidence of instability (fragile microvessels/inflammation) with adjacent areas containing mature vessels, fibrotic tissue, and minimal inflammation to represent the control material.
3. Cut the carefully dissected segments of artery into 4–7 smaller pieces to aid in the identification of lesions for LCM catapulting, embed in OCT, and cut serial sections to a thickness of 6–8 μm and mount on slides (*see* Fig. 1).
4. Cut the arterial specimens in a cryostat in RNase-free conditions and mount on membrane slides for total extraction of LCM-captured regions or subsequent to adventitial/media or intimal dissection and RNA extraction, all under RNase-free conditions.

3.3 Identification of Microenvironments and Vascular Regions for Subsequent LCM

The aim here is to compare several protocols to visualize the important areas including rapid immunofluorescence, rapid UEA-1, rapid IHC, and analysis of unstained serial sections in relation to IHC stained adjacent ones (*see* **Notes 4** and **5**).

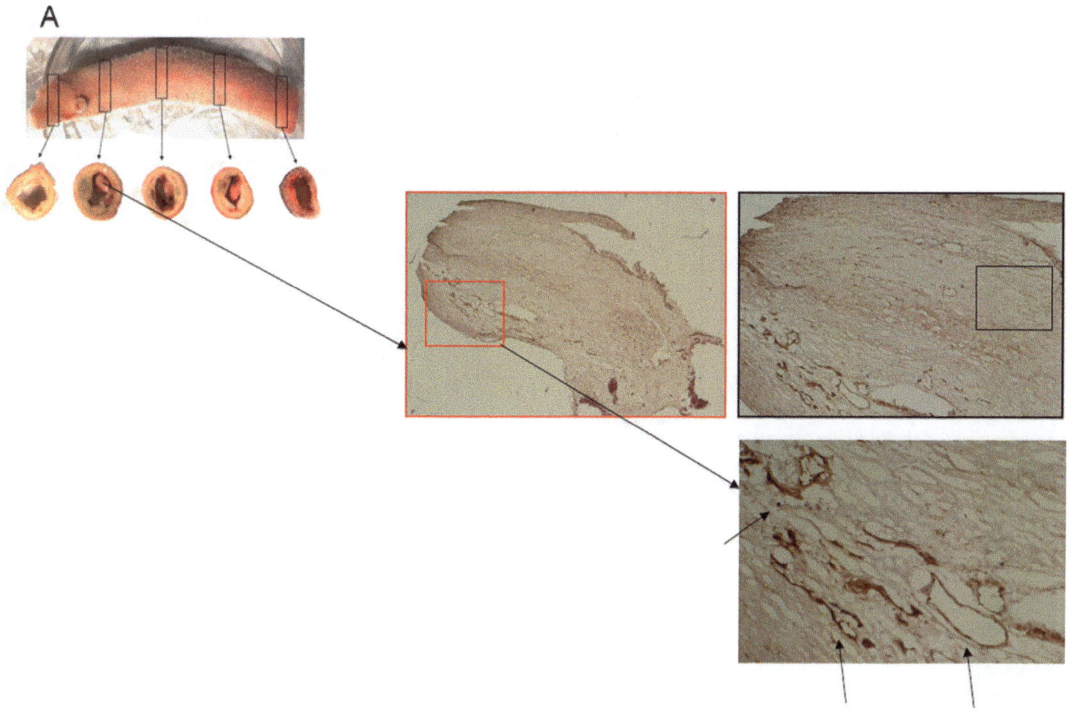

Fig. 1 Carefully dissected segments of artery cut into between 4 and 7 smaller pieces and mounted on SuperFrost® slides

1. Stain the frozen sections for CD31 and CD105 to identify active and quiescent microvessels, respectively, and H&E to study general morphology (Fig. 2).
2. Equilibrate the sections to 22 °C and fix in cold acetone. Other antibodies recognizing other plaque components can be used as alternatives to the ones we have applied in this example.
3. Wash the sections in phosphate-buffered saline (PBS), and apply 0.3 % peroxidase in methanol for 30 min to eliminate background DAB staining.
4. Apply 5 % BSA in PBS for 1 h to inhibit nonspecific antigen. Incubate with monoclonal mouse antihuman CD105 and mouse antihuman CD31 as primary antibodies to demonstrate the active and total microvessels, respectively.
5. Wash the sections in PBS and apply the secondary antibody (rabbit anti-mouse biotinylated) at a concentration of 0.5 %.
6. Wash the sections again in PBS and then apply the avidin-biotin-peroxidase complex.
7. Develop the color using 3,3′-diaminobenzidine (DAB) until color appears and then rinse.
8. Wash the sections in PBS and stain with Acustain Harris Hematoxylin (Sigma-Aldrich, Dorset, UK) and then dehydrate through 30, 50, 75, and 100 % ethanol and finally xylene.
9. Mount the sections using VectaMount permanent mounting media.
10. View the slides on a Leica light microscope and capture images of the sections.

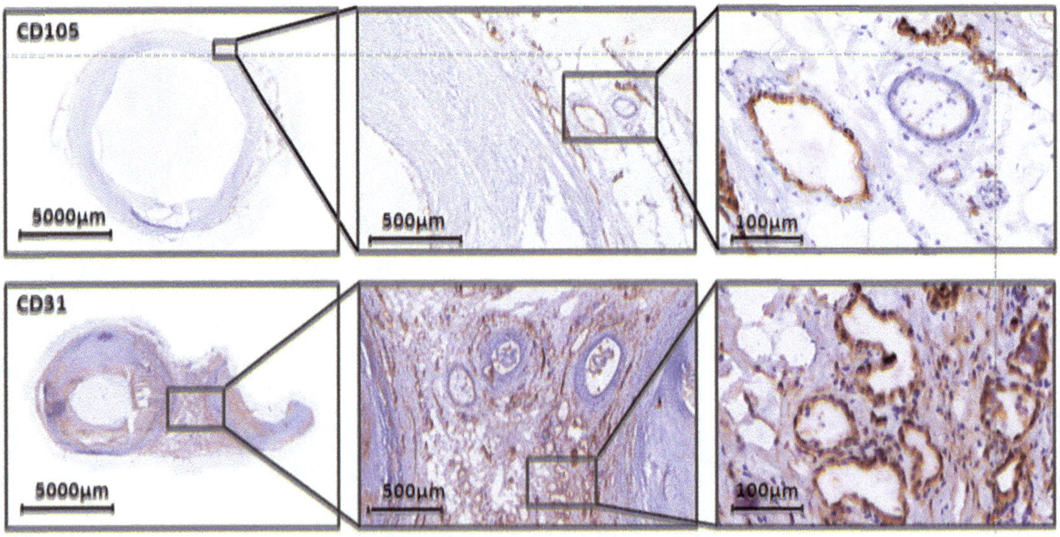

Fig. 2 Frozen sections of artery stained for CD31 and CD105. These markers identify active and quiescent microvessels. The H&E staining is carried out to study general morphology

3.4 Image Processing

Slide scanning is employed to obtain high-quality high-magnification images of adjacent, stained sections. Using the Mirax Histology System, scan the whole slides (200×) using a predefined threshold. This magnification equates to 0.23 μm/pixel. Briefly, the system utilizes tile-based stitching of the images. The system visualizes thousands of field of view (FOV) and seamlessly links them together.

1. Acquire images using a high-speed Marlin F-146C camera. Crop the image edges so as to exclude the spherical aberration of the lens. Each FOV is imaged in milliseconds and an average tissue section takes around 5–10 min to scan.
2. Visualize the resulting data via 3DHistech's Mirax Viewer software. Export the images of interest in JPEG format.
3. Create hybrid artificially colored maps (hybrid mapping) by imaging whole serial sections at 10× magnification and merged using Adobe Photoshop.
4. Overlay serial sections to create a hybrid of CD31 and CD105 staining, and artificially color to highlight the distinction between immature microvessel-rich/leaking vessels and mature microvessels in stable regions.
5. Using these maps as a guide, select serial-adjacent sections (previously dissected and divided into micro portions and mounted on PEN membrane slides) and subject to LCM the micro-areas of interest, e.g., vessels or areas surrounding groups of vessels (Fig. 3).

3.5 Laser Capture of Samples and Preparation Isolation of RNA/ cDNA

In order to accurately identify tissue components/microvessels in a specific localization in unstained serial sections, the previous stained section and the one cut directly after the one retained for LCM need to be compared directly to the section prepared for LCM (*see* **Note 6**).

1. Retrieve sections from the −80 °C freezer, and after defrosting, fix for 1 min in 100 % ethanol at 4 °C.
2. After air-drying in the sterile cabinet, mark the adventitia (or cut) with a fine needle into 5–8 equal segments making it easier to identify and localize the specific regions of interest using the LCM microscope.
3. Laser microdissect the samples (vascular-rich areas of either active-plaque associated vessels or non-active areas opposite to the plaque growth) and capture from 6 μm thick tissue slices using the PALM system (Zeiss) from PEN membrane slides (*see* **Note 5**).
4. Collect the samples in RLT buffer and keep at −80 °C until processing for RNA extraction.

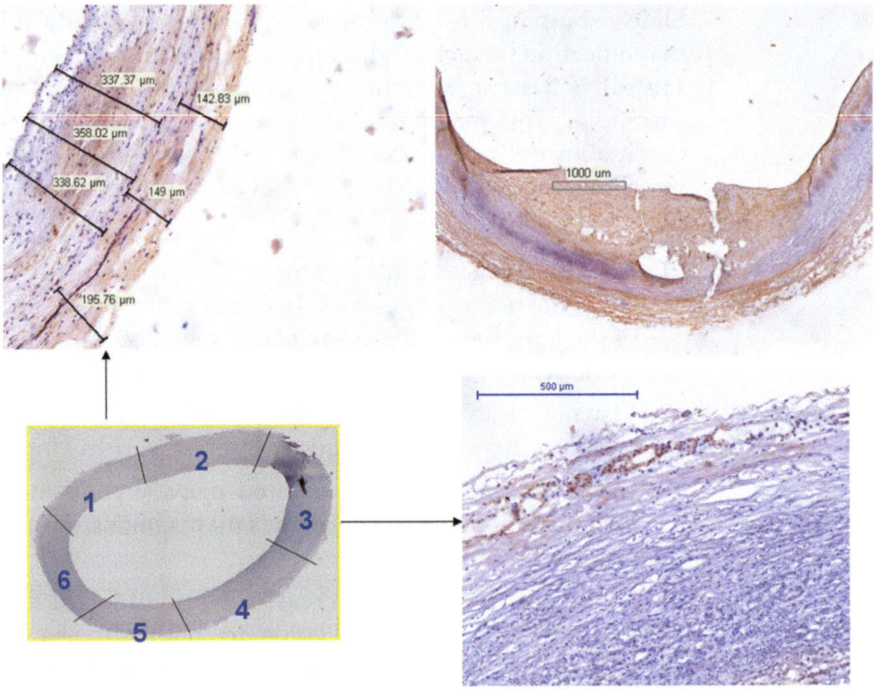

Fig. 3 Serial-adjacent sections, dissected and divided into micro portions, mounted on PEN membrane slides. These sections should be subsequently subjected to LCM and micro-areas of interest

3.6 RNA Extraction and Quantification

Perform preliminary studies in order to identify adventitial regions of potential interest using IHC and determine the overall adventitial RNA quality (methods shown above). Microdissect the chosen samples by LCM and extract the RNA using the Qiagen kit.

1. Homogenize the LCM-cut areas in 150 µL RLT buffer. Take four serial sections from each coronary artery specimen and mount on a single PEN membrane slide.
2. Pool the cut areas from two serial sections into the RLT buffer to provide sufficient RNA for analysis. Extract RNA using the RNeasy™ Micro Kit.
3. Quantify the RNA and assess for quality using RNA picochips and the Bioanalyzer 2100.
4. Test each sample for RNA quality and quantity in the picochip analyzer at least twice to avoid errors and possible false-negative results. Samples containing visible 18S and 28S peaks (or bands) and RIN values of greater than five should be considered for further investigation (Fig. 4).

3.7 Analysis and Selection of Housekeeping Genes

One of the major problems with micro-extraction and accurate analysis of gene deregulation, particularly in samples of arterial origin, is the heterogeneity of the tissue. Comparisons of gene expression throughout the different layers of the artery, within individual layers, and in particular in plaque tissue should be done

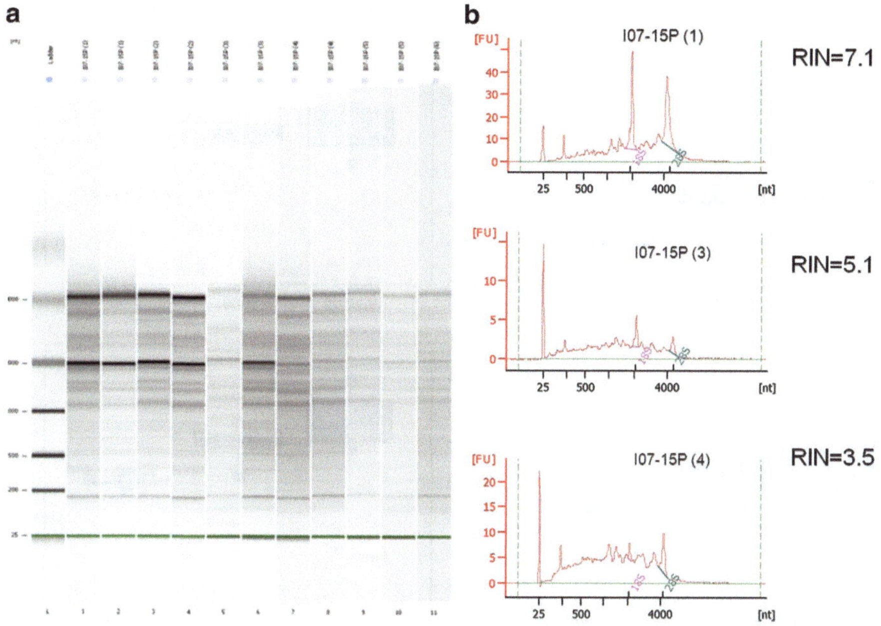

Fig. 4 RNA extracted from the adventitial regions of potential interest. Each sample of RNA was tested for quality and quantity in the picochip analyzer. Samples containing visible 18S and 28S peaks (or bands) (**a**) and RIN values (**b**) of greater than 5 should be considered for further investigation

carefully because some of the endogenous genes may show differences. For this particular study, the most important criteria were to find the best housekeeping genes stable when comparing microenvironmental regions of adventitial tissue in different grade lesions (*see* **Note 7**).

Here it is important to analyze an array of housekeeping genes and examine the following criteria in order to establish the best endogenous genes for a particular study:

1. Specificity: low detection limit (i.e., Ct < 32 as recommended for low-density arrays by Life Technologies).
2. Reproducibility: small standard deviation among replicates SD < 0.40 (recommended by Life Technologies).
3. Stability: similarity of threshold values (Cts) when the amount of starting material is the same. This can be tested with different programs (i.e., Data Assist from Life Technologies and GeNorm).
4. Sensitivity: capacity to detect small cDNA concentration differences in the total amount of starting total RNA.
5. Quality dependence: related to RNA quality (measured as RNA integrity number: RIN).
6. Consistency: are they affected by study conditions or factors to be analyzed (those will depend on every individual study).

Figures 5, 6, and 7 illustrate the findings.

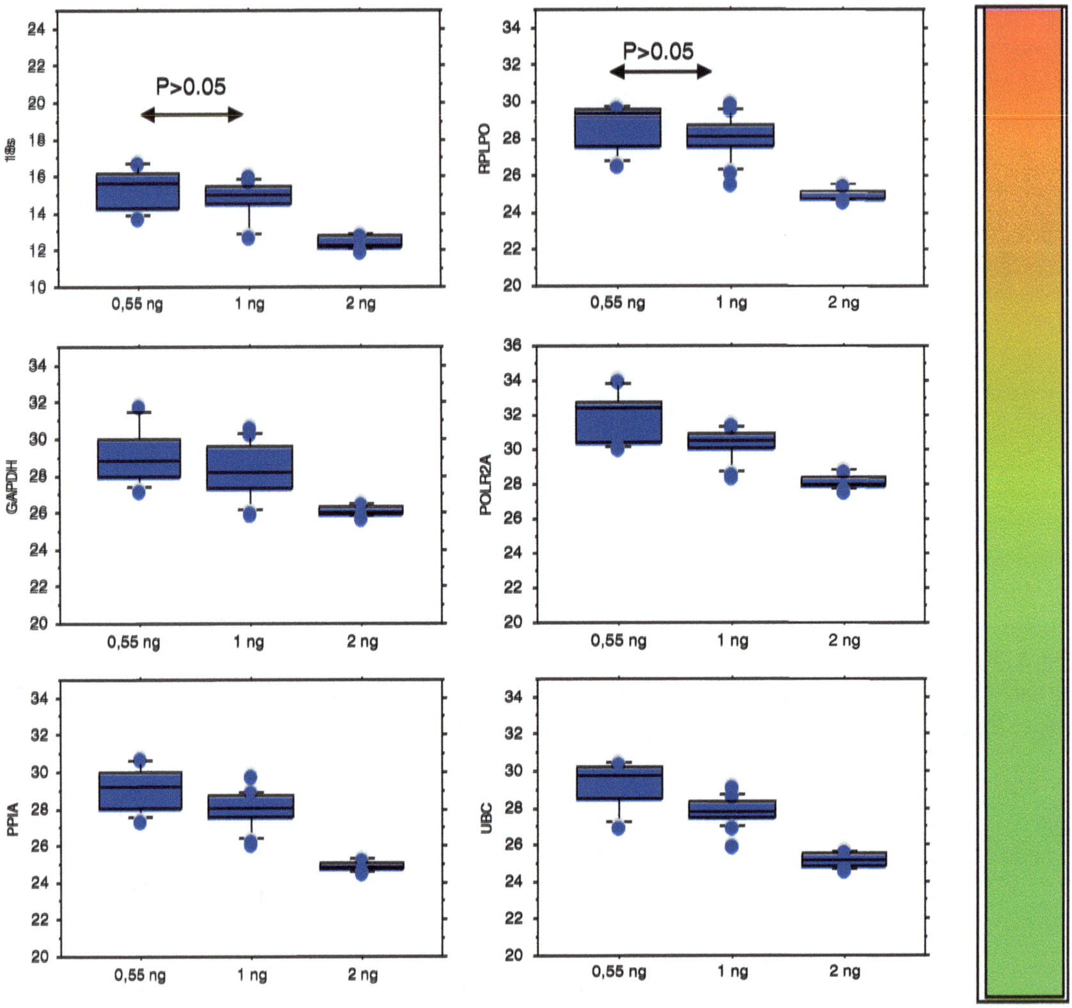

Fig. 5 Threshold values (Ct) for the most stable preselected genes and for the different concentration groups. Please note the overlapping between the first two groups as stated before. When analyzing (–Ct values with ANOVA, Partek) and correcting by RIN, differences between 0.55 and 1 ng were only obtained for GAPDH, PPIA, POLR2A, and UBC. Only nonsignificant differences are shown, and the remaining paired comparisons showed $p < 0.05$

These criteria can be applied in order to choose the most adequate candidate reference genes for any individual tissue- or cell-based assay/study. For example, TaqMan® Low-Density Endogenous Control Panels with 16 human reference genes in triplicate (Life Technologies) can be used to analyze samples from non-microdissected whole sections. High-capacity cDNA archive kit can be for converting total RNA into cDNA. The amount of starting total RNA and RNA quality will vary but should be acceptable: from 0.55 ng (RIN 0.5.0) and 1 ng (RIN 5.0) to 2 ng (only for whole sections RIN 6.0).

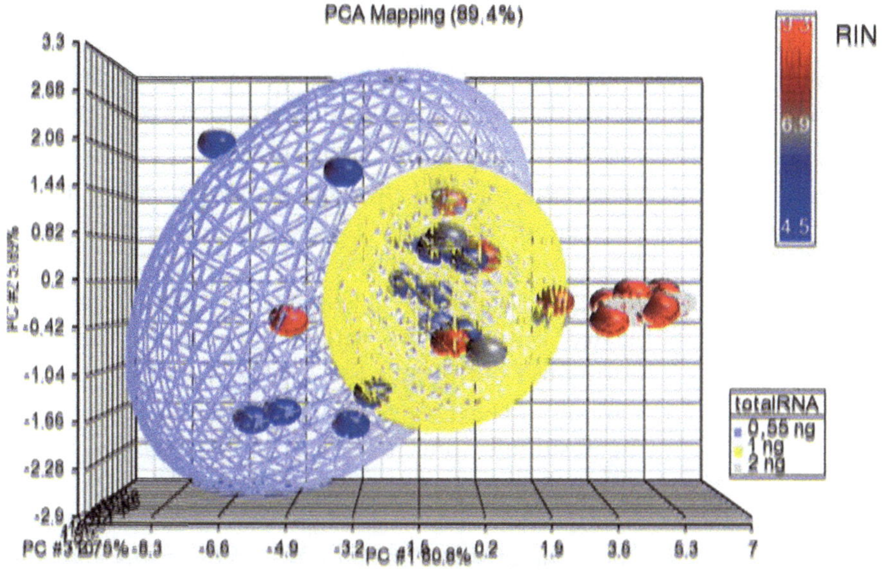

Fig. 6 Principal component analysis (PCA) of the expression of the eight preselected endogenous genes. Note differences for the 2 ng samples but certain overlapping among 0.55 and 1 ng samples. RNA quality expressed as RNA integrity number RIN (*color spheres*) and quantity of starting total RNA (*ellipsoids*)

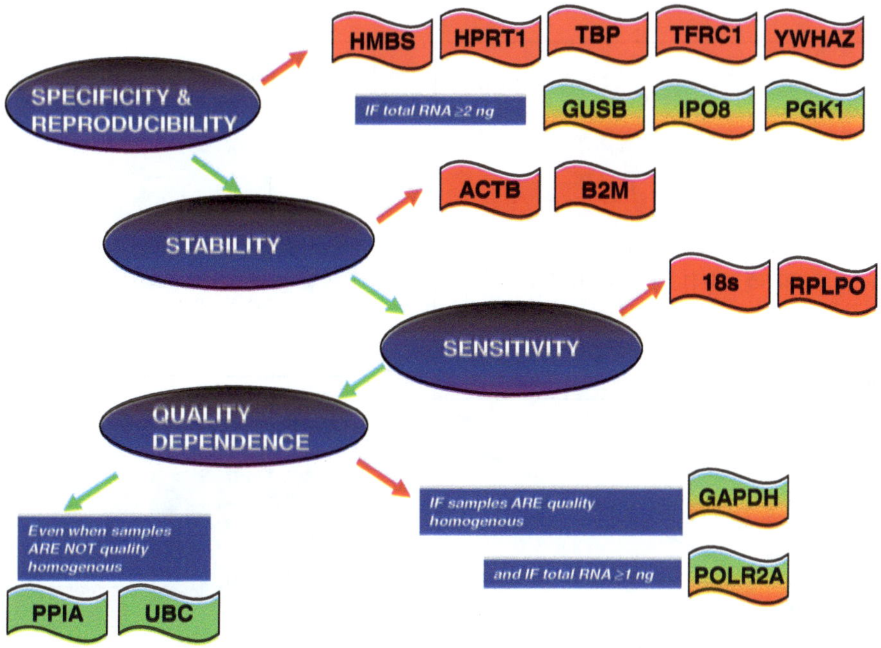

Fig. 7 Diagram showing the different criteria applied to choose the most adequate control endogenous genes and the results found for the particular conditions of our study on samples from human coronary arteries. *Red* for excluded candidates and *green* for those that adhere to the criteria; a gradation of both colors means that they can only be recommended to be used under the specified conditions

3.8 Pre-amplification of cDNA Prior to Array Analysis

Pre-amplification of the low quantities of RNA produced following direct micro-extraction is a requirement prior to microarray (microfluidity card) analysis. This pre-amplification is essentially an additional PCR cycling system which converts picogram quantities of RNA to nanogram amounts of cDNA.

1. Here, the Applied Biosystems amplification kit should be carried out control studies to ensure the equal amplification of all genes had taken place.
2. Amplification of the appropriate housekeeping genes will be carried out together with samples using primers that can distinguish one sample set from another.
3. Care should be taken to ensure that following pre-amplification, the resulting concentrations of each sample should remain within the limits set by the manufacturers (Applied Biosystems), i.e., <1.5ΔΔCt.
4. The samples should also be tested to ensure that they express the desired or expected differences in test gene thus ensuring the validity of the comparison groups (*see* Fig. 8 as an example CD105/CD31).

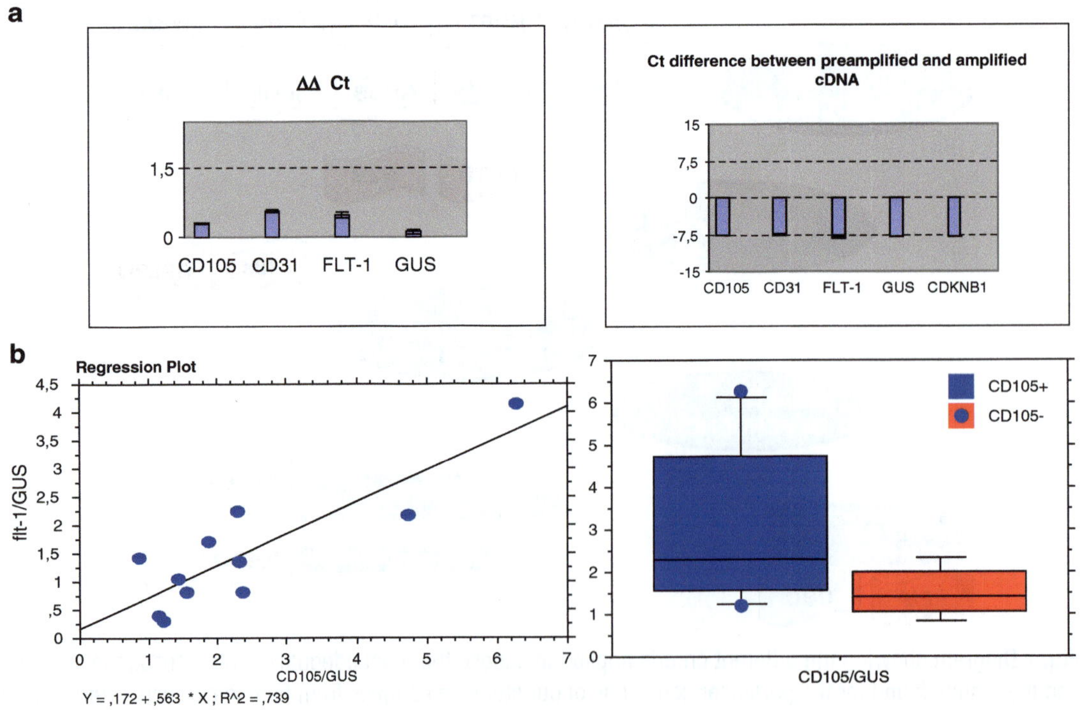

Fig. 8 (**a**) Pre-amplification did not affect the relative expression of test genes compared with the housekeeping control b-glucuronidase (GUS). (**b**) Pre-amplification did not affect the relative expression of test genes compared with the housekeeping control b-glucuronidase (GUS)

3.9 Employment of Microarrays and Data Analysis

A comparison of gene expression in paired microenvironments can now be carried out using targeted low-density arrays/microfluidity cards running comparative samples on the same cards.

As the data above has shown, a minimal quantity and quality of RNA is required in order to obtain reliable information, and when comparing samples, compared aliquots of RNA were chosen having similar RIN values (+/−1.5) and concentrations (e.g., 4–6 ng rather than comparing 1 ng with 5 ng).

3.10 Summary

The most important findings from our considerable investigations are that from a general point, consistent reproducible isolation of high-quality RNA from arterial tissue is very difficult to achieve. This paper is important as very few groups have published information relating to extraction of RNA from this tissue and the use of laser-capture microdissection to analyze the pathophysiological events occurring during development of atherosclerosis. It is often the case that published articles omit detailed information as to how the process has been carried out, and this is worrying considering the findings described in this paper. Gräbner et al. [9] used laser-capture microdissection to isolate adventitial areas from the abdominal aortas of mice (wild-type and apoE−/−), comparing by microarray the gene signatures; however, no information is supplied about the technology used to produce the samples. In 2006, Tiwari et al. [10] obtained human arterial carotid and femoral biopsy material and used laser-capture microdissection on rapidly immunostained frozen sections isolating vascular smooth muscle cells. Details of the procedure are not provided; however, our experiments suggest that use of this protocol may significantly reduce the RIN values of the samples making them of limited use for subsequent analysis. Pre-amplification of samples is also mentioned, but appropriate controls to check for equality of amplification or to confirm samples contain the material of interest are not shown.

Similarly, hematoxylin and eosin staining was used on sections used for laser capture where subsequent caspase-3-stained serial sections had identified caspase-positive regions in plaques from apoE−/− mice [11]. In this work, Trogan et al. described their rapid immunostaining protocol in adequate detail but suggested that this had no degradational effects on the isolated RNA. Clearly, in all these studies, satisfactory examination of the appropriate housekeeping controls was also not carried out leading to possible skewing of the final data. Papers published by Babaev et al. using LDL-null mice [12] and Bobryshev [13] examining human aortas and carotid artery segments used similar techniques but have failed to describe how they have confirmed the accuracy and reproducibility of their results.

While we are not prepared to suggest that the data presented in those papers is unreliable, we believe that a far more stringent set of parameters has to be set when conducting this type of analysis and that information alluding to this must be included within the manuscript or as additional data. We have summarized in the form of a table the steps we believe should be taken in order to ensure accurate, reproducible data is achieved when isolating individual cells or microenvironments from arterial tissue (Fig. 7).

4 Notes

1. The most useful method for identification of cellular components of the artery and plaque is a simple and standard IHC protocol, for example, using the ABC kit available from Thermo Scientific (Pierce, US). Antibodies used will be determined by the elements needed to be highlighted. Here we used anti-CD105 and anti-CD31 antibodies obtained from Abcam (Cambridge, UK) to discriminate between active immature and inactive mature vessels.
2. Target areas are identified by standard microscopy, photographs taken, and hard copies produced.
3. If necessary, layers within the artery can be individually dissected using a low-power stereo-dissecting microscope directly from the slide before LCM.
4. This was the method we found to produce the best quality samples and RNA.
5. By direct comparison of these small dissected areas, it is possible to accurately laser-dissect the microenvironments of interest.
6. Sample selection and LCM of paired samples (example used, arterial microvessels): Rapid staining with UEA-1 and IHC has been utilized previously in order to directly visualize microvessels prior to laser cutting and RNA analysis. Here, the results of our extensive experimental data demonstrate that exposure of the arterial tissue sections to DEPC water, PBS, or ethanol as well as excessive cutting time (>1/2 an hour) at room temperature can significantly reduce RNA quality measured as RNA integrity number (RIN) from between 5 and 9 (suitable for microarray analysis) to a value of less than 4. Also, we have found that rapid staining with UEA-1 cannot discriminate between active and inactive or mature and immature vessels and the contrast quality necessary in order to identify exact points of cut is difficult to visualize. The most important finding from the first part of this study shows that even simple exposure of thawed, hydrated sections to room temperature for 30 min reduced the RIN value from >6 to <2 making the samples of limited use for gene analysis. We showed that dehydrated samples can retain their RNA quality for longer (>2 h). Treating the sections immediately after thawing with RNA-Later (4 min) slowed down the process of degradation, giving

an RIN value of around 5 after 30 min. However, treatment with PBS reduced the RIN values from approximately 9 to 1. Ethanol, acetone, or DEPC treatment had a negative effect on the RIN value over the half-hour period. In summary, our final treatment protocol involved treating thawed sections with RNA-Later and LCM was carried out within 30 min in a cold room. Rapid staining procedures resulted in low RIN values not of sufficient quality for subsequent microarray analysis. Results are summarized in Table 1.

7. Housekeeping gene analysis: On the basis of the Specificity and Reproducibility criteria, we chose the first eight genes and excluded the remaining genes. Although excluded from downstream analysis, *GUSB*, *IPO8*, and *PGK1* fulfilled these criteria but only for samples with 2 ng of starting total RNA and good quality RNA. *POLR2A* was included since it was only for the 0.55 ng samples that we saw values higher than 32 Cts and 0.40 for SD.

The stability analysis was run for each starting amount of RNA, layer, and lesion category, and results were evaluated in a scoring system (1 for the most stable and 0.5 for the other candidate genes with an *M* value of maximum 0.6). B2M and ACTB are found to be the least stable genes for these conditions and therefore excluded from downstream analysis. These two were the least stable candidate genes when analyzed by starting quantity and by layer and plaque developmental stage using GeNorm and validated with Data Assist both when looking at the most extreme samples (0.55 ng) and all the data obtained.

Table 1
RIN after various extraction procedures

Protocol or treatment	RIN before	RIN after
Immediate cut	9.1	9.1
Rapid IHC	9.1	1.1
Rapid UEA-1	9.1	2.4
30 min RT	6.1	1.0
RNA-Later-DEPC 30 min	6.1	4.9
30 min PBS/RT	9.2	1.1
RNA-Later and DEPC/RT	6.7	5.0
RNA-Later and DEPC/UEA-1	6.7	3.8
RNA-Later and PBS/4 °C	6.7	4.8

All values represent an average of three readings from one tissue section
Each experiment was carried out at least three times

These results could be explained by a confounding effect for B2M being more expressed in media than in intima or adventitia (when only 1 ng data are analyzed; $p < 0.05$) and ACTB more expressed in earlier lesions (it applies to 0.55 and 1 ng samples; $p < 0.05$).

In terms of sensitivity to differences in amount of cDNA, the ribosomal 18S and RPLPO were less sensitive to small changes in starting total RNA (0.55 ng vs 1 ng $p > 0.05$) than the rest. It should be mentioned here that the only means to get an approximate quantification at these concentrations is using the picochip lab-on-a-chip system of the Bioanalyzer 2100 technology (Agilent Technologies) although it is not a specifically recommended system to quantify but for RNA quality control. Looking at the principal component analysis (PCA mapping), we observe an inherent variation evident in the endogenous housekeeping gene expression between samples at the lower concentration limits of 0.55 ng (RIN 6.17 ± 1.65) and 1 ng (RIN 6.48 ± 0.97) which did not vary according to their quality ($p > 0.05$ MANOVA) but showed a clear overlapping, as opposed to those with 2 ng of starting total RNA of higher quality (RIN 8.4 ± 0.5 $p < 0.05$ MANOVA).

Endogenous controls besides correcting for variations in the amount of starting total RNA are expected to correct for RNA quality. *GAPDH* and *POLR2A* were not significantly ($p > 0.05$) quality dependent in our RIN range, although PPIA and UBC expression were affected ($p < 0.05$). Therefore, we can conclude that for our particular study and taking into account the RNA restrictions in terms of quantity and quality, PPIA and UBC are the best endogenous controls because they showed the highest Specificity and Reproducibility, were more stable and sensitive to small variations in starting material, and also corrected for quality differences. If our study had homogenous RIN values, *GAPDH* would be a good choice as well as *POLR2A* if we had more than 1 ng of RNA for first strand annealing (and the resulting cDNA loaded in the same kind of microfluidity card to determine, e.g., 21 or 22 target genes and 2 or 3 reference genes by duplicate).

Acknowledgments

We would like to thank the PIV and the Foundation BBVA for their generous support of Professor Mark Slevin through the award of BBVA Chair in Clinical Biomedicine at the ICCC, St Pau Hospital, Barcelona. This work was also supported by the Spanish Ministry of Science (SAF2006-10091 to L.B.), Instituto de Salud Carlos III (CIBEROBN-CB06/03 to L.B.), and RyC (RyC2007-01466 to M.B.-P.). Maribel Baldellou and Lina Badimon have received funding by the network REDINSCOR RD06/0003/0015.

References

1. Nichols M, Townsend N, Luengo-Fernandez R, Leal J, Gray A, Scarborough P, Rayner M (2012) European cardiovascular disease statistics 2012. European Heart Network/ European Society of Cardiology, Brussels/ Sophia Antipolis
2. Mukherjee D, Patil CG (2011) Epidemiology and the global burden of stroke. World Neurosurg 76:85–90
3. Chowdhury M, Ghosh J, Slevin M, Smyth JV, Alexander MY, Serracino-Inglott F (2010) A comparative study of carotid atherosclerotic plaque microvessel density and angiogenic growth factor expression in symptomatic versus asymptomatic patients. Eur J Vasc Endovasc Surg 39(4):388–395
4. Slevin M, Turu MM, Rovira N, Luque A, Baldellou M, Krupinski J, Badimon L (2010) Identification of a 'snapshot' of co-expressed angiogenic markers in laser-dissected vessels from unstable carotid plaques with targeted arrays. J Vasc Res 47(4):323–335
5. Slevin M, Krupinski J, Badimon L (2009) Controlling the angiogenic switch in developing atherosclerotic plaques: possible targets for therapeutic intervention. J Angiogenes Res 1(4):1–10
6. Bergers G, Song S (2005) The role of pericytes in blood-vessel formation and maintenance. Neuro Oncol 7(4):452–464
7. Rupp C, Dolznig H, Puri C, Schweifer N, Sommergruber W, Kraut N, Rettig WJ, Kerjaschki D, Garin-Chesa P (2006) Laser capture microdissection of epithelial cancers guided by antibodies against fibroblast activation and endosialin. Diagn Mol Pathol 15: 35–42
8. Slevin M, Turu MM, Rovira N, Luque A, Baldellou M, Krupinski J et al (2009) Identification of a 'Snapshot' of co-expressed angiogenic markers in laser-dissected vessels from unstable carotid plaques with targeted arrays. J Vasc Res 47:323–335
9. Gräbner R, Lötzer K, Döpping S, Hildner M, Radke D, Beer M, Spanbroek R, Lippert B, Reardon CA, Getz GS, Fu YX, Hehlgans T, Mebius RE, van der Wall M, Kruspe D, Englert C, Lovas A, Hu D, Randolph GJ, Weih F, Habenicht AJ (2009) Lymphotoxin beta receptor signaling promotes tertiary lymphoid organogenesis in the aorta adventitia of aged ApoE-/- mice. J Exp Med 206:233–248
10. Tiwari S, Zhang Y, Heller J, Abernethy DR, Soldatov NM (2006) Atherosclerosis-related molecular alteration of the human CaV1.2 calcium channel alpha1C subunit. Proc Natl Acad Sci U S A 103:17024–17029
11. Trogan E, Choudhury RP, Dansky HM, Rong JX, Breslow JL, Fisher EA (2002) Laser capture microdissection analysis of gene expression in macrophages from atherosclerotic lesions of apolipoprotein E-deficient mice. Proc Natl Acad Sci U S A 99:2234–2239
12. Babaev VR, Ishiguro H, Ding L, Yancey PG, Dove DE, Kovacs WJ, Semenkovich CF, Fazio S, Linton MF (2007) Macrophage expression of peroxisome proliferator-activated receptor-alpha reduces atherosclerosis in low-density lipoprotein receptor-deficient mice. Circulation 116:1404–1412
13. Bobryshev YV (2005) Subset of cells immunopositive for neurokinin-1 receptor identified as arterial interstitial cells for Cajal in human large arteries. Cell Tissue Res 321:45–55

Part V

Stimulation of Cerebral Angiogenesis by Gene Delivery

Chapter 25

Induction of Brain Arteriovenous Malformation in the Adult Mouse

Wanqiu Chen, William L. Young, and Hua Su

Abstract

Brain arteriovenous malformations (bAVM) are tangles of abnormal, dilated vessels that directly shunt blood between the arteries and veins. The pathogenesis of bAVM is currently unknown. Patients with hereditary hemorrhagic telangiectasia (HHT) have a higher prevalence of bAVM than the general population. Animal models are important tools for dissecting the disease etiopathogenesis and for testing new therapies. Here, we introduce a method that induces the bAVM phenotype through regional deletion of activin-like kinase 1 (*Alk1*, the causal gene for HHT2) and vascular endothelial growth factor (VEGF) stimulation.

Key words Arteriovenous malformation, Activin-like kinase 1, Angiogenesis, VEGF, Hereditary hemorrhagic telangiectasia, Mouse model

1 Introduction

Arteriovenous malformations (AVM) are tangles of abnormal, dilated vessels that directly shunt blood between the arteries and veins without a true capillary bed. It is a common phenotype in patients with hereditary hemorrhagic telangiectasia (HHT). The two main subtypes of HHT are as follows: (a) HHT1, caused by loss-of-function mutations in the endoglin (*ENG*), a type III transforming growth factor-beta (TGFβ) coreceptor, and (b) HHT2, caused by mutations in activin-like kinase 1 (*ALK1* or *ACVRL1*) gene, a type I TGFβ signaling receptor. Compared to the general population, the prevalence of bAVM in HHT1 (*ENG*) is 1,000-fold higher, and in HHT2 (*ALK1*), 100-fold higher (10/100,000) [1].

Brain AVM accounts for 1–2 % of all strokes [2]. The malformed vessels are fragile and prone to rupture, and rupture of bAVM can lead to intracranial hemorrhage and serious neurological disability or death. The risk of intracranial hemorrhage of bAVM is about 2–4 % per year [2]. Patients with unruptured bAVM can develop many neurological symptoms, e.g., seizures and headache. The pathogenesis of bAVM is currently unknown [3].

Richard Milner (ed.), *Cerebral Angiogenesis: Methods and Protocols*, Methods in Molecular Biology, vol. 1135, DOI 10.1007/978-1-4939-0320-7_25, © Springer Science+Business Media New York 2014

To better understand the underlying mechanisms of bAVM formation, it is critical to establish reproducible and reliable animal models that mimic both macroscopic and microscopic morphological features of the human bAVM lesional phenotype, including large dysplastic, tangled vessels, and arteriovenous shunting.

Previous studies have shown that homozygous deletion of *Alk1* or endoglin results in embryonic lethality [4, 5]. Interestingly, induction of *Alk1* gene deletion at the adult stage leads to AVM formation only in small intestinal, pulmonary, and uterine vessels, but not in the brain [6]. Our group has previously shown that vascular endothelial growth factor (VEGF) stimulation in adult *Alk1* or *Eng* heterozygous mice results in vascular abnormalities at the capillary level [7–9], which supports the response-to-injury paradigm. Thus, other than genetic modification, environmental stimulus might also be involved in the onset of bAVM formation in adults. We have successfully induced the AVM phenotype in the adult mouse brain through regional *Alk1* deletion in combination with VEGF stimulation [10].

2 Materials

2.1 Animals

Adult $Alk1^{2f/2f}$ mice (exons 4–6 flanked by loxP sites) [11].

2.2 Viral Vectors

1. Adenoviral vector with cytomegalovirus (CMV) promoter driving Cre recombinase expression (Ad-Cre, Vector Biolabs, Philadelphia, PA, USA).
2. Control adenoviral vector with green fluorescent protein expression (Ad-GFP, Vector Biolabs).
3. Adeno-associated viral vectors expressing vascular endothelial growth factor (AAV-VEGF) packaged in AAV serotype 1 capsid [12–14].
4. Adeno-associated viral vectors expressing β-galactosidase (AAV-LacZ) [12–14].

2.3 Viral Vector Injection

1. Isoflurane.
2. 30 % oxygen/70 % nitrogen.
3. Anesthetic vaporizer and flowmeter.
4. Stereotactic frame (David Kopf Instruments, Tujunga, CA, USA).
5. Homeothermic temperature system (Harvard Apparatus, Holliston, MA, USA).
6. Hot Bead Sterilizers.
7. Rechargeable Cordless Micro Drill (Stoelting, Wood Dale, IL, USA).

2.4 Assorted Surgical Instruments

1. FS-2, 4-0 black silk suture.
2. Microdissecting tweezers.
3. McPherson-Vannas microdissecting scissors.
4. Hamilton syringe.
5. Forceps.

2.5 Latex Vessel Casting

1. 25-gauge 5-mL syringe.
2. Blue latex dye (Connecticut Valley Biological Supply Co., Southampton, MA, USA).
3. 4 % paraformaldehyde.
4. Methanol.
5. Benzyl alcohol.
6. Benzyl benzoate.
7. Microscope.

2.6 Determining Dysplasia Index

1. Fluorescein-lycopersicon esculentum-lectin (Vector Laboratory, Burlingame, CA, USA).
2. Leica CM1900 cryostat.
3. Fluorescent microscope.

3 Methods

3.1 Stereotactic Injection of Viral Vectors into the Basal Ganglia (See Note 1)

1. This protocol was approved by the Animal Care Committee of the University of California, San Francisco.
2. Following induction of anesthesia with 5 % isoflurane, 8-week-old $Alk1^{2f/2f}$ mice were placed in a stereotactic frame (David Kopf Instruments) with a mouth holder. The anesthesia was maintained with 1.5 % isoflurane.
3. A 1-cm midline skin incision at the top of the head was made to expose the sagittal suture and bregma.
4. A burr hole was drilled in the pericranium 2 mm lateral to the sagittal suture and 1 mm posterior to the coronal suture.
5. A 10-μL Hamilton syringe was slowly inserted into the left caudate nucleus at a depth of 3.0 mm under the dura (*see* **Note 2**).
6. A 2-μL viral suspension containing 2×10^7 plaque forming unit (PFU) Ad-Cre and 1 μL containing 2×10^9 genome copies (gcs) of AAV-VEGF were injected into the right basal ganglia at a rate of 0.2 μL/min using a ultramicro pump (World Precision Instruments). For the control group, Ad-GFP and AAV-LacZ were used.
7. The needle was retained in the brain for 10 min and then slowly withdrawn over a 5-min period (*see* **Note 3**).

8. The skin wound was closed with a 4-0 suture.
9. Brain sections were collected for analysis 8 weeks later (*see* **Note 4**).

3.2 Analysis

Visualizing malformed vessels in the brain is the first step to confirm the successful induction of the AVM phenotype. There are different techniques to visualize cerebral vasculature in small animals, including casting vessels with microfil [10, 15] or latex. Latex perfusion is the usual choice to display the AVM-like phenotype, since the particles in the latex dye are too big to pass through the capillaries [6]. Latex will present in the veins after intra-artery infusion when there is direct flow between arteries and veins (i.e., AV shunting, a major characteristic of AVM). Latex perfusion was first described by Coyle and Jokelainen [16] and later modified by Maeda et al. [17]. It is the liquid rubber that solidifies after the vasculature is filled.

Lesion volume measurement and dysplastic vessel index quantification (the number of vessels larger than 15 μm/200 vessels) [8–10, 18–20] are methods to evaluate the severity of the phenotype, enabling quantitative measurement of cerebrovascular abnormality, which could provide critical outcome evaluation in innovative brain vascular malformation therapies.

3.3 Vessel Casting with Latex Perfusion

1. Deep-anesthetize the mice with isoflurane. Open the abdominal and thoracic cavities and expose the heart.
2. Cut both the left and right atria. Slowly inject 1-mL blue latex dye into the left ventricle with a 25-gauge 5-mL syringe (*see* **Note 5**).
3. Remove the brain and fix with 4 % paraformaldehyde overnight.
4. Dehydrate the brain tissue using methanol series (50, 75, 95, and 100 % methanol, 24 h for each concentration).
5. Clarify the brain with benzyl alcohol/benzyl benzoate (1:1 ratio).
6. Cut the brain coronally using a razor blade. Image the brain AVM vessels under the microscope (Fig. 1).

3.4 Lesion Volume Quantification

1. After imaging the latex-perfused brain slide, rehydrate the brain tissue and snap-freeze in dry ice.
2. Coronally section the brain serially into 50-μm-thick sections using a cryostat. Section the entire brain region containing the AVM lesion.
3. Under a 5× microscope objective lens, image both the AVM lesion side and the contralateral corresponding region.
4. Quantify the latex area using NIH Image 1.63 software after binary (Fig. 2).

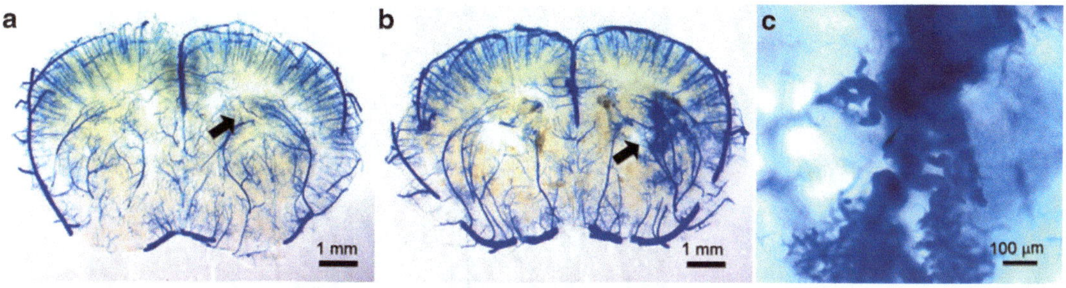

Fig. 1 Vessel casting by latex perfusion shows the AVM lesion. (**a**) No abnormal vessels were detected in the brain of wild-type mouse around the vector injection site (*arrow*). Scale bar: 1 mm. (**b**) AVM phenotype was detected around the vector injection site of 8-week-old *Alk1*$^{2f/2f}$ mouse (*arrow*). Scale bar: 1 mm. (**c**) High magnification of the injection area shows the abnormal vascular structure. Scale bar: 100 μm

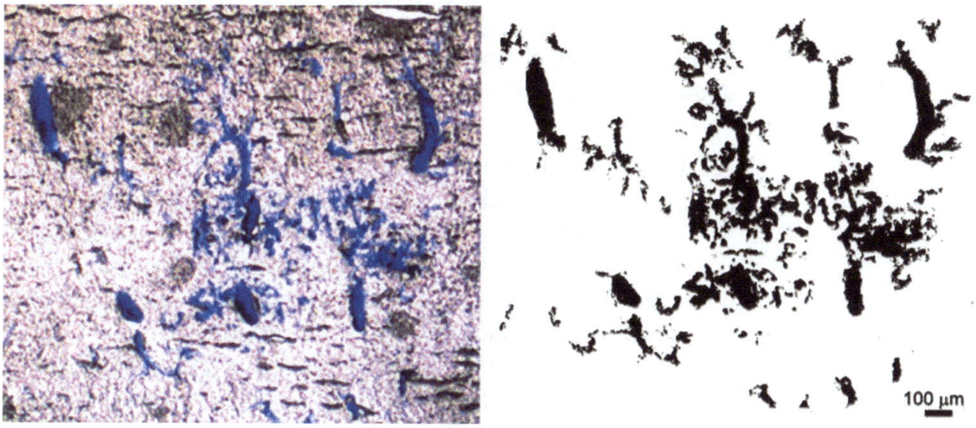

Fig. 2 An example of blue latex section image (*left*) and an image after binary (*right*). Scale bar: 500 μm

5. Based on known section thickness, estimate the volume by summing serial volumes. This method is adapted from what we have used for quantifying infarct volume in mice with middle cerebral artery occlusion (MCAO) [21].

3.5 Capillary Density and Dysplasia Index Quantification (See Note 6)

1. 8 weeks following viral injection, perfuse the mice with cold 1× PBS through the left ventricle of the heart, using a Masterflex Pump Controller (Cole Parmer Instrument, Vernon Hills, IL, USA) at 4 mL/min.
2. Harvest the brains and snap-freeze in dry ice.
3. Coronally slice the brain serially at 20-μm thickness using a cryostat.
4. Choose two sections per mouse, one mm rostral and one mm caudal of the injection site, for vessel staining using fluorescein-lycopersicon esculentum-lectin (1:200, Vector Laboratory) (Fig. 3; *see* **Note 7**).

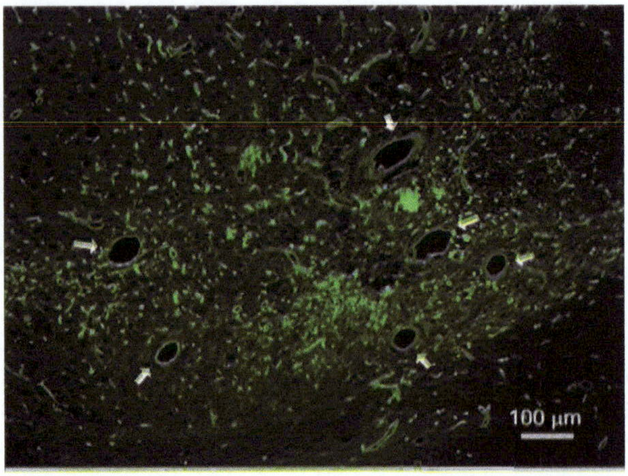

Fig. 3 Representative image of a brain section with lectin staining. *Alk1*$^{2f/2f}$ mouse received 2×10^7 PFU Ad-Cre and 2×10^9 gcs of AAV-VEGF in the basal ganglia for 8 weeks. A cluster of dysplastic vessels formed in the viral injection site (*white arrows*). Scale bar: 100 μm

5. Under a 20× microscope objective lens, capture three areas (to the right and left of and below the injection site) of each section. In each image, count total vessel number and vessels with diameter larger than 15 μm using NIH Image 1.63 software. Calculate the vascular density for each animal as the mean of the total vessel number obtained from the six images taken under 20× microscope objective lens. Calculate the dysplasia index as the number of vessels with a diameter larger than 15 μm per 200 vessels.

4 Notes

1. The vectors can also be injected into other brain regions, such as the cortex, to induce the AVM phenotype.
2. During viral injection, fix the mouse head in a horizontal position, and insert the needle perpendicularly into the surface of the dura. Needle insertion at an angle will affect the lesion location or may even inject into the brain ventricle.
3. After viral injection, it is critical to retain the needle in the brain for 10 min and then slowly withdraw the needle. Pulling the needle out too fast could lead to the virus leaking out.
4. Other than Ad-Cre-mediated regional gene deletion, *Alk1* can be conditionally deleted in adult mice, either systemically or tissue-/cell-specifically by crossbreeding *Alk1*$^{2f/2f}$ mice with transgenic mouse lines that express inducible Cre recombinase. The brain AVM phenotype can be induced in mice that have systemic/endothelial-specific *Alk1* gene deletion through intra-brain injection of AAV-VEGF.

5. For latex perfusion, make sure that the latex is fresh and that no solidified clusters have formed.
6. Vessels can also be visualized by perfusing fluorescent-labeled lectin through intravenous injection.
7. The sections can be stained with endothelial-specific antibodies, such as the anti-CD31 antibody.

Acknowledgements

This work was supported by grants from the NIH National Institute of Neurological Disorders and Stroke—R01 NS027713 (to WLY), P01 NS044155 (to WLY and HS), R21 NS070153 (to HS), and GM008440 (to WC)—and the American Heart Association, AHA 10GRNT3130004 (to HS).

References

1. Kim H, Marchuk DA, Pawlikowska L et al (2008) Genetic considerations relevant to intracranial hemorrhage and brain arteriovenous malformations. Acta Neurochir Suppl 105:199–206
2. Gabriel RA, Kim H, Sidney S et al (2010) Ten-year detection rate of brain arteriovenous malformations in a large, multiethnic, defined population. Stroke 41:21–26
3. Kim H, Su H, Weinsheimer S et al (2011) Brain arteriovenous malformation pathogenesis: a response-to-injury paradigm. Acta Neurochir Suppl 111:83–92
4. Urness LD, Sorensen LK, Li DY (2000) Arteriovenous malformations in mice lacking activin receptor-like kinase-1. Nat Genet 26: 328–331
5. Sorensen LK, Brooke BS, Li DY et al (2003) Loss of distinct arterial and venous boundaries in mice lacking endoglin, a vascular-specific TGFbeta coreceptor. Dev Biol 261:235–250
6. Park SO, Wankhede M, Lee YJ et al (2009) Real-time imaging of de novo arteriovenous malformation in a mouse model of hereditary hemorrhagic telangiectasia. J Clin Invest 119:3487–3496
7. Xu B, Wu YQ, Huey M et al (2004) Vascular endothelial growth factor induces abnormal microvasculature in the endoglin heterozygous mouse brain. J Cereb Blood Flow Metab 24:237–244
8. Hao Q, Su H, Marchuk DA et al (2008) Increased tissue perfusion promotes capillary dysplasia in the ALK1-deficient mouse brain following VEGF stimulation. Am J Physiol Heart Circ Physiol 295:H2250–H2256
9. Hao Q, Zhu Y, Su H et al (2010) VEGF induces more severe cerebrovascular dysplasia in Endoglin$^{+/-}$ than in Alk1$^{+/-}$ mice. Transl Stroke Res 1:197–201
10. Walker EJ, Su H, Shen F et al (2011) Arteriovenous malformation in the adult mouse brain resembling the human disease. Ann Neurol 69:954–962
11. Park SO, Lee YJ, Seki T et al (2008) ALK5- and TGFBR2-independent role of ALK1 in the pathogenesis of hereditary hemorrhagic telangiectasia type 2 (HHT2). Blood 111:633–642
12. Su H, Lu R, Kan YW (2000) Adeno-associated viral vector-mediated vascular endothelial growth factor gene transfer induces neovascular formation in ischemic heart. Proc Natl Acad Sci U S A 97:13801–13806
13. Su H, Huang Y, Takagawa J et al (2006) AAV serotype-1 mediates early onset of gene expression in mouse hearts and results in better therapeutic effect. Gene Ther 13:1495–1502
14. Shen F, Su H, Liu W et al (2006) Recombinant adeno-associated viral vector encoding human VEGF165 induces neomicrovessel formation in the adult mouse brain. Front Biosci 11:3190–3198
15. Walker EJ, Shen F, Young WL et al (2011) Cerebrovascular casting of adult mouse for 3D imaging and morphological analysis. J Vis Exp (57): e2958
16. Coyle P, Jokelainen PT (1982) Dorsal cerebral arterial collaterals of the rat. Anat Rec 203:397–404
17. Maeda K, Hata R, Hossmann KA (1998) Differences in the cerebrovascular anatomy of C57black/6 and SV129 mice. Neuroreport 9:1317–1319

18. Choi EJ, Walker EJ, Shen F et al (2012) Minimal homozygous endothelial deletion of Eng with VEGF stimulation is sufficient to cause cerebrovascular dysplasia in the adult mouse. Cerebrovasc Dis 33:540–547
19. Choi EJ, Walker EJ, Degos V et al (2013) Endoglin deficiency in bone marrow is sufficient to cause cerebrovascular dysplasia in the adult mouse after vascular endothelial growth factor stimulation. Stroke 44:795–798
20. Chen W, Guo Y, Walker EJ et al (2013) Reduced mural cell coverage and impaired vessel integrity after angiogenic stimulation in the *Alk1*-deficient brain. Arterioscler Thromb Vasc Biol 33:305–310
21. Shen F, Walker EJ, Jiang L et al (2011) Coexpression of angiopoietin1 with VEGF increases the structural integrity of the blood-brain barrier and reduces atrophy volume. J Cereb Blood Flow Metab 31:2343–2351

Chapter 26

Stimulation of Cerebral Angiogenesis by Gene Delivery

Yaohui Tang, Yaning Li, Xiaojie Lin, Peng Miao, Yongting Wang, and Guo-Yuan Yang

Abstract

Angiogenesis, an important process for long term neurological recovery, could be induced by ischemic brain injury. In this chapter, we describe a system to deliver adeno-associated viral (AAV) vector-mediated gene therapy for ischemic stroke. This includes the methods to construct, produce, and purify an AAV vector expressing target gene and an approach to quantify the number of microvessels and capillary density with synchrotron radiation angiography (SRA) imaging.

Key words Adeno-associated virus, Angiogenesis, Capillary density, Cerebral ischemia, Netrin-1, Synchrotron radiation angiography

1 Introduction

Blood vessels are formed during development while maintained at a basal level in the normal mature brain, critically regulated by concerted angiogenic factors [1, 2]. Pathological processes such as cerebral ischemia induce endogenous angiogenesis and show great benefits for motor functional recovery [3–5]. Clinical studies show that stroke patients with a higher density of microvessels are associated with less morbidity and longer survival [3, 6–8]. These observations indicate that focal angiogenesis plays an important role in postischemic brain repair and remodeling.

Angiogenesis is a significant event resulting in changes of the brain microvasculature following cerebral ischemia. We and other colleagues showed that active angiogenesis is induced after 7 and 14 days of focal cerebral ischemia in rodents [9, 10]. In human brain, active angiogenesis takes place as early as 3 days following ischemic insult [11]. Clinical observation demonstrates a strong correlation between neovascularization and increased cerebral blood flow (CBF), suggesting that newly formed vessels are functional [12]. Ischemia-evoked angiogenesis not only supplies nutrients

Richard Milner (ed.), *Cerebral Angiogenesis: Methods and Protocols*, Methods in Molecular Biology, vol. 1135,
DOI 10.1007/978-1-4939-0320-7_26, © Springer Science+Business Media New York 2014

and oxygen to the ischemic lesion area but also removes necrotic brain tissue [13, 14]. Besides angiogenesis, cerebral ischemia also evokes collateral artery perfusion and growth in the ischemic border zone, restoring perfusion, which may improve long-term functional recovery together with angiogenesis [15–17].

The adeno-associated viral (AAV) vector belongs to the helper-dependent member of the parvovirus family. It currently represents a promising vector for gene therapy and is widely used in various cerebral nervous system (CNS) disease models [18]. The AAV vector has several advantages over other viral vectors, including low immunogenicity, and the ability to mediate long-term target gene expression and to infect both dividing and nondividing cells [19–21]. AAV is a principal vector of choice in experimental ischemic stroke studies [22–24]. In addition, the usage of various serotypes and their different preferences to cells and tissues significantly broadened their application [25–27].

Netrin-1 (NT-1), a conserved protein initially purified from chick embryos, provides key guidance cues for the development of commissural axons [28]. NT-1 can either attract or repel axonal growth cones through binding to different receptors [29, 30]. In vitro studies show that NT-1 also stimulates the proliferation and adhesion of endothelial and vascular smooth muscle cells, acting as a survival factor for endothelial cells [31, 32]. Studies demonstrate that NT-1 promotes neovascularization and improves limb perfusion in hind limb ischemia [33]. Our previous studies showed that NT-1 overexpression increased angiogenesis in normal and ischemic mouse brain [24, 34, 35].

Gene transfer technology, which has been progressed to clinical trails [42, 43], has been widely used to explore the molecular mechanisms and therapeutic strategies of a number of neurological diseases, including Alzheimer's disease [36–38], Parkinson's disease [39–41], and glioblastoma multiforme. Although gene therapy for cerebrovascular diseases including ischemic stroke has shown promising results, it is still in its infancy stage. While overexpression of beneficial genes by exogenous gene transfer has shown to be effective for the treatment of experimental cerebral ischemia, several concerns for gene therapy remain to be addressed. First, in many animal studies, gene transfer is performed before artery occlusion, which is not applicable for clinical trials [9, 35, 44]. Further studies should focus on introducing promising genes post-artery occlusion, as this is more clinically relevant. The second concern related to the transduction efficiency of gene therapy. Although the intraventricular or multiple parenchymal injection method effectively increases target protein expression in the experimental rat brain, the vast size difference between the rodent brain and the human brain challenges efficient expression of genes in human subjects [45–48]. Recently, convection-enhanced AAV gene delivery and multiple AAV serotypes were developed to enhance widespread transduction and broad specificity [49–52]. Applying gene therapy protocols to larger animals for transduction and therapeutic efficiency is important. However, uncontrolled target gene expression can cause some unwanted side effects, such as hemangioma formation. Therefore, it is crucial to develop an efficient technique to control target gene expression [53, 54]. Our lab developed conditional expression of VEGF, using tissue hypoxia as a trigger (under the HIF-1α promoter), to increase cerebral angiogenesis without causing hemangioma formation [9].

Although cerebral ischemia-induced angiogenesis is well documented, currently, very few methods allow monitoring of cerebral vascular changes in real time or the quantitative characterization of cerebral vasculature in rodents. Histological analysis requires sacrificing the animals and is time consuming. In recent years, synchrotron radiation angiography (SRA) has been developed as a powerful tool for characterization of the rodent cerebral vasculature. SRA is known for its high resolution compared to conventional X-ray angiography and magnetic resonance angiography (MRA). Using the Shanghai Synchrotron Radiation Facility (SSRF), our group established an SRA technique to quantitatively study the cerebral vascular morphology in the suture middle cerebral artery occlusion (MCAO) model in real time [55, 56]. Using this technique we can dynamically monitor the changes of cerebral blood flow and diameter of blood vessels in deep brain [35, 57, 58]. Thus, SRA may represent a novel solution to directly and dynamically monitor blood vessel changes.

2 Materials

2.1 AAV-Virus Construction and Production

2.1.1 AAV-Virus Plasmid Construction

1. AAV Helper-Free Gene Delivery and Expression System (Stratagene Inc., La Jolla, CA).
2. Plasmid extraction kit (Axygen, Hangzhou, China).
3. Restriction enzymes and T4 DNA ligase.
4. Agarose gel electrophoresis equipment.
5. Competent cells: *E. coli* strain DH5α.

2.1.2 AAV-Virus Transfection, Purification, and Titer Determination

1. Sterile glassware: 15-cm tissue culture dishes, 24-well tissue culture dishes, 50-mL Corning tubes, 1-mL pipettes.
2. PBS: 137 mM NaCl,2.6 mM KCl,10 mM Na_2HPO_4, and 1.8 mM KH_2PO_4, pH 7.4.
3. 2× HBS: 280 mM NaCl, 1.5 mM $Na_2HPO_4{\cdot}7H_2O$, and 50 mM HEPES, pH 7.1.
4. 0.3 and 0.5 M $CaCl_2$.
5. Lysis buffer: 100 mM Tris–HCl, 150 mM NaCl, pH 8.0.

6. 3 M NaCl.
7. Dialysis buffer: 10 mM HEPES pH 7.4, 140 mM NaCl, 0.1 % Tween-80, 5 % sorbitol.
8. 2× SSC: 300 mM NaCl, 30 mM sodium citrate, pH 7.0.
9. Chloroform.
10. $CsCl_2$.
11. Polyethylene glycol (PEG).
12. DNase.
13. RNase.
14. Proteinase K (New England Biolabs, Ipswich, MA).
15. Dulbecco's modified Eagle's medium with 4.5 g/L glucose and l-glutamine.
16. Fetal bovine serum.
17. Trypsin-EDTA solution.
18. HEK 293 cells (American Type Culture Collection, Gaithersburg, MD).
19. SYBR Premix Ex Taq II (Takara, Dalian, China).
20. High-speed centrifuge and ultra speed centrifuge.
21. Cell culture room equipped with incubators and biological hood.
22. Primers of interest gene for real-time PCR titer determination.

2.2 AAV-Virus Intracerebral Injection and SRA Imaging

Optimally the small animal surgery room needs to be kept quiet and pathogen-free with the room temperature maintained at 25 °C. The following items are recommended for small animal surgery:

1. Sprague Dawley rats, weighing 250–300 g.
2. Operational microscope.
3. Exercise Physiological System (AD Instruments, PowerLab/4SP, Castle Hill, Australia).
4. Temperature controller (Homeothermic Blanket Control Unit, Harvard Apparatus, Cambridge, MA).
5. Dry sterilizer.
6. Stereotaxic frame: Model 900 (David Kopf Instruments, Tujunga, CA).
7. pH/blood gas analyzer (Bayer, Radiolab 248, Tarrytown, NY).
8. High-speed micro drill.
9. Surgical equipment kit (nylon suture, bipolar forceps, micro-scissors for small vessels, surgical scissors for animal skin and tissue use, microforceps, needle holder, skin hook for the exposure, ruler).

10. Hamilton syringe with replaceable beveled needle (World Precision Instruments, Sarasota, FL).
11. Ketamine and xylazine.
12. Contrast agent solution: 300 μL Omnipaque (300 mg I/mL, GE Healthcare), 150 μL saline.
13. Injection pump.
14. PE-tube: PE-10 tube.

3 Methods

3.1 AAV Vector Construction

1. Prepare enough pAAV-MCS vectors for restriction enzyme double digestion using the plasmid extraction kit.
2. The pAAV-NT1 vector can be generated by inserting the chicken NT-1 cDNA into multiple cloning sites in the pAAV-MCS vector. Then, validate the successful ligation by sequencing the plasmid.
3. The plasmid AAV-MCS is included in the AAV Helper-Free Gene Delivery and Expression System. The expression cassette includes the CMV promoter, NT-1 cDNA, and HGH polyadenylation signal (poly A).

3.2 AAV-Virus Production

3.2.1 Transfection

1. Seed 4.5×10^6 HEK 293 cells in 15-cm tissue culture dishes 2 days before transfection. Cells in one 90 % confluent dish can be distributed into 5–8 dishes and cultured in 25 mL DMEM plus 10 % FBS.
2. Co-transfect the AAV vector containing the target gene with pAAV-RC and pHelper vectors into the HEK293 cells by using the calcium phosphate precipitation method as described below. pAAV-RC contains the AAV rep and cap genes (*see* **Note 1**). pHelper has the adenoviral VA, E2A, and E4 regions that mediate AAV vector replication. To transfect one 15-cm dish, mix a total of 50 μg DNA (17 μg DNA of each plasmid) with 1 mL of 0.3 M $CaCl_2$. A mixture for transfection of four dishes can be prepared in one 50-mL Corning tube.
3. Place a sterile pipette into the tube. Gently bubble air while 1 mL (4 mL for four plates) of 2× HBS is added to the tube drop by drop.
4. Distribute the transfection mixture to each plate (2 mL mixture solution/15-cm plate).
5. Incubate the plates at 37 °C for 6 h.
6. Change to medium DMEM containing 2 % FBS. Culture the cells at 37 °C for 48 h.

3.2.2 Purification

1. Dislodge the cells from the dishes by gently pipetting and transfer to 50-mL Corning tubes.
2. Remove the media by centrifugation (1,000 × *g* for 5 min at 4 °C).
3. Resuspend cells (1×10^7 cells/mL) and then lyse with three freeze-and-thaw cycles (alternating between dry ice-ethanol and 37 °C water baths).
4. Centrifuge the lysates at 10,000 × *g* for 15 min at 4 °C to remove the cell debris.
5. Precipitate the cleared supernatant with 0.5 M $CaCl_2$ (final concentration 25 mM) on ice for 1 h.
6. Remove the precipitates by centrifuging at 10,000 × *g* for 15 min at 4 °C.
7. Add NaCl and PEG (8000) to the supernatant to make a final concentration of 620 mM NaCl and 8 % (w/v) PEG.
8. Incubate the supernatants on ice for 3 h.
9. Collect the AAV vector-containing precipitate by centrifugation at 300 × *g* for 30 min at 4 °C and then resuspend.
10. Remove the insoluble substances by centrifugation at 10,000 × *g* for 15 min at 4 °C.
11. Purify the AAV vectors by $CsCl_2$ gradient centrifugation. Add solid $CsCl_2$ to the supernatant to a final concentration of 1.4 g/mL. Centrifuge the samples at 15 °C for 16 h at 223,000 × *g*.
12. Fractionate the gradient (0.5–1 mL/fraction) and assay by real-time PCR to detect the viral particles (*see* Subheading 3.2.3).
13. Pool the AAV vector-containing fractions, put in dialysis cassettes, and dialyze against 1 L (1,000 times the volume of viral fraction) dialysis buffer, three times at 4 °C. Change the dialysis buffer every 2 h.
14. The vectors can also be purified by other methods (*see* **Note 2**).

Optional treatment to eliminate invalid viral nucleic acid and protein debris (*see* **Note 3**):

1. Add DNase and RNase to the final concentration of 1 μg/mL at room temperature.
2. Add an equal volume of chloroform into the viruses, and centrifuge at 10,000 × *g* for 15 min at 4 °C.
3. Collect the supernatant as the viral pool.

3.2.3 Titer Determination

Viral titers are determined by real-time PCR analysis of the DNA particles as described below (*see* **Note 4**). PCR was performed using SYBR Premix Ex Taq II.

Viral Sample Pretreatment

The viral sample can be pretreated to improve the purity of the sample as described below (*see* **Note 5**):

1. To pretreat the viral sample with DNase digestion, incubate 5 μL of the viral sample with 1 μL of DNase I (7,500 U/mL) in a final volume of 50 μL at 37 °C for 30 min. Afterward, inactivate DNase I by incubation at 65 °C for 10 min.
2. Following DNase I digestion, digest the sample with Proteinase K by adding 10 μg of Proteinase K to the viral sample at 50 °C for 60 min. Afterward, inactivate Proteinase K by incubation at 95 °C for 20 min.
3. Analyze 2 μL of the sample by PCR.

Quantitative Real-Time PCR

1. PCR was carried out in a final volume of 20-μL reaction system as below (*see* **Note 6**):

Template or standard plasmid	2 μL
Primer forward (0.4 μM)	0.8 μL
Primer reverse (0.4 μM)	0.8 μL
$MgCl_2$ (25 mM)	3.2 μL
Enzyme mix	2 μL
H_2O	11.2 μL

2. In each qPCR run, a standard curve was generated using a 5 log spanning serial dilution of the vector plasmid pAAV-MCS containing one CMV promoter per plasmid molecule. Serial dilution ranged from 0.01 to 100 pg of the vector plasmid. Each dilution step was measured in triplicate per cycle run. The standard curve was graphed by the average of cycle threshold and log of copies.

   Copies of 1 ng gene fragment can be calculated as follows (copies/mL):

   1×10^{-9}/length of the fragment (bp) $\times 660 \times 6 \times 10^{23} \times 2 \times 103$.
3. The copies of sample can be calculated by the generated formula from the standard curve.

3.2.4 *Toxicity*

The toxicity of the viral stock is checked by infection of HEK293 cells.

1. Seed HEK 293 cells in a 24-well tissue culture dishes, 1×10^5 cells/well. Culture cells in 0.5 mL DMEM with 10 % FBS for 24 h at 37 °C.
2. Then, add 10 μL of testing viral stock to each well and culture with the cells for 24 h.
3. Change the medium 24 h later.

4. Monitor the status of the infected cells everyday for 3–5 days and compare with uninfected control cells. The viral stock is considered toxic if cell death or slower growth is observed.
5. More chloroform precipitation steps can be used, if a viral stock is found to be toxic to cells.

3.3 Middle Cerebral Artery Occlusion (MCAO) Model, AAV-Virus Transplantation, SRA Imaging, and Blood Vessel Quantification

3.3.1 AAV-Virus Injection (for Mouse Model, See Notes 7 and 8)

1. The use of animals is approved by the Institutional Animal Care and Use Committee of Shanghai Jiao Tong University, Shanghai, China.
2. Anesthetize the rat with ketamine/xylazine (100 mg/10 mg/kg *see* **Note 9**) intraperitoneally.
3. Place the rat in a stereotactic frame with a mouth holder.
4. Incise a linear skin over the bregma, and drill a burr hole in the skull 3 mm lateral to the bregma using a handheld driller. Slowly insert a 10-μL syringe into the brain 6 mm under the dura and then slowly withdraw it by 1 mm.
5. Inject 10 μL AAV viral suspension containing 2×10^{10} genome copies of virus into the brain at a rate of 1 μL/min. Withdraw the needle after 15 min of injection.
6. Seal the craniotomy site with bone wax and close the wound.
7. Return the rat to its home cage after it recovers from the anesthesia.

3.3.2 Middle Cerebral Artery Occlusion (MCAO) Model

MCAO surgery is performed as detailed described previously by our lab [59].

3.3.3 SRA Imaging

1. The use of animals is approved by the Institutional Animal Care and Use Committee of Shanghai Jiao Tong University, Shanghai, China.
2. Fill the contrast agent solution into a 10-mL syringe and install on the injection pump.
3. Anesthetize adult rats with ketamine/xylazine. Carefully isolate the left common carotid artery (CCA), the external carotid artery (ECA), and the internal carotid artery (ICA). Then, ligate the pterygopalatine artery (PPA, *see* **Note 10**). Insert the PE-10 tube into the ECA stump and advance to the bifurcation of the CCA. Then, connect the PE-10 tube onto the 10-mL syringe. Make X-ray path vertical to sagittal plan of rat brain (Fig. 1).
4. Imaging setup: SRA was conducted at the X-ray imaging beamline BL13W in Shanghai Synchrotron Radiation Facility (SSRF). The energy of hard X-ray is 33.2 keV and a field of view is up to 40 mm (H)×4 mm (V). The resolution of the charge-coupled device (CCD) is 9 μm.

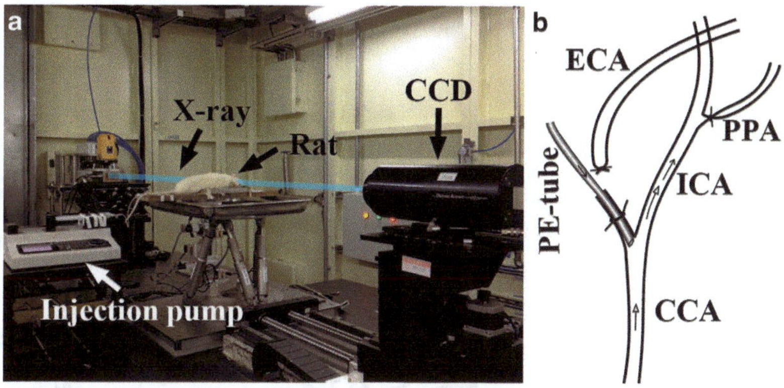

Fig. 1 The imaging system of synchrotron radiation angiography. (**a**) The setup of the imaging system. The X-ray path is vertical to the sagittal plane of rat's brain. (**b**) The schematic diagram of the cannula. The PE-tube was inserted into the external carotid artery (ECA) and then reversed and the PPA was ligated. *Arrows* indicate the direction of blood flow. *CCA* common carotid artery, *ICA* internal carotid artery

5. Imaging procedures and data recording: eight flat-field (baseline) images were recorded following with contrast agent (150 μL) injection at 133.3 μL/s. The frame rate of X-ray CCD camera is 4 fps. Single recording session takes 20 s and 80 images were aquired totally.

3.3.4 Blood Vessel Quantification

1. Data processing: The recorded absorption image was corrected by the flat-field data (Fig. 2a). The vasculature was manually segmented based on the fully perfused absorption images (Fig. 2c, d). Each layer of image was finally stitched together to form a whole angiographic image by Photoshop software.
2. Vascular parameter calculation: A binary segmented image of the vasculature was first used to calculate the vessel density (blood vessel area/total area). Then, the segmented image was skeletonized and Euclidean distance map (from foreground to background) was calculated. The vessel radius was extracted from distance map at the skeleton of each vessel. After that, the vessel length distribution respect to vessel radius was calculated (Fig. 2e).

4 Notes

1. The AAV Helper-Free Gene Delivery and Expression System was generated based on the AAV serotype 2 virus. Many new AAV serotypes have been cloned in recent years [60–63]. Recombinant cross-packaging of the AAV genome of one

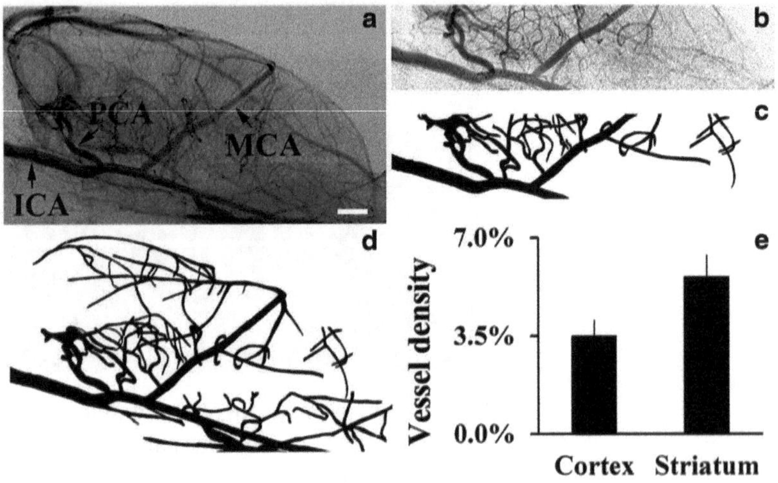

Fig. 2 Vessel density calculation. (**a**) Whole brain vascular image after background correction. *Arrows* indicate MCA (middle cerebral artery), ICA (internal carotid artery), and PCA (posterior cerebral artery), respectively. (**b**) One section of the brain vascular image. (**c**). The brain vascular image of (**b**) after manual segmentation. (**d**). The whole brain vascular image after manual segmentation. (**e**). The statistical graph of vessel density of cortex and striatum in the brain. Bar = 1 cm

serotype into capsids of other AAV serotypes to achieve optimal tissue-specific gene transduction is now possible. Studies by us and others [64] have shown that the AAV serotype 1 results in more efficient transduction of genes in the murine and human adult heart compared with serotypes 2, 3, 4, and 5. AAV serotype 2 has better efficient gene transduction in the brain [35].

2. AAV can be purified using other methods, such as nonionic iodixanol gradients, ion exchange, or heparin affinity chromatography by either conventional or high-performance liquid chromatography columns [65].
3. These optional steps aim to eliminate invalid viral nucleic acid and protein debris. We can repeat the chloroform precipitation step to remove all the protein debris until there are no white precipitates.
4. The AAV titer can also be determined by dot blot [49]. Infectious particles can be determined by using serially diluted viral stocks to infect HEK 293 or other cells and quantifying the infected cells or transgene expression, especially when the vector contains a GFP fraction. Assuming the viral preparations have particle-to-infectivity ratios between 56 and 1,000 [66], we can easily estimate how many valid particles are in a well and then back-calculate to determine how many particles are in each batch.

5. DNase I digestion can be used to eliminate unpackaged DNA in viral stocks and Proteinase K can be used to digest the viral capsid to release DNA [65].
6. A 10-μL reaction system is considered economical, but you can also double or triple your reaction system accordingly.
7. The method for *mouse* AAV-NT-1 gene transfer is similar to the rat model described above. The differences are (1) the bone hole location is situated on the pericranium 2 mm lateral to the sagittal suture and 1 mm posterior to the coronal suture and (2) the total inject volume of the AAV vector is 5 μL.
8. The volume and concentration of AAV-NT-1 can be adjusted depending on the experimental design.
9. Proper anesthesia is very important for animal studies. The optimal anesthetic and analgesic will vary depending on the surgical or experimental procedure and the strain of animal. In addition, anesthesia is neuroprotective, which should be considered carefully during animal models.
10. The PPA should be ligated to avoid disturbing the brain vascular map while performing SRA.

References

1. Gentile C, Muise-Helmericks RC, Drake CJ (2013) VEGF-mediated phosphorylation of eNOS regulates angioblast and embryonic endothelial cell proliferation. Dev Biol 373:163–175
2. McLeod DS, Hasegawa T, Baba T et al (2012) From blood islands to blood vessels: morphologic observations and expression of key molecules during hyaloid vascular system development. Invest Ophthalmol Vis Sci 53: 7912–7927
3. Krupinski J, Kaluza J, Kumar P et al (1993) Prognostic value of blood vessel density in ischaemic stroke. Lancet 342:742
4. Chopp M, Zhang ZG, Jiang Q (2007) Neurogenesis, angiogenesis, and MRI indices of functional recovery from stroke. Stroke 38:827–831
5. Zhu W, Fan Y, Hao Q et al (2009) Postischemic IGF-1 gene transfer promotes neurovascular regeneration after experimental stroke. J Cereb Blood Flow Metab 29:1528–1537
6. Lapchak PA, Araujo DM (2007) Advances in ischemic stroke treatment: neuroprotective and combination therapies. Expert Opin Emerg Drugs 12:97–112
7. Gursoy-Ozdemir Y, Yemisci M, Dalkara T (2012) Microvascular protection is essential for successful neuroprotection in stroke. J Neurochem 123(Suppl 2):2–11
8. Terpolilli NA, Kim SW, Thal SC et al (2012) Inhalation of nitric oxide prevents ischemic brain damage in experimental stroke by selective dilatation of collateral arterioles. Circ Res 110:727–738
9. Shen F, Su H, Fan Y et al (2006) Adeno-associated viral-vector-mediated hypoxia-inducible vascular endothelial growth factor gene expression attenuates ischemic brain injury after focal cerebral ischemia in mice. Stroke 37:2601–2606
10. Sbarbati A, Pietra C, Baldassarri AM et al (1996) The microvascular system in ischemic cortical lesions. Acta Neuropathol 92:56–63
11. Krupinski J, Kaluza J, Kumar P et al (1994) Role of angiogenesis in patients with cerebral ischemic stroke. Stroke 25:1794–1798
12. Gunsilius E, Petzer AL, Stockhammer G et al (2001) Serial measurement of vascular endothelial growth factor and transforming growth factor-beta1 in serum of patients with acute ischemic stroke. Stroke 32:275–278
13. Yu SW, Friedman B, Cheng Q et al (2007) Stroke-evoked angiogenesis results in a transient population of microvessels. J Cereb Blood Flow Metab 27:755–763

14. Marti HH (2005) Angiogenesis - a self-adapting principle in hypoxia. EXS 163–180
15. Bernaudin M, Marti HH, Roussel S et al (1999) A potential role for erythropoietin in focal permanent cerebral ischemia in mice. J Cereb Blood Flow Metab 19:643–651
16. Herzog S, Sager H, Khmelevski E et al (2002) Collateral arteries grow from preexisting anastomoses in the rat hindlimb. Am J Physiol Heart Circ Physiol 283:H2012–H2020
17. Wei L, Erinjeri JP, Rovainen CM et al (2001) Collateral growth and angiogenesis around cortical stroke. Stroke 32:2179–2184
18. McCown TJ (2011) Adeno-associated virus (AAV) vectors in the CNS. Curr Gene Ther 11:181–188
19. Shen F, Su H, Liu W et al (2006) Recombinant adeno-associated viral vector encoding human VEGF165 induces neomicrovessel formation in the adult mouse brain. Front Biosci 11:3190–3198
20. Bellomo M, Adamo EB, Deodato B et al (2003) Enhancement of expression of vascular endothelial growth factor after adeno-associated virus gene transfer is associated with improvement of brain ischemia injury in the gerbil. Pharmacol Res 48:309–317
21. Tsai TH, Chen SL, Chiang YH et al (2000) Recombinant adeno-associated virus vector expressing glial cell line-derived neurotrophic factor reduces ischemia-induced damage. Exp Neurol 166:266–275
22. Shen F, Kuo R, Milon-Camus M et al (2013) Intravenous delivery of adeno-associated viral vector serotype 9 mediates effective gene expression in ischemic stroke lesion and brain angiogenic foci. Stroke 44:252–254
23. Gao P, Shen F, Gabriel RA et al (2009) Attenuation of brain response to vascular endothelial growth factor-mediated angiogenesis and neurogenesis in aged mice. Stroke 40:3596–3600
24. Sun H, Le T, Chang TT et al (2011) AAV-mediated netrin-1 overexpression increases peri-infarct blood vessel density and improves motor function recovery after experimental stroke. Neurobiol Dis 44:73–83
25. Weber M, Rabinowitz J, Provost N et al (2003) Recombinant adeno-associated virus serotype 4 mediates unique and exclusive long-term transduction of retinal pigmented epithelium in rat, dog, and nonhuman primate after subretinal delivery. Mol Ther 7:774–781
26. Zabner J, Seiler M, Walters R et al (2000) Adeno-associated virus type 5 (AAV5) but not AAV2 binds to the apical surfaces of airway epithelia and facilitates gene transfer. J Virol 74:3852–3858
27. Davidson BL, Stein CS, Heth JA et al (2000) Recombinant adeno-associated virus type 2, 4, and 5 vectors: transduction of variant cell types and regions in the mammalian central nervous system. Proc Natl Acad Sci U S A 97: 3428–3432
28. Serafini T, Kennedy TE, Galko MJ et al (1994) The netrins define a family of axon outgrowth-promoting proteins homologous to C. elegans UNC-6. Cell 78:409–424
29. Fazeli A, Dickinson SL, Hermiston ML et al (1997) Phenotype of mice lacking functional Deleted in colorectal cancer (Dcc) gene. Nature 386:796–804
30. Hong K, Hinck L, Nishiyama M et al (1999) A ligand-gated association between cytoplasmic domains of UNC5 and DCC family receptors converts netrin-induced growth cone attraction to repulsion. Cell 97:927–941
31. Park KW, Crouse D, Lee M et al (2004) The axonal attractant Netrin-1 is an angiogenic factor. Proc Natl Acad Sci U S A 101:16210–16215
32. Castets M, Coissieux MM, Delloye-Bourgeois C et al (2009) Inhibition of endothelial cell apoptosis by netrin-1 during angiogenesis. Dev Cell 16:614–620
33. Wilson BD, Ii M, Park KW et al (2006) Netrins promote developmental and therapeutic angiogenesis. Science 313:640–644
34. Fan Y, Shen F, Chen Y et al (2008) Overexpression of netrin-1 induces neovascularization in the adult mouse brain. J Cereb Blood Flow Metab 28:1543–1551
35. Lu H, Wang Y, He X et al (2012) Netrin-1 hyperexpression in mouse brain promotes angiogenesis and long-term neurological recovery after transient focal ischemia. Stroke 43:838–843
36. Tuszynski MH (2007) Nerve growth factor gene therapy in Alzheimer disease. Alzheimer Dis Assoc Disord 21:179–189
37. Edelstein ML, Abedi MR, Wixon J et al (2004) Gene therapy clinical trials worldwide 1989–2004 - an overview. J Gene Med 6:597–602
38. Tuszynski MH, Thal L, Pay M et al (2005) A phase 1 clinical trial of nerve growth factor gene therapy for Alzheimer disease. Nat Med 11:551–555
39. Cederfjall E, Sahin G, Kirik D et al (2012) Design of a single AAV vector for coexpression of TH and GCH1 to establish continuous DOPA synthesis in a rat model of Parkinson's disease. Mol Ther 20:1315–1326
40. Muramatsu S, Fujimoto K, Kato S et al (2010) A phase I study of aromatic L-amino acid decarboxylase gene therapy for Parkinson's disease. Mol Ther 18:1731–1735

41. Koprich JB, Johnston TH, Huot P et al (2011) Progressive neurodegeneration or endogenous compensation in an animal model of Parkinson's disease produced by decreasing doses of alpha-synuclein. PLoS One 6:e17698
42. Colombo F, Barzon L, Franchin E et al (2005) Combined HSV-TK/IL-2 gene therapy in patients with recurrent glioblastoma multiforme: biological and clinical results. Cancer Gene Ther 12:835–848
43. Rainov NG (2000) A phase III clinical evaluation of herpes simplex virus type 1 thymidine kinase and ganciclovir gene therapy as an adjuvant to surgical resection and radiation in adults with previously untreated glioblastoma multiforme. Hum Gene Ther 11:2389–2401
44. Zhu W, Fan Y, Frenzel T et al (2008) Insulin growth factor-1 gene transfer enhances neurovascular remodeling and improves long-term stroke outcome in mice. Stroke 39: 1254–1261
45. Sun Y, Jin K, Clark KR et al (2003) Adeno-associated virus-mediated delivery of BCL-w gene improves outcome after transient focal cerebral ischemia. Gene Ther 10:115–122
46. Dayton RD, Wang DB, Klein RL (2012) The advent of AAV9 expands applications for brain and spinal cord gene delivery. Expert Opin Biol Ther 12:757–766
47. Colle MA, Piguet F, Bertrand L et al (2010) Efficient intracerebral delivery of AAV5 vector encoding human ARSA in non-human primate. Hum Mol Genet 19:147–158
48. Li J, Daly TM (2002) Adeno-associated virus-mediated gene transfer to the neonatal brain. Methods 28:203–207
49. Rabinowitz JE, Rolling F, Li C et al (2002) Cross-packaging of a single adeno-associated virus (AAV) type 2 vector genome into multiple AAV serotypes enables transduction with broad specificity. J Virol 76:791–801
50. Young LS, Searle PF, Onion D et al (2006) Viral gene therapy strategies: from basic science to clinical application. J Pathol 208:299–318
51. Hadaczek P, Kohutnicka M, Krauze MT et al (2006) Convection-enhanced delivery of adeno-associated virus type 2 (AAV2) into the striatum and transport of AAV2 within monkey brain. Hum Gene Ther 17:291–302
52. Cunningham J, Pivirotto P, Bringas J et al (2008) Biodistribution of adeno-associated virus type-2 in nonhuman primates after convection-enhanced delivery to brain. Mol Ther 16:1267–1275
53. Vilaboa N, Voellmy R (2006) Regulatable gene expression systems for gene therapy. Curr Gene Ther 6:421–438
54. Goverdhana S, Puntel M, Xiong W et al (2005) Regulatable gene expression systems for gene therapy applications: progress and future challenges. Mol Ther 12:189–211
55. Yuan F, Wang Y, Guan Y et al (2013) Real-time imaging of mouse lenticulostriate artery following brain ischemia. Front Biosci (Elite Ed) 5:517–524
56. Yuan F, Tang Y, Lin X et al (2012) Optimizing suture middle cerebral artery occlusion model in C57BL/6 mice circumvents posterior communicating artery dysplasia. J Neurotrauma 29:1499–1505
57. Cai J, Sun Y, Yuan F et al (2012) A novel intravital method to evaluate cerebral vasospasm in rat models of subarachnoid hemorrhage: a study with synchrotron radiation angiography. PLoS One 7:e33366
58. Cai J, He C, Yuan F et al (2012) A novel haemodynamic cerebral aneurysm model of rats with normal blood pressure. J Clin Neurosci 19:135–138
59. Su H, Yang GY (2011) Treatment of focal brain ischemia with viral vector-mediated gene transfer. Methods Mol Biol 686:429–446
60. Chiorini JA, Kim F, Yang L et al (1999) Cloning and characterization of adeno-associated virus type 5. J Virol 73:1309–1319
61. Chiorini JA, Yang L, Liu Y et al (1997) Cloning of adeno-associated virus type 4 (AAV4) and generation of recombinant AAV4 particles. J Virol 71:6823–6833
62. Rutledge EA, Halbert CL, Russell DW (1998) Infectious clones and vectors derived from adeno-associated virus (AAV) serotypes other than AAV type 2. J Virol 72:309–319
63. Xiao W, Chirmule N, Berta SC et al (1999) Gene therapy vectors based on adeno-associated virus type 1. J Virol 73:3994–4003
64. Du L, Sullivan CC, Chu D et al (2003) Signaling molecules in nonfamilial pulmonary hypertension. N Engl J Med 348:500–509
65. Zolotukhin S, Potter M, Zolotukhin I et al (2002) Production and purification of serotype 1, 2, and 5 recombinant adeno-associated viral vectors. Methods 28:158–167
66. Rohr UP, Wulf MA, Stahn S et al (2002) Fast and reliable titration of recombinant adeno-associated virus type-2 using quantitative real-time PCR. J Virol Methods 106:81–88

Chapter 27

Investigating the Role of Perlecan Domain V in Post-Ischemic Cerebral Angiogenesis

Aileen Marcelo and Gregory Bix

Abstract

Cerebral angiogenesis is an important process for physiological events such as brain development, but it also occurs in pathological conditions such as stroke. Defined as the generation of new blood vessels from preexisting vasculature, angiogenesis after ischemic stroke is important to limit the subsequent neuronal injury and death, as well as contribute to neurorepair. However, current therapies for ischemic stroke are largely focused on reestablishing uninterrupted blood flow, an important but inherently risky proposition. Furthermore, these therapies can have limited efficacy due to narrow therapeutic windows, and in the case of mechanical clot removal, are invasive procedures. Therefore, better stroke therapies are needed. Since the brain possesses mechanisms, including angiogenesis, to attempt self-repair after injury, it may prove beneficial to look at how such mechanisms are regulated to identify potential targets for new and improved stroke therapies. Perlecan domain V (DV), an endogenous extracellular matrix protein fragment, may represent one such therapeutic target. Key to its appeal is that perlecan DV is endogenously and persistently generated in the brain after stroke and has significant angio-modulatory properties. These, and other properties, have been therapeutically manipulated to improve experimental stroke outcomes, suggesting that DV could represent a promising new stroke therapy. Here we discuss a novel approach to studying DV-mediated angiogenesis in vitro using a coculture model.

Key words Perlecan, Domain V, Angiogenesis, Ischemia

1 Introduction

1.1 Background

Stroke is the fourth leading cause of death in the USA and the number one cause of significant morbidity [1]. It results from either reduced blood flow to areas of the brain via a blood clot, or by the rupture of a brain blood vessel. To date, tissue plasminogen activator (t-PA) is the only FDA-approved pharmaceutical treatment for ischemic stroke, and acts by dissolving the clot. However, its therapeutic window is very narrow (3–4.5 h), which reduces its "real world" effectiveness. If administered outside this window, there is potential for hemorrhagic transformation, making a bad situation that much worse. Mechanical clot retrieval, with a similar goal of cerebrovascular recanalization, may also be performed for

Richard Milner (ed.), *Cerebral Angiogenesis: Methods and Protocols*, Methods in Molecular Biology, vol. 1135,
DOI 10.1007/978-1-4939-0320-7_27, © Springer Science+Business Media New York 2014

particularly large clots that might not be otherwise pharmacologically dissolved. Therapy for hemorrhagic stroke includes surgery and other medical therapies, but with limited efficacy. Therefore, there is a very great need for better therapeutic interventions to treat both ischemic and hemorrhagic strokes.

Until quite recently, most experimental stroke therapies have focused on protecting neurons. However, these treatments have failed spectacularly to translate in clinical trials for a large number of reasons [2]. Interestingly, the brain does try to repair itself after stroke with, unfortunately, quite limited success. Angiogenesis, the formation of new blood vessels from preexisting ones, occurs to induce collateral circulation, and is also coupled with neurogenesis, the formation of new neurons. Potentially, the best stroke therapies might be those that focus on both neuroprotective and neuroreparative mechanisms. Current research in our laboratory has focused on enhancing endogenous neuroprotective and angiogenesis repair mechanisms. Here, the role of perlecan domain V, an extracellular matrix protein, is discussed as a regulator of postischemic cerebral angiogenesis. Furthermore, a method to study brain angiogenesis in vitro is described.

1.2 Stroke

According to the American Stroke Association [1], stroke is the fourth leading cause of death in the USA. Patients who survive stroke can have debilitating effects such as physical, emotional, and mental impairment, and indeed, stroke is the number one cause of serious morbidity. Strokes can be classified into two main groups: ischemic and hemorrhagic. Ischemic strokes occur when there is an obstruction, typically a blood clot in a cerebral blood vessel, reducing or eliminating blood flow to a part of the brain. These clots are due to either (1) cerebral thrombosis, where the clot forms in the blood vessel, or (2) cerebral embolism, in which the clot, or embolus, forms in a different part of the circulatory system, travels to the brain, and reaches a vessel that is too small for the clot to pass. Hemorrhagic strokes occur when a weakened blood vessel ruptures, and the blood compresses the surrounding brain tissue. This type of stroke can be a result of either an aneurysm, where the vessel wall balloons and eventually bursts, or an arteriovenous malformation (AVM), where there is abnormal blood vessel growth without a capillary bed [3].

Current treatment for ischemic strokes is limited to the FDA-approved drug, tissue plasminogen activator (t-PA). It works by breaking up the clot; however, the window of opportunity for effective treatment is limited to 3–4.5 h after stroke [4]. After this time frame, administration of t-PA is risky, with increased potential for brain hemorrhage, making a bad situation that much worse. The clinical problem is to accurately pin point the time of stroke onset as some strokes can occur during sleep (so called "wakeup strokes") or the patient may be unaware of the stroke event.

Other treatments have focused beyond the 3-h window. One method is the mechanical retrieval system, in which a cork screw-shaped device wraps around the clots to remove it and restore blood flow. Another method is the penumbra system, which allows for revascularization of the occluded vessel and reestablishment of blood flow. Unfortunately, cerebrovascular recanalization can lead to reperfusion injury. Hemorrhagic stroke is treated with surgery by placing a clip at the base of the aneurysm or removing the AVM. Additionally, there are also endovascular procedures that involve inserting a catheter that releases a mechanical agent to prevent the rupture of an aneurysm or AVM. These methods have been approved by the FDA, but they are invasive procedures with inherent risk. Collectively, treatment for both types of strokes is limited, risky, and flawed. Thus, new stroke therapies are desperately needed.

Most research on stroke treatments has focused on the excitotoxicity, oxidative stress, and apoptosis pathways that occur after stroke. Reduction of blood flows results in a loss of ATP in neurons, which affects the various ionic gradients that keep neurons functioning properly. Additionally, excessive neurotransmitter release can occur. Excessive glutamate (a neurotransmitter that is important for neuronal synaptic transmission) release can be excitotoxic and impair calcium homeostasis [5]. Mitochondrial mechanisms such as oxidative phosphorylation begin to fail and generation of reactive oxygen species (ROS) increases. ROS can lead to the activation of apoptotic pathways in neurons.

It is important to stress that stroke is not exclusively a neuronal or vascular deficit. Research has focused on integration and coordination of neurons, astrocytes, pericytes, and endothelial cells of the cerebrovasculature, which collectively has been termed the "neurovascular unit." With astrocytes acting as an intermediary, endothelial cells and neurons exhibit bidirectional communication. Neurons are highly metabolic, and their interaction with the cerebral endothelium ensures that neurons receive the appropriate nutrients and energy to survive. However, stroke events disrupt this interaction. Lack of blood flow results in energy failure in neurons. Neuronal cells in the core region of an ischemic stroke suffer severe deficits while the penumbra (ischemic but still viable region surrounding the ischemic core) exhibits a lesser insult due to the reperfusion of the region. Cell death pathways are activated, and disruption in the blood–brain barrier (BBB), a protective vascular interface between the brain and circulatory system, occurs in early neurovascular damage [6].

BBB function is dependent on interactions with the extracellular matrix (ECM). The matrix might seem like a static environment, acting like a scaffold for blood vessels and neurons. However, the matrix is highly dynamic, and its disruption can be detrimental to BBB integrity. The brain ECM consists primarily of fibronectin, laminin, type IV collagen, and heparan sulfate proteoglycans. Proteolysis of the matrix by various matrix metalloproteinases (MMPs) can result in its degradation, which disrupts the integrity of the BBB. Research has shown that an increase in MMP-9 correlates with an increase in infarct volume in patients who experience acute ischemic stroke [7]. In animal models of ischemia, MMP-9 increased and is related to BBB disruption [8]. MMP-9 also increased after t-PA administration and can lead to cell detachment from the extracellular matrix [9]. In addition to MMP-9, studies have demonstrated that MMP-2 also increases in during ischemia [10]. While matrix proteolysis during and after stroke clearly has its downside, it is also possible that such an event could release matrix fragments that could be of some benefit to the injured brain [11].

1.3 Angiogenesis and Perlecan Domain V

Angiogenesis is an important process in physiological and pathological events. In the postnatal brain, angiogenesis is down-regulated. However, angiogenesis is up-regulated after stroke and other forms of brain injury and is believed, along with neurogenesis, to be an important mechanism of the brain's attempt to repair itself [12].

One particular ECM component that has been studied as a potential candidate for stroke therapy is perlecan. Perlecan, a heparan sulfate proteoglycan, is found within the basement membrane of blood vessels (vascular basement membrane). It is also the major heparan sulfate proteoglycan secreted by endothelial cells. Perlecan is involved in cell proliferation, tumorigenesis, angiogenesis, and bone formation [13]. Lack of perlecan causes embryonic lethality at day 10.5 in mice [14], when vasculogenesis and angiogenesis occur. These embryos exhibit chondrodysplasia and defects in the heart outflow tract, and most die in utero. Perlecan-deficient mice have normal development of the basement membrane. However, the areas of the basement membrane that experience high mechanical stress will deteriorate [15]. Perlecan binds to various growth factors, such as fibroblast growth factor (FGF) and vascular endothelial growth factor (VEGF), as well as other ECM components such as collagen IV, laminin and fibronectin. When perlecan binds to FGF, it protects this growth factor from proteolytic degradation and also acts as a sink for FGF, which can be released later to interact with its receptor [16]. Additional studies have also shown that internalization of bFGF involves perlecan, which could play a role in maintaining BBB function [17]. Mice deficient in perlecan heparan sulfate (HS) demonstrated decreased wound healing and impaired tumor growth and angiogenesis [18].

Perlecan is highly conserved across species. It consists of five domains, with domain V (DV), its C-terminal domain, having both pro- and anti-angiogenic properties upon proteolysis from full length perlecan. Three heparan sulfate chains found at the N-terminus are important for growth factor interactions, and are able to modulate their signaling and serve as a growth factor sink. DV has three laminin globular (LG) domains separated by four EGF repeats [19]. Importantly, after experimental stroke in non-human primates, perlecan is the most sensitive ECM component to be proteolytically cleaved [20]. Both DV and LG3 are proteolytically cleaved by cathepsins and BMP-1/Tolloid-like MMPs [21]. These proteases are increased in the brain after stroke.

DV has been shown to possess both anti- and pro-angiogenic properties through interaction with integrin receptors. Upon binding to the $\alpha_2\beta_1$ integrin, DV inhibits angiogenesis. The interaction between DV and $\alpha_2\beta_1$ integrin receptor increases cAMP and activates protein kinase A [22]. Additionally, DV activates focal adhesion kinase, p38 MAPK and Hsp 27, of which the latter two proteins are down-regulated leading to disassembly of the actin cytoskeleton in human umbilical vein and dermal microvascular endothelial cells [22]. It has been demonstrated that perlecan domain V binds to $\alpha_2\beta_1$ integrin and to VEGF receptor 2 (VEGFR2) in a region other than the VEGF-A binding site [23]. Furthermore, it has been shown that LG1 and 2 domains of DV block pro-angiogenic signaling by VEGF-A by binding to VEGFR2 in endothelial cells [24].

Studies have shown that this is not the case in the brain. Interestingly, the brain microvasculature does not express $\alpha_2\beta_1$ [25]. Instead, DV has been shown to bind to the $\alpha_5\beta_1$ integrin expressed on activated brain endothelial cells after stroke. Expression of the α_5 and β_1 subunits is increased by collagen I, collagen IV, and fibronectin, but perlecan causes the most significant increase in integrin expression [26]. It is also interesting to point out that $\alpha_5\beta_1$ is expressed on brain endothelial cells during brain development, is largely absent in these cells in the postnatal brain, and re-expressed during brain injury such as a stroke. DV and $\alpha_5\beta_1$ are both increased post stroke [27]. In human stroke and several different stroke animal models, free DV is rapidly (within 24 h) and persistently (for at least 15 days after stroke) generated in the brain due to proteolytic cleavage of perlecan [27]. Figure 1 demonstrates that DV is present and/or is generated in the brains of human stroke patients. The importance of such post-stroke DV generation is underscored by worsened post-stroke brain injury and recovery in perlecan reduced mice (express 10 % of normal perlecan levels) [27]. Therefore, we reasoned that administering exogenous DV after stroke might be of therapeutic benefit. In vitro, DV interacts with the brain endothelial $\alpha_5\beta_1$ integrin receptor, leading to an increase in vascular endothelial growth factor (VEGF), a potent angiogenic factor. VEGF is up-regulated during physiological and pathological angiogenesis [28]. It is also

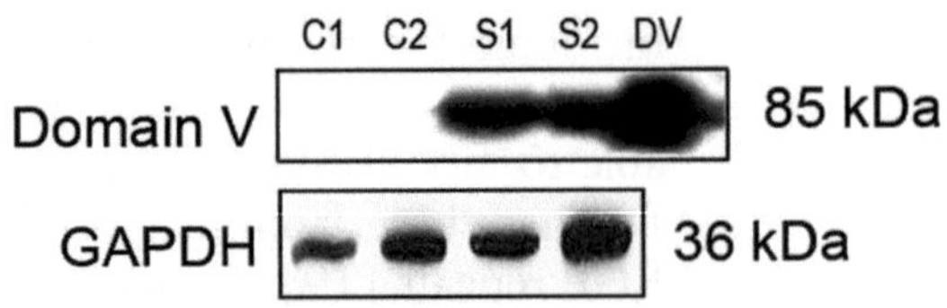

Fig. 1 DV is present and/or generated in human stroke patients. Western blot analysis strongly showed the presence of DV in human cerebral tissue samples (kindly provided by Dr. Kunlin Kim, Buck Institute, CA), whereas there was complete absence of DV in control tissues. C1 and C2 denote two independent non-stroked human brain tissue samples, and S1 and S2 denote samples from two hemorrhagic stroke patients (middle cerebral artery rupture, surgery performed within 24 h). Purified recombinant human DV (rhDV) was loaded as a positive control. GAPDH protein loading control is shown for the four brain tissue samples

important for neurogenesis, and decreased VEGF levels have been shown to lead to neurodegeneration [29]. Indeed, VEGF is able to activate angiogenesis in the ischemic penumbra, and also increase neuron survival in the dentate gyrus and subventricular zone [12]. Studies demonstrated that post DV treatment results in increased VEGF levels via an $\alpha_5\beta_1$-mediated mechanism, and the neuroprotective and pro-angiogenic activity was mediated by VEGFR2 [27]. This $\alpha_5\beta_1$-mediated mechanism leads to the phosphorylation of ERK and subsequent activation of eIF4E and HIF-1α, the regulator of VEGF expression [30].

Additionally, it has been shown that the C-terminal fragment of DV, laminin-like globulin domain 3 (LG3) is also increased in ischemic tissue [31]. Our laboratory has demonstrated that this LG3 increase occurs in neurons and pericytes after oxygen–glucose deprivation (OGD) [31, 32]. Thus, LG3 is neuroprotective and pro-angiogenic following OGD. This increase in LG3 is likely the result of increased perlecan synthesis and secretion [33]. Furthermore, cathepsin-L and cathepsin-B are increased, which is important for LG3 cleavage under normal conditions and following OGD, respectively [33].

Why should we care about perlecan domain V being a candidate for stroke therapy? Aside from the evidence supporting its role in angiogenesis, DV is generated by stroke, experimentally in both animal and human models. Furthermore, as mentioned previously, DV is present in ischemic stroke both acutely and chronically [11]. Proteases are released after stroke and break down ECM, which may seem detrimental, but as previously mentioned may also lead to release of beneficial ECM-derived protein fragments such as DV [11]. Furthermore, studies in our lab have demonstrated that DV administered systemically is able to reach the stroke region of the brain in transient middle cerebral artery occlusion models [27].

Surprisingly, DV confers neuroprotection when administered by intraperitoneal (I.P.) injection 24 h after stroke and every other day.

DV treatment decreased expansion of the ischemic stroke lesion, and resulted in motor function recovery to pre-stroke levels [27]. Furthermore, LG3 was neuroprotective in fetal cortical neurons subjected to oxygen-glucose deprivation in vitro [31]. Preliminary data in our lab also demonstrates that DV promotes neurogenesis, neuroblast migration, and synaptic connectivity in vitro and in vivo after stroke. DV treatment also decreased astrogliosis, and inhibited glial scar formation [34]. Lastly, DV administration increased peri-infarct angiogenesis by increasing VEGF through $\alpha 5\beta 1$ integrin [27]. Interestingly, despite this increase in VEGF, acute DV administration does not promote increased BBB breakdown and subsequent hemorrhage [27], a problem with acute post-stroke VEGF administration [35] suggesting that DV may have other, as yet uncharacterized, therapeutic mechanisms of action.

2 Materials

Since angiogenesis is modulated by DV and various cell types in the neurovascular unit, the following protocol outlines a novel assay to measure tube formation, an indicator of angiogenesis, in a coculture system of brain endothelial cells and neurons.

1. Dulbecco's Modified Eagle Medium (DMEM) containing 10 % fetal bovine serum (FBS), 10 % horse serum, glucose (6 mM), insulin (10 mg/mL), glutamine (200 nM), 50 U/mL penicillin, 50 μg/mL streptomycin, 10 μg/mL human transferrin, 10 μg/mL insulin, and 10 ng/mL selenium.
2. 4 % Paraformaldehyde (w/v), permeabilized with 0.1 % Triton-X100.
3. Laminin.
4. 1× Phosphate buffer saline (PBS).
5. Polystyrene centrifuge tubes.
6. Bovine serum albumin.
7. Conical tubes.
8. 20-, 74-, and 149-μm nylon fiber filters that fit over 50 mL conical tubes.
9. Collagenase/dispase.
10. Papain.
11. Iscove's Modified Dulbecco's Medium (IMDM) supplemented with 20 μg/mL endothelial cell growth supplement (ECGS), 50 μg/mL heparin, and 10 % heat-inactivated FBS.
12. Antibiotic–Antimycotic solution (100×).
13. VEGF.

14. Primary antibodies raised against von Willebrand factor and Tuj-1.
15. Fluorochrome-conjugated secondary antibodies.
16. Potential pro-angiogenic factor to be tested.
17. Rat cerebellar granule neurons.
18. Mouse or rat brain endothelial cells (BECs).

3 Methods

3.1 Primary Culture of Cerebellar Neurons [36]

1. Anesthetize 8-day-old pups and carefully remove the cerebella.
2. Digest cerebellar tissue with papain and triturate to dissociate cells.
3. Place the dissociated cells from a single rat cerebellum into a sterile 50 mL polystyrene centrifuge tube and incubate overnight at 37 °C in 5 % CO_2 in 20 mL DMEM. Loosen the cap of the centrifuge tube to allow for proper aeration.
4. *See* Subheading 3.3 for continuation of coculture.

3.2 Brain Endothelial Cell (BEC) Isolation

1. The protocol is performed based on a previously published protocol [37].
2. After removing brains from anesthetized 3-month-old mice or rats, remove the meninges and mince the brains in PBS supplemented with 0.1 % bovine serum albumin (PBSA).
3. Centrifuge the minced brains at 250 × *g* for 10 min, and resuspend the resulting pellet in 15 mL PBSA.
4. Homogenize the suspension with a Dounce tissue grinder, and pass the suspension successively through 149-, 74-, and 20-μm nylon fiber filters.
5. Wash the filtrate with PBSA at 450 × *g* for 5 min and discard the supernatant.
6. Digest the pellet and the 20- and 74-μm filters in separate 50 mL conical tubes with PBS containing 0.2 % collagenase/dispase for 3 h at 37 °C. Manually shake the tubes three times per hour.
7. Centrifuge the digested material at 150 × *g* for 5 min. Resuspend the pellets from both filters and from the filtrate in 40 mL each of IMDM.
8. Add antibiotic–antimycotic solution (to 1× final concentration) to the medium for the first eight passages.
9. Plate cells on coated dishes and incubate at 37 °C with 5 % CO_2 for 4 h.

10. Wash the cell cultures two times with PBS or serum-free medium and add fresh growth medium. Change the medium twice a week.

3.3 Establishment of the Neuronal–Endothelial Coculture System

1. Seed neurons at 5×10^4 cells/well in an eight-chamber well slide that has been precoated with laminin.
2. The following day, add 3×10^4 brain endothelial cells (BECs) to the neuron-containing wells, and allow them to grow overnight in DMEM containing serum (*see* **Note 1**).
3. Vigorously wash the cells with warm (37 °C) PBS to remove the serum.
4. If testing a potential pro-angiogenic agent, incubate the cells in serum-free DMEM containing this agent (at different concentrations to be determined by the user). VEGF-treated cells are used as a positive control (*see* **Note 2**). If testing a potential anti-angiogenic agent, incubate the cells in serum-free media containing this agent and VEGF (necessary to add for endothelial cell tube formation to occur).

3.4 Tubulogenesis Assay Using Immunofluorescence

1. Fix the cells with 4 % paraformaldehyde, and incubate in primary antibodies raised against von Willebrand factor (to detect brain endothelial cells) or Tuj-1 (to detect neurons).
2. Wash the cells and incubate in fluorochrome-conjugated secondary antibodies to detect the von Willebrand factor and Tuj-1, using a different fluorochrome for each. The cells should be in 1× PBS after immunohistochemistry to avoid drying out.
3. Visualize the cells on a fluorescent microscope.
4. Quantify the tube formation as pixels per high-power field, using Photoshop CS. Measure tube formation from ten random fields per treatment condition.
5. Figure 2 shows the capillary tube-like formation of brain endothelial cells in coculture with neurons in the presence of DV.

4 Notes

1. Prior to the addition of endothelial cells, the neurons can be removed based on previously published protocol [38] so that only the extracellular matrix and attached growth factors remain.
2. Endothelial cells can be added to laminin-coated wells overnight, and then treated with test reagents +/–VEGF in cerebellar granule cell conditioned media.

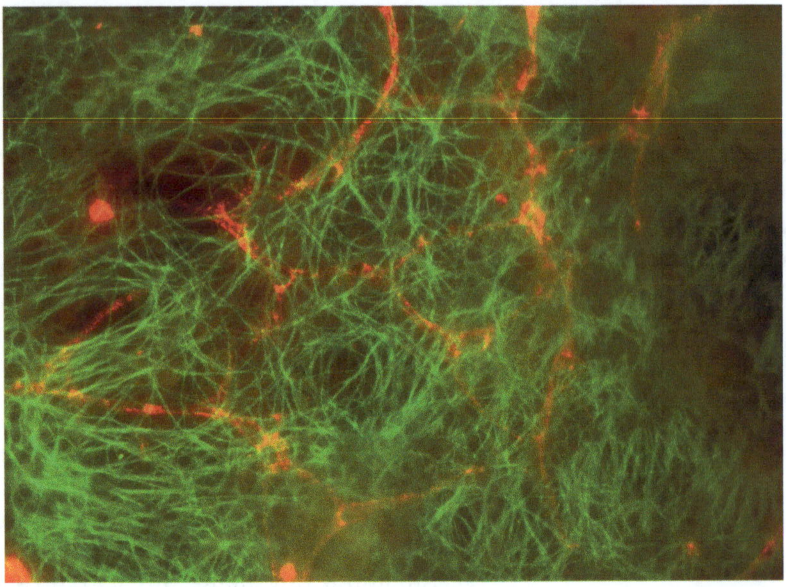

Fig. 2 DV enhances capillary tube-like formation of brain endothelial cells in a coculture system with neurons. Brain endothelial cells (*red* = von Willebrand factor) were cultured over cerebellar granule neurons (*green* = Tuj-1) and treated with DV for 6 h

References

1. American Stroke Association (2013). www.strokeassociation.org
2. Kahle M, Bix G (2013) Neuronal restoration following ischemic stroke: influences, barriers, and therapeutic potential. Neurorehabil Neural Repair 27(5):469–478
3. Kim H, Pawlikowska L, Chen Y, Su H, Yang G, Young W (2009) Brain arteriovenous malformation biology relevant to hemorrhage and implication for therapeutic development. Stroke 40:S95–S97
4. Davis S, Donnan G (2009) 4.5 hours: the new time window for tissue plasminogen activator in stroke. Stroke 40:2266–2267
5. Mehta A, Prabhakar M, Kumar P, Deshmukh R, Sharma P (2013) Excitotoxicity: bridge to various triggers in neurodegenerative disorders. Eur J Pharmacol 698:6–18
6. Arai K, Lok J, Guo S, Hayakawa K, Xing C, Lo E (2011) Cellular mechanism of neurovascular damage and repair after stroke. J Child Neurol 26(9):1193–1198
7. Demir R, Ulvi H, Ozel L, Ozdemir G, Guzelcik M, Aygul R (2012) Relationship between plasma metalloproteinase-9 levels and volume and severity of infarct in patients with acute ischemic stroke. Acta Neurol Belg 112:351–356
8. Ramos-Fernandez M, Bellolio M, Stead L (2011) Matrix metalloproteinase-9 as a marker for acute ischemic stroke: a systemic review. J Stroke Cerebrovasc Dis 20(1):47–54
9. Sumii T, Lo E (2002) Involvement of matrix metalloproteinase in thrombolysis-associated hemorrhagic transformation after embolic focal ischemia in rats. Stroke 33:831–836
10. Horstmann S, Kalb P, Koziol J, Gardner H, Wagner S (2003) Profiles of matrix metalloproteinases, their inhibitors, and laminin in stroke patients: influence of different therapies. Stroke 34:2165–2170
11. Bix G (2013) Perlecan domain V therapy in stroke: a beacon of hope? ACS Chem Neurosci 4(3):370–374
12. Fan Y, Yang G-Y (2007) Therapeutic angiogenesis for brain ischemia: a brief review. J Neuroimmune Pharmacol 2:284–289
13. Mongiat M, Fu J, Oldershaw R, Greenhalgh R, Gown A, Iozzo R (2003) Perlecan protein core interacts with extracellular matrix protein 1 (ECM1), a glycoprotein involved in bone formation and angiogenesis. J Biol Chem 278(19):17491–17499
14. Arikawa-Hirasawa E, Watanbe H, Takami H, Hassell J, Yamada Y (1999) Perlecan is essential

for cartilage and cephalic development. Nat Genet 23:254–358
15. Costell M, Gustafsson E, Aszodi A, Morgelin M, Bloch W, Hunziker E, Addicks K, Timpl R, Fassler R (1999) J Cell Biol 147(5):1109–1122
16. Segev A, Nili N, Strauss B (2004) The role of perlecan in arterial injury and angiogenesis. Cardiovasc Res 63:603–610
17. Deguchi Y, Okutsu H, Okura T, Yamada S, Kimura R, Yuge T, Furukawa A, Morimoto K, Tachikawa M, Ohtsuki S, Hosoya K, Terasaki T (2002) Internalization of basic fibroblast growth factor at the mouse blood-brain barrier involves perlecan, a heparan sulfate proteoglycan. J Neurochem 83:381–389
18. Zoeller J, Whitelock J, Iozzo R (2009) Perlecan regulates developmental angiogenesis by modulating to VEGF-VEGFR2 axis. Matrix Biol 28:284–291
19. Bix G, Iozzo R (2008) Novel interactions of perlecan: unraveling perlecan's role in angiogenesis. Microsc Res Tech 71:339–348
20. Fukuda S, Fini C, Mabuchi T, Koziol J, Eggleston L, de Zoppo G (2004) Focal cerebral ischemia induces active proteases that degrade microvascular matrix. Stroke 35:998–1004
21. Gonzalez E, Reed C, Bix G, Fu J, Zhang Y, Gopalakrishnan B, Greenspan D, Iozzo R (2005) BMP-1/tolloid-like metalloproteases process endorepellin, the angiostatic C-terminal fragment of perlecan. J Biol Chem 280(8): 7080–7087
22. Bix G, Fu J, Gonzalez E, Macro L, Barker A, Campbell S, Zutter M, Santoro S, Kim J, Hook M, Reed C, Iozzo R (2004) Endorepellin causes endothelial cell disassembly of actin cytoskeleton and focal adhesions through $\alpha 2\beta 1$ integrin. J Cell Biol 166(1):97–109
23. Goyal A, Pal N, Concannon M, Paul M, Doran M, Poluzzi C, Sekiguchi K, Whitelock J, Neill T, Iozzo R (2011) Endorepellin, the angiostatic module of perlecan, interacts with both the $\alpha 2\beta 1$ integrin and vascular endothelial growth factor receptor 2 (VEGFR2): a duel receptor antagonism. J Biol Chem 286(29): 25947–25962
24. Willis C, Poluzzi C, Mongiat M, Iozzo R (2013) Endorepellin LG 1/2 domains bind Ig3-5 of VEGFR2 and block proangiogenic signaling by VEGFA in endothelial cells. FEBS J 280(10):2271–2284
25. McGeer P, Zhu S, Dedhar S (1990) Immunostaining of human brain capillaries by antibodies to very late antigens. J Neuroimmunol 26(3):218–231
26. Milner R, Hung S, Wang X, Berg G, Spatz M, del Zoppo G (2007) Responses to endothelial cell and astrocyte matrix-integrin receptors to ischemia mimic those observed in the neurovascular unit. Stroke 39:191–197
27. Lee B, Clarke D, Al Ahmad A, Kahle M, Parham C, Auckland L, Shaw C, Fidanboylu M, Orr A, Ogunshola O, Fertala A, Thomas S, Bix G (2011) Perlecan domain V is neuroprotective and proangiogenic following ischemic stroke in rodents. J Clin Invest 121(8):3005–3023
28. Ferrera N, Gerber H-P, LeCouter J (2003) The biology of VEGF and its receptors. Nat Med 9(6):669–676
29. Storkebaum E, Carmeliet P (2004) VEGF: a critical player in neurodegeneration. J Clin Invest 113:14–18
30. Clarke D, Al Ahmad A, Lee B, Parham C, Auckland L, Fertala A, Kahle M, Shaw C, Roberts J, Bix G (2012) Perlecan domain V induces VEGF secretion in brain endothelial cells through integrin $\alpha\beta 51$ and ERK-dependent signaling pathways. PLoS One 7(9):e45257
31. Saini M, Pinteaux E, Lee B, Bix G (2011) Oxygen-glucose deprivation and interleukin-1α trigger the release of perlecan LG3 by cells of neurovascular unit. J Neurochem 119:760–771
32. Saini M, Bix G (2012) Oxygen-glucose deprivation (OGD) and interleukin-1 (IL-1) differentially modulate cathepsin B/L mediated generation of neuroprotective perlecan LG3 by neurons. Brain Res 1438:65–74
33. Roberts J, Kahle M, Bix G (2012) Perlecan and the blood-brain barrier: beneficial proteolysis? Front Pharmacol 3:155
34. Al Ahmad A, Lee B, Saini M, Bix G (2011) Perlecan domain V modulates astrogliosis in vitro and after focal cerebral ischemia through multiple receptors and increased nerve growth factor release. Glia 59:1822–1840
35. Yang R, Bunting S, Ko A, Keyt B, Modi N, Zioncheck T, Ferrara N, Jin H (1998) Substantially attenuated hemodynamic responses to Escherichia coli-derived vascular endothelial growth factor given by intravenous infusion compared with bolus injection. J Pharmacol Exp Ther 284(1):103–110
36. Bix G, Clark G (1998) Platelet-activating factor receptor stimulation disrupts neuronal migration in vitro. J Neurosci 18:307–318
37. Sapatino B, Welsh C, Smith C, Bebo B, Linthicum D (1993) Cloned mouse cerebrovascular endothelial cells that maintain their differentiation markers for factor VIII, low density lipoprotein, and angiotensin-converting enzyme. In Vitro Cell Dev Biol 29A:923–928
38. Serebriskii I, Castello-Cros R, Lamb A, Golemis E (2008) Fibroblast-derived 3D matrix differentially conditions the growth and drug-responsiveness of human cancer cells. Matrix Biol 27:573–578

Part VI

Methods to Study Cerebral Angiogenesis In Vitro

Chapter 28

Isolation and Culture of Primary Mouse Brain Endothelial Cells

Jennifer V. Welser-Alves, Amin Boroujerdi, and Richard Milner

Abstract

Blood vessels in the central nervous system (CNS) are unique in forming the blood–brain barrier (BBB), which confers high electrical resistance and low permeability properties, thus protecting neural cells from potentially harmful blood components. Endothelial cells, which form the inner cellular lining of all blood vessels, play a critical role in this process by forming tight adhesive interactions between each other. To study the properties of primary brain endothelial cells (BECs), a number of different methods have been described. In this chapter, we present a relatively simple method that produces high numbers of primary mouse BECs that are highly pure (greater than 99 % CD31-positive). In addition, we also describe an immunocytochemical approach to demonstrate the endothelial purity of these cultures.

Key words Brain endothelial cells (BECs), Papain, Collagen I, Puromycin

1 Introduction

All blood vessels are lined with an internal layer of endothelial cells. Blood vessels in the central nervous system (CNS) are unique in forming the blood–brain barrier (BBB), which confers high electrical resistance and low permeability properties, thus protecting neural cells from potentially harmful blood components [1–5]. Evidence suggests that the molecular basis of the BBB depends on the tight interactions that form between adjacent endothelial cells, as well as the vascular basal lamina that endothelial cells rest on, and the influence of astrocyte end-feet and pericytes [5–7]. Vascular remodeling and BBB breakdown play an integral part in the evolution of many neurological diseases, including ischemic stroke, vascular dementia, multiple sclerosis (MS), and Alzheimer's disease (AD) [8–14]. BBB breakdown is almost always a pathogenic event, resulting in cerebral edema and leukocyte infiltration [15–18]. However, vascular remodeling can either be beneficial, resulting in the production of new high-integrity vessels capable of increasing

Richard Milner (ed.), *Cerebral Angiogenesis: Methods and Protocols*, Methods in Molecular Biology, vol. 1135,
DOI 10.1007/978-1-4939-0320-7_28, © Springer Science+Business Media New York 2014

blood supply to a hypoxic region [19–21], or detrimental, resulting in the production of unstable leaky vessels [18, 22].

As brain endothelial cells (BECs) play such a central role in the processes of vascular remodeling, angiogenesis and establishment and maintenance of the BBB, it is clearly important to study the function and behavior of these important cells. In vivo studies offer the direct advantage of studying these cells in their natural environment, and so have obvious relevance to the study of molecular mechanisms that underlie vascular remodeling or pathological breakdown. However, studying the contribution of individual cell types in their complex three-dimensional context in vivo can be difficult to interpret, and so to examine cellular behavior at a more reductionist level, in vitro systems can provide a useful parallel approach. Indeed, for BECs, many different isolation and cell cultivation approaches have been described. BECs have been isolated from many different species (bovine, rat, mouse, human), from brains of different developmental ages (neonatal, adolescent, and elderly), and using many different types of approaches (selective centrifugation through percoll gradients, selective filtration through different pore-size filters, and selective adhesion) [23–31]. Preparation of a reliable consistent culture of primary BECs is an important prerequisite of any biological study of this cell type, and this becomes even more important when considering the value of using in vitro models of the BBB. While many in vitro BBB studies use immortalized BECs, we believe that such cell lines have inherent limitations and can never entirely faithfully recapitulate the properties of primary BECs. Thus, the value of good culture methods for obtaining primary BECs is paramount.

In this chapter, we describe a method of preparing pure cultures of mouse BECs that is relatively simple and which produces high numbers of cells that are highly pure (greater than 99 % CD31-positive) [30, 32]. It consists of four basic steps: (1) standard enzyme-based tissue disruption, (2) selective adhesion of BECs to collagen I, (3) culture in defined endothelial-optimized medium, and (4) selective survival over non-endothelial cells when cultured with the protein synthesis inhibitor, puromycin. We also describe an immunocytochemical approach to demonstrate the endothelial purity of these cultures.

2 Materials

2.1 Cell Culture

2.1.1 Dissection Equipment

1. Large scissors, straight.
2. Small scissors, curved.
3. Curved forceps.
4. Plastic 100 mm petri dishes, sterile, bacteriological grade (to contain the isolated brains).

5. 15 ml polypropylene centrifuge tubes, sterile.
6. 70 % Ethanol in a spray bottle.

2.1.2 General Equipment

1. Biological safety cabinet.
2. Inverted microscope with ×10 and ×20 objectives.
3. A 37 °C Incubator with humidity and gas control to maintain >95 % humidity and an atmosphere of 5 % CO_2 in air.
4. Pipetman.
5. Low-speed centrifuge.

2.1.3 Tissue Culture Equipment

1. Tissue culture 6-well plates.
2. 15 and 50 ml polypropylene centrifuge tubes.
3. 10 ml plastic serological pipettes.
4. Plastic pipette tips: 1 ml and 200 μl.
5. 10 ml syringes.
6. 19 gauge needles.
7. 21 gauge needles.

2.1.4 Media and Reagents

1. MEM-HEPES with penicillin/streptomycin.
2. Papain Cell Dissociation System (Worthington Biochemical Corp., Lakewood, NJ), consisting of: papain and DNase I.
3. 30 % Albumin solution from bovine serum.
4. Nutrient Mixture F12 Ham (Sigma-Aldrich), with sodium bicarbonate.
5. Fetal bovine serum (FBS). Separate into 50 ml aliquots and store at −20 °C (*see* **Note 1**).
6. 100× Penicillin/Streptomycin solution. Aliquot into 5 ml lots and store at −20 °C. One aliquot of penicillin/streptomycin is added to 500 ml bottles of MEM-HEPES and Nutrient Mixture F12 Ham.
7. Endothelial cell growth supplement (Millipore 02-102): rehydrate one 15 mg vial with 1 ml Ham's F12 medium = 15 mg/ml stock, aliquot into 100 μl aliquots and freeze at −20 °C. Add one aliquot (100 μl) to 50 ml base medium (Ham's F12 containing 10 % FBS and pen/strep) to give working conc. = 30 μg/ml.
8. 400 mM L-glutamine solution: add 5 ml F12 medium into the L-glutamine bottle, mix well (keep shaking the bottle during the aliquoting so as to keep the glutamine evenly distributed), and separate into 0.5 ml aliquots. Freeze at −20 °C (*see* **Note 2**). Add one aliquot to 50 ml base medium (1:100 dilution).
9. Ascorbate: add 10 ml water to 5 mg ascorbate = 0.5 mg/ml stock, 0.25 ml aliquots, freeze at −20 °C, use at 1:200, add one aliquot to 50 ml base medium to give working conc. = 2.5 μg/ml.

10. Heparin: add 9 ml F12 medium into a bottle containing 10,000 U Heparin (Sigma-Aldrich H3149) (approx 72 mg) to give approx 0.8 % stock (8 mg/ml), separate into 0.25 ml aliquots, freeze at −20 °C. Add one 0.25 ml aliquot to 50 ml base medium (1:200 dilution) to give working conc. = 40 μg/ml.
11. Stock F12 medium with penicillin/streptomycin: add one 5 ml aliquot of 10× penicillin/streptomycin to 500 ml Nutrient Mixture F12 Ham medium (with sodium bicarbonate).
12. Stock F12 medium with 10 % FBS and penicillin/streptomycin: add one 50 ml aliquot FBS and one 5 ml aliquot of 100× penicillin/streptomycin to 500 ml Nutrient Mixture F12 Ham medium (with sodium bicarbonate) to make stock F12 medium with FBS and antibiotics.
13. Endothelial cell growth medium (ECGM): F12 medium with 10 % FBS and penicillin/streptomycin, endothelial cell growth supplement, ascorbate, L-glutamine, and heparin. Aliquot 50 ml F12 stock medium with FBS and antibiotics into a 50 ml conical tube. Add 100 μl of endothelial cell growth supplement, 250 μl ascorbate, 250 μl heparin, and 500 μl L-GLUTAMINE (*see* **Note 3**).
14. Collagen coating solution: collagen from calf skin Type I diluted 1:5 in sterile water.
15. 0.05 % Trypsin with EDTA.
16. Phosphate buffered saline (PBS): Dilute 10× PBS to 1× PBS.
17. Puromycin hydrochloride: Puromycin hydrochloride in a 10 mg vial (Alexis Biochemicals, Grunberg, Germany) should be diluted in 1 ml sterile water to make a 10 mg/ml stock solution. Separate into 50 μl aliquots and freeze at −20 °C. Dilute to 4 μg/ml in ECGM + FBS medium before adding to the cultures.

2.2 Immunohistochemistry

1. 24-well tissue culture plates.
2. Microscope cover glass circles No. 1–0.13 to 0.17 mm thick; Size: 12 mm (Fisher Scientific).
3. Ice-cold acetone–methanol: Acetone and methanol (50:50 mixture). The acetone–methanol should be stored at −20 °C.
4. Phosphate buffered saline (PBS): Dilute 10× PBS to 1× PBS.
5. Dako Antibody Diluent.
6. CD31 antibody: The FITC rat anti-mouse CD31 antibody (clone MEC 13.3 from BD Pharmingen) should be used at a dilution of 1:100.
7. Alexa Fluor 488 goat anti-rat secondary antibody: The Alexa Fluor 488 goat anti-rat IgG (H+L) antibody (Invitrogen) should be diluted to 1:200 in antibody diluent.
8. Hoechst stain: Hoechst stain solution diluted 1:2 in PBS.
9. Aqua-Poly/Mount (Polysciences).

3 Methods

3.1 Preparation of Mouse Brain Endothelial Cells

1. Make up fresh ECGM + FBS medium (add aliquots of ECGS, Ascorbate, L-glutamine, and Heparin) (*see* **Note 3**). Coat 4 wells of a 6-well plate with collagen type I diluted 1:5 in water (2 ml per well) for 2 h at 37 °C. Wash with PBS three times and keep the final wash in the well and transfer to the incubator to ensure gas equilibration before adding the endothelial cells.
2. Prepare the brain endothelial cells (BEC) from the brains of six mice (6–8 weeks of age preferred) euthanized by CO_2 inhalation (*see* **Notes 4** and **5**).
3. Flame all the tools and the razor blade to sterilize.
4. Ethanol spray the head and then use scissors to remove the skin on top of the skull. Next, use scissors to cut through the skull from the back of the head on both sides of the head towards the eyes. Carefully lift up the skull plate and remove the brain; place upside down in a 100 mm dish containing enough MEM-HEPES medium to totally cover the brain. Repeat this process for all six brains.
5. Transfer the brains to the lid of the 100 mm dish and then dice up the tissue using a flamed razor blade. Be thorough; chopping in both axes to ensure no tissue chunks remain.
6. Pour on some MEM-HEPES medium and aspirate the tissue using a 10 ml pipette into a 15 ml universal tube. Centrifuge this at 290 × *g* for 5 min.
7. At this stage, prepare the papain dissociation solution by adding 5 ml MEM-HEPES to the vial of papain and 0.5 ml MEM-HEPES to the DNase tube. Incubate these solutions in the 37 °C incubator for 10 min to ensure they have fully dissolved.
8. Aspirate the supernatant from the brain tissue and add 5 ml papain solution and 250 μl of the DNase solution onto the brain tissue. Mix well and incubate at 37 °C for 1 h 10 min.
9. After 1 h and 10 min incubation, transfer the contents of the tube to a 50 ml tube to allow access of the syringe for trituration. Break up the tissue using a 10 ml syringe, first using a 19-gauge needle, ten times up and down, then with a 21-gauge needle. Be careful not to over-titurate (*see* **Note 6**). The aim is to break up the microvessels from the brain tissue, but to maintain the vascular tubes intact.
10. After trituration, transfer the brain homogenate (approximately 6–7 ml) to a new 15 ml tube and add approximately 7 ml of 22 % BSA to the tube and mix well. Centrifuge the 15 ml tube at 1360 × *g* for 10 min. Following this spin, the myelin is retained at the top of the tube, while the vascular tubes and liberated cells pellet at the bottom of the tube.

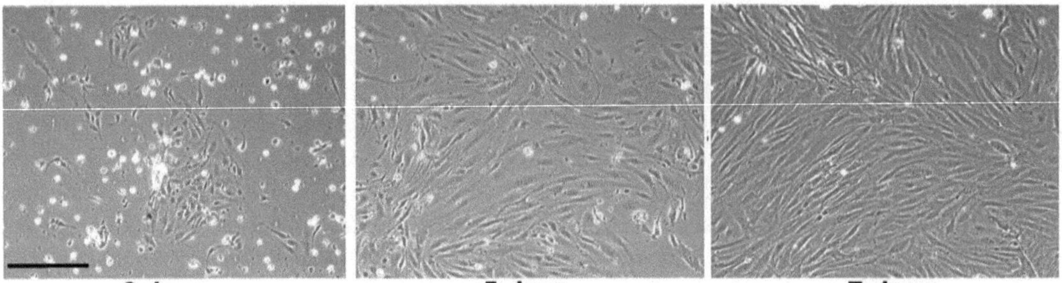

Fig. 1 Phase micrograph of brain endothelial cell (BEC) cultures at different stages of culture. Scale bar = 100 μm

11. After the spin, tilt the tube at 45° to free the myelin from the edge of the tube, and then pour the myelin plus BSA into a waster container. Remove the remainder of the myelin with a pipette being careful not to allow the myelin to mix with the pellet. Suspend the blood vessel pellet with 1 ml ECGM and transfer into a new 15 ml centrifuge tube. Add an additional 1 ml ECGM medium to the suspension.
12. Wash the blood vessels by spinning down at 290 × *g* for 5 min.
13. Resuspend the cell pellet in 8 ml ECGM and plate this onto 4 wells of a 6-well plate coated with type I collagen (*see* above). Incubate at 37 °C. Over the next few hours the vascular tubes attach and form colonies that continue to grow and expand with time (*see* Fig. 1).
14. Change all medium the next morning (day 1). Wash the cells with F12 medium (lacking FBS) until most of the debris is removed (seven to eight washes are recommended). Add new ECGM + FBS medium to the cells and leave until the early evening. In the evening, remove the medium and replace with fresh ECGM + FBS medium prior to adding puromycin.
15. Add 4 μg/ml puromycin to the cultures for 2.5 days to prevent the growth of non-endothelial cells (*see* Subheading 3.2 below).
16. After 2.5 days incubation with puromycin, remove the medium and replace with 2 ml fresh ECGM + FBS per well. Mouse BECs grow as swirling groups of cells and will reach confluence between 5 and 7 days (*see* Fig. 1).

3.2 Purification of BEC Using Puromycin (Very Important Step)

To obtain 99 % pure cultures of BEC, it is important to add puromycin on Day 1 of the culture. After dissociation of brain tissue, it is best to allow the cells 24 h to attach and spread on the culture plate before treating them with puromycin. Puromycin effectively kills all cells in the culture with the exception of BECs. BECs are protected because they express high levels of the multidrug resistance protein,

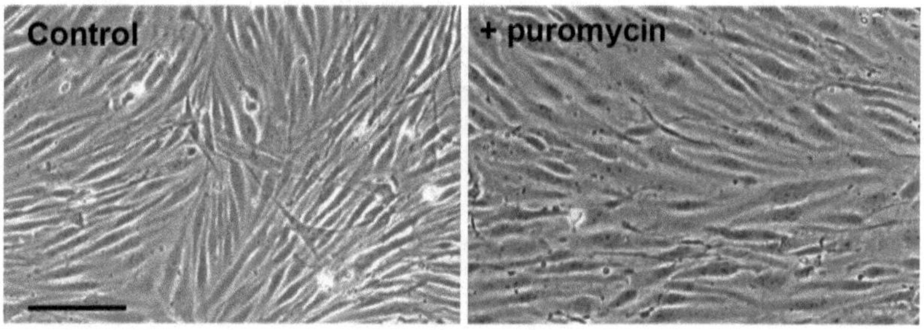

Fig. 2 Phase micrograph of BEC cultures demonstrating the effect of puromycin to improve BEC purity. Scale bar = 50 μm

MDR1 also known as P-glycoprotein. Thus, BECs pump out puromycin and are resistant to its toxic effects to block protein synthesis. The outcome is that the vast majority of non-endothelial cells (predominantly pericytes) in the culture are killed during the 2.5 day exposure. Importantly, BECs survive puromycin treatment only if added to the culture early, as extended time in culture causes BECs to lose expression of MDR1 and thus the ability to shunt out toxic substances. For this reason, puromycin treatment should only be used for 2.5 days before changing back to puromycin free ECGM + FBS, otherwise the BEC will also start to die (*see* Fig. 2 for comparison of BEC cultures cultured with or without puromycin).

3.3 Harvesting and Seeding Primary BEC on a 24-Well Plate

1. Coat a 24-well plate with 500 μl collagen type I per well (other extracellular matrix proteins can be used instead depending on the experiment).
2. Remove medium from the 6-well plate and wash with F12 medium.
3. Add 2 ml trypsin to each well and place the plate in the 37 °C incubator for 5 min (*see* **Note** 7).
4. Look at the cells under the microscope to ensure that all the cells have detached from the plate. Once all of the cells have detached, add 2 ml F12 + FBS medium to each well. Remove the cells and medium into a 15 ml tube. All four wells can be pooled together at this time.
5. Spin down the cells at 290 × *g* in a centrifuge for 5 min to pellet the cells.
6. Remove the supernatant and resuspend the pellet in 1 ml ECGM containing FBS.
7. Seed the cells using the desired amount of ECGM so that each well gets 500 μl cell suspension. It is best to make sure that the BECs are not seeded too thinly otherwise they will start to die.

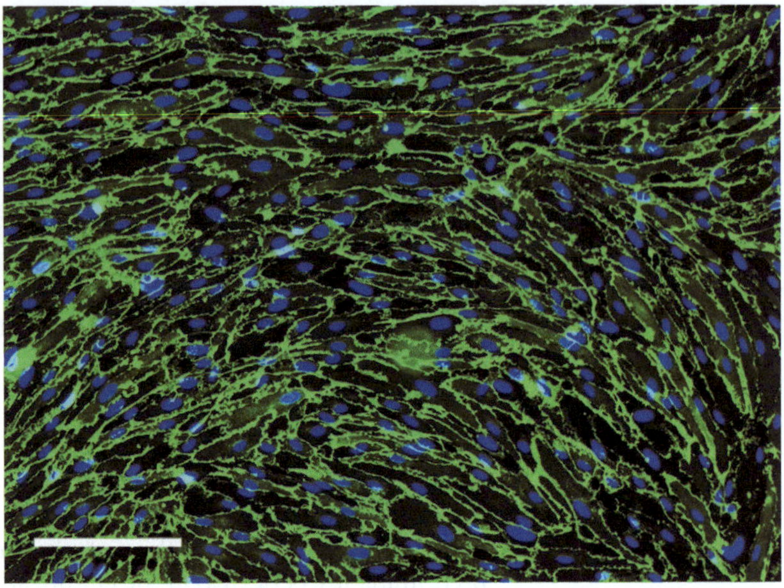

Fig. 3 Dual IF of BEC culture stained with the endothelial-specific marker CD31 (*green*) and the nuclear stain Hoechst (*blue*). Scale bar = 100 μm. Note that the culture is highly pure

3.4 Immunohistochemistry to Validate the Purity of BEC Cultures

In our research, we routinely use immunohistochemistry to determine the purity of BEC cultures. We employ CD31 (endothelial-specific)/Hoechst (nuclei) dual-immunofluorescence to quantify the proportion of cells expressing CD31 (*see* Fig. 3). Typically, endothelial cell cultures treated with puromycin will be >99 % pure.

1. Sterilize glass coverslips by dipping in 70 % ethanol and placing into the wells of a 24-well plate. Air-dry in the T. C. hood.
2. Prepare the collagen I solution to be coated onto coverslips. Collagen I is used at approximately 0.2 mg/ml (1 in 5 of stock in H_2O). Place 25 μl drops of collagen I onto the center of the coverslips. Leave to coat for 2 h at 37 °C (*see* **Note 8**).
3. Prepare the cells by harvesting (follow **steps 1–5** from Subheading 3.3) and centrifuge the cells then resuspend in a small volume of medium. Approximate volume of medium required = no. of drops × 25 μl.
4. Just before plating the cells onto the coverslips, remove the collagen coating solution and wash the ECM-coated areas 1× with PBS.
5. Apply the 25 μl cell drops to the substrates, taking great care to add the cell drops exactly over the wetted area previously coated with collagen. Mix the cell suspension well between each plating of cells.
6. Allow the cells to attach in the 37 °C incubator for 1 h. Check under the microscope that the majority of cells have attached and spread. If not, then extend this period for up to another hour.

7. Add 0.5 ml ECGM + FBS medium to the well, and leave the cells to grow in the 37 °C incubator.
8. When the cells are approximately 50 % confluent, remove the coverslips from the wells using a curved needle and curved forceps, and fix in ice-cold acetone–methanol (50:50) in a 24-well plate for 5 min at −20 °C.
9. Wash the coverslips very well by dipping in three PBS-filled 50 ml tubes. Take care to remove all of the acetone–methanol (*see* **Note 9**).
10. Incubate in FITC-rat CD31 anti-mouse antibody diluted 1:100 in antibody diluent for 30 min in a 37 °C incubator.
11. Wash in PBS by dipping in a PBS-filled 50 ml tubes three times.
12. Incubate with Alexa Fluor 488 goat anti-rat antibody (diluted 1:200 in antibody diluent) for 30 min in a 37 °C incubator.
13. Wash in PBS by dipping in a PBS-filled 50 ml tubes three times.
14. To stain nuclei, use Hoechst stain solution (diluted 1:2 in PBS) and incubate for 10 min.
15. Wash in PBS by dipping in a PBS-filled 50 ml tubes three times.
16. Apply Aqua-Poly/Mount mounting medium and cover with a glass coverslip.
17. To remove excess Aqua-Poly/Mount, lay a piece of tissue paper on the top of the coverslips and apply gentle pressure to squeeze out the excess medium which is then absorbed by the tissue paper.
18. Store the coverslips in the dark until ready for analysis by fluorescent microscopy.

4 Notes

1. It is worth batch-testing several different sources of FBS. Most companies are happy to let you have free samples of their FBS to test. Once you have found the batch that supports the optimal growth of your cultures, then stick with it and order several 500 ml bottles and store at −20 °C.
2. When dissolving L-glutamine in small volumes of medium, it does not fully dissolve, but forms precipitates. Thus, it is important to mix this concentrated solution very well and ensure all the L-glutamine is transferred to the 500 ml bottle, where it fully dissolves in the larger volume. If dividing into 0.5 ml aliquots, shake the L-glutamine solution well in between each dispense to ensure equal distribution.

3. For every new BEC isolation, make up a fresh aliquot of 50 ml ECGM + FBS medium.
4. We find that it is important to dissect the brains from mice immediately after euthanization while the mice are still warm.
5. We usually prepare BEC cultures with six brains going into 4 wells of a 6-well plate (1.5 brains per well). This protocol can be performed using fewer mice. However, it is critical that elements such as papain and DNase are downsized. For example, if using only one mouse brain (for instance when numbers of mice are limited such as with specific knockout strains), papain should be reduced to 833.3 μl and DNase to 41.7 μl and cells should be plated into smaller sized wells, e.g., use wells of a 12-well plate.
6. When triturating the brain suspension, the aim is to break up the brain tissue with a slow methodical use of the syringe, but do not over-triturate or move the tissue too violently as this will greatly reduce cell viability. BECs survive much better when they are contained within still intact vascular tubes; isolated BECs tend to die in the early stages of culture.
7. Trypsin incubation times vary with different cultures, and are influenced by: age of culture (the older more established cultures tend to need longer to detach), density of culture, batch of trypsin, and how long the bottle of trypsin has been at 4 °C for (the longer the trypsin has been thawed, the lower the activity).
8. When moving the plates containing the small 25 μl drops around, take great care not to jolt or jar the plate, as this will move the droplet off the wetted area and will create a larger ECM-coated area, which will distort the size of the prepared area, and thus disrupt the experiment.
9. It is imperative to thoroughly remove all of the fixative from the coverslip by washing several times in PBS-filled 50 ml tubes. If all fixative is not removed, this will subsequently denature the added antibodies and prevent effective staining of cells.

Acknowledgements

This work was supported by the National Multiple Sclerosis Society; Harry Weaver Neuroscience Scholar Award (R.M.), and Postdoctoral Fellowship (J.V.W.-A.), by the American Heart Association (Western State Affiliate) Postdoctoral Fellowship (A.B.), and by the NIH grant RO1 NS060770.

References

1. Ballabh P, Braun A, Nedergaard M (2004) The blood–brain barrier: an overview structure, regulation and clinical implications. Neurobiol Dis 16:1–13
2. Brett FM, Mizisin AP, Powell HC, Campbell IL (1995) Evolution of neuropathologic abnormalities associated with blood–brain barrier breakdown in transgenic mice expressing interleukin-6 in astrocytes. J Neuropathol Exp Neurol 54:766–775
3. Huber JD, Egleton RD, Davis TP (2001) Molecular physiology and pathophysiology of tight junctions in the blood–brain barrier. Trends Neurosci 24:719–725
4. Pardridge WM (2003) Blood–brain barrier drug targetting: the future of brain drug development. Mol Med 3:90–105
5. Wolburg H, Lippoldt A (2002) Tight junctions of the blood–brain barrier; development, composition and regulation. Vascul Pharmacol 38: 323–337
6. del Zoppo GJ, Milner R (2006) Integrin-matrix interactions in the cerebral microvasculature. Arterioscler Thromb Vasc Biol 26:1966–1975
7. Daneman R, Zhou L, Kebede AA, Barres BA (2010) Pericytes are required for blood–brain barrier integrity during embryogenesis. Nature 468:562–566
8. Faraci FM (2011) Protecting against vascular disease in brain. Am J Physiol Heart Circ Physiol 300:H1566–H1582
9. Grammas P (2011) Neurovascular dysfunction, inflammation and endothelial activation: implications for the pathogenesis of Alzheimer's disease. J Neuroinflammation 8:26
10. Holley JE, Newcombe J, Whatmore JL, Gutowski NJ (2010) Increased blood vessel density and endothelial cell proliferation in multiple sclerosis cerebral white matter. Neurosci Lett 470:65–70
11. Holman DW, Klein RS, Ransohoff RM (2011) The blood–brain barrier, chemokines and multiple sclerosis. Biochim Biophys Acta 1812:220–230
12. Krupinski J, Kaluza J, Kumar P, Kumar S, Wang JM (1994) Role of angiogenesis in patients with cerebral ischemic stroke. Stroke 25:1794–1798
13. Lin TN, Sun SW, Cheung WM, Chang C (2002) Dynamic changes in cerebral blood flow and angiogenesis after transient focal cerebral ischmeia in rats. Evaluation with serial magnetic resonance imaging. Stroke 33:2985–2991
14. Ludwin SK, Henry JM, McFarland HF (2001) Vascular proliferation and angiogenesis in MS: clinical and pathogenic implications. J Neuropathol Exp Neurol 60:505
15. del Zoppo GJ, Hallenbeck JM (2000) Advances in the vascular pathophysiology of ischemic stroke. Thromb Res 98:V73–V81
16. Dirnagl U, Iadecola C, Moskowitz MA (1999) Pathophysiology of ischemic stroke: an integrated view. Trends Neurosci 22:391–397
17. Hamann GF, Okada Y, del Zoppo GJ (1996) Hemorrhagic transformation and microvascular integrity during focal cerebral ischemia. J Cereb Blood Flow Metab 16:1373–1378
18. Zhang ZG, Zhang L, Jiang Q, Zhang R, Davies K, Powers C, Bruggen NV, Chopp M (2000) VEGF enhances angiogenesis and promoted blood–brain barrier leakage in the ischemic brain. J Clin Invest 106:829–838
19. LaManna JC, Kuo NT, Lust WD (1998) Hypoxia-induced brain angiogenesis. Signals and consequences. Adv Exp Med Biol 454: 287–293
20. Milner R, Hung S, Erokwu B, Dore-Duffy P, LaManna JC, del Zoppo GJ (2008) Increased expression of fibronectin and the $\alpha5\beta1$ integrin in angiogenic cerebral blood vessels of mice subject to hypobaric hypoxia. Mol Cell Neurosci 38:43–52
21. Persson AB, Buschmann IR (2011) Vascular growth in health and disease. Front Mol Neurosci 4:1–15
22. Goel S, Duda DG, Xu L, Munn LL, Boucher Y, Fukumura D, Jain RK (2011) Normalization of the vasculature for treatment of cancer and other diseases. Physiol Rev 91:1071–1121
23. Abbott NJ (2002) Astrocyte-endothelial interactions and blood–brain barrier permeability. J Anat 200:629–638
24. Abbruscato T, Davis T (1999) Protein expression of brain endothelial cell E-cadherin after hypoxia/aglycemia: influence of astrocyte contact. Brain Res 842:277–286
25. Greenwood J, Walters CE, Pryce G, Kanuga N, Beraud E, Baker D, Adamson P (2003) Lovastatin inhibits brain endothelial cell Rho-mediated lymphocyte migration and attenuates experimental autoimmune encephalomyelitis. FASEB J 17:905–907
26. Haudenschild CC, Zahniser D, Folkman J, Klagsbrun M (1976) Human vascular endothelial cells in culture. Lack of response to serum growth factors. Exp Cell Res 98:175–183
27. Milner R, Campbell IL (2002) Developmental regulation of $\beta1$ integrins during angiogenesis in the central nervous system. Mol Cell Neurosci 20:616–626

28. Risau W, Esser S, Engelhardt B (1998) Differentiation of blood–brain barrier endothelial cells. Pathol Biol 46:171–175
29. Risau W, Hallman R, Albrecht U, Henke-Fahle S (1986) Brain astrocytes induce the expression of an early cell surface marker for blood–brain barrier specific endothelium. EMBO J 5: 3179–3183
30. Sapatino BV, Welsh CJ, Smith CA, Bebo BF, Linthicum DS (1993) Cloned mouse cerebrovascular endothelial cells that maintain their differentiation markers for factor VIII, low density lipoprotein, and angiotensin-converting enzyme. In Vitro Cell Dev Biol Anim 29:923–928
31. Wang J, Milner R (2006) Fibronectin promotes brain capillary endothelial cell survival and proliferation through α5β1 and αvβ3 integrins via MAP kinase signaling. J Neurochem 96:148–159
32. Milner R, Hung S, Wang X, Berg G, Spatz M, del Zoppo G (2008) Responses of endothelial cell and astrocyte matrix-integrin receptors to ischemia mimic those observed in the neurovascular unit. Stroke 39:191–197

Chapter 29

Purification of Endothelial Cells from Rat Brain

Jinhua Luo, Xiangling Yin, Alma Sanchez, Debjani Tripathy, Joseph Martinez, and Paula Grammas

Abstract

Endothelial cells make up a minor population of cells in a tissue, but play a major role in tissue homeostasis, as well as in diverse pathologies. To understand the biology of cerebral endothelium, purification and characterization of the cerebrocortical endothelial cell population is highly desirable. For this purpose, rat brains are mechanically minced and subsequently digested enzymatically with collagenase. In this protocol, the capillary fraction (microvessels) and the fraction enriched in small arterioles and arteries (resistance vessels) are separated. Each produces its own homogenous endothelial culture, namely, MV-EC and RV-EC. The endothelial origin of these cells is identified by positive immunofluorescent staining for the endothelial cell surface antigen Factor VIII. Unlike MV-EC, RV-EC cultures are capable of serial cultivation for up to 15 passages. Primary MV-ECs are able to retain their characteristic endothelial morphology for 6–8 weeks.

Key words Brain, Resistance vessels, Microvessels, Endothelial cell culture, Factor VIII

1 Introduction

Vascular endothelial cells, located at the interior surface of blood vessels, form the thin layer that lines the entire circulatory system, from the heart to the smallest capillaries. Endothelial cells of the microvasculature are key regulators of metabolite exchange and communication at the blood–tissue interface. The study of endothelial cell physiology and pathobiology has increased exponentially as the key contribution of endothelial dysfunction to cardiovascular, metabolic, and neurodegenerative diseases is increasingly recognized [1, 2]. Endothelial cells derived from different vasculatures display morphological heterogeneity as well as regional and organ-specific functional properties [3]. The differentiated endothelial cell phenotype is maintained even in vitro as global expression profiling of cultured endothelial cells documents vascular bed-specific endothelial cell types with characteristic and distinct gene expression programs [4].

Richard Milner (ed.), *Cerebral Angiogenesis: Methods and Protocols*, Methods in Molecular Biology, vol. 1135,
DOI 10.1007/978-1-4939-0320-7_29, © Springer Science+Business Media New York 2014

Among the most specialized endothelial cells are the brain microvascular endothelial cells. Brain endothelial cells have highly developed tight junctions and little pinocytotic activity, which protects neural tissue from fluctuations in blood composition [5]. Brain-derived endothelial cells express unique features that serve their blood–brain barrier function including surface expression of enzymes such as monoamine oxidase that prevent passive entry of neuroactive molecules into the brain [6] as well as energy-dependent carrier-mediated nutrient transporters [7]. Because of the metabolic demands of energy transport, cerebrovascular endothelial cells have a higher concentration of mitochondria relative to endothelia in other vascular beds [8].

To study the properties and functions of the cerebromicrovasculature requires the isolation and/or culture of endothelial cells derived from isolated brain blood vessels. In contrast to the relative ease of isolating and culturing endothelium from peripheral vessels, the inaccessibility of intraparenchymal brain vessels as well as the less robust endothelial cell growth pattern of brain-derived endothelial cells present challenges. Moreover, there are different vessel types, such as capillaries, venules, and arterioles present in the brain microcirculation. The procedure presented herein is a modification of previously published protocols [9, 10]. The microvessels (capillaries) are clearly distinguished from the resistance vessels (small arterioles and arteries). From each fraction, endothelial cells (MV-EC or RV-EC) are isolated for long-term in vitro cultivation.

2 Materials

Sterilize all solutions and equipment. Keep all reagents as well as tissues on ice or as close to 4 °C as possible throughout the vessel isolation process to minimize degradation unless indicated otherwise. Coat all glass tools that come into contact with vessels including beakers and homogenizers with silicon (*see* **Note 1**). Diligently follow all waste disposal regulations when disposing of waste materials.

1. Hanks' balanced salt solution (HBSS) without Ca^{2+} and Mg^{2+}.
2. Homogenizer (Wheaton Overhead Stirrer 222-44-382; Fisher Scientific, Hampton, NH) fitted with a serrated Teflon pestle (0.13–0.18 mm clearance).
3. 15 % dextran solution: HBSS containing 15 % dextran (MW 150,000–200,000), 5 % fetal bovine serum (FBS).
4. Dulbecco's modified Eagle's medium (DMEM) containing 1 g/L D-glucose.
5. Growth medium: DMEM containing 20 % FBS, 2 mM glutamine, 1 % antibiotic–antimycotic solution (10,000 U/mL

penicillin G, 10 mg/mL streptomycin sulfate, 25 μg/mL amphotericin B).

6. Freezing medium: growth medium with 10 % DMSO.
7. Endothelial cell growth supplement (ECGS, E0760; Sigma Aldrich, St. Louis, MO).
8. Cloning Peni Cylinder (Fisher Scientific).

3 Methods

3.1 Isolation of Cerebral Resistance Vessels (RVs) and Microvessels (MVs)

1. For each preparation use at least 24 rats each weighing 150–200 g.
2. Euthanize the rats and remove the brains. Rinse the brains with HBSS to remove blood and debris (*see* **Note 2**). Use Kimwipes moistened with HBSS to gently wipe off pial membrane containing blood vessels, and tweezers to remove the cerebellum and white matter.
3. Place the cleaned cerebral cortices (*see* **Note 3**) in a beaker containing 150 mL HBSS, then scissor mince and homogenize the cortices using 25 up and down strokes in a homogenizer set at 350 rpm.
4. Transfer the homogenate to two conical centrifuge tubes (250 mL) and centrifuge at 1,000 × *g* for 15 min.
5. Decant and discard the supernatant and resuspend the pellet in 200 mL of 15 % dextran solution per 250 mL conical centrifuge tube and mix thoroughly. Centrifuge again at 4,000 × *g* for 20 min.
6. Carefully aspirate the supernatant including the floating white fatty layer from the centrifuge tube and discard (*see* **Note 4**).
7. Resuspend the pellet from both centrifuge tubes thoroughly in 100 mL HBSS and transfer using a siliconized pipette onto a 150 μm nylon mesh sieve.
8. To collect RVs, use a rubber policeman to remove vessels off the 150 μm sieve and into a pre-chilled cryovial containing 1 mL freezing medium. Aliquot the vessels into approximately four cryovials (250 μL each) and add to each vial 750 μL freezing medium, and store in liquid nitrogen until use (*see* **Note 5**).
9. To isolate MVs, collect the filtrate from the 150 μm sieve into a 250 mL siliconized beaker and transfer onto a 53 μm nylon mesh sieve using a siliconized pipette.
10. Use a rubber policeman to remove the vessels off the 53 μm sieve and into a pre-chilled cryovial containing 1 mL freezing medium. Aliquot the vessels into approximately four cryovials (250 μL each) and add to each vial 750 μL freezing medium, and store in liquid nitrogen until use (*see* **Note 6**).

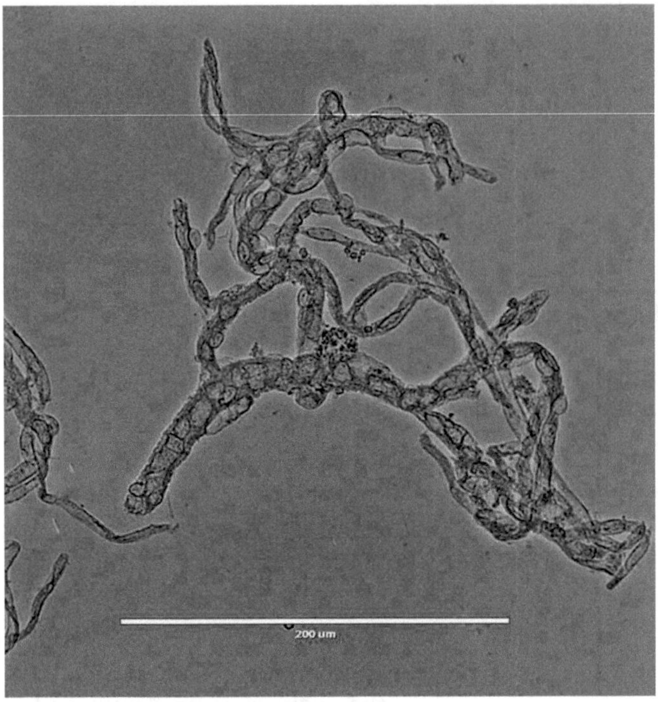

Fig. 1 Phase-contrast micrograph of an isolated resistance vessel (RV) fraction collected from a 150 μm sieve. Scale bar = 200 μm

11. Monitor the procedure and assess the purity of the vessels by phase-contrast microscopy. RV and MV preparations are shown in Figs. 1 and 2, respectively.

3.2 Initiation, Cloning, and Maintenance of RV-EC Cell Cultures

1. Resuspend the RVs in 10 mL HBSS containing 0.1 % collagenase/dispase and 2 % FBS with a siliconized pipette tip.
2. Incubate the vessels in a 37 °C water bath for 1 h with periodic shaking.
3. Centrifuge the RVs at 200 × *g* for 10 min, and resuspend in 2 mL growth medium.
4. Seed the cultures into a 35 mm (tissue culture dish), and maintain at 37 °C in a humidified 5 % CO_2 incubator.
5. Change the growth medium after 24 h, and remove the large vessel fragments with a beaded tip Pasteur pipette.
6. Change the growth medium every 3–4 days thereafter.
7. Cells emerge in about 1–2 days, and by day 7 the primary cultures consist of both smooth muscle and endothelial cells. Small colonies of endothelial cells of cobblestone-like morphology develop interspersed within the mixed cell monolayer (*see* **Note 7**).

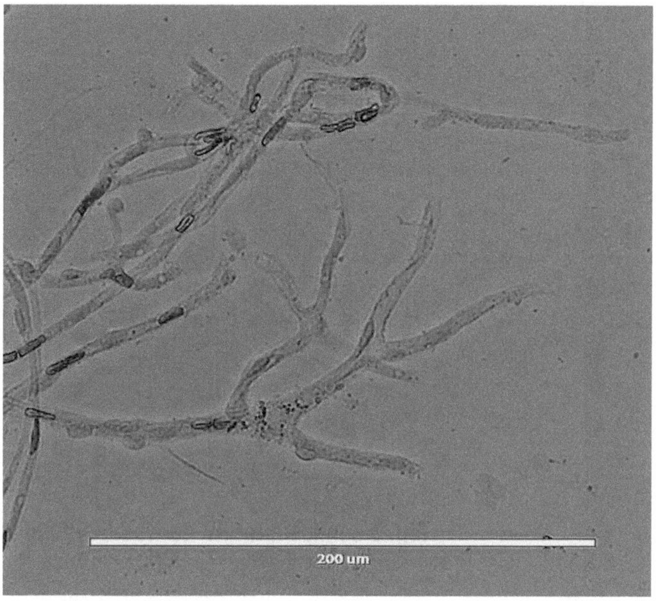

Fig. 2 Phase-contrast micrograph of an isolated microvessel (MV) fraction collected from a 53 μm sieve. Scale bar = 200 μm

8. By 7–14 days isolate 6–8 distinct endothelial cell colonies using cloning Peni Cylinder.
9. Wash the cultures with HBSS and add enough HBSS to form a thin film on cultures. Use the microscope to find endothelial cell colonies and draw a line on the outside bottom of the culture dish with a marker under the colonies. Grease the bottom of the Peni Cylinder with silicon and place over the endothelial cell colonies (using mark as a guide). Add trypsin (0.25 %) to the Peni Cylinder. Incubate for 3–5 min at 37 °C (using microscope to check that cells are detached) and remove the cells with a siliconized pipette.
10. Pool and seed the trypsinized cells into 35 mm tissue culture dishes containing growth medium.
11. Change the growth medium every 3–4 days thereafter.
12. Upon reaching confluency, transfer RV-ECs into 60 mm and subsequently into 100 mm tissue culture dishes. Once a homogenous culture is established, routinely split the endothelial cells (1:6) every 7–9 days. RV-ECs can be successfully maintained for up to 15 passages. An RV-EC monolayer is shown in Fig. 3.
13. Confirm the purity of the endothelial cell cultures using antibodies to the endothelial cell surface antigen factor VIII (*see* Fig. 4).

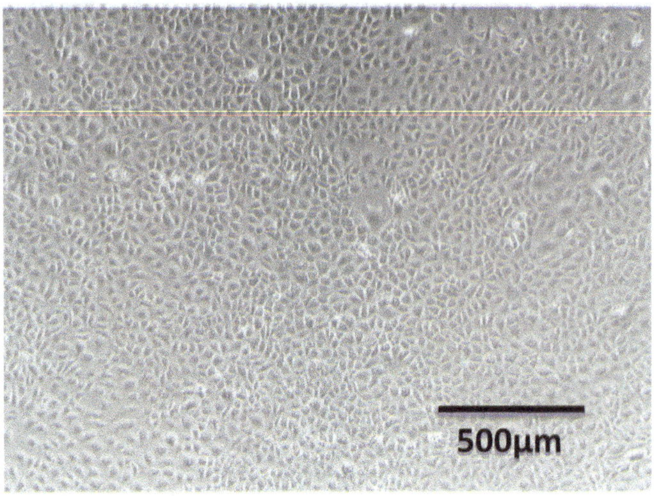

Fig. 3 Confluent rat RV-derived EC monolayer. Scale bar = 500 μm

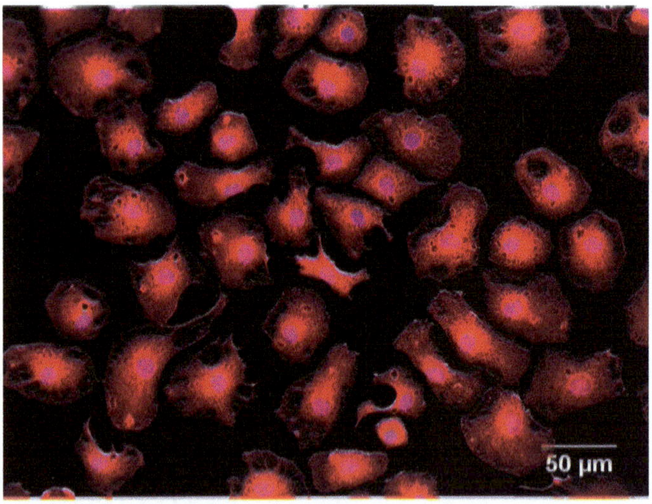

Fig. 4 Rat brain RV-derived endothelial cell cultures stained with an antibody to endothelial cell surface antigen Factor VIII. Scale bar = 50 μm

3.3 Initiation and Maintenance of Primary MV-EC Cell Cultures

1. Resuspend the MVs in 10 mL HBSS containing 0.1 % collagenase/dispase and 2 % FBS with a siliconized pipette tip.
2. Incubate the vessels in a 37 °C water bath for 15–20 min with periodic shaking.
3. Centrifuge the microvessels at 200×*g* for 10 min, and resuspend in 2 mL growth medium.
4. Seed the cultures into 35 mm tissue culture dishes with densities of 20–40 fragments per cm^2 (*see* **Note 8**), and maintain at 37 °C in a humidified 5 % CO_2 incubator.

5. Change growth medium after 24 h, and remove large vessel fragments with a beaded tip Pasteur pipette.
6. At 72 h post-seeding, change the medium to growth medium containing 150 μg/mL ECGS.
7. Change the growth medium containing 150 μg/mL ECGS every 3–4 days thereafter.
8. Migration occurs 1–2 days after plating, and marked proliferation begins after 5–7 days when endothelial cell colonies are evident. Primary MV-EC cultures form a confluent monolayer by 12–14 days (*see* **Note 9**), and retain their characteristic endothelial morphology for 6–8 weeks (*see* **Note 10**).
9. Confirm the purity of endothelial cell cultures using antibodies to the endothelial cell surface antigen factor VIII.

4 Notes

1. Microvessels have the characteristic of adhering to surfaces. Coating the glassware with silicon minimizes vessel loss during isolation procedure.
2. 24 rats yield approximately 30 g of total brain tissue.
3. Cleaned cortices from 24 rats yield approximately 15 g of tissue.
4. After centrifugation, the tube contains three distinct layers: a thin white fatty top layer; clear liquid middle layer; and a reddish solid pellet on the bottom. Repeat **steps 5** and **6** if there is excessive fatty material in the pellet.
5. Instead of freezing, to directly culture RVs, use a siliconized pipette to transfer RVs to approximately two centrifuge tubes (50 mL) containing 10 mL HBSS with 0.1 % collagenase/dispase and 2 % FBS, and go to the procedure in Subheading 3.2 (**step 2**).
6. Instead of freezing, to directly culture MVs, use a siliconized pipette to transfer MVs to approximately two centrifuge tubes (50 mL) containing 10 mL HBSS with 0.1 % collagenase/dispase and 2 % FBS, and go to the procedure in Subheading 3.3 (**step 2**).
7. Early passage resistance vessel-derived smooth muscle cells exhibit a markedly slow recovery rate from the frozen stock. To enrich mixed cultures for endothelial cells, freeze cultures in freezing medium and store in liquid nitrogen. (Thaw and replace medium with growth medium and go to the procedure in Subheading 3.2 (**step 6**)).
8. Approximately 50 % of the seeded fragments are capable of initiating cell colonies.

9. Upon reaching confluency, proliferation ceases and endothelial cells do not overgrow into multilayer aggregates. Phase-contrast and electron microscopic depiction of MV-ECs is presented in ref. 9.
10. After this, primary endothelial cells lose their phenotypic appearance and gradually deteriorate.

Acknowledgments

Sources of support: This work was supported in part by grants from the National Institutes of Health (AG020569 and AG028367). Dr. Grammas is the recipient of the Shirley and Mildred Garrison Chair in Aging.

References

1. Cai H, Harrison DG (2000) Endothelial dysfunction in cardiovascular diseases: the role of oxidant stress. Circ Res 87:840–844
2. de la Torre JC (2002) Alzheimer disease as a vascular disorder: nosological evidence. Stroke 33:1152–1162
3. Aird WC (2012) Endothelial cell heterogeneity. Cold Spring Harb Perspect Med 2(1):a006429
4. Chi JT, Chang HY, Haraldsen G et al (2003) Endothelial cell diversity revealed by global expression profiling. Proc Natl Acad Sci U S A 100:10623–10628
5. Simionescu M, Gafencu A, Antohe F (2002) Transcytosis of plasma macromolecules in endothelial cells: a cell biological survey. Microsc Res Tech 57:269–288
6. Kalaria RN, Harik SI (1987) Blood–brain barrier monoamine oxidase: enzyme characterization in cerebral microvessels and other tissues from six mammalian species, including human. J Neurochem 49:856–864
7. Hawkins RA, O'Kane RL, Simpson IA et al (2006) Structure of the blood–brain barrier and its role in the transport of amino acids. J Nutr 136:218S–226S
8. Olendorf WH, Cornford ME, Brown WJ (1977) The apparent work capability of the blood–brain barrier: a study of the mitochondrial content of capillary endothelial cells in brain and other tissues of the rat. Ann Neurol 1:409–417
9. Diglio CA, Grammas P, Giacomelli F et al (1982) Primary culture of rat cerebral microvascular endothelial cells. Lab Invest 46:554–563
10. Diglio CA, Liu W, Grammas P et al (1993) Isolation and characterization of cerebral resistance vessel endothelium in culture. Tissue Cell 25:833–846

Chapter 30

Generation of Primary Cultures of Bovine Brain Endothelial Cells and Setup of Cocultures with Rat Astrocytes

Hans C. Helms and Birger Brodin

Abstract

In vitro models of the blood–brain barrier are useful tools to study blood–brain barrier function as well as drug permeation from the systemic circulation to the brain parenchyma. However, a large number of the available in vitro models fail to reflect the tightness of the in vivo blood–brain barrier. The present protocol describes the setup of an in vitro coculture model based on primary cultures of endothelial cells from bovine brain microvessels and primary cultures of rat astrocytes. The model displays a high electrical tightness and expresses blood–brain barrier marker proteins.

Key words Blood–brain barrier, In vitro model, Bovine brain endothelial cells, Primary cell isolation, Cell culture, Astrocyte coculture

1 Introduction

Numerous in vitro models of the blood–brain barrier (BBB) have been established based on primary cells isolated from a range of different species (for review, *see* [1]). However, many of these models have low paracellular tightness compared to the in vivo barrier. Common measures of tightness are transendothelial electrical resistance (TEER) and permeability of hydrophilic low-molecular-weight compounds such as mannitol (182 g/mol), sucrose (342 g/mol), or inulin (~5,000 g/mol). The in vivo TEER of the blood–brain barrier has been estimated to be ~1,900 Ω cm^2 [2]; however, values up to 6,000 Ω cm^2 have been reported [3]. The in vivo permeability of radiolabelled mannitol has been estimated to ~8.4×10^{-8} cm/s, sucrose estimates range from 1- to 2.7×10^{-8} cm/s, and inulin permeability has been estimated to ~1×10^{-8} cm/s [4, 5]. In vitro models rarely match the in vivo values for TEER and paracellular permeability. Models based on human endothelial cells have typically shown TEER values below 100 Ω cm^2 and high variation [6–9], whereas rat and mice models in some instances develop TEER values up to around 600 Ω cm^2

Richard Milner (ed.), *Cerebral Angiogenesis: Methods and Protocols*, Methods in Molecular Biology, vol. 1135,
DOI 10.1007/978-1-4939-0320-7_30,

[10, 11] but in most cases display TEER values in the region of 100–300 Ω cm^2 [12–14]. Models based on primary cells of porcine and bovine origin have been reported to display TEER values in the range of 600–1,800 Ω cm^2 and mannitol or sucrose permeabilities in the range of 10^{-7}–10^{-6} cm/s [15–20]. Therefore, these species seem to be the most promising for the development of a tight BBB model usable for small molecule transport studies.

Bovine brain endothelial cells were originally isolated by Bowman et al. and used in monoculture as BBB models [21, 22]. The model was further developed into a coculture model by Dehouck et al. who cocultured the bovine brain endothelial cells with rat astrocytes and observed improved paracellular tightness across the model [15]. A number of research groups have since applied this culture setup, but techniques to isolate and culture the endothelial cells vary greatly between laboratories. In this protocol, we describe our technique, which is a modification of the isolation and culture techniques developed by Gaillard et al. [16]. Our isolation protocol is based upon filtration steps to isolate the brain microvessels followed by enzymatic digestion and culture of the digested microvessels [17]. The resulting endothelial cells are cocultured with rat cortical astrocytes, and a specific differentiation medium with high buffer concentration is added for the last 3 days of coculture to induce the tightness of the model. The protocol results in cocultures with an average TEER of ~1,000 Ω cm^2 and mannitol permeabilities below 10^{-6} cm/s, which is usable for studies of vectorial transport of small molecules across the blood–brain barrier (*see* **Note 1**) [17, 23].

2 Materials

2.1 Acquisition of Bovine Brains

1. Dulbecco's Phosphate-Buffered Saline (PBS).
2. Plastic bags.
3. Transport box with ice.
4. Meat saw, hammer, chisel, axe.
5. Sharpened meat knife and sharpening steel.

2.2 Isolation of Brain Microvessels

1. Dulbecco's Modified Eagles Medium (DMEM), 2–3 bottles per brain.
2. Complete culture medium (DMEM-comp): To 500 ml, DMEM is added: 50 ml fetal bovine serum (FBS), 5 ml MEM nonessential amino acids, and 5 ml penicillin/streptomycin (0.1 g/l streptomycin sulfate and 100,000 U/l penicillin G sodium).
3. Trypsin TRL stock solution. Prepare a 900 U/ml solution in PBS. Sterilize the solution by passage through a 0.2 μm filter and store in aliquots of 2 ml at −20 °C.

4. Collagenase type III stock solution. Prepare a 2,000 U/ml solution in PBS. Sterilize the solution by passage through a 0.2 μm filter. Prepare 2 ml per brain on the day of isolation, and store at 4 °C until use.
5. DNAse I stock solution. Prepare a 3,400 U/ml solution in PBS. Sterilize the solution by passage through a 0.2 μm filter and store in aliquots of 1 ml at −20 °C.
6. Digestion medium: Prepare immediately before use by mixing 1 aliquot Trypsin TRL, 1 aliquot collagenase type III, 1 aliquot DNAse I, and 15 ml DMEM-comp.
7. Freezing mix: Prepare immediately before use by mixing 1 ml of dimethyl sulfoxide, molecular biology grade (DMSO) with 9 ml FBS.
8. Dounce tissue grinder, 40 ml, with loose and tight fitting pestles (Sigma-Aldrich, Cat. D9188).
9. Filter holders—ø 47 mm (Millipore, Copenhagen, Denmark, Cat. SX0004700).
10. Nylon mesh filters are cut out of the fabric in small circles matching the size of the filter holder (ø 45 mm) and autoclaved. Approximately 15 filters of 160 μm (Millipore, Cat. NY6H00010) and 1 filter of 200 μm (Merrem & la Porte, Zaltbommel, Netherlands) are needed pr. brain.

2.3 Culture of Brain Microvessels and Endothelial/Astrocyte Coculture Models

1. Collagen stock solution: Prepare by dissolving 5 mg collagen type IV in 50 ml PBS overnight at 4 °C. Store the collagen stock solution in aliquots of 5 ml at −20 °C.
2. Fibronectin stock solution: Prepare by dissolving 5 mg fibronectin in 5 ml sterile Milli-Q water. Store the fibronectin stock solution in aliquots of 500 μl at −20 °C.
3. Heparin stock solution: Prepare a 5 mg/ml solution by dissolving heparin sodium salt in PBS. Sterilize the solution by passage through a 0.2 μm filter and store at 4 °C.
4. Growth medium$^+$ (GM^+): Prepare immediately before use. Mix DMEM-comp with astrocyte-conditioned medium (1:1) (*see* **Note 2**) and add heparin stock solution (25 μl/ml GM^+).
5. Growth medium$^-$ (GM^-): Prepare immediately before use. Add heparin stock solution (25 μl/ml GM^-) to DMEM-comp.
6. Trypsin-EDTA for endothelial cells.
7. 50 mM TES-buffered culture medium (DMEM-TES) (TES: *N*-[Tris(hydroxymethyl)methyl]-2-aminoethanesulfonic acid): Dissolve 6.65 g powdered DMEM and 5.80 g TES-free acid in 500 ml Milli-Q water and adjust pH to 7.4 (*see* **Note 3**). Sterilize the solution by passage through a 0.2 μm filter and add 5 ml MEM nonessential amino acids and 5 ml penicillin/streptomycin. The medium can be stored at 4 °C in aliquots of 45 ml. Before use, add 5 ml FBS to one aliquot.

8. Cyclic AMP stock solution: Prepare by dissolving 8-(4-CPT)-cyclic adenosine monophosphate in Milli-Q water (12.3 mg/ml). Sterilize the solution by passage through a 0.2 μm filter and store in aliquots of 255 μl at –20 °C.
9. RO 20-1724 stock solution: Dissolve 4-(3-Butoxy-4-methoxybenzyl)imidazolidin-2-one in DMSO (9.7 mg/ml). Store the solution in aliquots of 25 μl at –20 °C.
10. Dexamethasone stock solution: Prepare by dissolving dexamethasone in absolute ethanol (3.9 mg/ml). Dilute the solution ten times in absolute ethanol to 5 μM. Store the solution in aliquots of 25 μl at –20 °C.
11. TES-buffered differentiation medium (DM-TES): Prepare immediately before use by mixing DMEM-TES with cyclic AMP stock solution (12.5 μl/ml DMEM-TES), RO 20-1724 stock solution (0.5 μl/ml DMEM-TES), and dexamethasone stock solution (0.5 μl/ml DMEM-TES).

2.4 Viability Test of Isolation Products

1. Hanks Balanced Salt Solution (HBSS) with calcium and magnesium, without sodium bicarbonate and phenol red.
2. Calcein-acetoxymethyl ester (Calcein-AM): Prepare a 5 mM stock solution by dissolving 1 vial (50 μg) (Life Technologies, Cat. C3100MP) in 10 μl DMSO. Store at –20 °C for up to 2 weeks.
3. Propidium iodide stock solution: Prepare a 1.5 mM stock solution by dissolving 10 mg propidium iodide in 10 ml PBS. Store at room temperature.
4. Viability staining solution: Prepare the staining solution immediately before use by mixing 1 μl Calcein-AM stock solution and 1 μl propidium iodide stock solution into 1 ml HBSS.

3 Methods

Figure 1 shows a flowchart of the most important steps in the isolation of the bovine brain microvessels. The microvessel isolation procedure described in Subheading 3.2 takes approximately 7–8 h if one person is working with one brain. The most time-consuming steps are the isolation of gray matter (Subheading 3.2.1, **step 5**) and the homogenization of brain tissue (Subheading 3.2.2, **steps 1–3**).

3.1 Removal of the Brain

Brains are acquired from calves below 12 months of age, directly at the slaughterhouse. The brains should be taken from the skull and cooled on ice as fast as possible and no longer than an hour after the animal is sacrificed.

1. Remove the skin and fat from the top of the skull using a sharp knife.

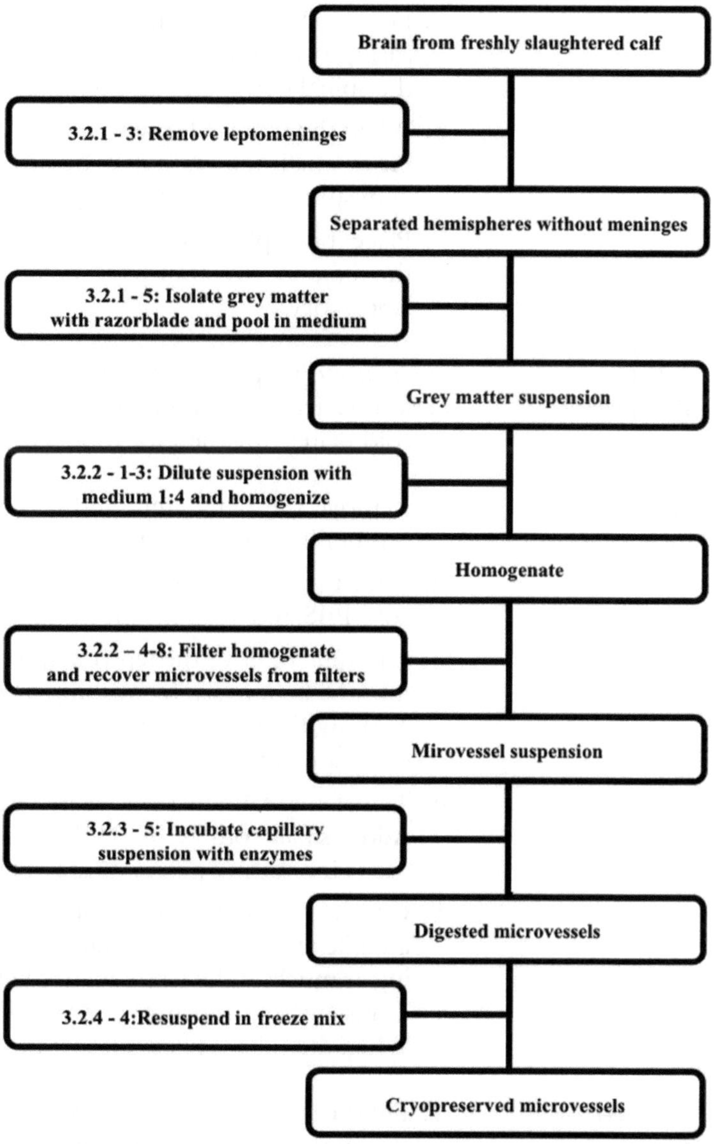

Fig. 1 Flowchart showing the method for isolating bovine brain microvessels. *Boxes* on the *left* refer to the different protocol steps; *boxes* on the *right* show the state of the tissue during the isolation of microvessels

2. Saw the skull 2/3 through, just below the horns using the meat saw.
3. Crack open the skull, using the axe or the chisel as leverage to expose the brain.
4. Cut away the dura mater using the knife to expose the brain. Remove small bone protrusions in the skull using the hammer and chisel.

5. Loosen the brain, while still in the skull, by gently destructing (using your fingers) the connection to the olfactory nerve and the spinal cord. The cerebellum can also be gently removed.
6. When the brain is loosened, gently lift it out. If the skull opening is small, the brain may be parted in two, and each cerebral hemisphere lifted out separately.
7. Place it in a plastic bag with approx 250 ml ice-cold PBS. Place the bag in the transport box with ice and take it to the laboratory.

3.2 Isolation of Brain Capillaries

Preparation: Prepare the laminar airflow bench with cover paper, a suction device, and aluminum trays for the ice bath. Alternatively, use cold plates designed for food storage instead of the ice baths.

3.2.1 Cleaning and Initial Processing

1. Remove the brain stem and cerebellum and separate the two cerebral hemispheres using the hands.
2. Place the hemispheres in a Petri dish (ø 15 cm) containing ice-cold PBS and keep on the cooling tray.
3. Carefully remove the leptomeninges with the hands. The brain must be kept moist by washing it regularly in PBS. Finish up by washing the brain carefully in PBS to flush off remaining pieces of leptomeninges (*see* **Note 4**).
4. Transfer the brain to a clean Petri dish containing DMEM. Add a small volume of DMEM to the lid of the Petri dish and use it for storing pieces of gray matter later on.
5. Cut the hemispheres in smaller lumps (for easier handling) using a sterile razor blade. Carefully slice off the gray matter while leaving as much of the white matter behind as possible (*see* **Note 5**). Transfer the gray matter to a 50 ml tube with 15–20 ml DMEM. When the tube is filled, place it in an ice bath.

3.2.2 Homogenizing and Filtration

1. Fill the slim part of the homogenizer approximately 1/5 with the gray matter suspension and add DMEM to fill up the slim part.
2. Start homogenizing by carefully moving the "Loose/A" pestle up and down 8 times (*see* **Note 6**). Then, switch to the "Tight/B" pestle and homogenize 8 times.
3. Transfer the homogenate to a clean 500 ml bottle (on ice) and repeat **steps 1–3** until all of the suspension has been homogenized. Optionally, collect a small sample of homogenate (approximately 500 μl) for viability test (*see* Subheading 3.4, Fig. 3a) (*see* **Note 7**).
4. Assemble the filter holders containing a rubber ring and a 160 μm nylon net filter. Place the filter on top of a 500 ml bottle for waste (as the filtrate is not going to be used further), and attach a 20 ml syringe on top of the filter.

5. Filter the suspension through the nylon filters by simply pouring it into the syringe and let the filtrate run through by itself. Filtrate approximately 30–40 ml suspension, and then flush with 10 ml DMEM (*see* **Note 8**).
6. Take out the filter with microvessels and transfer to a small Petri dish (ø 10 cm) containing 10 ml DMEM. Repeat the **steps 4–6** until the whole suspension has been filtered (*see* **Note 9**).
7. Move one filter containing microvessels to a clean Petri dish (ø 10 cm) and flush the microvessels into the dish with 7.5 ml DMEM-comp. Repeat the flushing until the filter appears mostly clean of microvessels.
8. Transfer the microvessel suspension to a 50 ml centrifuge tube in an ice bath. Repeat **steps 7–8** until all filters are flushed. Fill the centrifuge tubes to a maximum of 40 ml. (Normally, the final yield corresponds to three tubes or 120 ml capillary suspension.)

3.2.3 Purification and Incubation

1. Centrifuge the tubes with microvessel suspension for 5 min at 500 × *g*, room temperature.
2. Aspire the medium and, at the same time, remove any contaminating white matter, which can be seen as a distinct white band atop the microvessel pellet.
3. Resuspend the pellet in 40 ml DMEM-comp. Optionally, collect a small sample (200 μl) for viability test (*see* Subheading 3.4, Fig. 3b) and repeat **steps 1–2** (*see* **Note 7**).
4. While centrifuging, prepare the digestion medium as described in Subheading 2.2, **item 6** (20 ml per brain).
5. Resuspend and pool the pellets in the digest mix and incubate for 1 h in the 37 °C water bath while shaking the suspension every 15 min. During the incubation, label cryovials (10 per brain) and prepare freezing mix (10 ml per brain) as described in Subheading 2.2, **item 7** (*see* **Note 10**).

3.2.4 Final Processing and Storage

1. Assemble the filter holders containing a rubber ring and a 200 μm nylon net filter. Place the filter on top of a 50 ml centrifuge tube.
2. Filter the digest suspension followed by 30 ml DMEM-comp. Use the medium to rinse the suspension tube to minimize microvessel loss. Optionally, collect a small sample (200 μl) for viability test (*see* Subheading 3.4, Fig. 3c) (*see* **Note 7**).
3. Spin the tube for 5 min (500 × *g*, room temperature) and remove any remaining white matter with the suction device.
4. Resuspend the pellet in the freezing mix and mix thoroughly to obtain a fully dispersed suspension.

5. Transfer the suspension to cryovials immediately, 1 ml in each, and freeze the vials in a −80 °C freezer overnight in a cell-freezing unit. Perform **steps 4** and **5** as fast as possible.
6. On the following day, move the vials to liquid nitrogen storage. Microvessels can be stored for at least 1 year and thawed for culturing endothelial cells (*see* **Note 11**).

3.3 Culture of Brain Microvessels and Endothelial/Astrocyte Coculture Models

Figure 2 shows the expected growth of endothelial cells after seeding microvessels.

3.3.1 Day 1: Coat Culture Flasks and Seed Microvessels

1. Mix 0.7 ml collagen stock solution and 6.3 ml PBS and add the resulting collagen solution to a T75 culture flask. Leave the flask for a minimum of 2 h at room temperature or overnight at 4 °C.
2. Aspire the collagen and rinse three times with PBS (*see* **Note 12**).
3. Mix 0.5 ml fibronectin stock solution with 49.5 ml PBS. Add 7 ml of this to the collagen-coated flask for a minimum of 30 min at room temperature, and then aspire and use directly. The remaining diluted fibronectin solution can be stored at 4 °C for 6 weeks.
4. While the flasks are incubating with fibronectin, thaw 1 vial of microvessels in a 37 °C water bath (*see* **Note 13**).
5. Wipe the vial with 70 % ethanol and transfer the content of the vial to 30 ml DMEM-comp. Centrifuge for 5 min at 500 × *g*, room temperature.
6. Aspire the culture medium and resuspend the pellets gently in 10 ml DMEM-comp. Optionally, collect a small sample (200 μl) for viability test (*see* Subheading 3.4, Fig. 3d) (*see* **Note** 7).
7. Pipette the cell suspension gently into the coated flask and place the flask in the incubator for 4–6 h (37 °C, 10 % CO_2) (*see* **Note 14**).
8. Prepare GM+ before the end of the incubation period as described in Subheading 2.3, **item 4**.

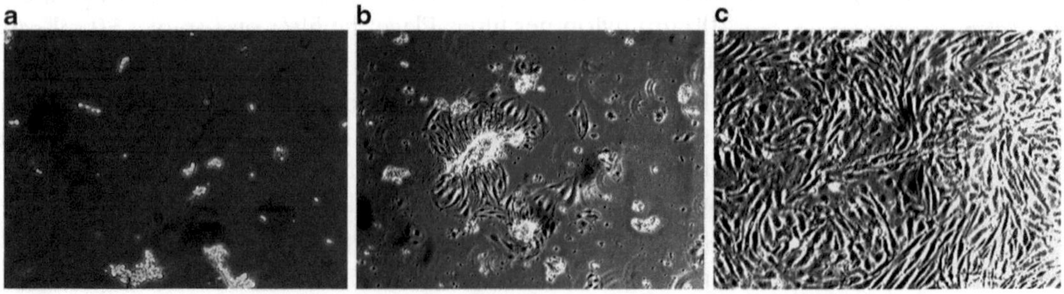

Fig. 2 Micrograph showing brain microvessels and endothelial cell outgrowth at the day of seeding: day 1 (**a**), day 3 (**b**), and day 5 (**c**)

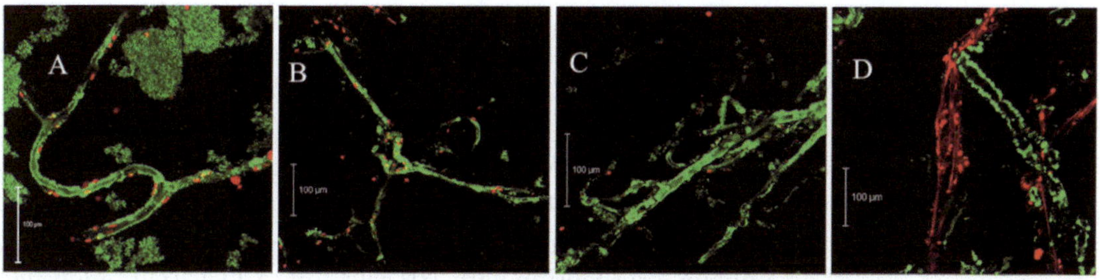

Fig. 3 Viability staining of bovine brain microvessel samples withdrawn after homogenization (**a**), filtration and resuspension (**b**), enzyme treatment (**c**), and a freeze/thaw cycle in liquid nitrogen (**d**). The samples were incubated with Calcein-AM (*green*) to visualize live cells and propidium iodide (*red*) to visualize dead cells. Bars = 100 μm. The figure demonstrates the presence of live capillaries but also several living tissue fragments in the homogenate (**a**). After filtration, most living tissue fragments have disappeared, whereas viable capillaries remain (**b**). Dead single cell nuclei are visible, some attached to the microvessels, which could be pericytes. After the enzyme treatment, most cells adherent to the microvessels have disappeared, whereas the microvessels remain viable (**c**). The structure is altered after the breakdown of the basement membrane and adhesion proteins. After freezing the microvessels in liquid nitrogen and re-thawing, some microvessels have died; however, approximately 50 % are still viable and endothelial cells will sprout upon culture

9. After 4 h of incubation, take the flask from the incubator and inspect with light microscopy. Now, you should be able to see microvessels attaching to the flask in small colonies. Attached capillaries will be present in a focal plane close to the bottom of the culture flask and will not move when the medium is gently agitated. If only a few or no capillaries have attached, the flask may be transferred back to the incubator until attachment is observed. Carefully aspire the medium and add 10 ml GM+pr. flask gently to the bottom or down the side of the flask, in order not to flush away the attached capillaries (Fig. 2a).
10. Place the flasks in the incubator at 37 °C, 10 % CO_2.

3.3.2 Day 3: Seed Astrocytes on Filter Inserts and Change Medium on Endothelial Cells in Flasks (See ***Note 15****)*

1. Mix 0.7 ml collagen stock solution and 6.3 ml PBS. Add 500 μl of the solution to the apical surface of each filter insert from a T12 Transwell cell culture tray (surface area = 1.12 cm^2, pore size = 0.4 μm). Incubate for minimum 2 h at room temperature or overnight at 4 °C.
2. Aspire the collagen and rinse 3 times with PBS (*see* **Note 12**).
3. Add 500 μl of the diluted fibronectin solution from Subheading 3.3.1, **step 3**, to the apical side of each collagen-coated filter insert and coat for at least 30 min.
4. Aspire the fibronectin and move the coated filter inserts to a Petri dish (ø 15 cm), where they are placed upside down.
5. Add 1.5 ml DMEM-comp into each of the wells in the T12 tray (without the filter inserts) and place the tray in the incubator (37 °C, 10 % CO_2).

6. Thaw 1 vial of astrocytes (approximately 2×10^6 cells) in a water bath, 37 °C. See Hertz et al. for astrocyte isolation protocol [24].
7. Wipe the vial with 70 % ethanol and transfer the content of the vial to 30 ml DMEM-comp. Centrifuge for 5 min at $500 \times g$, room temperature.
8. Aspire the supernatant and resuspend in 2 ml DMEM-comp to obtain a suspension with 10^6 cells/ml. Add 120 µl (equivalent to 120,000 cells) astrocyte suspension to the bottom of each coated filter insert. Distribute the 120 µl so it covers the entire filter insert bottom.
9. Place the Petri dish with astrocyte covered filter inserts in the incubator for 15 min, in order to allow the astrocytes to adhere to the permeable supports.
10. Transfer the filter inserts back into the T12 tray prefilled with DMEM-comp from **step 5**.
11. Incubate the astrocytes on the bottom of the filter inserts for 2 days (37 °C, 10 % CO_2).
12. Carefully aspirate the GM^+ from the brain microvessels seeded on day 1 and apply fresh GM^+. Endothelial cell sprouting and division should be clearly visible by now (Fig. 2b).

3.3.3 Day 5: Passage of Endothelial Cells and Seeding on the Filter Inserts

1. Prepare GM^- as described in Subheading 2.3, **item 5**. Prepare 20 ml per T12 tray.
2. Aspire the medium on the astrocytes and replace with freshly mixed GM^-. Save a small amount of GM^- (approximately 2 ml) for later resuspension of endothelial cells (**step 8**).
3. Incubate the astrocytes in the GM^- for at least 2 h before proceeding with the passage of endothelial cells.
4. Take the endothelial cells from the incubator and inspect under the microscope. They should be at least 50–60 % confluent in order to provide enough cells for 12 filter inserts (*see* **Note 16**) (Fig. 2c).
5. Rinse the endothelial cells twice with PBS.
6. Add 2 ml Trypsin-EDTA for endothelial cells. Immediately place the flask in the incubator for 1 min, and then take out the flask and observe continuously under the microscope. As soon as most endothelial cells have started to round up and detach, bang the flask gently against the table to detach the cells. Then stop the trypsinization by adding 10 ml DMEM-comp. It is important to avoid prolonged trypsinization, as this can detach pericytes growing from the microvessels (*see* **Note 17**). Preferably stop the trypsinization within 3 min. Flush the bottom of the flask a few times with the medium to detach the cells.

7. Transfer the cell suspension to a 50 ml centrifuge tube and adjust the volume to 30 ml with DMEM-comp.
8. Centrifuge the cell suspension for 5 min at 500 × *g*, room temperature and resuspend the pellet thoroughly but gently in 1 ml GM^-.
9. Count the cells in the FastRead counting plate, by adding 12 μl of cell suspension to the counting chamber, and count the amount of cells in a 4 × 4 grid, under microscope. Count at least 3 of the 4 × 4 grids and use the average cell count per 4 × 4 grid.

$$Cell\ /\ ml = Count \times 10^4 \Leftrightarrow Count = cells\ /\ (ml\ /\ 10^4)$$

$$V_{per\,filter} = \frac{100{,}000\,cells}{cells\ /\ ml} = \frac{100{,}000\,ml}{Count \times 10^4} = \frac{10\,ml}{count} = X\,ml$$

10. Calculate the volume to be added to each filter ($V_{\text{per filter}}$), using the equation above.
11. Apply the calculated volume of cell suspension to each filter insert (100,000 cells per T12 filter insert, or 9×10^4 cells/cm^2) by adding the suspension to the medium present in the culture wells (*see* **Notes 16** and **18**).
12. Place the T12 tray in the incubator at 37 °C, 10 % CO_2.

3.3.4 Day 8: Change to Differentiation Medium

1. Prepare TES-buffered differentiation medium (DM-TES) as described in Subheading 2.3, **item 11** (*see* **Note 19**).
2. Remove the T12 tray from the incubator and gently aspirate the medium. Take care when aspirating the medium to avoid damaging the cells. The medium is first aspirated from the basolateral compartment and then the apical. Change the Pasteur pipette between basolateral and apical liquid removal, in order to avoid apical contamination with astrocytes. Do not completely aspire the medium on the apical side, but leave a small volume (approximately 50 μl) to avoid damaging the endothelial cells.
3. Add 500 μl DM-TES to the apical side and 1,000 μl DM-TES to the basolateral side. Use a micropipette to apply accurate volumes.
4. Place the tray in the incubator at 37 °C, 10 % CO_2, and culture for 3 days.

3.3.5 Day 11: Experimental Day

1. Three days after the addition of DM, the transendothelial electrical resistance (TEER) will normally have stabilized, and the cocultures will be ready for experiments (*see* **Note 20**). Measure the TEER early on the day of the study, allowing the cocultures to stabilize in the incubator before the experiment.

Handle the T12 trays and the filters with caution (you may use a Styrofoam transport box in order to avoid temperature changes, when moving trays from the tissue culture/cell room to the lab).

3.4 Viability Test of Isolated Cells

1. Briefly centrifuge samples from Subheadings 3.2.2, 3.2.3, 3.2.4, and 3.3.1, to pellet the tissue.
2. Resuspend the pellets in 200 μl HBSS and pellet again.
3. Resuspend the pellets in 200 μl staining solution each (*see* Subheading 2.4, **item 4**) and incubate for 30 min at room temperature.
4. Pellet the samples and remove most of the staining solution, leaving behind approximately 50 μl.
5. Resuspend the samples in the remaining staining solution and apply a sample (approximately 10 μl) to a glass coverslip.
6. Inspect the samples with a confocal laser scanning microscope using two channels—Calcein-AM (excitation 488 nm, emission 520 nm) and propidium iodide (excitation 543 nm, emission 620 nm). Figure 3 shows sample pictures from the various isolation steps, indicated in the figure text.

4 Notes

1. All isolations present variations in the final yield of endothelial cells and the conservation of their paracellular tightness. Even when the protocol is meticulously followed, approximately only half of the isolations result in endothelial cells which are able to reach transendothelial resistance values above 1,000 Ω cm^2. It is therefore an advantage to perform two simultaneous isolations to improve the chance of success.
2. Astrocyte-conditioned medium is collected after medium change during the third week of culture of primary rat astrocytes (*see* Hertz et al. [24]). The medium can be stored in aliquots of 10–15 ml at −20 °C for at least 1 year.
3. Instead of TES-free acid, HEPES (4-(2-Hydroxyethyl) piperazine-1-ethanesulfonic acid) or MOPS (3-(*N*-Morpholino) propanesulfonic acid) buffer can be used to buffer the medium with similar effects on the endothelial cells [17].
4. It is important to be thorough when removing the leptomeninges from the cerebral hemispheres. Remaining material will cause problems during the homogenization (Subheading 3.2.2, **steps 1–3**). The homogenizer may stop when meningeal remnants get caught between the pestle and the beaker. This will cause large fluctuations in homogenization speed, detrimental

to the microvessels (*see* also **Note 6**). Furthermore, if meningeal blood vessels are carried through the procedure, they will contaminate the final capillary yield.

5. This is a very laborious step in the protocol and probably the most time-consuming part of the isolation. The gray matter is a very thin layer (approximately 2 mm) outside the white matter, and it is practically impossible to slice it off without removing some white matter as well. As there are several purification steps later on to get rid of white matter, it is not essential at this step, but still aim to keep it to a minimum.
6. Be cautious when performing this part of the isolation, as previous experience has shown that the homogenization is one of the essential steps of the isolation. Do not fill the homogenizer more than the stated 1/5 with brain pieces, and make sure to perform the homogenization with slow and steady movements as rapid movements are detrimental to microvessel survival.
7. Microvessel isolation is a long process, where several steps may affect the isolation outcome. Thus, it can be difficult to identify the reason for a failed isolation if microvessel viability is not assessed at several points through the procedure. It is therefore beneficial, especially when performing the procedure for the first time, to withdraw samples of homogenate and microvessel pellets at the indicated steps to pinpoint where the isolation might have gone wrong. The samples can either be stored at 4 °C for examination the following day, or they can be examined in parallel by another person.
8. The homogenate will not be able to run through the filter by gravitational force alone, if it is too concentrated or if too much homogenate is applied to one filter. The volume to apply per filter (e.g., 30–40 ml) should thus be corrected if the filters continue to clog. Sometimes, the homogenate is prevented from running by air bubbles and may thus be able to run through the filter after a brief pressure with the syringe stamp. However, avoid pressuring the entire homogenate through the filters with the stamp, as this will result in more contaminating tissue fragments in the final microvessel pellet.
9. The filter holders eventually clog. Remove large pieces of tissue stuck in the holder between each filtration, and clean the filter holder with the suction device. A filter partly covered with material indicates that the filter holder is clogged. This causes more small tissue remnants and free cells to adhere to the filters and yields microvessel preparations with more impurities.
10. Different batches of enzymes may vary in activity. The incubation time may therefore have to be modified when applying new enzyme batches, and it is recommended to determine an

optimal incubation time in the laboratory. However, 1 h is usually an appropriate time frame in our experience. It may be beneficial to withdraw small samples during the incubation to observe tissue degradation under the microscope. When finished, the capillaries should be broken into smaller fragments, where the basal membrane structure is degraded and the endothelial cells are visible as rounded cells like pearls on a string. However, the endothelial cells should still be connected in smaller capillary-like structures.

11. As stated in **Note 1**, endothelial cell yield varies between individual isolation batches. However, the yield is relatively stable within one isolation batch. One cryovial from a batch can be used to determine expected endothelial cell yield from all the vials from the isolation. The endothelial cell yield is determined after 4 days of culture (microvessels seeded Monday and cultured until Friday).
12. Culture flasks and/or Transwell filter inserts coated with collagen can also be stored at 4 °C for up to 2 weeks. Rinse the collagen-coated filter 3 times with PBS and allow the surface to dry before storing.
13. Some isolations yield few viable capillaries. If seeding one vial of microvessels in a T75 culture flask results in few endothelial cells, then attempt to pool 2 or even 3 vials in 1 T75 culture flask, which may result in a significantly higher yield of endothelial cells.
14. We use 10 % CO_2 when culturing the endothelial cells to obtain a more stable culture pH. Small fluctuations in CO_2 concentrations may occur in pulsed incubators. When 10 % CO_2 is used, relative variations in pCO_2, and thereby culture media pH variations, will be minimized.
15. The endothelial cells can also be established in monocultures or in noncontact cocultures with astrocytes seeded at the bottom of the well instead of the bottom of the inserts. This can be preferential for ease of culture or for specific experiments, where astrocytes may interfere with the interpretation of the results. However, the endothelial monocultures, noncontact cocultures, and contact cocultures differ in morphology. We have observed a spindle-shaped morphology of the endothelial cells when cultured in monoculture, a cobblestone-like morphology in contact cocultures, and an intermediate morphology in noncontact cocultures (Fig. 4). The cell morphology can thus also be used as an indicator of differentiation. Endothelial cell cultures in coculture should display the morphology shown in Fig. 4c, f.

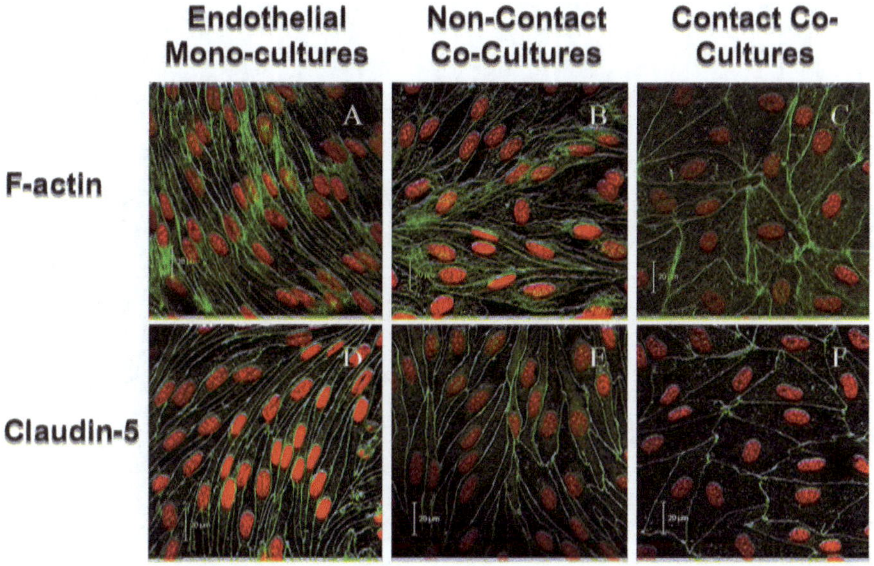

Fig. 4 Blood–brain barrier models at day 6 of mono- (**a**, **d**), noncontact (**b**, **e**), and contact coculture (**c**, **f**) were stained with phalloidin to visualize filamentous actin (**a–c**) as well as with antibodies against claudin-5 (**d–e**) (*green*). All samples were incubated with propidium iodide to visualize cell nuclei (*red*). Bars = 20 μm

16. It is important to maintain the seeding density of 9×10^4 cells/cm^2. Lower seeding densities lead to low paracellular tightness in the resulting models, even if the cells are cultured for longer time than the recommended 6 days. If the endothelial cell yield is smaller than expected, then prepare fewer filter inserts with cocultures instead of lowering the seeding density. Seed 2 vials of microvessels in 1 T75 flask for any future models with this isolation batch (*see* **Notes 1** and **11**).
17. The endothelial cells will sprout from the capillary and dominate the cell population in the culture flasks for the first 3–4 days of culture. On days 4–5, pericytes may grow from the capillary fragments. The trypsinization procedure will, if carefully performed, cause a detachment of the endothelial cells yielding a pure endothelial primary culture. The purity of the culture should be validated regularly, as described in **Note 20**. Pericytes are firmly attached to the surface of the flask and can be subcultured further if a primary pericyte culture is wanted.
18. Variations in seeding density can be a cause for large variations in transendothelial electrical tightness across the individual filter inserts. Resuspend the endothelial cell suspension with the pipette between adding cells to the filter inserts to minimize sedimentation of the cells. The best seeding results are achieved by seeding the endothelial cells in 2 rounds: half of the volume is added in the first round, and in the consecutive round, the

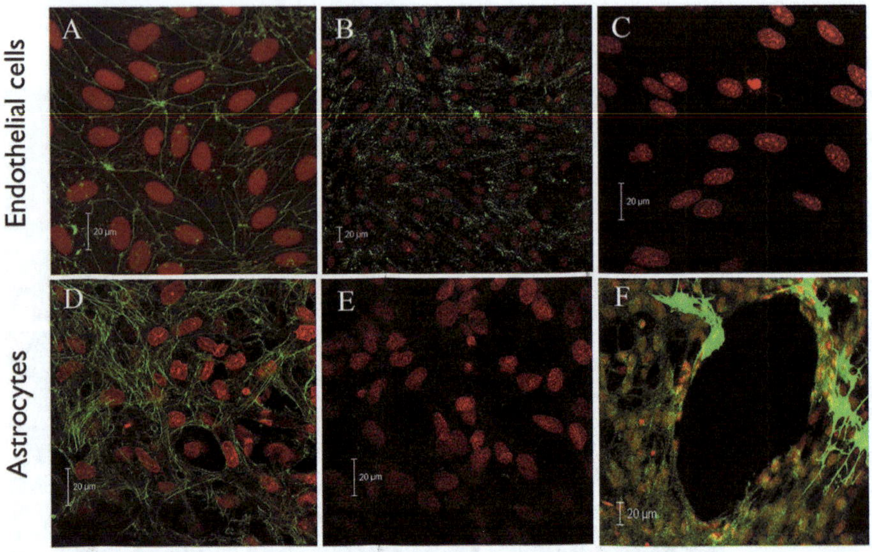

Fig. 5 Blood–brain barrier models at day 6 of coculture were stained with phalloidin to visualize filamentous actin (**a** + **d**) as well as with antibodies against von Willebrand factor (**b** + **e**) and glial fibrillary vacidic protein (**c** + **f**) (*green*). All samples were incubated with propidium iodide to visualize cell nuclei (*red*). Both the *top* of the filter supports containing endothelial cells (**a–c**) and the bottom containing astrocytes (**d–f**) were examined. Bars = 20 μm

last half of the volume is added in the opposite order (if wells 1–12 were filled in the first round, well 12–1 will be filled in the second round).

19. TES, MOPS, or HEPES is added to the differentiation medium to obtain a high buffer capacity, which has shown to promote high paracellular tightness in the coculture models [17]. Also, this medium maintains a relatively stable pH when removed from the incubator, whereby experiments on live cells can be performed outside the incubator without a prior change of medium. This gives, in our hands, the most robust setup to perform transport experiments, as change of the apical medium prior to experiments results in an immediate and significant drop in paracellular tightness.

20. It is important to characterize the endothelial cell/astrocyte coculture models regarding cell morphology and expression of cell-specific markers, especially when implementing the model in the laboratory but also as a routine validation. Sometimes, pericytes may be passaged along with endothelial cells, which will result in a significantly lower transendothelial electrical resistance across the filter inserts. If this is observed, it may be a good idea to perform the immunocytochemical validation again to confirm or rule out the presence of pericytes. Typical cell morphologies and expression of cell-specific markers are shown in Fig. 5.

References

1. Deli MA, Abraham CS, Kataoka Y, Niwa M (2005) Permeability studies on in vitro blood–brain barrier models: physiology, pathology, and pharmacology. Cell Mol Neurobiol 25(1):59–127
2. Crone C, Olesen SP (1982) Electrical resistance of brain microvascular endothelium. Brain Res 241(1):49–55. doi:0006-8993(82)91227-6 [pii]
3. Butt AM, Jones HC, Abbott NJ (1990) Electrical resistance across the blood–brain barrier in anaesthetized rats: a developmental study. J Physiol 429:47–62
4. Rapoport SI, Ohno K, Fredericks WR, Pettigrew KD (1978) Regional cerebrovascular permeability to [14C]sucrose after osmotic opening of the blood–brain barrier. Brain Res 150(3):653–657
5. Amtorp O (1980) Estimation of capillary permeability of inulin, sucrose and mannitol in rat brain cortex. Acta Physiol Scand 110(4): 337–342
6. Gu X, Zhang J, Brann DW, Yu FS (2003) Brain and retinal vascular endothelial cells with extended life span established by ectopic expression of telomerase. Invest Ophthalmol Vis Sci 44(7):3219–3225
7. Urich E, Lazic SE, Molnos J, Wells I, Freskgard PO (2012) Transcriptional profiling of human brain endothelial cells reveals key properties crucial for predictive in vitro blood–brain barrier models. PLoS One 7(5):e38149. doi:10.1371/journal.pone.0038149
8. Megard I, Garrigues A, Orlowski S, Jorajuria S, Clayette P, Ezan E, Mabondzo A (2002) A co-culture-based model of human blood–brain barrier: application to active transport of indinavir and in vivo–in vitro correlation. Brain Res 927(2):153–167. doi:S000689930 1033376 [pii]
9. Kannan R, Chakrabarti R, Tang D, Kim KJ, Kaplowitz N (2000) GSH transport in human cerebrovascular endothelial cells and human astrocytes: evidence for luminal localization of Na+-dependent GSH transport in HCEC. Brain Res 852(2):374–382
10. Abbott NJ, Dolman DE, Drndarski S, Fredriksson SM (2012) An improved in vitro blood–brain barrier model: rat brain endothelial cells co-cultured with astrocytes. Methods Mol Biol 814:415–430. doi:10.1007/978-1-61779-452-0_28
11. Yu XY, Lin SG, Zhou ZW, Chen X, Liang J, Yu XQ, Chowbay B, Wen JY, Duan W, Chan E, Li XT, Cao J, Li CG, Xue CC, Zhou SF (2007) Role of P-glycoprotein in limiting the brain penetration of glabridin, an active isoflavan from the root of Glycyrrhiza glabra. Pharm Res 24(9):1668–1690. doi:10.1007/s11095-007-9297-1
12. Nakagawa S, Deli MA, Nakao S, Honda M, Hayashi K, Nakaoke R, Kataoka Y, Niwa M (2007) Pericytes from brain microvessels strengthen the barrier integrity in primary cultures of rat brain endothelial cells. Cell Mol Neurobiol 27(6):687–694. doi:10.1007/s10571-007-9195-4
13. Perriere N, Yousif S, Cazaubon S, Chaverot N, Bourasset F, Cisternino S, Decleves X, Hori S, Terasaki T, Deli M, Scherrmann JM, Temsamani J, Roux F, Couraud PO (2007) A functional in vitro model of rat blood–brain barrier for molecular analysis of efflux transporters. Brain Res 1150:1–13. doi:10.1016/j.brainres.2007.02.091
14. Deli MA, Abraham CS, Niwa M, Falus A (2003) N, N-diethyl-2-[4-(phenylmethyl)phenoxy]ethanamine increases the permeability of primary mouse cerebral endothelial cell monolayers. Inflamm Res 52(Suppl 1):S39–S40
15. Dehouck MP, Meresse S, Delorme P, Fruchart JC, Cecchelli R (1990) An easier, reproducible, and mass-production method to study the blood–brain barrier in vitro. J Neurochem 54(5):1798–1801
16. Gaillard PJ, Voorwinden LH, Nielsen JL, Ivanov A, Atsumi R, Engman H, Ringbom C, de Boer AG, Breimer DD (2001) Establishment and functional characterization of an in vitro model of the blood–brain barrier, comprising a co-culture of brain capillary endothelial cells and astrocytes. Eur J Pharm Sci 12(3):215–222. doi:S0928-0987(00)00123-8 [pii]
17. Helms HC, Waagepetersen HS, Nielsen CU, Brodin B (2010) Paracellular tightness and claudin-5 expression is increased in the BCEC/astrocyte blood–brain barrier model by increasing media buffer capacity during growth. AAPS J 12(4):759–770. doi:10.1208/s12248-010-9237-6
18. Hoheisel D, Nitz T, Franke H, Wegener J, Hakvoort A, Tilling T, Galla HJ (1998) Hydrocortisone reinforces the blood–brain properties in a serum free cell culture system. Biochem Biophys Res Commun 247(2): 312–315
19. Malina KCK, Cooper I, Teichberg VI (2009) Closing the gap between the in-vivo and in-vitro blood–brain barrier tightness. Brain Res 1284:12–21. doi:10.1016/J.Brainres.2009.05.072
20. Patabendige A, Skinner RA, Abbott NJ (2013) Establishment of a simplified in vitro porcine blood–brain barrier model with high transendothelial electrical resistance. Brain Res 1521:1–15. doi:10.1016/j.brainres.2012.06.057

21. Bowman PD, Betz AL, Ar D, Wolinsky JS, Penney JB, Shivers RR, Goldstein GW (1981) Primary culture of capillary endothelium from rat brain. Vitro 17(4):353–362
22. Bowman PD, Ennis SR, Rarey KE, Betz AL, Goldstein GW (1983) Brain microvessel endothelial cells in tissue culture: a model for study of blood–brain barrier permeability. Ann Neurol 14(4):396–402. doi:10.1002/ana.410140403
23. Helms HC, Madelung R, Waagepetersen HS, Nielsen CU, Brodin B (2012) In vitro evidence for the brain glutamate efflux hypothesis: Brain endothelial cells cocultured with astrocytes display a polarized brain-to-blood transport of glutamate. Glia 60(6):882–893. doi:10.1002/glia.22321
24. Hertz L, Juurlink BHJ, Hertz E, Fosmark H (1989) Preparation of primary cultures of mouse (rat) astrocytes. In: Shahar A, Vellis JVAD, Haber B (eds) A dissection and tissue culture. Manual of the nervous system. Alan R. Liss Inc., New York, pp 105–108

Chapter 31

Isolation and Culture of Primary Pericytes from Mouse Brain

Amin Boroujerdi, Ulrich Tigges, Jennifer V. Welser-Alves, and Richard Milner

Abstract

Pericytes are perivascular cells that play an important role in the development, maturation, and remodeling of blood vessels. However, studies of this important cell type on vascular remodeling have been hindered due to the difficulty of culturing pericytes in adequate numbers to high purity. In this chapter, we present a novel yet simple method to isolate and culture large numbers of pure pericytes from the mouse central nervous system (CNS). In our approach, vascular cells obtained from adult mice brains are cultured initially under conditions optimized for endothelial cells. Following two passages, the medium is switched over to optimize pericyte growth. After growing the cells for 2–3 additional passages, this approach produces a largely homogeneous population of cells that express the pericyte markers NG2, PDGFβ receptor, and CD146 but are negative for markers of endothelial cells (CD31), astrocytes (GFAP), and microglia (Mac-1), demonstrating a highly pure pericyte culture. Thus, our technique provides an effective method to culture CNS pericytes that is easy to establish and provides large numbers of highly pure pericytes for extended periods of time. This system provides a useful tool for those wishing to study pericyte behavior.

Key words Pericyte, Brain endothelial cell (BEC), Culture, Purity, Coculture, Matrigel

1 Introduction

Pericytes are specialized mesenchymal cells associated with the walls of capillaries and other small diameter vessels [1]. Their function is still not entirely understood, but studies have shown they significantly contribute to the development, maturation, stabilization, and remodeling of blood vessels [2]. Strong evidence indicates that pericytes participate in the regulation of the blood–brain barrier (BBB) and in the process of angiogenesis [3–5]. Pericyte dysfunction has also been shown to be linked to several vascular pathologies, including hypertension [6, 7], diabetic microangiopathy, and various pathological diseases in which disordered angiogenic remodeling is a major part [2, 8]. Therefore, it is important that we enhance our understanding of pericyte

Richard Milner (ed.), *Cerebral Angiogenesis: Methods and Protocols*, Methods in Molecular Biology, vol. 1135,
DOI 10.1007/978-1-4939-0320-7_31, © Springer Science+Business Media New York 2014

biology in hopes of generating better therapeutics for the treatment of different vascular diseases.

The study of pericytes has been greatly facilitated by the isolation and culture of primary pericytes. Most techniques are aimed at the isolation of pericytes from the retina [9] or the brain [10], two tissues whose capillaries are rich in pericytes. Pericyte isolation strategies generally start with enzymatic digestion of tissue, usually followed by the isolation of microvessel fragments via successive filtration steps. Fragment outgrowth strategies are relatively easy to perform but the cultures generated often contain several different cell types, resulting in impure pericyte populations (i.e., also containing endothelial and glial cells). Pure pericyte cultures can be obtained through either positive or negative immunoselection via the use of magnetic beads or flow cytometry; the downside of this approach is that it is expensive and complicated and often leads to very low cell yields.

In our studies of pericytes, we recently developed a modified approach of the fragmented outgrowth strategy that allows for pericyte isolation from the mouse CNS and ultimately results in very pure pericyte cultures [11]. In this approach, vascular cells obtained from adult mice brains are cultured initially under conditions optimized for brain endothelial cells (BECs) but, after two passages, switched to a medium optimized for pericyte growth. After growing the cells for 1–2 passages, the BECs and other cell types are gradually outgrown by pericytes, to produce high-density cultures composed of very pure pericytes. We validated the pericyte purity in these cultures by performing immunocytochemistry for the pericyte markers NG2, PDGFβ receptor, and CD146 and excluded the possibility of contaminating alternative cell types using markers of endothelial cells (CD31), astrocytes (GFAP), and microglia (Mac-1). These pericyte cells can be maintained for extended periods of time at high purity making this a powerful system for analyzing pericyte behavior.

2 Materials

2.1 Cell Culture

2.1.1 Dissection Equipment

1. Large scissors, straight.
2. Small scissors, curved.
3. Curved forceps.
4. Plastic 100-mm Petri dishes, sterile, bacteriological grade (to contain the isolated brains).
5. 15-ml polypropylene centrifuge tubes, sterile.
6. 70 % ethanol in a spray bottle.

2.1.2 General Equipment

1. Biological safety cabinet.
2. Inverted microscope with 10× and 20× objectives.
3. A 37 °C incubator with humidity and gas control to maintain >95 % humidity and an atmosphere of 5 % CO_2 in air.
4. Pipetman.
5. Low-speed centrifuge.

2.1.3 Tissue Culture Equipment

1. Tissue culture 6-well plates.
2. 15- and 50-ml polypropylene centrifuge tubes.
3. 10-ml plastic serological pipettes.
4. Plastic pipette tips: 1 ml and 200 μl.
5. 10-ml syringes.
6. 19-gauge needles.
7. 21-gauge needles.

2.1.4 Media and Reagents

1. MEM-HEPES with penicillin/streptomycin.
2. Papain Cell Dissociation System (Worthington Biochemical Corp., Lakewood, NJ), consisting of papain and DNase I.
3. 30 % albumin solution from bovine serum.
4. Ham's Nutrient Mixture F12 (Sigma-Aldrich), with sodium bicarbonate.
5. Fetal bovine serum (FBS). Separate into 50 ml aliquots and store at −20 °C (*see* **Note 1**).
6. 100× penicillin/streptomycin solution. Aliquot into 5 ml lots and store at −20 °C. One aliquot of penicillin/streptomycin is added to 500-ml bottles of MEM-HEPES and Ham's Nutrient Mixture F12.
7. Endothelial cell growth supplement (Millipore 02-102): Rehydrate one 15-mg vial with 1 ml Ham's F12 medium = 15 mg/ml stock, aliquot into 100 μl aliquots, and freeze at −20 °C. Add one aliquot (100 μl) to 50 ml base medium (Ham's F12 containing 10 % FBS and pen/strep) to give working conc. = 30 μg/ml.
8. 400 mM L-glutamine solution: Add 5 ml F12 medium into the L-glutamine bottle, mix well (keep shaking the bottle during the aliquoting so as to keep the glutamine evenly distributed), and separate into 0.5 ml aliquots. Freeze at −20 °C (*see* **Note 2**). Add one aliquot to 50 ml base medium (1:100 dilution).
9. Ascorbate: Add 10 ml water to 5 mg ascorbate = 0.5 mg/ml stock, 0.25 ml aliquots, freeze at −20 °C, use at 1:200, and add one aliquot to 50 ml base medium to give working conc. = 2.5 μg/ml.

10. Heparin: Add 9 ml F12 medium into a bottle containing 10,000U heparin (Sigma-Aldrich H3149) (approx. 72 mg) to give approx. 0.8 % stock (8 mg/ml), separate into 0.25 ml aliquots, and freeze at –20 °C. Add one 0.25 ml aliquot to 50 ml base medium (1:200 dilution) to give working conc. = 40 μg/ml.
11. Stock F12 medium with penicillin/streptomycin: Add one 5 ml aliquot of 10× penicillin/streptomycin to 500 ml Ham's Nutrient Mixture F12 medium (with sodium bicarbonate).
12. Stock F12 medium with 10 % FBS and penicillin/streptomycin: Add one 50 ml aliquot FBS and one 5 ml aliquot of 100× penicillin/streptomycin to 500 ml Ham's Nutrient Mixture F12 medium (with sodium bicarbonate) to make stock F12 medium with FBS and antibiotics.
13. Endothelial cell growth medium (ECGM): F12 medium with 10 % FBS and penicillin/streptomycin, endothelial cell growth supplement, ascorbate, L-glutamine, and heparin. Aliquot 50 ml F12 stock medium with FBS and antibiotics into a 50-ml conical tube. Add 100 μl of endothelial cell growth supplement, 250 μl ascorbate, 250 μl heparin, and 500 μl L-glutamine (*see* **Note 3**).
14. Pericyte growth medium (PGM): Add 500 μl of Pericyte Growth Supplement (ScienCell 1252, San Diego, CA) into 50 ml of Pericyte Medium (ScienCell 1201).
15. Collagen coating solution: Collagen from calf skin type I diluted 1:5 in sterile water.
16. 0.05 % trypsin with EDTA.
17. Phosphate-buffered saline (PBS): Dilute 10× PBS to 1× PBS.

2.2 Immunohistochemistry

1. 24-well tissue culture plates.
2. Microscope cover glass circles No. 1–0.13 to 0.17 mm thick; size: 12 mm (Fisher Scientific).
3. Ice-cold acetone/methanol: The acetone/methanol (50:50 mixture). The acetone/methanol should be stored at –20 °C.
4. Phosphate-buffered saline (PBS): Dilute 10× PBS to 1× PBS.
5. Dako antibody diluent.
6. NG2 polyclonal anti-rabbit (AB5320, Millipore) should be used at 1:100 dilution.
7. PDGFβR polyclonal anti-rabbit (ab5511, Abcam) should be used at 1:100 dilution.
8. The FITC rat anti-mouse CD31 antibody (clone MEC 13.3 from BD Pharmingen) should be diluted 1:100.
9. GFAP polyclonal anti-rabbit (ab7260, Abcam) should be used at 1:100 dilution.

10. The rat anti-mouse Mac-1 antibody (clone M1/70 from BD Pharmingen) should be diluted 1:100.
11. Alexa Fluor 488 goat anti-rabbit secondary antibody: The Alexa Fluor 488 goat anti-rabbit IgG (H+L) antibody (Invitrogen) should be diluted 1:200 in antibody diluent.
12. Alexa Fluor 488 goat anti-rat secondary antibody: The Alexa Fluor 488 goat anti-rat IgG (H+L) antibody (Invitrogen) should be diluted to 1:200 in antibody diluent.
13. Hoechst stain: Hoechst stain solution diluted 1:2 in PBS.
14. Aqua-Poly/Mount (Polysciences).

3 Methods

3.1 Preparation of Mouse Brain Vascular Cultures (Containing BECs and Pericytes)

1. Make up fresh ECGM+FBS medium (add aliquots of ECGS, ascorbate, L-glutamine, and heparin) (*see* **Note 3**). Coat 4 wells of a 6-well plate with collagen type I diluted 1:5 in water (2 ml per well) for 2 h at 37 °C. Wash with PBS 3 times and keep the final wash in the well and transfer to the incubator to ensure gas equilibration before adding the endothelial cells.
2. Prepare the brain endothelial cells (BEC) from the brains of 6 mice (6–8 weeks of age preferred) euthanized by CO_2 inhalation (*see* **Notes 4** and **5**).
3. Flame all the tools and the razor blade to sterilize.
4. Ethanol spray the head and then use scissors to remove the skin on top of the skull. Next, use scissors to cut through the skull from the back of the head on both sides of the head towards the eyes. Carefully lift up the skull plate and remove the brain; place upside down in a 100-mm dish containing enough MEM-HEPES medium to totally cover the brain. Repeat this process for all 6 brains.
5. Transfer the brains to the lid of the 100-mm dish and then dice up the tissue using a flamed razor blade. Be thorough, chopping in both axes to ensure no tissue chunks remain.
6. Pour on some MEM-HEPES medium and aspirate the tissue using a 10-ml pipette into a 15-ml universal tube. Centrifuge this at 290×*g* for 5 min.
7. At this stage, prepare the papain dissociation solution by adding 5 ml MEM-HEPES to the vial of papain and 0.5 ml MEM-HEPES to the DNase tube. Incubate these solutions in the 37 °C incubator for 10 min to ensure they have fully dissolved.
8. Aspirate the supernatant from the brain tissue and add 5 ml papain solution and 250 μl of the DNase solution onto the brain tissue. Mix well and incubate at 37 °C for 1 h and 10 min.

9. After 1 h and 10 min incubation, transfer the contents of the tube to a 50-ml tube to allow access of the syringe for trituration. Break up the tissue using a 10-ml syringe, first using a 19-gauge needle, 10 times up and down, and then with a 21-gauge needle. Be careful not to over-triturate (*see* **Note 6**). The aim is to break up the microvessels from the brain tissue, but to maintain the vascular tubes intact.
10. After trituration, transfer the brain homogenate (approximately 6–7 ml) to a new 15-ml tube and add approximately 7 ml of 22 % BSA to the tube and mix well. Centrifuge the 15-ml tube at 1,360 × *g* for 10 min. Following this spin, the myelin is retained at the top of the tube, the vascular tubes, and liberated cell pellet at the bottom of the tube.
11. After the spin, tilt the tube at 45° to free the myelin from the edge of the tube, and then pour the myelin plus BSA into a waste container. Remove the remainder of the myelin with a pipette being careful not to allow the myelin to mix with the pellet. Suspend the blood vessel pellet with 1 ml ECGM and transfer into a new 15-ml centrifuge tube. Add an additional 1 ml ECGM medium to the suspension.
12. Wash the blood vessels by spinning down at 290 × *g* for 5 min.
13. Resuspend the cell pellet in 8 ml ECGM and plate this onto 4 wells of a 6-well plate coated with type I collagen (*see* **step 1** above). Incubate at 37 °C. Over the next few hours, the vascular tubes attach and form colonies that continue to grow and expand with time (*see* Fig. 1).
14. Change all medium the next morning (day 1) replacing the ECGM with fresh ECGM and allow the cells to grow to confluence.

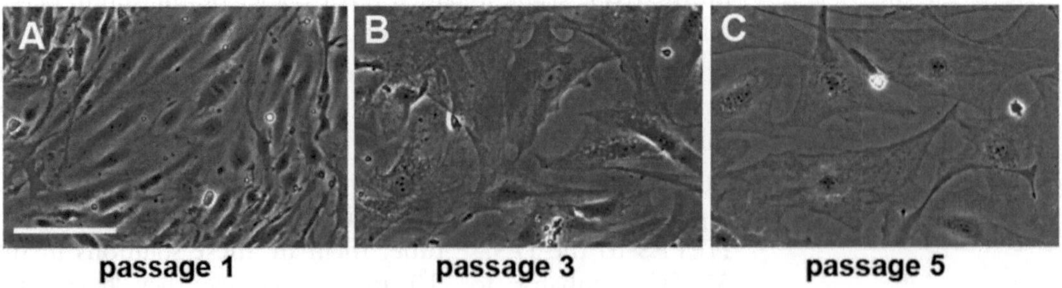

Fig. 1 Phase micrograph of pericyte cultures at different stages of culture: (**a**) passage 1, (**b**) passage 3, and (**c**) passage 5. Note that the passage 1 culture contains many cells with a fusiform morphology, typical of endothelial cells. In contrast in later passage cultures, the predominant cell type has a polygonal rhomboid morphology typical of pericytes in culture. Scale bar = 50 μm

3.2 Growing and Selecting for Pericytes

1. After the cells reach confluence, coat 4 wells of a 6-well plate with type I collagen for 2 h at 37 °C. Wash with PBS 3 times and keep the final wash in the well and transfer to the incubator to ensure gas equilibration before adding the endothelial cells (*see* Subheading 3.1, **step 1**).
2. Remove the medium from the 6-well plate containing the vascular cells and wash with F12 medium.
3. Add 2 ml trypsin to each well and place the plate in the 37 °C incubator for 5 min (*see* **Note 7**).
4. Look at the cells under the microscope to ensure that all the cells have detached from the plate. Once all of the cells have detached, add 2 ml F12 + FBS medium to each well. Remove the cells and medium into a 15-ml tube. All four wells can be pooled together at this time.
5. Spin down the cells at 290 × *g* in a centrifuge for 5 min to pellet the cells.
6. Remove the supernatant and resuspend the cell pellet in 1 ml ECGM, followed by an additional 3 ml ECGM for a total of 4 ml cell suspension.
7. Seed 1 ml of ECGM into each of the new collagenized 4-well plates that have 1 ml of ECGM already present in them (*see* **step 1** above) and incubate at 37 °C until confluent.
8. When confluent, passage the cells again. This time resuspend and grow the cells in pericyte growth medium (PGM) instead of ECGM and repeat **steps 1**–7.
9. From this point onwards, perform all passaging of cells with PGM. To passage pericytes onto a 24-well plate, follow **steps 1–4** and use the desired volume of PGM so that each well gets 500 μl of the cell suspension. *See* Fig. 1 which demonstrates the morphology of the cultures at different stages of passage.

3.3 Immunocytochemistry to Validate the Purity of Pericyte Cultures

In our research, we use immunocytochemistry to determine the purity of pericyte cultures (*see* **Note 8**). We employ NG2 or PDGFβ receptor antibodies to identify pericytes and CD31, GFAP, and Mac-1 antibodies to ensure the absence of endothelial cells, astrocytes, and microglia, respectively, in these cultures.

1. Sterilize glass coverslips by dipping in 70 % ethanol and placing into the wells of a 24-well plate. Air-dry in the T. C. hood.
2. Prepare the collagen I solution to be coated onto coverslips. Collagen I is used at approximately 0.2 mg/ml (1 in 5 of stock in H_2O). Place 25 μl drops of collagen I onto the center of the coverslips. Leave to coat for 2 h at 37 °C (*see* **Note 9**).
3. Prepare the cells by harvesting (follow **steps 1–5** from the section: growing and selecting of pericyte section) and centrifuge

the cells and then resuspend in a small volume of medium. Approximate volume of medium required = no. of drops × 25 μl.

4. Just before plating the cells onto the coverslips, remove the collagen coating solution and wash the ECM-coated areas 1× with PBS.
5. Apply the 25 μl cell drops to the substrates, taking great care to add the cell drops exactly over the wetted area previously coated with collagen. Mix the cell suspension well between each plating of cells.
6. Allow the cells to attach in the 37 °C incubator for 1 h. Check under the microscope that the majority of cells have attached and spread. If not, then extend this period for up to another hour.
7. Add 0.5 ml PGM medium to the well, and leave the cells to grow in the 37 °C incubator.
8. When the cells are approximately 50 % confluent, remove the coverslips from the wells using a curved needle and curved forceps, and fix in ice-cold acetone/methanol (50:50) in a 24-well plate for 5 min at −20 °C.
9. Wash the coverslips very well by dipping in three PBS-filled 50-ml tubes. Take care to remove all of the acetone/methanol (*see* **Note 10**).
10. Incubate with the appropriate primary antibody (NG2, PDGFβR, CD31, GFAP, or Mac-1) diluted 1:100 in antibody diluent for 30 min in a 37 °C incubator.
11. Wash in PBS by dipping in a PBS-filled 50-ml tubes three times.
12. Incubate with the appropriate corresponding secondary Alexa Fluor 488 (anti-rat or anti-rabbit) antibody (diluted 1:200 in antibody diluent) for 30 min in a 37 °C incubator.
13. Wash in PBS by dipping in a PBS-filled 50-ml tubes three times.
14. To stain nuclei, use Hoechst stain solution (diluted 1:2 in PBS) and incubate for 10 min.
15. Wash in PBS by dipping in a PBS-filled 50-ml tubes three times.
16. Apply Aquamount mounting medium and cover with a glass coverslip.
17. To remove excess Aquamount, lay a piece of tissue paper on the top of the coverslips and apply gentle pressure to squeeze out the excess medium which is then absorbed by the tissue paper.
18. Store the coverslips in the dark until ready for analysis by fluorescent microscopy.

4 Notes

1. It is worth batch-testing several different sources of FBS. Most companies are happy to let you have free samples of their FBS to test. Once you have found the batch that supports the optimal growth of your cultures, stick with it and order several 500-ml bottles to store at −20 °C.
2. When dissolving L-glutamine in small volumes of medium, it does not fully dissolve, but forms precipitates. Thus, it is important to mix this concentrated solution very well and ensure all the L-glutamine is transferred to the 500-ml bottle, where it fully dissolves in the larger volume. If dividing into 0.5 ml aliquots, shake the L-glutamine solution well in between each dispense to ensure equal distribution.
3. For every new brain vascular isolation, make up a fresh aliquot of 50 ml ECGM + FBS medium.
4. We find that it is important to dissect the brains from mice immediately after euthanization while mice are still warm—this gives the highest yield of cells.
5. We usually prepare brain vascular cultures with 6 brains going into 4 wells of a 6-well plate (1.5 brains per well). This protocol can be performed using fewer mice. However, it is critical that elements such as papain and DNase are downsized. For example, if using only one mouse brain (for instance, when numbers of mice are limited such as with specific knockout strains), papain should be reduced to 833.3 μl and DNase to 41.7 μl and cells should be plated into smaller-sized wells, e.g., use wells of a 12-well plate.
6. When triturating the brain suspension, the aim is to break up the brain tissue with a slow methodical use of the syringe, but do not over-triturate or move the tissue too violently as this will greatly reduce cell viability. BECs survive much better when they are contained within still intact vascular tubes; isolated BECs tend to die in the early stages of culture.
7. Trypsin incubation times vary with different cultures and are influenced by age of culture (the older more established cultures tend to need longer to detach), density of culture, batch of trypsin, and how long the bottle of trypsin has been at 4 °C (the longer the trypsin has been thawed, the lower the activity).
8. Using this approach, pericyte cultures become more pure with increasing passage. In the early passages, significant numbers of BECs still remain, but by the 4th or 5th passage, these pericyte cultures become highly pure. Therefore, to assure the purity of pericyte cultures, it is optimal to perform immunocytochemical analysis after this point.

9. When moving the plates containing the small 25-μl drops around, take great care not to jolt or jar the plate, as this will move the droplet off the wetted area and will create a larger ECM-coated area, which will distort the size of the prepared area and thus disrupt the experiment.
10. It is imperative to thoroughly remove all of the fixative from the coverslip by washing several times in PBS-filled 50-ml tubes. If all fixative is not removed, this will subsequently denature the added antibodies and prevent effective staining of cells.

Acknowledgements

This work was supported by the National Multiple Sclerosis Society—Harry Weaver Neuroscience Scholar Award (RM) and Postdoctoral Fellowship (JVW-A)—by the American Heart Association (Western State Affiliate) Postdoctoral Fellowship (AB), and by the NIH grant RO1 NS060770.

References

1. Armulik A, Abramsson A, Betsholtz C (2005) Endothelial/pericyte interactions. Circ Res 97: 512–523
2. Hirschi KK, D'Amore PA (1996) Pericytes in the microvasculature. Cardiovasc Res 32:687–698
3. Bergers G, Song S (2005) The role of pericytes in blood-vessel formation and maintenance. Neuro Oncol 7:452–464
4. Dore-Duffy P, LaManna JC (2007) Physiologic angiodynamics in the brain. Antioxid Redox Signal 9:1363–1371
5. Hamilton NB, Attwell D, Hall CN (2010) Pericyte-mediated regulation of capillary diameter: a component of neurovascular coupling in health and disease. Front Neuroenergetics 2:1–14
6. Kutcher ME, Herman IM (2009) The pericyte: cellular regulator of microvascular blood flow. Microvasc Res 77:235–246
7. Herman IM, Jacobson S (1988) In situ analysis of microvascular pericytes in hypertensive rat brains. Tissue Cell 20:1–12
8. Hammes HP (2005) Pericytes and the pathogenesis of diabetic retinopathy. Horm Metab Res 37:39–43
9. Gitlin JD, D'Amore PA (1983) Culture of retinal capillary cells using selective growth media. Microvasc Res 26:74–80
10. Dore-Duffy P (2003) Isolation and characterization of cerebral microvascular pericytes. Methods Mol Med 89:375–382
11. Tigges U, Welser-Alves JV, Boroujerdi A, Milner R (2012) A novel and simple method for culturing pericytes from mouse brain. Microvasc Res 84:74–80

Chapter 32

Assays to Examine Endothelial Cell Migration, Tube Formation, and Gene Expression Profiles

Shuzhen Guo, Josephine Lok, Yi Liu, Kazuhide Hayakawa, Wendy Leung, Changhong Xing, Xunming Ji, and Eng H. Lo

Abstract

Common methods for studying angiogenesis in vitro include the tube formation assay, the migration assay, and the study of the endothelial genome. The formation of capillary-like tubes in vitro on basement membrane matrix mimics many steps of the angiogenesis process in vivo and is used widely as a screening test for angiogenic or antiangiogenic factors. Other assays related to the study of angiogenesis include the cell migration assay, the study of gene expression changes during the process of angiogenesis, and the study of endothelial-derived microparticles. Protocols for these procedures will be described here.

Key words Angiogenesis, Endothelial cells, Matrigel matrix, Tube formation, Migration, Transcriptome, Microparticles

1 Introduction

The study of a complicated biological process such as angiogenesis requires the use of appropriate in vitro models to facilitate an understanding of the signaling pathways in angiogenesis. About 30 years ago, Montesano and Orci reported that bovine microvascular endothelial cells formed tubelike structures in a 3D collagen-I gel and that this process may serve as an in vitro angiogenesis model [1]. This model was expanded by George E. Davis to represent vasculogenesis or angiogenesis [2]; however, the assay required a period of several days, and was limited by the fact that not all the cells formed tubes, making it difficult to perform any type of quantification. In 1988 Kubota et al. showed that the use of a reconstituted basement membrane matrix enabled endothelial cells to attach and align rapidly and to form capillary-like tubes within hours, representing many of the steps of angiogenesis in vivo. These capillary-like tubes had characteristics of endothelial cells in vivo, including the ability to form tight cell-cell contacts and to take up acetylated

Richard Milner (ed.), *Cerebral Angiogenesis: Methods and Protocols*, Methods in Molecular Biology, vol. 1135, DOI 10.1007/978-1-4939-0320-7_32, © Springer Science+Business Media New York 2014

low-density lipoprotein [3]. The basement membrane matrix tube formation assay has been shown to be an easy, rapid, and reliable assay and is widely used to study the signaling pathways and mechanisms of angiogenesis, along with migration assays and proliferation assays [4].

Additional important assays for the study of endothelial cells include the analyses of the endothelial genome as well as the study of endothelial vesicles. A large amount of vesicles are found to originate from eukaryotic cells and are present in almost all body fluids [5]. These vesicles are called exosomes, microparticles, or microvesicles, according to their origins and characteristics [6]. These vesicular structures are pivotal in cellular communication both under physiological and pathological conditions [7, 8]. For instance, extracellular vesicles derived from platelets are reported to promote angiogenesis [9, 10]. Likewise, endothelial-derived microparticles containing proteins or mRNA of the matrix metalloproteinase family were also reported to contribute to angiogenesis [11, 12].

In this chapter we describe the Matrigel tube formation assay, the scratch migration assay [13], the transcriptome analysis of brain microvascular endothelial cells purified by PECAM-1 antibody [14], and the preparation of extracellular vesicles, including microparticles and microvesicles.

2 Materials

2.1 Growth Factor Reduced (GFR) Matrigel Matrix (BD Bioscience, San Jose, CA)

Matrigel is the most commonly used commercial basement membrane. A solubilized matrix extracted from the Engelbreth-Holm-Swarm (EHS) mouse sarcoma, it contains laminin, collagen IV, entactin, and heparin sulfate proteoglycan (perlecan). GFR BD Matrigel matrix contains reduced concentrations of growth factors compared to standard Matrigel matrix, resulting in a lower basal level of tube formation, which is advantageous for testing factors that increase angiogenesis.

2.2 Rat Brain Endothelial Cells

1. The rat brain microvascular endothelial cell line, RBE4.
2. Endothelial growth medium (EGM), as endothelial basal medium (EBM-2) with supplements (catalogue # CC-4147, Lonza, Cambridge, MA), which consist of 5 mL fetal bovine serum, hEGF, hydrocortisone, GA-1000 (gentamicin, amphotericin-B), VEGF, hFGF-B, R3-IGF-1, and ascorbic acid. The working medium is EBM-2 with 100 units/mL penicillin-streptomycin without serum and any supplements.
3. RBE4 cells were plated on collagen-I-coated flasks or plates.

2.3 Cell Detachment Solution

0.05 % trypsin-EDTA solution, stored at 4 °C.

2.4 Fixation Solution

4 % paraformaldehyde (PFA).

2.5 Animals for Brain Endothelial Cell Preparation

CD57 mice (6–8 weeks old) were used. For each preparation pool the cortexes from 3 mice.

2.6 Reagents for Endothelial Cell Purification

1. HBSS without calcium/magnesium, referred to as HBSS in this chapter.
2. Collagenase-A dissolved in H_2O and stored as 200 mg/mL stock in −20 °C. The working solution is 2 mg/mL diluted in HBSS with calcium/magnesium and warmed to 37 °C for use.
3. Rat antibody anti-mouse PECAM-1 (clone MEC13.3, Pharmingen, Franklin Lakes, NJ).
4. Dynabeads M-450 with sheep anti-rat IgG (Dynal Biotech).
5. Magnetic separator (Dynal Biotech).

2.7 Reagents for Total RNA Preparation and Measurement

1. 14.3 M β-mercaptoethanol (β-ME).
2. Ethanol (70, 80, and 96–100 %), molecular reagent grade.
3. RNeasy Plus Micro Kit (Qiagen).
4. RNA 6000 Pico kits including reagents and chips (Agilent Technologies, Santa Clara, CA).
5. Ovation Pico WTA system (NuGEN, San Carlos, CA).

2.8 Exosome-Free Fetal Bovine Serum (FBS)

The FBS used for conditioned culture medium to prepare extracellular vesicles is prepared by ultracentrifugation at 100,000 × *g* for 16 h prior to use, to remove bovine exosomes.

3 Methods

3.1 Matrigel Tube Formation

1. Thaw the Matrigel on ice or in cold room completely and slowly (*see* **Notes 1**, **2**). Keep it on ice until use.
2. Keep the multiwell plate on ice. Use precooled tips to add 100 μL soluble matrix into wells of 48-well plate, making sure that it covers the well evenly, with no bubbles in the gel mixture.
3. Incubate the plates with Matrigel at 37 °C for 30–60 min. This process causes the matrix to polymerize into insoluble gels. The Matrigel is now ready for use.
4. Detach RBE4 cells with detachment solution when they reach 80–90 % confluence. Collect and suspend the cells in EBM-2 with 0.5 % FBS.
5. Count the cells with a hemocytometer, diluting the cells in EBM-2 with 0.5 % FBS to 2.5×10^4 per cells/mL (*see* **Note 3**).

6. Seed the diluted RBE4 cells on to the surface of Matrigel-coated wells at a density of 5×10^3 cells/well (*see* **Note 4**).
7. Incubate the plate in the CO_2 incubator for 6–18 h, with temperature at 37 °C (*see* **Note 5**).
8. Examine the cells using a phase-contrast microscope, and take 2–3 images for each well using the light microscope (*see* **Note 6**).
9. Count the number of complete rings in each well. The number of rings is a measure of the capability to form tubes and a reflection of the angiogenic ability (*see* **Note 7**).

3.2 Scratch Migration Assay

1. Use RBE4 cells when confluent, wash twice with PBS, then add 3 mL cell detachment solution, and incubate at 37 °C for 5 min.
2. Add fresh 3 mL EGM, gently pipette to resuspend the cells, and centrifuge the suspension to get a cell pellet.
3. Resuspend the cell pellet with EGM, and plate 2×10^5 cells per well in a 6-well plate pre-coated with collagen-I or 5×10^4 cells per well in a 24-well plate.
4. Return the plates to the CO_2 incubator at 37 °C. Change to fresh EGM every 2–3 days until the cells are almost confluent (*see* **Note 8**).
5. Change to working medium, and incubate overnight for serum starvation (*see* **Note 9**).
6. Scrape the cell monolayer in a straight line with a p200 or p1000 pipette tip (*see* **Note 10**). Remove the debris by washing twice with warm working media, and then replace with fresh working medium.
7. To enable imaging the area with scratch injury, mark the location of the scratch with reference points on the outer bottom of the culture plate using a razor blade or an ultrafine tip marker.
8. Return the cells back to the CO_2 incubator for 8–24 h (*see* **Note 11**).
9. After the incubation, wash the cells twice with PBS gently to remove detached or dead cells, and fix cells with 4 % PFA for 10 min at 4 °C (*see* **Note 12**).
10. Place the culture plate under a phase-contrast microscope, locate the scratch injury site using the reference point, and acquire images from at least 4 random areas (*see* **Note 13**).
11. Quantify the number of migrated cells by counting cells that cross into the scratched area from their reference points (*see* **Note 14**).

3.3 The Transcriptome of Mouse Brain Microvascular Endothelial Cells (See Note 15)

3.3.1 Preparation of Anti-Mouse PECAM-1 Dynabeads

1. Take 12 μL of bead slurry; mix well in a 15 mL tube with 0.1 % BSA in PBS (*see* **Note 16**).
2. Place the tube in a magnetic separator for 1–2 min, then remove the supernatant.
3. Repeat washing as above for 3 times.
4. Resuspend the beads in 100 μL of 0.1 % BSA in PBS, then add 1 mL HBSS.
5. Add the rat anti-mouse PECAM-1 into the beads solution, at the ratio of 1 μg antibody for every 40 μL of beads. Incubate overnight on a rotator at 4 °C (or 2 h at RT) for antibody linkage.
6. Wash the beads/antibody mix with 2 mL 0.1 % BSA in PBS, for 3 times.
7. Resuspend the beads in original volume of 12 μL 0.1 % BSA to maintain beads at 4×10^8 beads/mL. Store the beads at 4 °C and they are ready for the purification. Use within 1–2 weeks.

3.3.2 Tissue Dissection and Dissociation

1. Sacrifice the mice by intracardiac perfusion with PBS, and wash the body with a generous amount of 70 % ethanol.
2. Harvest the brains in HBSS on ice; dissect the brains to remove the cerebral cortex.
3. Roll the cortical tissue on 3MM filter paper (autoclaved) to remove the big vessels.
4. Transfer the tissue to a new dish with a small amount of HBSS and finely mince the tissue with scissors.
5. Transfer the minced tissue to a 50 mL tube with warm collagenase-A solution (*see* **Note 17**).
6. Incubate the tissue in collagenase solution at 37 °C with gentle shaking for 20 min.
7. Using 14 gauge catheters attached to a syringe, triturate the enzyme suspension until there are no obvious big pieces.
8. Filter the enzyme suspension through a 70 μm cell strainer into a new 50 mL tube.
9. Centrifuge the cell suspension at 4 °C, 400 × *g* for 8 min.
10. Resuspend the pellet with 10 mL HBSS, centrifuge again for washing.

3.3.3 Purification of Endothelial Cells

1. Resuspend the pellet with 4 mL HBSS, and transfer into a new 15 mL tube.
2. Add anti-PECAM-1-coated Dynabeads, and mix by carefully rotating the tube up and down.

3. Incubate at room temperature for 10 min, or at 4 °C for 30 min, with gentle rotation for binding of endothelial cells to anti-PECAM-coated Dynabeads.
4. Mount the tube on a magnetic separator and wait for 1–2 min; then remove the supernatant (*see* **Note 18**).
5. Remove the tube from the magnetic separator and resuspend the Dynabeads with 4 mL HBSS by vigorous trituration (*see* **Note 19**).
6. Repeat **steps 4**, **5** for 3 times, then remove the supernatant.
7. After the last wash, directly add 400 μL Buffer RLT plus (with 1 % β-ME) for RNA preparation or freeze the Dynabead-bound cells in −80 °C for later preparation.

3.3.4 Preparation of RNA from Purified Endothelial Cells

Following the protocol suggested by the manufacturer:

1. Transfer the lysate of endothelial cells with Dynabeads to a 1.5 mL Eppendorf tube.
2. Homogenize the cell lysate with a 20 gauge needle attached to syringe, passing the cell lysate through the needle several times.
3. Transfer the lysate to the gDNA Eliminator spin column and centrifuge at 10,000 × *g* for 30 s.
4. Add 400 μL 70 % ethanol to the flow-through and mix well by gently pipetting.
5. Transfer the mixture to an RNeasy MinElute spin column in collection tube, and centrifuge at 10,000 × *g* for 30 s. Discard the flow-through.
6. Wash the RNeasy MinElute spin column with 700 μL Buffer RW1 and 500 μL Buffer RPE, respectively, by centrifuge at 10,000 × *g* for 30 s.
7. Wash again with 500 μL 80 % ethanol at ≥8,000 g for 2 min. And completely dry the column by centrifuge in a new collection tube at maximum speed for 5 min.
8. Transfer the column to a new collection tube; elute the RNA sample with 30 μL H_2O. Aliquot the RNA samples on ice and store at −80 °C.

3.3.5 Quantification and Qualification of RNA Samples for Microarray Service

1. Check the quantity of RNA samples. Apply 1 μL of RNA onto the NanoDrop, for quantification of RNA concentration, and the ratio of A_{260}/A_{280} and A_{230}/A_{280}.
2. Check the integrity of the RNA samples with Agilent 2100 Bioanalyzer. Apply 1 μL of RNA (diluted to concentrations less than 1 ng/mL) to Eukaryotic RNA Pico chip following the protocol of manufacturer. The respective ribosomal RNAs should appear as sharp peaks. The apparent ratio of 28S

rRNA to 18S rRNA should be approximately 2. The RIN (RNA integrity number) score should be larger than 7.0.

3. Use 500 pg–50 ng of RNA in a volume less than 5 μL for amplification to get cDNA with the NuGEN Ovation Pico WTA system, then fragmentation and labeling with Encore Biotin Module. The individual samples are hybridized to Affymetrix GeneChip Mouse Genome 430 2.0 microarray.

3.4 Preparation of Extracellular Vesicles

1. Use cells at 70 % confluence in several 15 cm culture dishes (*see* **Note 20**).
2. Discard culture media and wash thoroughly with pre-warmed PBS three times (*see* **Note 21**).
3. Add culture media with exosome-free FBS, and incubate 24–48 h to collect conditioned media with extracellular vesicles (*see* **Note 22**).
4. Collect the conditioned medium and centrifuge at 300 × *g*, 4 °C for 10 min, and followed by 2,000 × *g*, 4 °C for 20 min (*see* **Note 23**).
5. (a) To get microparticles, further centrifuge the supernatant at 20,000 × *g*, 4 °C for 20 min, resulting in a pellet containing the microparticles (*see* **Note 24**). (b) To get microvesicles, filter the supernatant with a 0.2 μm filter, and then ultracentrifuge at 100,000 × *g* (Rotor 70Ti) for 2 h, which results in a pellet containing the microvesicles (*see* **Note 25**).
6. Wash the pellets with filtered PBS, then resuspended in filtered PBS, and store at −80 °C (*see* **Note 26**).

4 Notes

1. It is recommended to use BD Matrigel matrix protein concentrations of 10 mg/mL or greater. Because the effect of Matrigel from different lots on tube formation may be different, it is preferable to use Matrigel from the same lot number for the entire experiment.
2. Thaw the soluble matrix slowly on ice, usually over night. Once thawed, swirl the bottle slowly on ice to make the solution even. Avoid repeated freezing and thawing. The unused Matrigel matrix should be dispensed into appropriate aliquots and refrozen immediately.
3. The angiogenic or antiangiogenic factors for testing can be added into the cell suspension before seeding.
4. The cell number seeded in each well is critical for this assay. Using an inadequate number of cells yields incomplete tubes, while using too many cells yields large areas of monolayers.

The optimum cell number is suggested to be approximately 4,800 cells/cm^2; however, this number may vary with different types of endothelial cells [4].

5. The cells could elongate and align to each other as early as 3 h after seeding. The images could be acquired at an early time (~6 h incubation time) or late time (~18 h incubation time) for analysis [4]. After longer incubation times, the proteinases secreted by the cells may digest the gels and destroy the supports for tubes.
6. For the thin coating of Matrigel matrix described here, the surface is not flat but crescent, which may make it difficult to get photos with good focus. There is a μ-Slide Angiogenesis chamber from ibidi GmbH (Planegg/Martinsried, Germany) used with only 10 μL of matrix and a flat surface to bring all cells into one focal plane.
7. The parameters for quantitative image analysis could be numbers, areas or perimeters of rings (tubes), or the numbers of branching points.
8. To maintain the cells in a monolayer, the endothelial cells should not be overgrown. Overgrown cells could be arranged in multiple layers, which cause higher amount of cell debris or more free-floating cells after scratching.
9. Serum starvation is an important step to minimize background signal before cell stimulation.
10. To create scratches of approximately similar size, it is critical to minimize any possible variation caused by a difference in the width of the scratches. For 24-well plates, the p200 pipette tip can be used more easily than the p1000 pipette tip.
11. Choose shorter incubation times under faster migrating conditions.
12. Before imaging, the washing step is important. Wash gently to remove all debris and free-floating cells to distinguish real migrating cells. Fixing cells helps to keep cell morphology and allows a longer imaging time.
13. RBE4 cells migrate into the scratched area under basal conditions (with working medium). Four random areas from both edges of the scratched area should be photographed, providing large sample sizes for statistical analysis.
14. Acquired images could be further analyzed quantitatively with other indexes, such as the migrating distance from reference point of at least 100 cells for each condition, or the area occupied by migrated cells. Some free software are available, such as Image J (http://rsb.info.nih.gov/ij/).

15. Great care should be taken during the preparation to eliminate possible RNase contamination and during the whole procedure to avoid introducing RNase into the samples. The use of sterile, disposable plastic wares (dishes and tubes) is recommended throughout the procedure. The reusable instruments should be cleaned and presoaked in 0.1 % DEPC for at least 12 h, rinsed with DW water, and autoclaved before use. Prepare all solutions using RNase-free water. Always wear latex or vinyl gloves while handling reagents and samples and change gloves frequently.
16. Prepare 4 μL of beads for each brain.
17. HBSS with calcium/magnesium is used to make collagenase-A solution, since the enzyme needs cations for its function. HBSS without calcium/magnesium is used in other steps, especially for the steps with magnetic beads.
18. The cell suspension here is cloudy; more HBSS can be added before placing the tube on the magnetic separator, making it easier to remove the supernatant while not losing beads.
19. Triturate vigorously here to wash out the contamination of other types of cells.
20. Media used for culture should be at room temperature before using. If unclear whether there is an adequate amount of extracellular vesicles production, collect as much conditioned culture medium as possible.
21. Cells should be washed thoroughly with PBS, especially if serum deprivation is required. This removes any residual nucleic acid and proteins from the serum.
22. Extracellular vesicle harvesting time varies from 24 to 48 h, depending on cell confluence.
23. During the preparation of vesicular structures, use 4 °C prechilled solutions. The centrifuge should be performed sequentially and immediately; frozen-thaw of conditioned medium even once would lead to a lot of unexpected pellets.
24. For microparticle preparation from blood, platelets can be easily stimulated to secrete more microparticles and can be an additional source of microparticles. It is highly recommended that one uses platelet-free plasma, along with citrate sodium as an anticoagulant.
25. Pure exosomes can be further purified by using continuous sucrose gradients.
26. The filtered PBS here should be prepared freshly and filtered through a 0.2 μm filter, for washing and dissolving the extracellular vesicles pellets.

References

1. Montesano R, Orci L (1985) Tumor-promoting phorbol esters induce angiogenesis in vitro. Cell 42(2):469–477
2. Koh W, Stratman AN, Sacharidou A, Davis GE (2008) In vitro three dimensional collagen matrix models of endothelial lumen formation during vasculogenesis and angiogenesis. Methods Enzymol 443:83–101
3. Kubota Y, Kleinman HK, Martin GR, Lawley TJ (1988) Role of laminin and basement membrane in the morphological differentiation of human endothelial cells into capillary-like structures. J Cell Biol 107(4):1589–1598
4. Arnaoutova I, George J, Kleinman HK, Benton G (2009) The endothelial cell tube formation assay on basement membrane turns 20: state of the science and the art. Angiogenesis 12(3): 267–274
5. Gallo A, Tandon M, Alevizos I, Illei GG (2012) The majority of microRNAs detectable in serum and saliva is concentrated in exosomes. PLoS One 7(3):e30679
6. Le Bihan MC, Bigot A, Jensen SS, Dennis JL, Rogowska-Wrzesinska A, Laine J, Gache V, Furling D, Jensen ON, Voit T, Mouly V, Coulton GR, Butler-Browne G (2012) In-depth analysis of the secretome identifies three major independent secretory pathways in differentiating human myoblasts. J Proteomics 77:344–356
7. Lai CP, Breakefield XO (2012) Role of exosomes/microvesicles in the nervous system and use in emerging therapies. Front Physiol 3:228
8. Skog J, Wurdinger T, van Rijn S, Meijer DH, Gainche L, Sena-Esteves M, Curry WT Jr, Carter BS, Krichevsky AM, Breakefield XO (2008) Glioblastoma microvesicles transport RNA and proteins that promote tumour growth and provide diagnostic biomarkers. Nat Cell Biol 10(12):1470–1476
9. Kim HK, Song KS, Chung JH, Lee KR, Lee SN (2004) Platelet microparticles induce angiogenesis in vitro. Br J Haematol 124(3): 376–384
10. Shai E, Varon D (2011) Development, cell differentiation, angiogenesis–microparticles and their roles in angiogenesis. Arterioscler Thromb Vasc Biol 31(1):10–14
11. Taraboletti G, D'Ascenzo S, Borsotti P, Giavazzi R, Pavan A, Dolo V (2002) Shedding of the matrix metalloproteinases MMP-2, MMP-9, and MT1-MMP as membrane vesicle-associated components by endothelial cells. Am J Pathol 160(2):673–680
12. Deregibus MC, Cantaluppi V, Calogero R, Lo Iacono M, Tetta C, Biancone L, Bruno S, Bussolati B, Camussi G (2007) Endothelial progenitor cell derived microvesicles activate an angiogenic program in endothelial cells by a horizontal transfer of mRNA. Blood 110(7): 2440–2448
13. Xing C, Lee S, Kim WJ, Wang H, Yang YG, Ning M, Wang X, Lo EH (2009) Neurovascular effects of CD47 signaling: promotion of cell death, inflammation, and suppression of angiogenesis in brain endothelial cells in vitro. J Neurosci Res 87(11):2571–2577
14. Guo S, Zhou Y, Xing C, Lok J, Som AT, Ning M, Ji X, Lo EH (2012) The vasculome of the mouse brain. PLoS One 7(12):e52665

Chapter 33

A Neurovascular Blood–Brain Barrier In Vitro Model

Christoph M. Zehendner, Robin White, Jana Hedrich, and Heiko J. Luhmann

Abstract

The cerebral microvasculature possesses certain cellular features that constitute the blood–brain barrier (BBB) (Abbott et al., Neurobiol Dis 37:13–25, 2010). This dynamic barrier separates the brain parenchyma from peripheral blood flow and is of tremendous clinical importance: for example, BBB breakdown as in stroke is associated with the development of brain edema (Rosenberg and Yang, Neurosurg Focus 22:E4, 2007), inflammation (Kuhlmann et al., Neurosci Lett 449:168–172, 2009; Coisne and Engelhardt, Antioxid Redox Signal 15:1285–1303, 2011), and increased mortality. In vivo, the BBB consists of brain endothelial cells (BEC) that are embedded within a precisely regulated environment containing astrocytes, pericytes, smooth muscle cells, and glial cells. These cells experience modulation by various pathways of intercellular communication and by pathophysiological processes, e.g., through neurovascular coupling (Attwell et al., Nature 468:232–243, 2010), cortical spreading depression (Gursoy-Ozdemir et al., J Clin Invest 113:1447–1455, 2004), or formation of oxidative stress (Yemisci et al., Nat Med 15:1031–1037, 2009). Hence, this interdependent assembly of cells is referred to as the neurovascular unit (NVU) (Zlokovic, Nat Med 16:1370–1371, 2010; Zlokovic, Neuron 57:178–201, 2008). Experimental approaches to investigate the BBB in vitro are highly desirable to study the cerebral endothelium in health and disease. However, due to the complex interactions taking place within the NVU in vivo, it is difficult to mimic this interplay in vitro.

Here, we describe a murine blood–brain barrier coculture model consisting of cortical organotypic slice cultures and brain endothelial cells that includes most of the cellular components of the NVU including neurons, astrocytes, and brain endothelial cells. This model allows the experimental analysis of several crucial BBB parameters such as transendothelial electrical resistance or tight junction protein localization by immunohistochemistry and live cell imaging of reactive oxygen species.

Key words Blood–brain barrier, Neurovascular unit model, In vitro blood–brain barrier model, Cortical organotypic slice cultures, bEnd.3

1 Introduction

Dysfunction of the blood–brain barrier plays a key role in various neurological disorders including inflammation [1], neurodegeneration [2, 3], cerebral ischemia [4, 5], and others [6–9]. The BBB is formed by brain endothelial cells (BEC) that are ensheathed by astrocytes and pericytes. Besides ensuring blood flow and glucose

Richard Milner (ed.), *Cerebral Angiogenesis: Methods and Protocols*, Methods in Molecular Biology, vol. 1135,
DOI 10.1007/978-1-4939-0320-7_33, © Springer Science+Business Media New York 2014

supply to the brain, BEC are required to maintain neural homeostasis. They have distinctive cellular features and express a complex regulated system of tight junction proteins [10–12] and various transporters [11] that essentially contribute to the adaptive sealing function of the BBB. Crucial aspects of an intact BBB model are reflected by the expression and physiological alignment of tight junction proteins forming a high transendothelial electrical resistance, which is a hallmark for the tightness of the BBB [13].

Within the NVU, perivascular cells have been shown to be capable of modulating the BBB [3, 14, 15] under physiological as well as pathological conditions. It has become evident that this interplay between endothelial cells, pericytes, astrocytes, smooth muscle cells, glia, and neuronal circuits is crucial in investigating BBB responses towards different stimuli. Numerous in vitro BBB models are based on a variety of different coculture systems that include combinations of astrocytes, pericytes, and BEC [16, 17]. However, these models lack a preserved neuronal compartment.

The Stoppini laboratory was the first to report a coculture of cortical slices and BEC which was investigated with electrophysiological recordings, electron microscopy, and microdialysis techniques [18]. Recently, our laboratory established a preparation that is composed of cortical organotypic slice cultures (COSC) and the BEC cell line bEnd.3 [19]. This system can be investigated with commonly used tools in BBB research, e.g., immunohistochemistry, evaluation of transendothelial electrical resistance, and live cell imaging. The procedure consists of preparing and culturing cortical organotypic slice cultures with the commercially available brain endothelial cell line bEnd.3.

2 Materials

2.1 Cell Culture

1. bEnd.3 cell line (American Type Culture Collection, Manassas, VA, USA).
2. bEnd.3 culture medium: DMEM GlutaMAX (with and without phenol red) with 1 % penicillin, 1 % streptomycin, 15 % fetal calf serum.
3. 0.25 % trypsin/EDTA.
4. Sterile glass coverslips.
5. 24-well culture plates.
6. Cell culture flasks.

2.2 Cortical Organotypic Slice Cultures

1. 70 % ethanol in water.
2. C57/Bl6 postnatal day 3 or 4 mouse pups.
3. Millicell-CM membrane filters 0.4 μm (Millipore, Bedford, MA, USA).

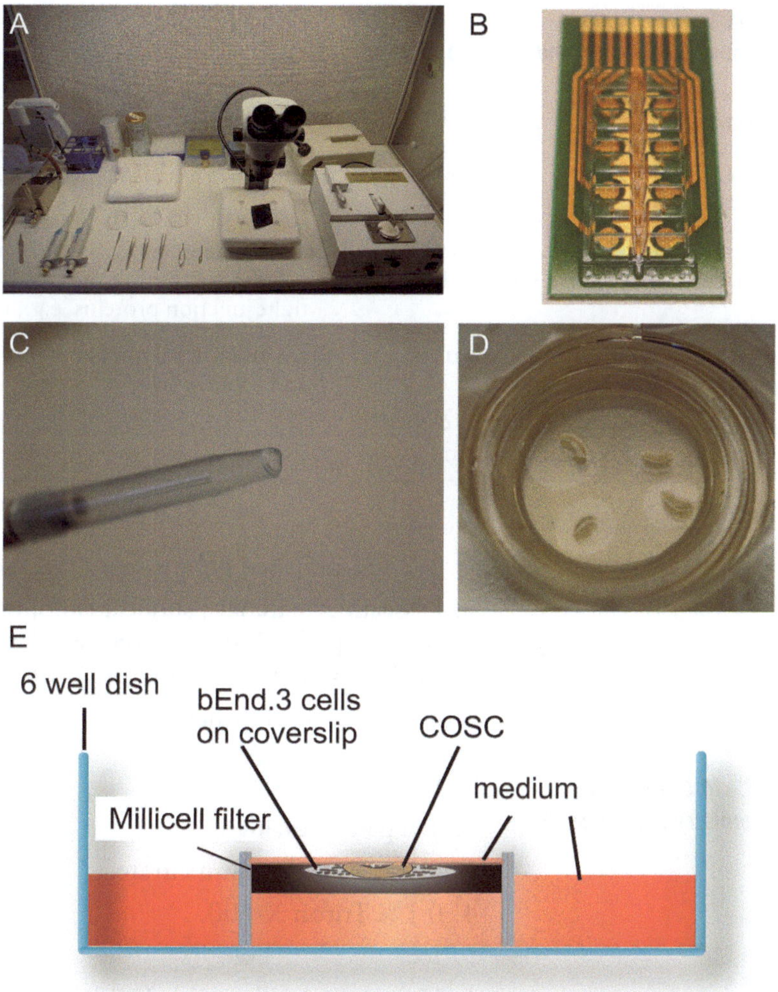

Fig. 1 Materials and tools. (**a**) Setup for COSC preparation with all required tools. To transfer COSC into an ECIS array (**b**) or onto Millicell membrane filters, a fire-polished pipette (**c**) is used. (**d**) Photograph of 4 COSC on a Millicell membrane in a 6-well dish. (**e**) Schematic drawing of COSC on top of bEnd.3 cells which are grown on a coverslip (*side view*). The coculture is placed on a Millicell filter and is covered with a film of bEnd.3 medium

4. 6-well culture plates, 35 mm diameter.
5. COSC culture medium (*see* **Notes 1**, **2**): prepare 200 ml of medium consisting of 100 MEM HEPES GlutaMAX supplemented with 1 mg/ml glucose, 50 ml horse serum, 43–45 ml HBSS containing calcium chloride (2 mmol/l), and magnesium chloride (1 mmol/l), 1 ml GlutaMAX. Add glucose dissolved in HBSS containing calcium (2 mmol/l) and magnesium (1 mmol/l) to reach a final concentration of 6–9 mg/ml. Warm the media up to 37 °C and adjust pH to 7.2.
6. Tissue chopper (e.g., from McIlwain, Mickle Laboratory Engineering, UK) (Fig. 1).

Table 1
Primary antibodies and their application

Primary antibody	Application
Mouse anti-claudin 5 (Life technologies #35-2500), 1:50	Claudin 5 is a transmembrane tight junction protein which is of major importance for BBB integrity
Rabbit anti-zonula occludens 1 (Life technologies #61-7300), 1:100	Zonula occludens 1 (ZO-1) is an intracellular tight junction protein connecting the cytoskeleton with transmembrane tight junction proteins, e.g., claudin 5 [11]
Rat anti-CD31, (Beckman Coulter #732116) 1:50–1:200	Brain endothelial marker of primary cortical microvessels and bEnd.3 cells
Rabbit anti-cleaved caspase-3 (Signaling Technology ASP 175, #9669), 1:200	Cleaved caspase-3 is a cellular marker for apoptosis

7. 2–3 fire-polished pipette tips (Fig. 1).
8. Dissecting materials: surgical lamp; benchtop microscope; 2 ice cooling pads, 3 sharp forceps; microsurgical scissors; scissors (3–5 mm blades, 0.1 mm tips); 1 scalpel; 1 scoop (Fig. 1).
9. 3–4 small petri dishes.

2.3 Immunohistochemistry (See Note 3)

1. Ice cold acetone.
2. 4 % paraformaldehyde.
3. 0.1 mol/l trisodium citrate.
4. 0.1 % Triton X-100.
5. Ice.
6. Phosphate buffer saline (PBS) 0.01 mol/l.
7. Normal goat serum.
8. Mounting medium.
9. Antibodies of interest. Table 1 gives an overview of some primary antibodies that we use to analyze our model.

2.4 Transendothelial Electrical Resistance

1. 8-well ECIS (Electric Cell-Substrate Impedance Sensing) device (8W10E), ECIS Z from ibidi (ibidi in cooperation with Applied BioPhysics, Martinsried, Germany) (*see* Fig. 1b).
2. Incubator in which the ECIS plates can be incubated at 37 °C; humidified atmosphere; 5 % CO_2.

2.5 Live Cell Imaging and ROS (Reactive Oxygen Species) Measurement

1. 2′,7′-Dichlorodihydrofluorescein diacetate (DCF, e.g., from Calbiochem or Sigma).
2. Calcein red AM (Life technologies).
3. HBSS containing 2 mmol/l calcium, 1 mmol/l magnesium, and 10 mmol/l glucose (without phenol red).

3 Methods

3.1 bEnd.3 Cell Culture

1. Grow bEnd.3 cells to confluence. Subculture cells with trypsin by adding a 0.25 % trypsin/0.02 % EDTA solution.
2. Subculture cells at a ratio of 1:10–1:40.
3. Exchange medium every 2–3 days.
4. For experiments seed 120,000 bEnd.3 cells per well in 500 μl culture medium on coverslips (in 24-well plates).
5. Confluence is usually reached within 48 h after seeding and should be monitored by light microscopy.

3.2 Preparation of COSC

1. Add 1 ml of COSC medium into a 35 mm well of a 6-well dish. Then, under sterile conditions, transfer the Millicell membranes into the wells with forceps. This procedure should be done well in advance of COSC preparation in order to equilibrate the membranes in an incubator before use (e.g., overnight).
2. Disinfect all scissors, scalpels, and forceps with 70 % ethanol prior to use. Place all instruments on an open airflow bench, using an arrangement of the instruments similar to Fig. 1a (photo of setup ready to use, **Notes 4**, **5**). One of the cooling packs is placed under the benchtop microscope, the other adjacent to it. Put one petri dish on the ice pack of the stage of the benchtop microscope, and place a dark piece of plastic sheet under it to improve the visibility of the tissue. The second ice pack should cool the second petri dish. Fill the dishes with cold (4 °C) COSC medium.
3. Rapidly decapitate one pup with scissors and discard the corpse. Disinfect the head with 70 % ethanol.
4. Grasp the head at the nose. Make a rostral midline incision in the skin starting caudally ranging from the foramen magnum to the nostrils using a pair of microsurgical scissors. The skull must be left intact at this step (Fig. 2a). Pull the skin aside to reveal the skull.
5. Make a rostral midline incision beginning at the foramen magnum towards the bregma to get access to the brain (Fig. 2b). With the help of a forceps, the soft skull can be flipped aside (Fig. 2c).
6. Remove the brain with a scoop and quickly transfer it into the cold COSC medium under the benchtop microscope.
7. Fix the brain with one pair of sharp forceps at the cerebellum. Remove the olfactory bulb and dissect both hemispheres with the other forceps. If possible at that stage remove the meninges carefully. Alternatively the meninges can be removed at **step 11**. Remove the cerebellum.

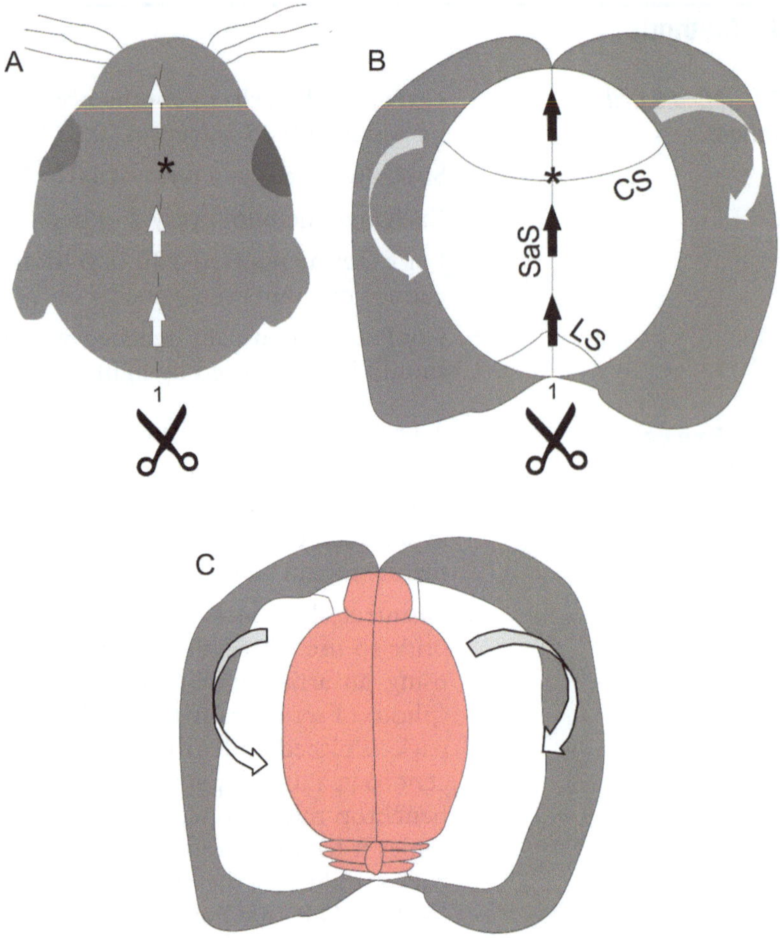

Fig. 2 Brain removal for COSC preparation. (**a**) To gain access to the mouse brain, a midline incision of the skin (*grey*) is performed with microsurgical scissors beginning at the foramen magnum [20] along the sagittal suture (SaS) directed towards the bregma (*). The skull must be left intact. (**b**) The skin is flipped aside to reveal the bone. Incise the skull carefully starting at the foramen magnum [20] towards the bregma (*). (**c**) The soft skull is then flipped aside to expose the brain. *CS* coronal suture, *LS* lambda suture

8. Place the brain onto the chopper, and remove remnants of media with a pipette.
9. Chop the brain at 350 μm thickness.
10. Fill a 1,000 μl pipette equipped with a fire-polished tip (*see* Fig. 1c) with medium and gently flush the sliced complete brain into the 2nd petri dish filled with medium. While pushing the slices off the Teflon plate, it is important to apply a continuous gentle flow with the pipette.
11. Separate the slices from each other with 2 pairs of sharp forceps. Afterwards use the scalpel to remove the hippocampi and remnants of the basal ganglia and the thalamus from the slices.

12. Place the 6-well dish containing the membranes next to the microscope.
13. Carefully aspirate the slice with a 1,000 μl pipette supplied with a fire-polished tip and transfer it onto the membranes.
14. Incubate the slices on the membranes (Fig. 1d) for at least 3 days after preparation to allow recovery from explantation. Exchange the medium 1 day after the preparation; thereafter medium is exchanged every 2nd day.

3.3 Combination of bEnd.3 and COSC on Coverslips

1. Fill a 6-well dish with 1 ml bEnd.3 medium per well, and insert one Millicell-CM membrane per well.
2. Put bEnd.3 grown coverslips on top of the Millicell-CM membrane (cells facing up, Fig. 1e), and cover the coverslip with a drop of bEnd.3 medium to avoid desiccation of cells.
3. Check COSC for necrosis by light microscopy. Only use COSC that display an intact morphology (*see* Fig. 3).
4. Gently remove COSC from the membranes using a scalpel or by flushing them off the membrane with a pipette. For that purpose add approximately 400 μl of bEnd.3 medium on top of the membrane.
5. Transfer the COSC carefully on top of the bEnd.3 coverslip.
6. Apply enough bEnd.3 medium so that the coverslip and the COSC are covered with a thin film of bEnd.3 medium (Fig. 1e).
7. Carefully put the 6-well plates back into the incubator (*see* **Note 6**).
8. Any movement of the plates should be avoided for the next following 48 h to let the COSC adhere to the bEnd.3 cells.
9. Wait for 48 h of coculture before manipulations can be performed (e.g., incubation in a hypoxic atmosphere in glucose-free medium to simulate ischemic conditions).

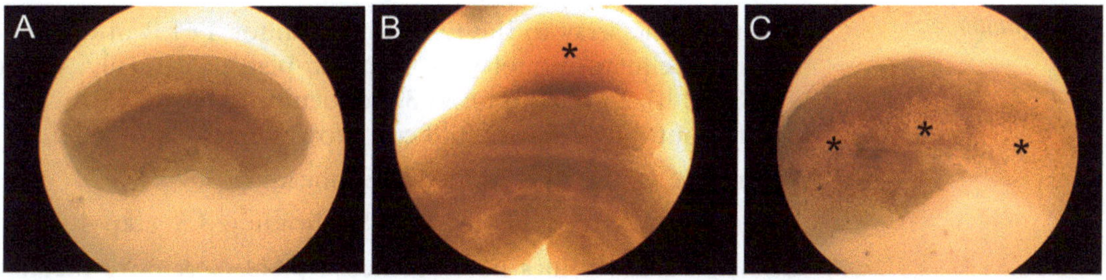

Fig. 3 Morphology of COSC evaluated by light microscopy. (**a**) An intact COSC. Note the smooth surface of the cortex and the homogeneous appearance of the cortical layers. (**b**) A microorganism-contaminated (*) COSC. (**c**) A COSC with inhomogeneous structure demonstrating loss of tissue possibly due to necrosis (marked with *asterisks*). Only specimens that display a proper morphology as shown in (**a**) should be used for further experiments

3.4 Live Cell Imaging of Cocultures to Analyze ROS

1. To visualize oxidative stress within bEnd.3 cells of the coculture, remove the medium very carefully with a pipette.
2. Transfer the Millicell filter into a 6-well dish containing 1 ml bEnd.3 medium supplemented with 10 μmol/l DCF. Next, add bEnd.3 medium (about 400 μl) supplemented with 10 μmol/l DCF on top of the coverslips including the COSC (*see* **Note 7**). We usually use an incubation time of 30 min.
3. Carefully exchange the medium containing DCF with colorless (without phenol red) HBSS or colorless bEnd.3 medium. A colorless buffer or medium is necessary to avoid interference with the fluorescent DCF signal. The sample is now ready for live cell imaging with the instrument setup of choice. We recommend a microscope setup that allows gassing of the coculture with 5 % CO_2, 20 % O_2, the rest N_2 or carbogen (a mixture of 5 % CO_2, the rest N_2).

3.5 Fixation of Coculture for Immunohistochemistry

1. For immunohistochemical staining, fix the cocultures with ice cold acetone (which to our experience works best with most antibodies).
2. Transfer the filters containing cocultures on coverslips with forceps into a glass petri dish.
3. Mount the petri dish onto an ice cold cooling pack and apply approximately 2 ml of ice cold acetone onto the Millipore filters.
4. Cover the petri dish to avoid evaporation of acetone.
5. Take care that the samples remain submerged in acetone for 10 min.
6. Transfer the fixed coverslips into a dish containing PBS. Proceed with immunohistochemical analysis or store the samples at 4 °C making sure the samples are submerged in PBS.

3.6 Combination of bEnd.3 and COSC for TEER Evaluation

1. Seed bEnd.3 cells at a density of 50,000–80,000 per 300 μl/well in an 8 ECIS array. Change the medium every 2–3 days.
2. Attach the array to the ECIS and monitor TEER in real time until cellular monolayers reach a plateau (usually within 48–72 h). In our lab bEnd.3 monocultures have TEER values of about 800–1,000 Ω (0.8 cm^2 growth area).
3. Prepare COSC as described in Subheading 3.3, **step 2**.
4. Remove COSC as described in Subheading 3.3, **step 4**.
5. When ECIS values have reached a plateau, remove medium from the wells with a pipette. During this procedure ECIS recording should be paused.
6. Transfer COSC into ECIS wells at a total volume of 300 μl bEnd.3 medium. No exchange of medium is necessary until maximum TEER values are reached.

7. Monitor TEER in real time with ECIS until TEER values reach a plateau. The coculture usually has higher TEER values than a monoculture of bEnd.3. Coculture TEER values reach about 1,000–1,150 Ω (0.8 cm^2 growth area). Usually a plateau is observed within 24–48 h after combining COSC with bEnd.3 cells. However, we recommend monitoring TEER changes in real time as growth conditions may vary.
8. After the plateau is reached, one can perform the desired experimental manipulations.

4 Notes

1. Culture medium for COSC may differ. In our lab we use different kinds of culture media. It is known that components such as cortisone, nonselective antibiotics (e.g., puromycin [17]), or growth factors may affect growth and differentiation of cell populations in vitro [17]. Therefore, we have recently switched to media free of antibiotics and growth factors to avoid selecting certain cell populations within COSC. This way we have observed that pericytes survive in COSC [21]. However, in our opinion the use of prokaryotic-specific antibiotics is also possible, and a composition of COSC media including antibiotics can be found in our recent publications [5, 19].
2. As already described in the original publication of the organotypic slice culture preparation by Stoppini et al. [22], it is crucial to adjust the pH to 7.2 with the antibiotic-free medium. At higher pH we have observed a bad quality of COSC.
3. For immunohistochemistry we obtained best results by fixing the tissue in ice cold acetone for 10 min. Caspase-3 stainings can also be done with paraformaldehyde (PFA)-fixed tissue. However, especially at concentrations of 4 % PFA, demasking procedures may be necessary. CD31 and ZO-1 usually work best with acetone fixation. Claudin 5 and CD105 also work fine after fixation in 2 % PFA for 2 h. One needs to figure out what kind of fixation works best for each specific purpose.
4. All experimental manipulations should be carried out under sterile conditions, ideally using a laminar flow cabinet.
5. Note that the laminar flow needs time to be properly established. Wait at least 15 min after turning on the airflow of the bench. Wipe off all instruments with 70 % alcohol before use.
6. Be sure that your incubator is fully humidified. The coculture is very sensitive to desiccation as it is only covered by a very thin film of culture medium. Check every day that there is enough medium on the cells.

7. At this step COSC may dislocate from the coverslip. Only cocultures with a firm adhesion on the coverslip can be used for live cell imaging.

Acknowledgements

This work is supported by grants of the Deutsche Forschungsgemeinschaft. CMZ is supported by a stage 1 grant of the University Medical Center of the Johannes Gutenberg-University Mainz. RW is financially aided by the MAIFOR program of the University Medical Center of the Johannes Gutenberg-University Mainz.

References

1. Argaw AT, Asp L, Zhang J et al (2012) Astrocyte-derived VEGF-A drives blood–brain barrier disruption in CNS inflammatory disease. J Clin Invest 122:2454–2468. doi:10.1172/JCI60842
2. Attwell D, Buchan AM, Charpak S et al (2010) Glial and neuronal control of brain blood flow. Nature 468:232–243. doi:10.1038/nature09613
3. Bell RD, Winkler EA, Sagare AP et al (2010) Pericytes control key neurovascular functions and neuronal phenotype in the adult brain and during brain aging. Neuron 68:409–427. doi:10.1016/j.neuron.2010.09.043
4. Rosenberg GA, Yang Y (2007) Vasogenic edema due to tight junction disruption by matrix metalloproteinases in cerebral ischemia. Neurosurg Focus 22:E4
5. Zehendner CM, Librizzi L, De Curtis M et al (2011) Caspase-3 contributes to ZO-1 and Cl-5 tight-junction disruption in rapid anoxic neurovascular unit damage. PLoS One 6:e16760. doi:10.1371/journal.pone.0016760
6. Gursoy-Ozdemir Y, Qiu J, Matsuoka N et al (2004) Cortical spreading depression activates and upregulates MMP-9. J Clin Invest 113:1447–1455. doi:10.1172/JCI21227
7. Yemisci M, Gursoy-Ozdemir Y, Vural A et al (2009) Pericyte contraction induced by oxidative-nitrative stress impairs capillary reflow despite successful opening of an occluded cerebral artery. Nat Med 15:1031–1037. doi:10.1038/nm.2022
8. Zlokovic BV (2010) Neurodegeneration and the neurovascular unit. Nat Med 16:1370–1371. doi:10.1038/nm1210-1370
9. Stanimirovic DB, Friedman A (2012) Pathophysiology of the neurovascular unit: disease cause or consequence? J Cerebr Blood Flow Metabol 32:1207–1221. doi:10.1038/jcbfm.2012.25
10. Coisne C, Engelhardt B (2011) Tight junctions in brain barriers during CNS inflammation. Antioxid Redox Signal 15(5):1285–1303. doi:10.1089/ars.2011.3929
11. Zlokovic BV (2008) The blood–brain barrier in health and chronic neurodegenerative disorders. Neuron 57:178–201. doi:10.1016/j.neuron.2008.01.003
12. Sandoval KE, Witt KA (2008) Blood–brain barrier tight junction permeability and ischemic stroke. Neurobiol Dis 32:200–219. doi:10.1016/j.nbd.2008.08.005
13. Deli MA, Abrahám CS, Kataoka Y et al (2005) Permeability studies on in vitro blood–brain barrier models: physiology, pathology, and pharmacology. Cell Mol Neurobiol 25:59–127
14. Armulik A, Genové G, Betsholtz C (2011) Pericytes: developmental, physiological, and pathological perspectives, problems, and promises. Dev Cell 21:193–215. doi:10.1016/j.devcel.2011.07.001
15. Daneman R, Zhou L, Kebede AA et al (2010) Pericytes are required for blood–brain barrier

integrity during embryogenesis. Nature 468:562–566. doi:10.1038/nature09513

16. Kuhlmann CRW, Zehendner CM, Gerigk M, Closhen D et al (2009) MK801 blocks hypoxic blood–brain-barrier disruption and leukocyte adhesion. Neurosci Lett 449:168–172. doi:10.1016/j.neulet.2008.10.096
17. Nakagawa S, Deli MA, Kawaguchi H et al (2009) A new blood–brain barrier model using primary rat brain endothelial cells, pericytes and astrocytes. Neurochem Int 54:253–263. doi:10.1016/j.neuint.2008.12.002
18. Duport S, Robert F, Muller D et al (1998) An in vitro blood–brain barrier model: cocultures between endothelial cells and organotypic brain slice cultures. Proc Natl Acad Sci U S A 95:1840–1845
19. Zehendner CM, Luhmann HJ, Kuhlmann CR (2009) Studying the neurovascular unit: an improved blood–brain barrier model. J Cereb Blood Flow Metab 29:1879–1884
20. Abbott NJ, Patabendige AAK, Dolman DEM et al (2010) Structure and function of the blood–brain barrier. Neurobiol Dis 37:13–25. doi:10.1016/j.nbd.2009.07.030
21. Zehendner CM, Wedler HE, Luhmann HJ (2013) A Novel In Vitro Model to Study Pericytes in the Neurovascular Unit of the Developing Cortex. PLoS ONE 8(11): e81637. doi:10.1371/journal.pone.0081637
22. Stoppini L, Buchs PA, Muller D (1991) A simple method for organotypic cultures of nervous tissue. J Neurosci Meth 37:173–182

Chapter 34

In Vitro Models of the Blood–Brain Barrier

Cathrin J. Czupalla, Stefan Liebner, and Kavi Devraj

Abstract

The blood–brain barrier (BBB) proper is composed of endothelial cells (ECs) of the cerebral microvasculature, which are interconnected by tight junctions (TJs) that in turn form a physical barrier restricting paracellular flux. Tight control of vascular permeability is essential for the homeostasis and functionality of the central nervous system (CNS). In vitro BBB models have been in use for decades and have been of great benefit in the process of investigating and understanding the cellular and molecular mechanisms underlying BBB establishment. BBB integrity changes can be addressed in vitro by determining cell monolayer permeability (Pe) to different solutes and measuring trans-endothelial electrical resistance (TEER).

This chapter describes procedures that can be utilized for both freshly isolated mouse brain microvascular ECs (MBMECs) and murine or human brain EC lines (bEnd5 or hCMEC/D3), cultivated either as a single monolayer or in cocultivation with primary mouse astrocytes (ACs). It starts with detailed information on how to perform transwell cell culture, including coating of inserts and seeding of the ECs and ACs. Moreover, it encompasses instructions for electrical assessment of the in vitro BBB using the more recent cellZscope® device, which was traditionally performed with chopstick electrodes of voltohmmeter type (EVOM). From continuous impedance measurements, the cellZscope® device provides TEER (paracellular resistance) and cell membrane capacitance (C_{cl}—transcellular resistance), two independent measures of monolayer integrity. Additionally, this chapter provides guidance through subsequent experiments such as permeability analysis (Pe, flux), expression analysis (qRT-PCR and Western blotting), and localization analysis of BBB junction proteins (immunocytochemistry) using the same inserts subjected earlier to impedance analysis.

As numerous diseases are associated with BBB breakdown, researchers aim to continuously improve and refine in vitro BBB models to mimic in vivo conditions as closely as possible. This chapter summarizes protocols with the intention to provide a collection of BBB in vitro assays that generate reproducible results not only with primary brain ECs but also with EC lines to open up the field for a broader spectrum of researchers who intend to investigate the BBB in vitro particularly aiming at therapeutic aspects.

Key words Blood–brain barrier, In vitro model, Microvascular brain endothelial cells, Astrocytes, Coculture, cellZscope®, TEER, C_{cl}, Permeability

Cathrin J. Czupalla and Kavi Devraj contributed equally.

Richard Milner (ed.), *Cerebral Angiogenesis: Methods and Protocols*, Methods in Molecular Biology, vol. 1135,
DOI 10.1007/978-1-4939-0320-7_34, © Springer Science+Business Media New York 2014

1 Introduction

The blood–brain barrier (BBB) protects the brain milieu from circulating toxins and pathogens and maintains the brain homeostasis. The principal sites of the BBB are the microvascular endothelial cells (ECs) with support from astrocytes (ACs) and pericytes (PCs). The tight junctions (TJs) formed by neighboring ECs restrict paracellular diffusion, while efflux transporters expressed on their plasma membranes prevent transcellular flux. However, transporters for vital nutrients such as glucose and iron are expressed at the BBB [1, 2]. The protective nature of the BBB is disadvantageous for the delivery of neurotherapeutics as the majority of drugs delivered do not reach the brain at all or do so with negligible bioavailability [3, 4]. Thus, there is a significant investment into development and characterization of in vitro BBB models for optimizing brain delivery of candidate drugs.

There is a plethora of in vitro models already in use, and several newer ones have emerged recently that comprise single culture of brain ECs from various species such as murine, bovine, and human. There are also models that utilize coculture between ECs and ACs and also triple cultures between ECs, ACs, and PCs [5]. Recently a model derived from human pluripotent stem cells has also been described [6]. The optimal model would possess BBB characteristics that would match the in vivo situation. The current chapter is therefore dedicated to describing protocols associated with the characterization of the BBB models in vitro to better aid in the assessment of their utility. All protocols have been tested primarily in murine ECs but are equally applicable to other species as indicated in the relevant sections. Many experimental disease models as well as transgenic models are available in mice from which the yield of brain ECs is meager. Therefore, we describe herein a protocol that utilizes the very same cells for different assays, which is not only a robust design but also minimizes the amount of cells needed.

Several in vitro BBB models rely on transwell chambers for functional measurements including permeability (Pe), trans-endothelial electrical resistance (TEER), and transmigration of inflammatory cells. However, the characterization of the same model is often either performed in dishes or flasks due to ease of handling. BBB ECs are known to dedifferentiate in vitro, and there are also treatments that help retain the native characteristics for a longer period of time [7, 8]. But no matter what treatment is utilized, BBB ECs behave differently according to their growth surface [9]. Classically, plastic has been the surface of choice, but as more functional assays are being performed in a transwell setup, semipermeable filter membranes (made of PET) have replaced plastic. We therefore propose that, if applicable, all assays ideally should be performed on the same growth surface.

In the current chapter, we describe such a protocol that can be utilized for both freshly isolated mouse brain microvascular ECs (MBMECs) and mouse brain EC lines either alone or in cocultivation with primary mouse ACs. We first give detailed procedures for handling transwell inserts and culturing ECs alone and in coculture with primary murine ACs. Furthermore, we describe the functional assay of impedance measurements using the cellZscope® device (nanoAnalytics, Münster, Germany) to assess monolayer tightness. Electrophysiological measurements that correlate with cell monolayer permeability have been established for some time [10, 11]. Traditionally, chopstick electrodes of the voltohmmeter type have (EVOM) been used, where an AC current is passed and the voltage is measured followed by applying Ohm's law to calculate the resistance of the monolayer. These kinds of measurements are very unreliable as they depend on the position and angle of electrodes. Further, the overall resistance of the monolayer is often expressed as a single value without any separation between paracellular and transcellular components. Impedance spectroscopy using the cellZscope® device separates these two components and hence gives a better indication of the monolayer barrier properties [10, 11]. The cellZscope® device provides noninvasive continuous monitoring of cell monolayers grown on semipermeable inserts with the advantage of using the inserts in subsequent experiments such as permeability analysis or expression analysis. Barrier integrity based upon impedance measurements, that is, reflective of ion flux, cannot be extrapolated to other solutes due to charge and size selectivity of paracellular permeability [12, 13]. Thus, a better assessment of barrier integrity is one that involves several solutes such as dextrans of varying molecular sizes in addition to the ion flux measured electrically [14]. Therefore, we provide a protocol for a fluorescence-based dextran permeability assay. For the same ECs subjected earlier to the functional assays, in the second part we describe the qRT-PCR expressional analysis and finally a detailed immunocytochemistry protocol for several BBB-related marker proteins (*see* Fig. 1).

2 Materials

2.1 TEER/Permeability

1. Freshly isolated mouse brain microvascular endothelial cells (MBMECs) [15] 3–4 days in culture (passage P0) or brain endothelial cell lines such as murine bEnd5 [16] (kindly provided by Prof. Dr. Britta Engelhardt, Theodor Kocher Institute, University of Bern, Switzerland) and human hCMEC/D3 [17] (kindly provided by Dr. Pierre–Olivier Couraud, Cochin Institute INSERM, Université Paris Descartes, France).
2. Primary astrocytes (ACs) isolated from murine C57/Bl6 wild-type cortices at postnatal day P4. Cells should be kept 3–4 days in culture before starting the assay (use passages P0-2) [18].

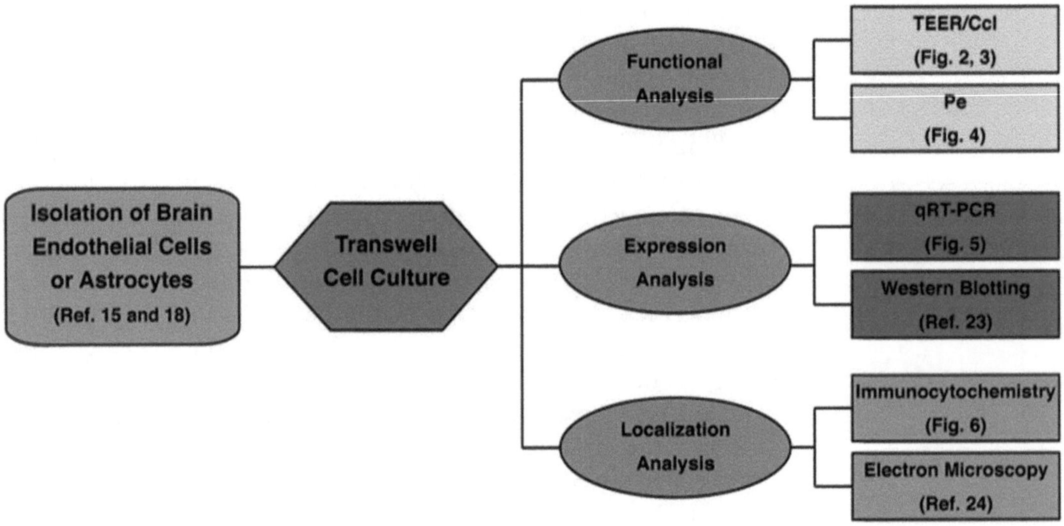

Fig. 1 Schematic representation of different assays that can be performed with the same transwell inserts. After initial isolation of brain ECs and ACs as described previously, we describe seeding the cells and maintaining these cultures on transwell inserts followed by functional assays of impedance and permeability measurements. This is followed up by expression analysis via qRT-PCR, while Western blotting is referred to previous work. We then describe immunocytochemistry for ECs on insert membranes followed by reference to previous work for electron microscopy

3. Endothelial complete growth medium: MCDB-131 (Gibco), 20 % fetal bovine serum (FBS) (*see* **Note 1**), 2 mM L-GLUTAMINE solution, 2 mM penicillin–streptomycin solution, 0.05 mg/ml ECGS (homemade; alternatively #E2759 Sigma-Aldrich), 1× MEM nonessential amino acids, 1 mM sodium pyruvate, 0.1 mg/ml heparin salt, 1 mg/ml sodium bicarbonate.
4. Astrocyte complete growth medium: DMEM GlutaMAX™, high glucose (Gibco), 10 % FBS, 2 mM L-glutamine solution, 2 mM penicillin–streptomycin solution.
5. Fibronectin (FN, # F1141; Sigma).
6. 1× phosphate buffered saline (PBS).
7. Poly-L-lysine solution, 0.01 % (PLL, #P4831; Sigma).
8. Trypsin (0.05 %)/EDTA.
9. 24-well ThinCert™, transparent, PET 1.0 μm pore size, growth surface 0.336 cm² (#662610; Greiner Bio One).
10. Companion plates for 24-well transwell inserts and 12-well plates (Greiner Bio One).
11. Tweezers for handling inserts.
12. cellZscope® system (nanoAnalytics, Münster, Germany) including windows 7 computer for continuous display and analysis.
13. Dedicated CO_2 (5 %) incubator (37 °C) for holding the cellZscope® device.

14. Lucifer yellow CH 0.45 kD (#L0259; Sigma), TXR 3 kD-dextran (#D3328; Invitrogen), 20 kD TMR-dextran (#73766; Sigma), and 70 kD FITC-dextran (#FD70S; Sigma).
15. Black 96-well plates (#655086; Greiner Bio One) for fluorescence measurements.
16. Infinite® M200 PRO multi-fluorescence plate reader.

2.2 qRT-PCR

1. 1× phosphate buffered saline (PBS).
2. RNeasy mini kit (#74106; Qiagen) for RNA extraction.
3. RNAse-free DNAse 1 set (#79254; Qiagen).
4. Experion™ automated electrophoresis station and associated RNA analysis kit (Bio-Rad) for RNA quality and quantity analysis.
5. RevertAid H Minus First Strand cDNA Synthesis Kit (#K1632; Thermo Scientific).
6. ABsolute QPCR SYBR Green Fluorescein Mix (Thermo Scientific).
7. Gene specific sense and antisense primer pairs (*see* Table 1).
8. MyiQ™ Single-Color Real-Time PCR Detection System (Bio-Rad) and iQ5 2.1 software (Bio-Rad).

2.3 Immunocytochemistry

1. 1× phosphate buffered saline (PBS).
2. Methanol, −20 °C.
3. Parafilm.
4. Blocking buffer: 20 % normal serum (derived from the species in which the secondary antibody was raised), 0.5 % Triton X-100, PBS, adjust pH to 7.4–7.6.
5. Antibody incubation buffer: 0.5 % BSA, 0.25 % Triton X-100, PBS. Adjust pH to 7.2.
6. Primary antibodies (*see* Table 2).
7. Fluorescently-labeled secondary antibodies.

Table 1
Primer sequences for qRT-PCR

Gene ID	Sense primer sequence	Antisense primer sequence
Axin-2	5′-gccgacctcaagtgcaaactc-3′	5′-ggctggtgcaaagacatagcc-3′
Claudin-3[a]	5′-cgtacaagacgagacggccaag-3′	5′-cacgtacaacccagctcccatc-3′
VE-cadherin	5′-gcccagccctacgaacctaaa-3′	5′-gggtgaagttgctgtcctcgt-3′
G6PDX	5′-gggtccaccactgccacttt-3′	5′-tttgcgttcattcagggcttt-3′

[a]Not intron spanning. Include −RT controls

Table 2
Primary antibodies for immunocytochemistry

Protein ID	Localization	Dilution	Distributer	Catalogue #
VE-cadherin	Adherens junctions	1:100	Santa Cruz	sc-6458
Claudin5	Tight junctions	1:200	Invitrogen	35–2,500

8. DAPI nucleic stain (4′,6-diamidino-2-phenylindole, dihydrochloride).
9. Aqua-Poly/Mount mounting medium (#18606; Polyscience Inc.).
10. Scalpel blade.
11. Glass slides and cover slips.
12. Confocal fluorescence imaging device.

3 Methods

3.1 Transwell Cell Culture

To assess TEER and cell permeability in vitro, transwell inserts have been used since many years. Thus, there are a variety of inserts available from different companies with diverse properties varying in their geometry and their filter membranes. The membranes vary in material (e.g., polycarbonate [PC] and polyethylene terephthalate [PET]), optical properties (translucent or transparent), growth surface area, pore size (typically 0.4, 1.0 and 3.0 μm), and pore density. Depending on the assay and cell type, one has to carefully choose an appropriate membrane. For all assays described below, transparent PET 24-well ThinCert™ transwell inserts with 1.0 μm pore size and 0.336 cm^2 surface area from Greiner Bio One have been used.

This section outlines the cultivation of brain microvascular ECs on the apical side of a filter membrane as single culture or in combination with basolaterally cultivated primary mouse ACs. For endothelial single culture, skip the PLL coating described in Subheading 3.2.1 and astrocyte coculture described in Subheading 3.2.2.

3.2 EC and Astrocyte Coculture

Preparation of the transwell inserts and seeding of the cells should be done in a laminar flow hood using sterile plasticware and solutions to avoid contamination. Always use sterile tweezers to handle the inserts and pre-warm PBS, trypsin, and media in a 37 °C water bath. Cells are incubated throughout the assays in 37 °C, 5 % CO_2 incubators. A vacuum pump can be used to aspirate liquids from the upper transwell compartment. We recommend attaching a yellow tip when glass Pasteur pipettes are used and to be very careful to avoid touching the membrane and/or cell layer with the tip.

Please note that depending on the cell line or cell type used in this assays, cell numbers and seeding time might need to be adapted. We recommend quadruplicate inserts per condition.

3.2.1 Coating Transwell Inserts

1. Slowly thaw a fresh FN solution aliquot at 2–8 °C. Carefully mix the FN solution by gently pipetting up and down (*see* **Note 2**). Each membrane is coated with 5 μg/cm² FN diluted in 150 μl sterile PBS (= 1.68 μl of a 1 mg/ml stock solution in 150 μl PBS per 24-well ThinCert™). Mix gently using a pipette.
2. Prepare a 0.001 % PLL solution by diluting 0.01 % PLL (stored at 4 °C) 1:10 in sterile distilled H_2O. For coating 24-well ThinCerts™ 70 μl per insert will be needed. Mix briefly by vortexing and keep at 4 °C for no longer than 2 weeks.
3. Sterilize tweezers with 70 % EtOH for 5 min and let air-dry in flow hood.
4. Open the lid of a 12-well plate and carefully remove one 24-well ThinCert™ using the sterilized tweezers from its sterile packaging. Invert the insert and place it bottom up in one well of the 12-well plate. Proceed with one insert after the other until you reach the desired number of wells you need for your experiment.
5. Add 70 μl of the 0.001 % PLL solution on the bottom of the ThinCerts™ facing now upwards.
6. Incubate for 45 min at room temperature (RT). To prevent drying, horizontally rotate the lid of the 12-well plate 180° and place it onto the plate. Now the lid does not fit properly but still prevents drying of the PLL.
7. In the meantime, prepare a 24-well plate with 450 μl PBS per well. Keep at RT.
8. Carefully aspirate the PLL using a pipette. Flip the ThinCerts™ back to their original orientation and place them in the 24-well plate with PBS prepared earlier.
9. Apply 150 μl of the previously prepared FN solution per ThinCerts™ and incubate for 45 min at RT with the lid closed.
10. Carefully aspirate the FN solution and add 250 μl PBS in the upper compartment. To balance liquid levels apply an additional 500 μl PBS to the lower compartment. The inserts can either be used immediately or be kept at 37 °C up to 7 days.

3.2.2 Seeding Astrocytes on the Basolateral Side of Transwell Inserts

1. Wash the ACs twice with PBS and detach the cells at 37 °C for 2–3 min with trypsin (e.g., 500 μl per T25 flask). Shake the flask horizontally until cells lift off.
2. Resuspend the cells in AC complete growth medium (add to 3 ml total volume) and determine the number of cells.
3. Per 24-well ThinCert™ 4.5×10^4 cells are seeded (=1.485×10^5/cm²). Make sure you have the required amount of cells (always

calculate the number of inserts +1) in not more than 70 µl of growth medium per insert. If necessary spin down the cells at 300×*g* for 3 min and resuspend in a smaller volume.

4. Remove PBS from the pre-coated inserts and invert them as described in Subheading 3.2.1, **step 4**.
5. Resuspend the ACs carefully and apply them on the bottom of the inserts facing now upwards.
6. Allow the ACs to attach to the surface for 45 min at RT. To prevent drying of the cells, follow **step 6** in Subheading 3.2.1.
7. In the meantime, prepare a 24-well plate with 950 µl AC complete growth medium per well and insert. Keep the plate in a 37 °C, 5 % CO_2-incubator.
8. When the cells are attached, remove the residual medium and cells using a 100 µl pipette. Flip the inserts back to their original orientation and place them immediately in the prepared 24 wells. Carefully add 250 µl of pre-warmed AC complete growth medium to the upper compartment of the ThinCert™.
9. Use a microscope and check for proper cell attachment and uniform seeding. At this point the cells are still round shaped and nicely visible. Later on they will stretch and flatten, which will render them more difficult to be seen.
10. Keep the plate in a 37 °C, 5 % CO_2-incubator for 24 h prior to seeding of the ECs.

3.2.3 Seeding ECs on the Apical Side of Transwell Inserts

1. Add 950 µl EC complete growth medium (*see* **Note 3**) per well to a new 24-well plate and incubate at 37 °C.
2. Wash the ECs twice with PBS and incubate 5 min in a 37 °C, 5 % CO_2-incubator. Aspirate the PBS and detach the ECs at 37 °C for 3–5 min with trypsin. Shake the flask horizontally until the cells lift off.
3. Resuspend the cells in EC complete growth medium (add to 3 ml total volume) and determine cell number.
4. Per 24-well ThinCert™ 3.3×10^4 MBMECs ($\approx10^5/cm^2$), 5.7×10^4 bEnd5 ($\approx1.7\times10^5/cm^2$) or 5.0×10^4 hCMEC/D3 ($\approx1.5\times10^5/cm^2$) cells should be seeded. Make sure you have the required amount of cells (always calculate number of inserts +1) in not more than 250 µl of growth medium per insert. If necessary spin down the cells at 300×*g* for 3 min and resuspend in a smaller volume.
5. Remove the plate containing the FN-coated ThinCert™ (with or without ACs) and 24-well plate containing the medium from the incubator and get ready to transfer the inserts.
6. Use sterilized tweezers to transfer the ThinCert™ and remove the medium or PBS from the apical compartment using a vacuum pump system. To prevent drying of membranes, transfer the inserts sequentially (e.g., only 4–6 inserts at a time).

7. Place the ThinCerts™ in the prepared 24-well plate, and seed the ECs in 250 μl EC complete growth medium on the apical side of the filter membrane.
8. Repeat **steps 6** and 7 until you have transferred all inserts.
9. Incubate the 24-well plate in a 37 °C, 5 % CO_2-incubator.

3.3 Functional Analysis

3.3.1 TEER/C_{cl}

The brain EC monolayer functions as a barrier that controls the para- and transcellular crossing of solutes from the blood to the brain parenchyma. The main components of this barrier are the junction molecules such as the adherens and tight junctions that are formed between adjacent ECs. These junction molecules respond to autocrine and paracrine signals that selectively trigger the opening and closing of the barrier [19].

Decades ago, it was shown that there is a direct correlation between EC permeability and its electrical resistance [20]. Thus, as a physiological readout, in vitro assays that measure the so-called trans-endothelial electrical resistance (TEER) are used to monitor the establishment or modulation of barrier-forming cell-to-cell contacts. The cellZscope® device offers a noninvasive way to electrically assess barrier integrity for cell monolayers grown on transwell filters. At several different frequencies of the applied AC voltage, the phase and amplitude of the resulting AC current is measured [13, 14]. The overall impedance of the monolayer at each frequency is then derived by applying an equivalent parallel circuit comprising paracellular resistance (TEER) governed by gaps at the cell–cell junctions and transcellular resistance governed by cell membrane capacitance (C_{cl}). Reduction in the paracellular clefts by TJs increases the TEER values. Morphological characteristics of cell membranes that increase the membrane surface area such as microvilli increase the C_{cl}, whereas increased alignment of the cells with smooth cell borders resulting from better organization or higher expression of the junction proteins reduces the effective membrane surface area and hence lowers the C_{cl}. Thus, tight barriers have high TEER values and low C_{cl} values.

When performing specific treatments, it is important to include the treatment compound both in apical and basal sides, as cells do not polarize correctly in vitro. It is also important to treat from both sides to avoid osmotic pressure differences across cell layer that might impact the ion flux and hence impedance measurements. Please note that the TEER values obtained in $\Omega \cdot cm^2$ using cellZscope® cannot be compared directly to measurements in other laboratories using different instruments such as ECIS system (Applied Biophysics Inc.) or EVOM (WPI Inc.). Several parameters such as height and surface area of the inserts along with the AC frequencies for TEER measurements are different, and in the ECIS system, there is no basolateral side as the cells are plated onto the electrode wells. Thus, there is a need for mathematical normalization of the above variables before TEER can be regarded as a truly absolute measure:

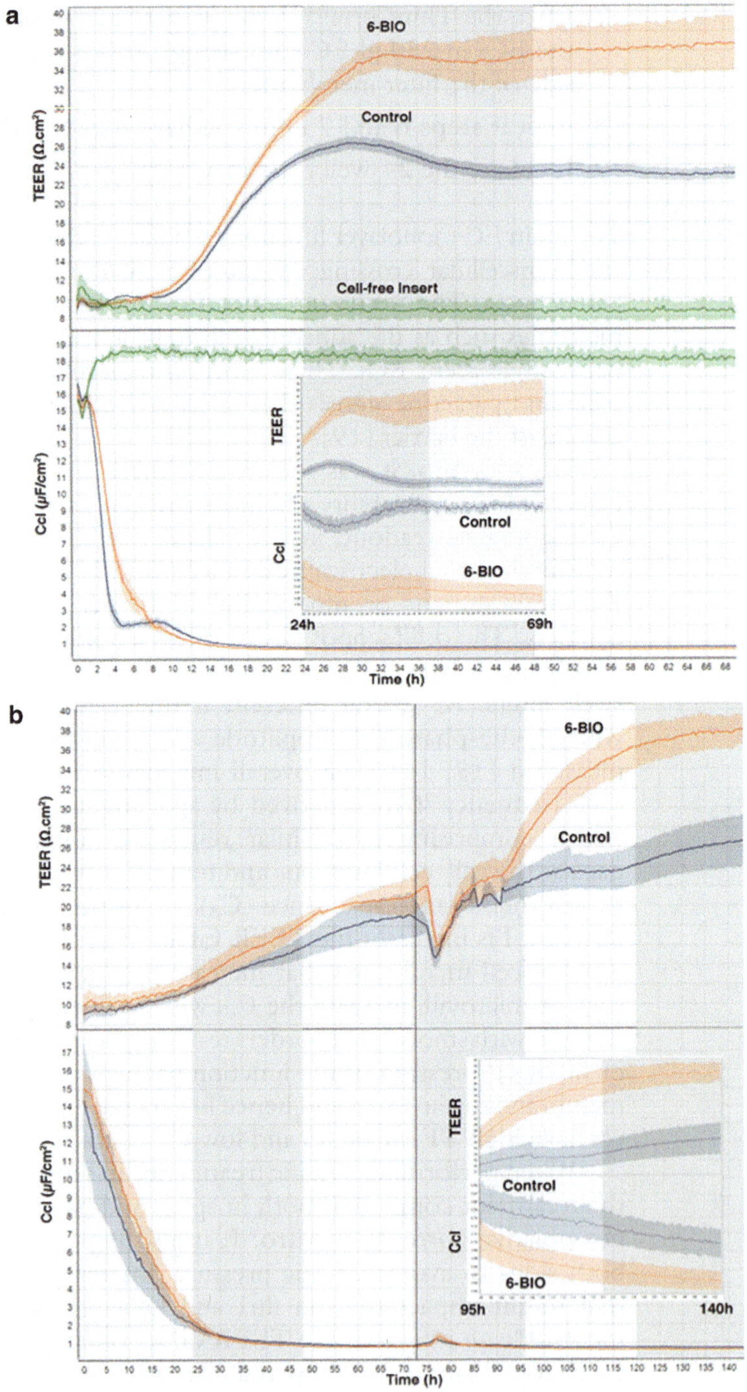

Fig. 2 Impedance measurements in cellZscope® device for primary mouse brain microvascular ECs (MBMECs) treated with 6-BIO, a canonical Wnt stimulator. (**a**) In this paradigm cells were treated at the time of seeding by adding 2.5 μM 6-BIO (Calbiochem), a GSK-3 inhibitor that stimulates canonical Wnt/β-catenin signaling. In both conditions MBMECs show gradual increase in the slope for TEER and decrease in the slope for C_{cl} until they reach a plateau, indicating monolayer confluency, whereas the cell-free inserts show a TEER of 8 $\Omega\cdot cm^2$ and C_{cl} of 17 $\mu F/cm^2$ that remain constant through the measurement. It can be noted that the control-treated cells reach a plateau for TEER around 23 $\Omega\cdot cm^2$, and upon canonical Wnt treatment using 6-BIO, there is an

1. Prepare the cellZscope® device for measurement by aspirating the ethanol that is applied during storage at RT. Add 1.5 ml PBS per 24-well (*see* also manufacturers description) to flush away residual ethanol and to equilibrate the pH. Finally, prewarm the device in a 37 °C, 5 % CO_2-incubator overnight or at least 6 h.
2. Before transferring the ThinCerts™, start the cellZscope® software and select File > New.
3. Check spectrum settings: start frequency, 1 Hz; end frequency, 100 kHz; points per decade: 9; spacing, linear (default cellZscope® settings).
4. Choose the correct type of insert (Greiner Bio One 24-well transparent).
5. Set up well matrix: set up > set up well matrix.
6. Adjust wait time: we recommend 5–15 min depending on the number of wells.
7. Aspirate PBS from the wells of the cellZscope® cell module.
8. Add 950 μl EC complete growth medium per well and transfer the ThinCerts™ filled with exactly 250 μl medium from the 24-well plate to the cellZscope® cell module (*see* **Note 4**).
9. Make sure the ThinCerts™ in the cellZscope® device are orientated correctly. Close the lids carefully; replace the cell module in a 37 °C, 5 % CO_2-incubator; and connect it to the controller.
10. Check electrode performance by hitting the "test electrode" button. If all wells pass the test indicated by a corresponding green signal, start the measurement by pressing the START button (*see* **Note 5**).
11. Depending on the duration of the measurement, medium might be exchanged after 24–72 h (*see* **Note 6**). Start treatment of the cells within 24–72 h, when TEER and C_{cl} values have reached the plateau phase indicating a confluent monolayer (*see* Figs. 2 and 3).

Fig. 2 (continued) increase in TEER to about 37 Ω·cm^2, whereas the capacitance (C_{cl}) goes down from control value of 0.7 to 0.6 μF/cm^2 in Wnt-treated cells. The increase in TEER and decrease in C_{cl} by Wnt treatment mean improved paracellular and transcellular resistance, respectively, and hence improved BBB properties by Wnt signaling as described previously [15]. (**b**) In this paradigm, the monolayer was first allowed to reach confluency (typically 48–72 h) followed by 6-BIO treatment. Again increase in TEER can be noted with a corresponding decrease in C_{cl} values with Wnt stimulation by 6-BIO treatment, whereas the control curves stay at the same level as before treatment. The decrease in capacitance could mean a decrease in the surface area of the cellular membranes resulting from better organization or change in the expression of tight-junction components. As the capacitance of the cell-free inserts and of cell-seeded inserts at early time points is high, zoomed version of C_{cl} curves and corresponding TEER curves after monolayer reached confluency has been shown in the bottom part of the graph in order to visualize the differences in C_{cl} with 6-BIO treatment. The figures (**a**) and (**b**) show representative experiments using quadruplicate wells with standard error in corresponding colored area

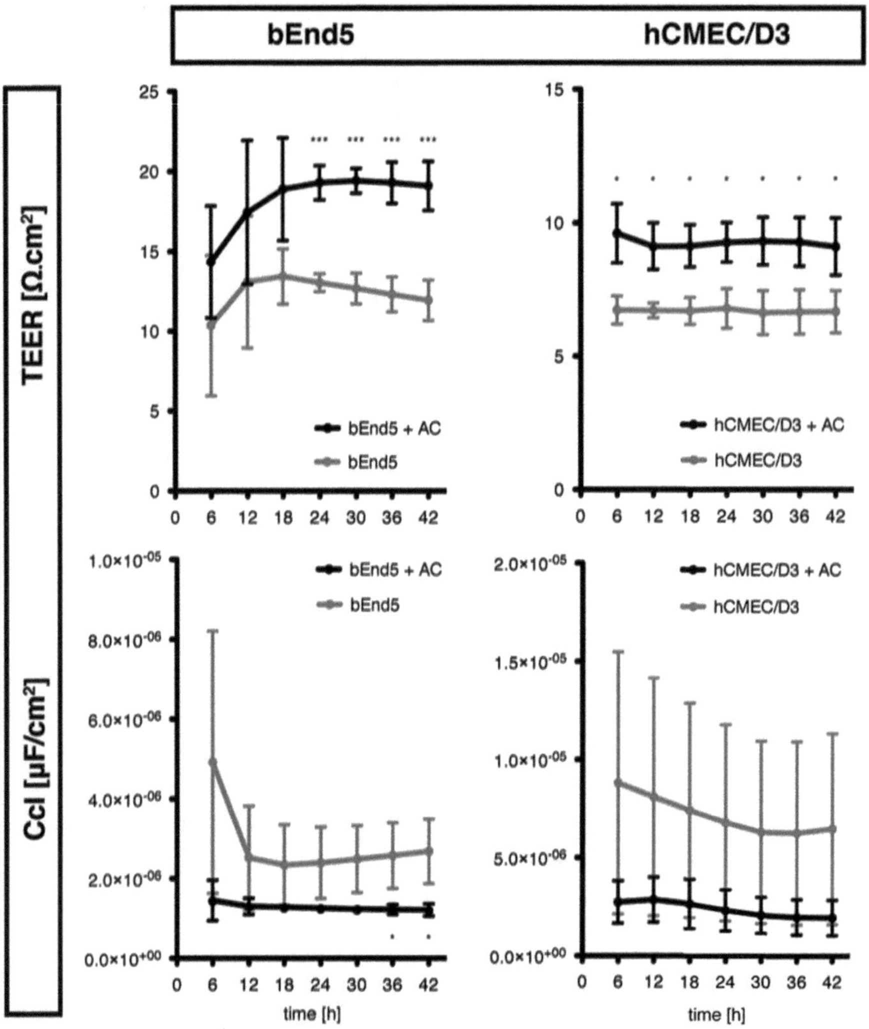

Fig. 3 Continuous TEER and C_{cl} measurements of brain endothelial cell lines—single culture versus AC coculture (cellZscope®). Brain ECs were either grown alone on the apical side of 24-well ThinCerts™ (bEnd5, hCMEC/D3; *grey*) or in cocultivation with primary mouse astrocytes from C57/Bl6 WT mice (+ AC; *black*) on the basolateral side. Measurement was started directly after seeding of the ECs. Typically, after 24–36 h a plateau phase is reached, and astrocyte cocultivation induces significant differences compared to the single monolayer condition. In general, bEnd5 cells show higher TEER and lower C_{cl} values in comparison to hCMEC/D3 cells ($n=3$; $^*P \leq 0.05$; $^{***}P \leq 0.005$)

3.3.2 Pe

The same transwell inserts subjected to TEER measurements can also be analyzed using classic tracer permeability assay, e.g., as an end point measurement (*see* Fig. 4). Barrier integrity based purely on ion flux such as the impedance measurements do not always correspond to changes to paracellular permeability as they are also sensitive to transcellular ion channels. Furthermore, the ionic composition of the media also strongly effects the impedance measurements. Thus, it is critical to assess cell monolayers also via other

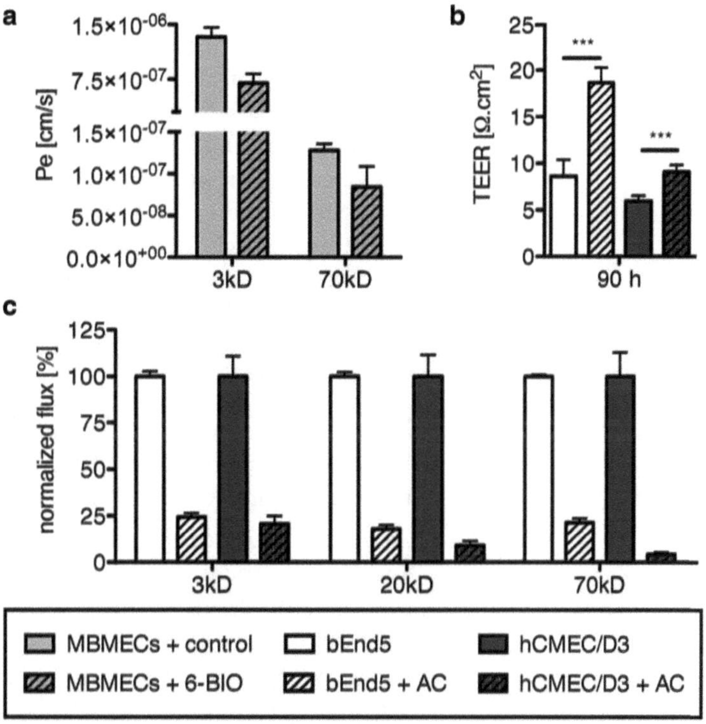

Fig. 4 Permeability (Pe) and relative flux measurements. (**a**) Permeability coefficients of control and Wnt-stimulated MBMECs to TXR-3 kD and FITC-70 kD-dextrans. The same transwell inserts from the cellZscope® were also subjected to permeability assay to obtain the permeability coefficients for MBMECs to 3 and 70 kD molecular weight dextrans having a fourfold difference in their hydrodynamic or Stokes radii. The Pe value for MBMECs is reduced from 1.3 to 0.7 (10^{-6} cm/s) for 3 kD-dextran and from 1.28 to 0.84 (10^{-7} cm/s) for 70 kD-dextran upon Wnt stimulation with 6-BIO. The Pe values correlate with the impedance measurements in the cellZscope® thus indicating that the barrier is tighter also to bigger solutes in addition to ions (primarily Na^+, Mg^{2+} measured in the cellZscope®) with about threefold lower hydrated radius compared to the 3 kD-dextran [11, 21]. (**b**) ACs induce elevated TEER values when cocultivated with bEnd5 and hCMEC/D3 ECs. Cells were grown for 48 h in serum-reduced (2 % FCS; w/o ECGS) DMEM–GlutaMAX-based endothelial complete growth medium until the plateau phase was reached. Then medium was exchanged to completely deplete cells from exogenous growth factors (w/o FCS and ECGS). After 90 h of measurement (in total), single cultures of bEnd5 and hCMEC/D3 showed further reduced TEER values in comparison to ECs cocultivated with ACs ($n=3$; $*P \leq 0.05$; $***P \leq 0.005$). The TEER measurement was stopped and the inserts were transferred into a 24-well plate containing medium. (**c**) To correlate permeability with TEER results, the flux of differently sized and labeled dextrans (3 kD, 20 kD, and 70 kD) was measured at 91 h. The flux rate of ECs grown in single culture was set as 100 %. In the presence of ACs, the permeability of both cell lines was reduced for all tracers ($n=1$; triplicates/condition)

uncharged solutes of varying sizes such as dextrans, which are highly inert and nontoxic and which are not taken up by the cells. In a typical triple tracer assay, TXR 3 kD-dextran, TMR 20 kD-dextran, and FITC 70 kD-dextran are used. Additionally, for very tight monolayers 0.45 kD Lucifer yellow (LY) can be included. The molecular weights of these solutes correspond to hydrodynamic or Stokes radii of 0.4, 1.4, 3.2, and 6.5 nm, respectively, for LY, 3, 20, and 70 kD-dextrans. In endothelial models in vitro, the permeability of these solutes correlates well with their Stokes radii in accordance with Stokes–Einstein equation [11, 21]. Thus, choosing a range of solutes is critical in a comprehensive assessment of the monolayer integrity:

1. Prepare the medium containing different tracers by adding 10 µl of each tracer (stock 1 mM in H_2O) per ml of the medium to obtain a final concentration of 10 µM for each tracer.
2. Stop the cellZscope® measurement at the desired time point, and transfer the inserts in the same order as in the cellZscope® cell module carefully into a 24-well plate containing 1 ml medium in each well.
3. From the inserts containing 250 µl medium, aspirate 200 µl using a 200 µl pipette so as to leave about 50 µl per insert to avoid drying.
4. Add 200 µl of the medium containing the tracers into each well going one well per condition followed by the duplicate well and so on.
5. At 30, 60, 90, 120, and 150 min, collect 100 µl from the bottom chamber without media replacement in a 96-well black plate. At the last time point, also collect 100 µl from the top chamber from all the inserts (*see* **Notes 7** and **8**).
6. Add 100 µl of tracer-free medium for background subtraction and the one containing the tracers (initial top chamber medium) also in quadruplicates into the 96-well plate (*see* **Note 9**).
7. Use the following order of excitation/emission (nm) for reading the fluorescence in the Tecan plate reader which can be auto-programmed in the plate reader software:
   (a) TXR 595/625
   (b) TMR 550/580
   (c) FITC 490/520
   (d) LY 425/525
8. Using the above-obtained raw fluorescence units (RFUs), the final Pe in cm/s is calculated from $\text{Pe} = (B/T)\,(V_B/(A \times 60 \times t))$, where B is the bottom chamber RFUs at time t and T is the top chamber fluorescence in RFUs, assumed constant as it does

not change much. V_B is the volume of media in the bottom chamber at time t (ml or cc) before collecting the sample aliquot. A is the cross-sectional area of the insert (sq.cm) and t is the time in minutes [22].

9. The value of Pe is obtained as the slope of the linear regression line (note: R^2 should be a minimum of 0.80 using at least 6 time points as above including the initial 0 min. time point) between $(B/T)\times(V_B)$ on y-axis (cc) and t on x-axis (min) and dividing the slope by $A\times 60$. The slope of the empty-coated inserts is subtracted from the slope of cells + insert to get the final Pe using: (1/Pe) = (1/slope cells + insert)–(1/slope insert) (*see* Fig. 3a).
10. When Pe calculation is not required, permeability flux percentage can be used to compare different conditions. This is obtained as a ratio ($R = B/T$) of bottom to top chamber fluorescence for the desired time points. Multiplying $(100/R_{\text{control}})$ with R-value for a particular condition will provide an easy readout of permeability flux with control set to 100 % (*see* Fig. 3c).

3.4 Expression Analysis

In order to better assess the functional permeability and impedance data, it is crucial to obtain expression changes in permeability-related genes. Gaps in the monolayer resulting from improper cell seeding or handling and change in ionic composition of the medium with specific treatments cause artifacts in the permeability data. Several treatments at high doses are toxic to the cells and thus manifest an increase in monolayer permeability. Indeed the cellZscope® device has been recommended for continuous monitoring of cellular toxicity upon a specific treatment [10]. Therefore, expression analysis using the same cells increases the confidence in the functional data as it permits comparison between barrier integrity and gene expression of barrier-related proteins. Due to a low-protein yield of cells grown on 24-well transwell inserts, qRT-PCR approach is the best approach for expression analysis at the mRNA level.

3.4.1 qRT-PCR

We recommend utilizing two of the quadruplicate inserts for each condition that went into the functional analysis. The RNA yield (about 1 μg) from duplicate inserts is sufficient to perform qRT-PCR of several target genes. When applied to brain ECs isolated from transgenic animal models with a low availability or treated with expensive compounds, qRT-PCR analysis becomes a robust tool to correlate with the functional data:

1. Rinse two inserts from the above functional analysis with cold 1× PBS.
2. Prepare RNA lysis buffer by adding 10 μl 2-mercaptoethanol per 1 ml of RLT buffer supplied in the RNA extraction kit and store on ice.

3. To extract RNA from ECs cocultured with ACs, invert the insert and scrape off the ACs with a pipette tip applying 50 µl cold 1× PBS. Repeat this two times.
4. Add 200 µl of the above RLT lysis buffer per insert and resuspend about 10 times up and down.
5. Pool the lysates from duplicate inserts, which will be about 350 µl due to losses during mixing and combining the inserts. Store this on ice.
6. After preparing the lysates from all the conditions in the above manner, RNA extraction can be started either immediately or the lysates stored at −80 °C until the extraction procedure.
7. Vortex the samples for 1 min at highest setting. If the samples were stored at −80 °C, they should be thawed at room temperature followed by vortexing the samples as above.

   Steps 8–14 (below) are exactly as listed in the Qiagen RNeasy handbook.
8. Add an equal volume of 70 % ethanol (about 350 µl) and mix well by pipetting. Do not centrifuge.
9. Transfer up to 700 µl into an RNeasy spin column and centrifuge at RT for 15 s at ≥8,000 × *g*. Discard the flow through.
10. Add 350 µl buffer RW1 to the spin column and centrifuge as in **step 8**. Discard the flow through again.
11. Add 10 µl DNase I stock solution to 70 µl buffer RDD and mix by gently inverting the tube. Do not vortex DNase I as it is quite sensitive to physical denaturation.
12. Add the above 80 µl DNase I mix directly to the spin column and incubate at RT for 15 min. Care should be taken to add the mix to the center of the spin column, as the digestion will be incomplete if part of the mix sticks to the column walls.
13. Add 350 µl buffer RW1 to the column and centrifuge as above discarding the flow through.
14. Add 500 µl buffer RPE to the column and centrifuge as above and discard flow through. Repeat by centrifugation for additional 2 min to ensure minimal ethanol carryover.
15. Place the spin column in a new 2.0 ml collection tube and add 25 µl RNase-free water to the column membrane. Centrifuge for 1 min as above to elute the RNA. Repeat this step with the eluted RNA to minimize losses on the membrane.
16. An aliquot of the RNA is then subject to quality analysis in the Bio-Rad Experion station according to the manufacturer's protocol. The typical yield is about 1 µg total RNA which can be reverse transcribed into cDNA or stored at −80 °C until use.
17. Utilizing 1 µg total RNA per reaction, prepare cDNA using First Strand cDNA Synthesis Kit (Fermentas) according to the manufacturer's protocol.

18. Apply RNase H digestion to the cDNA at 37 °C for 30 min (20:1 v/v) and store the sample at −80 °C until use.
19. Perform qRT-PCR using ABsolute QPCR SYBR Green Fluorescein Mix (Thermo Fisher Scientific, USA) according to the manufacturer's protocol applying the following conditions: 15 min at 95 °C, 45 cycles of 30 s at 95 °C, 30 s at 61 °C, and 35 s at 72 °C in MyiQ Real-Time PCR Detection System (Bio-Rad).
20. Analyze the qRT-PCR data with iQ5 2.1 software (Bio-Rad) (*see* Fig. 5).

3.4.2 Western Blotting

Standard Western blotting can be applied to cell lysates from transwell inserts [23]. Cell scraping can be performed with a 200 μl pipette tip for 24-well inserts or regular cell scraper for bigger inserts. However, the total protein yield is minimal (about 50 μg per insert for MBMECs) from 24-well inserts described in the current chapter. For increased yield we recommend pooling quadruplicate inserts or switching to 12-well or 6-well inserts.

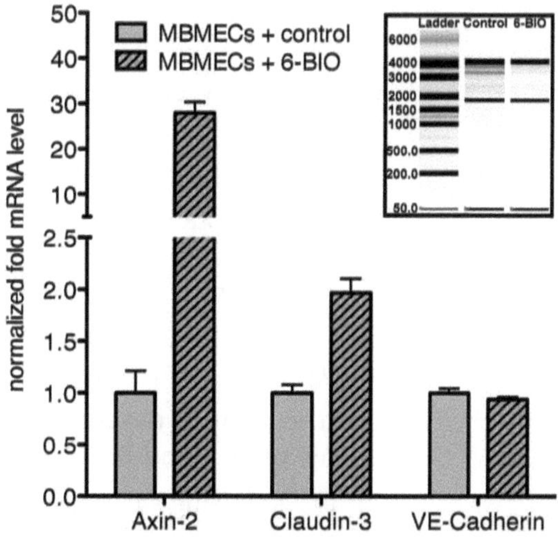

Fig. 5 qRT-PCR expression analysis of MBMECs upon canonical Wnt stimulation with 6-BIO. MBMECs show a 25-fold increase in Axin-2, a Wnt target gene indicative of canonical Wnt stimulation with 6-BIO using cDNA isolated from the same inserts subjected to the functional analysis. There is also a two-fold increase in claudin-3 levels as reported previously indicating better barrier properties observed in impedance and permeability assays as a result of increased expression of tight-junction (TJ) components such as claudin-3 [15]. There was however no difference in VE-cadherin, an adherens-junction (AJ) marker. Housekeeping gene G6PDX was used for normalization. The *top-right* portion of the graph is a virtual gel showing RNA quality information for MBMECs on transwell inserts. Both control and treated samples had high RNA quality index (RQI index > 9.0) indicative of intact RNA, with the RQI cutoff for intact RNA being 7.0 (Bio-Rad handbook)

3.5 Localization Analysis

3.5.1 Immunocytochemistry

To determine specific protein localization and patterning, cell monolayers grown on filter membranes can be fluorescently labeled. We recommend subjecting the remaining duplicate inserts from each condition that went into functional analysis for immunocytochemistry analysis. This is of importance as TEER and permeability changes are commonly linked to alterations in junctional protein expression and localization. In case of specific treatments, this technique aids to obtain additional confidence on the functional and mRNA expression data. In particular TJ proteins such as occludin, ZO-1 (zonula occludens 1), and members of the claudin family are usually affected. Furthermore, this method enables you to evaluate the purity and integrity of primary brain EC cultures (e.g., VE-cadherin staining). Depending on the antibody, methanol, acetone, or paraformaldehyde fixation is required. Thus, fixation has to be individually adapted to the antibody.

The following steps can be conducted at non-sterile conditions. Make sure that the membranes never dry during the procedure. To prevent drying, transfer the inserts sequentially (e.g., only 4–6 inserts at a time). Do not put the plates on a shaker as this might cause detachment of cells.

Fixation of Cells Grown on Filter Inserts

1. Prepare 2 × 24-well plates with 950 μl PBS per well.
2. Sterilize tweezers with 70 %EtOH for 5 min and let air-dry in flow hood.
3. Stop the TEER measurement.
4. Remove one transwell insert using the sterile tweezers and carefully aspirate the medium from the insert with glass pipette attached to a vacuum pump system.
5. Transfer the insert to the corresponding well in the multiwell plate and add 250 μl PBS to wash off the residual medium.
6. Proceed with the remaining inserts one after the other following **steps 4** and **5**.
7. The following steps have to be conducted quickly to avoid warming of the MeOH. Make sure you have everything prepared.
8. Put a third 24-well plate on ice. Fill each well with 950 μl −20 °C cold MeOH.
9. Remove the first insert from the plate with the tweezers and aspirate the PBS and transfer it immediately to the MeOH filled plate.
10. Move on to the next inserts until you have no more than 8 in total transferred. Adjust the number of wells processed at any time to your work speed.
11. Immediately add 250 μl of −20 °C cold MeOH to each well and set the timer for 3 min.
12. Proceed with aspirating the PBS from the next 8 inserts and continue as described in **steps 8–10**.

13. After 3 min remove the first 8 inserts from the MeOH plate and aspirate the MeOH from the top chamber. Place the inserts in the second 24-well filled with PBS and add 250 μl PBS in the upper compartment.
14. Continue with the residual inserts as the timer indicates to move on.
15. Either immediately continue with staining the cells or seal the lid with parafilm and store at 4 °C (inserts can be stored up to 2 weeks).

Staining of ECs Grown on Filter Inserts

1. Aspirate PBS from the inserts and wash cells once with fresh PBS for 5 min at RT.
2. Prepare a second 24-well plate with 450 μl blocking buffer per 24-well. Remove PBS of the washing step, and transfer the inserts to the new 24-well plate containing the block buffer. Add 100 μl of blocking buffer per apical compartment. Incubate at RT for 45 min.
3. In the meantime, prepare the primary antibody dilutions in antibody incubation buffer (*see* Table 2). Calculate for 100 μl per transwell insert.
4. Remove PBS from the initial 24-well plate (used for washing), and add 450 μl antibody incubation buffer (without antibodies) per 24-well instead.
5. When blocking is complete, carefully aspirate the blocking buffer from the inserts and transfer them directly into the plate with antibody incubation buffer. Immediately add 100 μl of primary antibody solution per insert and incubate for 1 h at RT.
6. Remove the blocking buffer from the second 24-well plate and add 950 μl PBS per 24-well instead.
7. Then wash off the primary antibody from the upper compartment using 250 μl PBS per insert and transfer them to the 24-well plate prepared in **step 6**. Discard the 24-well plate filled with antibody incubation buffer. Once finished with all wells, start over and repeat washing 2 times (3 times in total).
8. Prepare a fresh 24-well plate and add 450 μl antibody incubation buffer (without antibodies) per 24-well.
9. Aspirate PBS from the inserts and add 100 μl of secondary antibody solution (1:200) per insert and transfer them to the 24-well plate containing antibody incubation buffer. Incubate for 45 min at RT in the dark.
10. During the incubation time, set up and label the cover slides: cut parafilm in approximately 5 mm wide spacers (length should be adjusted to cover slide width). Apply the spacers to the upper side of the glass slide, one positioned at the top and one at the bottom of the transparent glass slide (*see* Fig. 6a).

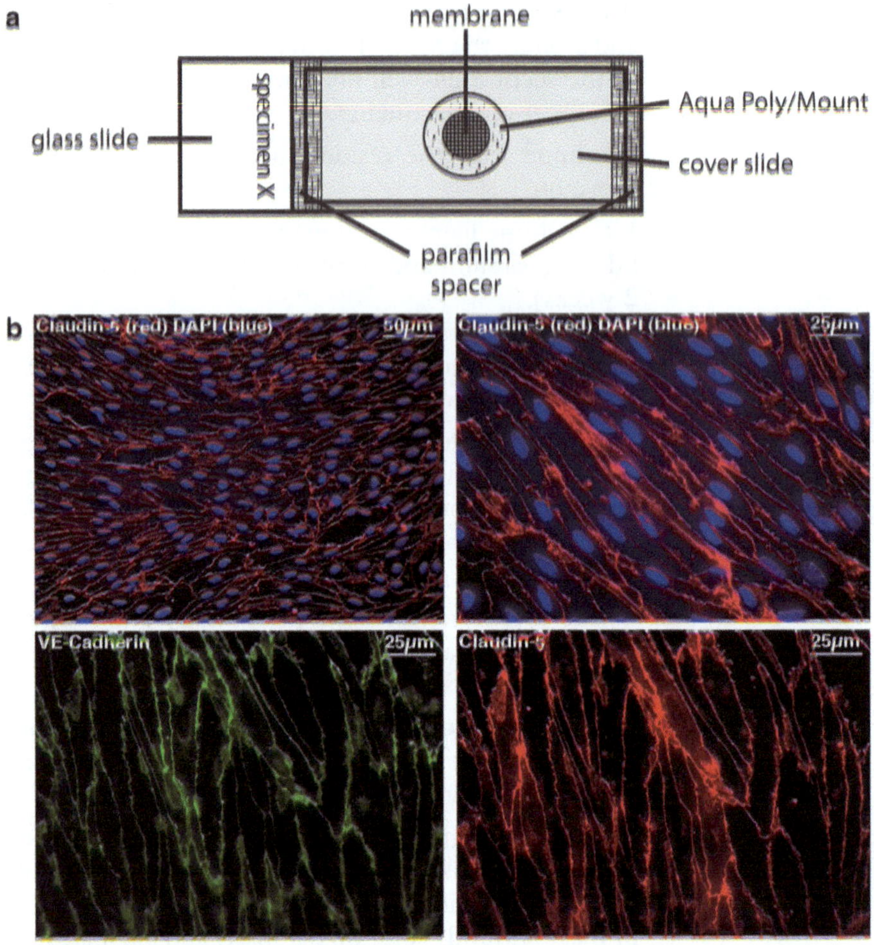

Fig. 6 (**a**) Schematic of a filter membrane mounted on a glass slide. Label the glass slide and apply the parafilm spacers on the transparent *top* and *bottom sites*. To mount filter membranes on glass slides after immunocytochemistry, apply one drop of Aqua-Poly/Mount mounting medium in the center of a glass slide. Cut out the membrane from the insert; invert it so that the EC cell monolayer is facing upwards; and position it onto the drop of Aqua-Poly/Mount mounting medium. Gently cover the glass slide with a cover slip avoiding embedding air bubbles and putting too much pressure. (**b**) Immunofluorescence for claudin-5 and VE-cadherin of MBMECs grown on transwell inserts. The same MBMECs from the functional assays were also subjected to immunocytochemistry analysis for claudin-5, a TJ member, and VE-cadherin, an AJ member, both being well-characterized endothelial markers. The ×20 image (*top left*) shows highly pure endothelial culture as claudin-5 staining at the membrane can be noticed on all DAPI-stained cells. In the higher magnification (×40) for claudin-5 (*top right*), well-organized and continuous cell–cell borders are evident with very few gaps. In the bottom panel, co-staining for VE-cadherin and claudin-5 is shown which indicates a clear co-localization at the cell–cell borders. Thus, staining the inserts can be very useful to correlate with functional permeability and impedance analysis. Treatment with 6-BIO resulted in tightly packed and highly aligned MBMECs that support the functional and expression analysis (data not shown)

11. Repeat the washing procedure described in **step 7** (use the same 24-well plate). Make sure all following steps are carried out in the dark.
12. Counter stain nuclei with DAPI nucleic acid stain solution diluted 1:1,000 in PBS. Use 100 μl per insert. Incubate at RT for 10 min.
13. In the meantime prepare for mounting: fill a small beaker with PBS. Turn the Aqua-Poly/Mount mounting medium upside down that bubbles can float up and equilibrate it to RT.
14. Mount one or two membranes per glass slide. Accordingly apply one drop Aqua-Poly/Mount mounting medium per membrane onto the glass slide.
15. Aspirate DAPI stain and put the insert inverted on a working bench. Excise the membrane using a scalpel blade. Clip the membrane carefully with thin tweezers and dip it in the PBS beaker to wash of the DAPI stain.
16. Invert the membrane so that the cell monolayer is facing upwards and position it onto the drop of Aqua-Poly/Mount mounting medium. Cover the glass slide with a cover slip avoiding embedding of air bubbles.
17. Remove excess Aqua-Poly/Mount mounting medium by gently pressing the cover slide on a paper tissue. Let Aqua-Poly/Mount mounting medium dry overnight in an appropriate specimen casing.
18. Next day acquire pictures using a confocal fluorescence imaging device (*see* Fig. 6b).

3.5.2 Electron Microscopy

Transwell inserts are also amenable to electron microscopy analysis. Refer to transwell scanning electron microscopy (SEM) protocol from Corning available on their website. Additionally refer to Zhang Y and colleagues for EM analysis on brain endothelial cells [24].

4 Notes

1. For EC and AC cultivation, FBS with low endotoxin levels should be used.
2. When handling FN, do not vortex, shake, or spin the tube, as this will cause FN to "precipitate," which is irreversible. Please stick to manufacturers guidelines.
3. After thawing ECs or isolation of primary MBMECs, ECs were cultivated in MCDB-131-based endothelial complete growth medium. ACs do not grow well in MCDB-131 medium. Thus, for astrocyte and EC coculture, ECs needed to be equilibrated to DMEM GlutaMAX-based complete growth medium.

Therefore, 24 h prior to TEER measurement EC growth medium was exchanged.

4. To avoid larger hydrostatic pressure on the cell layer(s) cultured on filter membranes, ensure equilibrium between the apical and basolateral compartment. When using the cellZscope®, please refer to chapter 4.1 of the operation manual.
5. If some of the electrodes do not pass the test, first restart the test to check if the corresponding electrodes pass the test. If they do not pass the test again, transfer the cell module back to the hood and ensure there are no air bubbles in the medium or debris on the electrodes. Also ensure enough media volume in the top chamber as the top electrodes need to be submerged in the liquid medium. The recommended insert volume is 250 μl, but due to pipetting errors or leakage from the bottom well (O-ring), the volume might be lower.
6. Medium exchange will lead to disturbances in cellZscope® measurement. Thus, short-term effects up to 12 h usually cannot be monitored properly.
7. The time points of the Pe assay can be adjusted based on the size of the tracer. For example, 70 kD FITC-dextran has minimal flux in the first hour. Thus, to obtain better signal to noise for fluorescence measurements, the assay time points can be hourly time points up to 5 h. Since Pe is dependent on the slope of the linear regression line, the time points chosen have no bearing on the final value. However, they should be chosen so as to obtain better signal to noise for the tracer fluorescence.
8. There is also a change in the hydrostatic pressure, as media is not replaced in the current protocol with tracer-free medium. If this is a concern for the solute flux for a particular cell type, the aliquot size can be reduced from 100 to 25 μl and then made up to 100 μl with tracer-free medium if read in a 96-well plate or 25 μl aliquots read directly in a 384-well plate.
9. Eliminate bubbles before reading fluorescence using 22-gauge needle to break bubbles if needed.

References

1. Abbott NJ, Patabendige AA, Dolman DE, Yusof SR, Begley DJ (2010) Structure and function of the blood-brain barrier. Neurobiol Dis 37:13–25
2. Pardridge WM (2003) Molecular biology of the blood-brain barrier. Methods Mol Med 89:385–399
3. Pardridge WM (2005) The blood-brain barrier: bottleneck in brain drug development. NeuroRx 2:3–14
4. Neuwelt E, Abbott NJ, Abrey L, Banks WA, Blakley B, Davis T, Engelhardt B, Grammas P, Nedergaard M, Nutt J, Pardridge W, Rosenberg GA, Smith Q, Drewes LR (2008) Strategies to advance translational research into brain barriers. Lancet Neurol 7:84–96
5. Deli MA, Abraham CS, Kataoka Y, Niwa M (2005) Permeability studies on in vitro blood-brain barrier models: physiology, pathology, and pharmacology. Cell Mol Neurobiol 25:59–127
6. Lippmann ES, Azarin SM, Kay JE, Nessler RA, Wilson HK, Al-Ahmad A, Palecek SP, Shusta EV (2012) Derivation of blood-brain barrier

endothelial cells from human pluripotent stem cells. Nat Biotechnol 30:783–791

7. Lyck R, Ruderisch N, Moll AG, Steiner O, Cohen CD, Engelhardt B, Makrides V, Verrey F (2009) Culture-induced changes in blood-brain barrier transcriptome: implications for amino-acid transporters in vivo. J Cereb Blood Flow Metab 29:1491–1502
8. Calabria AR, Shusta EV (2008) A genomic comparison of in vivo and in vitro brain microvascular endothelial cells. J Cereb Blood Flow Metab 28:135–148
9. Tilling T, Korte D, Hoheisel D, Galla HJ (1998) Basement membrane proteins influence brain capillary endothelial barrier function in vitro. J Neurochem 71:1151–1157
10. Benson K, Cramer S, Galla HJ (2013) Impedance-based cell monitoring: barrier properties and beyond. Fluids Barriers CNS 10:5
11. Gunzel D, Zakrzewski SS, Schmid T, Pangalos M, Wiedenhoeft J, Blasse C, Ozboda C, Krug SM (2012) From TER to trans- and paracellular resistance: lessons from impedance spectroscopy. Ann N Y Acad Sci 1257:142–151
12. Nitta T, Hata M, Gotoh S, Seo Y, Sasaki H, Hashimoto N, Furuse M, Tsukita S (2003) Size-selective loosening of the blood-brain barrier in claudin-5-deficient mice. J Cell Biol 161:653–660
13. Van Itallie C, Rahner C, Anderson JM (2001) Regulated expression of claudin-4 decreases paracellular conductance through a selective decrease in sodium permeability. J Clin Invest 107:1319–1327
14. Van Itallie CM, Anderson JM (2011) Measuring size-dependent permeability of the tight junction using PEG profiling. Methods Mol Biol 762:1–11
15. Liebner S, Corada M, Bangsow T, Babbage J, Taddei A, Czupalla CJ, Reis M, Felici A, Wolburg H, Fruttiger M, Taketo MM, von Melchner H, Plate KH, Gerhardt H, Dejana E (2008) Wnt/beta-catenin signaling controls development of the blood-brain barrier. J Cell Biol 183:409–417
16. Rohnelt RK, Hoch G, Reiss Y, Engelhardt B (1997) Immunosurveillance modelled in vitro: naive and memory T cells spontaneously migrate across unstimulated microvascular endothelium. Int Immunol 9:435–450
17. Weksler BB, Subileau EA, Perriere N, Charneau P, Holloway K, Leveque M, Tricoire-Leignel H, Nicotra A, Bourdoulous S, Turowski P, Male DK, Roux F, Greenwood J, Romero IA, Couraud PO (2005) Blood-brain barrier-specific properties of a human adult brain endothelial cell line. FASEB J 19:1872–1874
18. Uliasz TF, Hamby ME, Jackman NA, Hewett JA, Hewett SJ (2012) Generation of primary astrocyte cultures devoid of contaminating microglia. Methods Mol Biol 814:61–79
19. Liebner S, Czupalla CJ, Wolburg H (2011) Current concepts of blood-brain barrier development. Int J Dev Biol 55:467–476
20. Crone C, Olesen SP (1982) Electrical resistance of brain microvascular endothelium. Brain Res 241:49–55
21. Li G, Simon MJ, Cancel LM, Shi ZD, Ji X, Tarbell JM, Morrison B 3rd, Fu BM (2010) Permeability of endothelial and astrocyte cocultures: in vitro blood-brain barrier models for drug delivery studies. Ann Biomed Eng 38: 2499–2511
22. Antonetti DA, Wolpert EB (2003) Isolation and characterization of retinal endothelial cells. Methods Mol Med 89:365–374
23. Kroll S, El-Gindi J, Thanabalasundaram G, Panpumthong P, Schrot S, Hartmann C, Galla HJ (2009) Control of the blood-brain barrier by glucocorticoids and the cells of the neurovascular unit. Ann N Y Acad Sci 1165:228–239
24. Zhang Y, Li CS, Ye Y, Johnson K, Poe J, Johnson S, Bobrowski W, Garrido R, Madhu C (2006) Porcine brain microvessel endothelial cells as an in vitro model to predict in vivo blood-brain barrier permeability. Drug Metab Dispos 34:1935–1943

ERRATUM TO

Induction of Brain Arteriovenous Malformation in the Adult Mouse

Wanqiu Chen, William L. Young, and Hua Su

Richard Milner (ed.), *Cerebral Angiogenesis: Methods and Protocols*, Methods in Molecular Biology, vol. 1135,
DOI 10.1007/978-1-4939-0320-7_25, © Springer Science+Business Media New York 2014

DOI 10.1007/978-1-4939-0320-7_35

The paperback and online versions of the book contain some errors, and the corrections to these versions are given below:

Chen W, Sun Z, Han Z et al (2014) De novo cerebrovascular malformation in the adult mouse after endothelial Alk1 deletion and angiogenic stimulation. Stroke 45:900–902 was not credited for Figure 1 at the time of publication. The updated caption of figure 1 is given below.

Fig. 1 Vessel casting by latex perfusion shows the AVM lesion. (**a**) No abnormal vessels were detected in the brain of wild-type mouse around the vector injection site (*arrow*). Scale bar: 1 mm. (**b**) AVM phenotype was detected around the vector injection site of 8-week-old *Alk12f/2f* mouse (*arrow*). Scale bar: 1 mm. (**c**) High magnification of the injection area shows the abnormal vascular structure. Scale bar: 100 μm. Credit: Chen W, Sun Z, Han Z et al (2014) De novo cerebrovascular malformation in the adult mouse after endothelial Alk1 deletion and angiogenic stimulation. *Stroke* 45:900–902.

The online version of the original chapter can be found at
http://dx.doi.org/10.1007/978-1-4939-0320-7_25

Index

Richard Milner (ed.), *Cerebral Angiogenesis: Methods and Protocols*, Methods in Molecular Biology, vol. 1135, DOI 10.1007/978-1-4939-0320-7, © Springer Science+Business Media New York 2014

L

M

N

O

P

MIX
Papier aus verantwortungsvollen Quellen
Paper from responsible sources
FSC® C105338
FSC
www.fsc.org

If you have any concerns about our products,
you can contact us on
ProductSafety@springernature.com

In case Publisher is established outside the EU,
the EU authorized representative is:
Springer Nature Customer Service Center GmbH
Europaplatz 3, 69115 Heidelberg, Germany

Printed by Libri Plureos GmbH
in Hamburg, Germany